Understanding Ima

STUDIES IN HISTORY AND PHILOSOPHY OF SCIENCE

VOLUME 33

For further volumes:
http://www.springer.com/series/5671

Dennis L. Sepper

Understanding Imagination

The Reason of Images

Springer

Dennis L. Sepper
Department of Philosophy
University of Dallas
Irving, Texas, USA

ISSN 0929-6425
ISBN 978-94-007-9389-7 ISBN 978-94-007-6507-8 (eBook)
DOI 10.1007/978-94-007-6507-8
Springer Dordrecht Heidelberg New York London

Softcover re-print of the Hardcover 1st edition 2013

Printed on acid-free paper

Springer is part of Springer Science+Business Media (www.springer.com)

Acknowledgments

The first intimation of the project that led to this book emerged three decades ago, when in teaching a graduate philosophy course about scientific method I realized that what secondary works said about Descartes did not make sense. Descartes's early writings about method and mathematics focused on imagination, but the secondary accounts said or implied it was all about rationalizing intellect. In a sense the present book continues to draw out the consequences of that intimation.

It was inevitable that, in the course of decades of gestation, I would incur more debts than it is possible to enumerate. The ones I include here are the most outstanding of all.

At a crucial moment the Getty Research Institute awarded me a year-long residential research fellowship that allowed me to get this project moving. The interaction with artists, historians, art historians, philosophers, and art restorers gave real substance to my thinking about imagination. Many thanks, then, to the fellows and the staff of the Institute, and to the J. Paul Getty Trust for its support. A 2007–2008 National Endowment for the Humanities research fellowship (FB-53295-07) made it possible for me to turn many hundreds of pages of notes and thousands of pages of annotated photocopies into the conceptual and historical heart of this book. Thanks to the University of Dallas for awarding a concurrent sabbatical, and to the King and Haggar foundations for their funding of the development grants for University of Dallas faculty that have also supported my research.

I have benefited from the ministrations of librarians all over the world—even, and especially, in this era of universal electronic connectivity. The staffs of the Getty Research Institute Library in 2001–2002, the Bibliothèque nationale de France (François Mitterand) over the last decade, and the University of Dallas Blakley Library for thirty years, have granted my every bibliographic wish. As always I need to express a special thanks to Mrs. Alice Puro, the interlibrary loan librarian of the University of Dallas, for her tireless work in securing copies of even the most obscure publications.

Many colleagues and students have taken an interest in this work and offered valuable criticisms and perspectives. I hope I have been as generous in return. Friends at

St. John's College, in Annapolis and in Santa Fe, have long given encouragement, especially Harvey Flaumenhaft. And of course there is Eva Brann, who has been an inspiration. Without the example of her magisterial study of imagination I might not have found my way to this project. I have often benefited from the words of Katherine Pandora of the University of Oklahoma's Department of the History of Science. Thanks to Philipp Rosemann, chair of the Department of Philosophy at the University of Dallas, for his sage advice. A special word of appreciation goes to members of the Department of Psychology, especially Scott Churchill and Bob Kugelmann. If I am not yet enough of a phenomenologist, it is my fault, not theirs.

Heartfelt thanks to Stephen Gaukroger, the editor of the series and a friend for almost two decades. His invitation to Sydney many years ago not only gave me a first look at the Southern Cross but also opened my eyes to new approaches to the world of early modern thought and the emergence of the sciences. His wide interests and unfailing philosophical tact are exemplary.

I want to express my appreciation to Lynn Rohm for her promptness and skill in designing and producing the figures. Thank you to Lucy Fleet of Springer-Verlag, who has been generous with help and patient with questions, and to Gounasegarane Shanthy's guidance of the book's production.

I thank Blackwell Publishing, which has granted permission for the reproduction, in Chapter 7, of materials from two essays of mine that originally appeared in *A Companion to Rationalism*, edited by Alan Nelson (Oxford: Blackwell, 2005): "Cartesian Imaginations: The Method and Passions of Imagining" (pp. 156–176) and "Spinoza, Leibniz, and the Rationalist Reconceptions of Imagination" (pp. 322–342).

I would be remiss if I did not mention the La Madeleine Restaurant at Midway and Northwest Highway in Dallas. I have spent literally thousands of hours there reading and writing, drinking their coffee, and enjoying the good cheer of the staff. The major part of the first draft of this book was written there. I want also to express gratitude to the staffs of two Paris cafés: the Royal-Jussieu in the 5th, and La Terrasse in the 7th. They provided uninterrupted morning writing and revising time, with excellent café crème and croissants, every summer since 2000.

My children used to say that my wife (who is professor of history at Southern Methodist University) and I never provided them with much of a model for work life. They saw us sitting with a book or a notepad or a laptop, reading, writing, and drinking coffee. Matthew is now an attorney, and Elizabeth a law professor. They spend much of their time sitting with a book or a notepad or a laptop, reading, writing, and drinking their beverages of choice! I thank them for helping make the journey a rewarding one—and hope they find this book more approachable and interesting than the others! My mother, Marge, and my sister, Kathy, continue as always to offer their support.

As for my wife, Kathleen Wellman: she read to me, I read to her, and thus we corrected the page proofs. As you, reader, can tell, that meant many hours together! If proofreading is not exactly romantic, it still turns out to be an act of love. I dedicate the book to her.

Contents

1 Beginning in the Middle of Things 1
 1.1 Constellations of Questions About Imagination 1
 1.2 The Occluded-Occulted Tradition of Intelligent Imagining 2
 References 15

2 Locating Emergent Appearance 17
 2.1 Some Practice of Imagining, and Thoughts About It 20
 2.2 Psychologism, Antipsychologism, and the Persistence of the Visual Model 27
 2.3 Limits of the Visual Model 31
 2.4 Elementary and Complex Imagining 37
 2.5 Listening to Images 40
 2.6 Can Philosophers Sing? 44
 2.7 Simple Imagining and Beyond 47
 References 51

3 Locating Imagination: The Inceptive Field Productivity and Differential Topology of Imagining (Plus What It Means to Play a Game) 53
 3.1 Hume's Blue 55
 3.2 From Resemblant Production to Schematized Activity in Fields 58
 3.3 Imagination as a Release in/of/from the Conditions of Perception 64
 3.4 The Repositioning of Imagination and the Problem of Reifying Consciousness 68
 3.5 Fields 72
 3.6 Imaginative Topology and Topographies 75
 3.7 Placing the Topological Dynamics of Imagination 83
 3.8 From Basketball Practice to the Biplanarity of Imagining 87
 3.9 From the Biplanarity of Imagining to the Practice of Art 91
 3.10 Transition: Reversing the Occlusion and Occultation of Tradition 96
 References 100

4 Plato and the Ontological Placement of Images 103
4.1 Pre-Platonic Philosophy and the Emergence of the Image–Bearer 104
4.2 Image–Bearers, Figures, and Images in Plato's *Meno* 114
4.3 The Use and Abuse of Images 121
4.4 Speech as Image, Reason as Imaginative, and the Platonic Ontology of Imaging 127
4.5 The Multilevel Look of Things in the *Republic* 137
4.6 The Paradoxes of Imaging 142
4.7 The Ontology of Images and the Psychology of Scenario–Imagining 148
4.8 The Grand Image–Sequence of the *Republic*: From the Good Itself to the Dialectical Education of the Philosopher 154
4.9 Singing and Hearing the *logos* 163
4.10 Forming an Equable Icon of the Cosmos 169
4.11 The Perfect Image of the Cosmos as the Goal of Dialectic 173
4.12 Conclusion 180
References 183

5 Aristotle's *phantasia*: From Animal Sensation to Understanding Forms of Fields 185
5.1 Aristotle's Physiologically Based Psychology of Imagination 186
5.2 Placing Soul in Aristotelian Context 190
5.3 Aristotle's Imagination Conventionalized 197
5.4 *Phantasia* Beyond the Conventions 204
5.5 The Perplexities of Imagination in *On the Soul* III: An Overview 205
5.6 The Imagination of *On the Soul* III.3: What It Is and What It Isn't 210
5.7 Imagination, Sensation, Motion 217
5.8 What the Physics of Motion Implies 222
5.9 From Motions of Sensation to Structures of Imagining 226
5.10 What Aristotle's Definition of Imagination Means 233
5.11 Is Imagination the Same as Intellect? 240
5.12 Parsing the Phenomenon of Thinking 242
5.13 Thinking Imagination 252
5.14 Conclusion 262
References 264

6 The Dynamically Imaginative Cognition of Descartes 267
6.1 Imagination After Aristotle and Before Descartes 270
6.2 Descartes's Starting Point 273
6.3 The Imagination of Notebook C 279
6.4 Imaginative Representation and Manipulation 284
6.5 The Dynamism of Imaginative Ingenuity 289
6.6 How the Knowing Power Recognizes Itself in Imagining 298

6.7 The Limits of Imagination 303
6.8 Imagining the Cosmos 307
6.9 Imagination in the *Meditations* 313
6.10 Willing, Images, and Passions 321
References 328

7 The Cartesian Heritage: Kant and the Conceptual Topology of Imagination and Reason 331
7.1 How Imagination Got Misplaced, Part 1: The Way of Ideas 339
7.2 How Imagination Got Misplaced, Part 2: The Psychological Reification of the Idea 346
7.3 The Rationalist Loss of Confidence in Imagining, and the Rise of Aesthetics 352
7.4 Aesthetics, Genius, and the Ordinary Mystery of Sensation: The Need for an Expansive Concept of Imagination 359
7.5 Kant's Response to the Challenge: Transcendental Psychology 361
7.6 How Previous Philosophy Failed to See the Syntheses of Imagination 372
7.7 The Higher Syntheses 377
7.8 Aesthetics, Ethics, and the Limits of Kantian Imagination 383
References 397

8 After Kant: Appropriating the Conceptual Topology of Imagination 399
8.1 The Idealist-Romantic Appropriation of Infinite Imagination in Art 400
8.2 Tendencies of the Post-Kantian Topology 407
8.3 The Signitive Placement of Imagination 414
8.4 If Signification Is Imaginative, Can Reason Leave Imagination Behind? 426
8.5 Wittgenstein and the Imaginative Supports of *logos* 430
8.6 Semiotics: Thinking the Signification of Sequenced Phantasms 440
8.7 Semiology: Signs as the Fusion of Imaginative Planes 454
8.8 Language as the Social Imagination of the World 464
8.9 Conclusion: The Ontology of Language 476
References 479

9 The Ethos of Imagining 483
9.1 Delimiting Imagination Rationally 486
9.2 The Ethos of Imagining Found; or, Topological Topics of Placing Imagination 502
9.3 Conclusion: Toward a New Beginning 512
References 525

Index 527

Chapter 1
Beginning in the Middle of Things

> *The history of thought is also the history of its successive closures—and that is what renders ineliminable a critical attitude with regard to thinkers of other times. But it is also true that, among the forms thus created, some possess a mysterious and marvelous permanence. And the truth of thought is this movement itself in and through which the already created permanent part finds itself differently placed and illuminated by the new creation of which it has need in order not to sink into the silence of the simply ideal. (Castoriadis 1997, 336)*

1.1 Constellations of Questions About Imagination

We begin with four constellations of questions about imagination.

(1) Since the emergence of European Romanticism at the end of the eighteenth and the beginning of the nineteenth century, both popular and philosophical understanding have associated imagination with creativity. Creativity came to be understood as good for the individual human being and for the arts; more recently, it has come to be perceived as essential for social, economic, and scientific progress. But most religious and philosophical traditions, going back to Greek, Roman, and Jewish antiquity, have been suspicious of imagination as harboring not just falsity but even delusion and evil. Are these traditions reconcilable with the modern conception of imagination? Is imagination a solution to our conception of ourselves and our world, or is it a source of the difficulty of conceiving them?

(2) There was already a decisive break in the conception of imagination before Romanticism. In Greek antiquity, Aristotle, the first to give a careful delineation of the power of imagination as part of a complex theory of human and animal psychology, had claimed that for human beings there is no thinking without phantasms, or, as we would say, no thinking without images. That is, there would be no intelligent human activity, productivity, or morality without imagining. What is more, this applied to *scientific knowing* as well. In Western thought, this basic notion was

D.L. Sepper, *Understanding Imagination: The Reason of Images*, Studies in History and Philosophy of Science 33, DOI 10.1007/978-94-007-6507-8_1,

widely, though variously, accepted for nearly 2,000 years. Yet, since the seventeenth and eighteenth centuries, modern science has come to be understood as methodical rationality taking control of factual experience. As such, it has seemed not to need imagination. Imagination may well be used as a tool for conceiving new possibilities, but it must quickly give way to rational analysis and testing. In the final reckoning, science is indifferent to how and why theoretical and experimental innovations come about; only results count. Imagination is something for artists and for children, not for the sober rationality of adults.

How, why, and *exactly* when did this conceptual break occur? More importantly: was the break justified? Is imagination as irrelevant and accidental to knowledge as this modern scheme makes it appear?

(3) Are the first and second sets of questions related? Is the contemporary emphasis on imagination's creativity perhaps the *consequence* of the earlier split between its scientific/cognitive and its artistic/aesthetic functions? More generally, does the modern emphasis on creativity distort our understanding of imagination—so that, for example, it has become nearly impossible to recognize how and why it is essential to both science and art, as well as to all other kinds of thoughtful human action?

(4) Do historical shifts in the conception of imagination correspond to larger changes, for example to a change in the conception of what it is to be a human being and what human flourishing is about? Aristotle may have established the basic framework for imagination more than 2,000 years ago, but does what philosophers think about imagination make any real difference today? More generally, does what people *think* about imagination make a difference to its role in their lives?

****

We can summarize the concerns expressed above in the following four questions: Where does our idea of imagination come from? Is imagination the name of anything real? How can we arrive at an understanding of what it is? Will it, does it, make a difference to who or what we are?

There *are* answers to these questions. This book intends to track them down by a kind of philosophical archeology. The answers, or at least their elements, have been available for a very long time. But the tradition that ought to have handed them down to us has been repeatedly interrupted and obscured. The tradition has, again and again, been occluded and occulted. The time has come to recover it.

1.2 The Occluded-Occulted Tradition of Intelligent Imagining

How can there be intelligent imagining or imaginative intelligence when intelligence and imagination may be fundamentally at odds with one another? Intelligence is about the real, whereas imagination is irrealizing, fictive, untrue. Imagination serves not just the hopes of humanity but also its fears and superstitions. In the dark (says Shakespeare) it is what makes us take a bush for a bear; in credulous society (says Spinoza) it produces fabulous tales of gods and demons. Knowledge, by contrast, is disciplined intelligence. If it is not exactly necessary to *banish* imagination from the hard, logically driven work of understanding the world, it is important to keep it under control.

But what about when the real is defective or wrong? The real can be changed—and isn't it imagination that lets us see beyond the limits of the merely real? Moreover, what about when intelligence and reason try to control imagination by playing the tyrant? In their claim to rule and measure all things, they have often tried to suppress imaginative powers as unruly. In Western thought this starts with Plato (ca. 428–347 B.C.E.), most strikingly in his dialogue the *Republic*, which banishes all influences that do not measure up to austere reason's demands. Two thousand years later, the (at least perceived) tyranny of reason was behind Romanticism's furious repudiation of Enlightenment. More recently, it was behind the political rebellions that swept through the West in May 1968 after a student uprising in France against the stupidities of bureaucratic reason. It echoes in a favorite slogan of the day: "L'imagination au pouvoir!", "Put imagination in power!" Imagination by its nature rebels against restraint, and when fully engaged it is *creative*. If it is unreal, it is because the object aimed at *does not yet exist.* Imagination anticipates the new, the not-yet. Real-and-rational standards cannot lead us into the future because they follow the guide of what is established and even sclerotic. In a world that more and more depends on the ability to innovate, whether in politics, in business, in science, or in everyday life, imagination must take the lead. It is rationality that must learn how to follow. Yet, in the long run, it is not clear at all that imagination can live up to the hopes it inspires. It often loses its way in dreams; dreams of the new can become rigid; their force can peter out in the face of real problems.

These conflicting, and sometimes schizophrenic, conceptions about imagination are widespread, even popular, but that does not make any of them true. They are all notionally weak and historically underinformed stereotypes. Yet even the learned and historically informed think in these terms—that is, when they think about the questions at all. In the West, both philosophers and scientists have for more than a century deliberately turned their backs on imagination; and, when reason seems to need defending, they trot out a shabby theory of the nobility of enlightened reason and the irrationality of its opponents. This is in part a reaction against the Romantic elevation of imagination above all other human powers. The Romantic hyperelevation of imagination was itself a reaction against an eighteenth-century tendency to entrust truth exclusively to the rationality of science and to confine imagination to the fictions of children and art. It is not just a pun to say that the effective history of imagination has for the past several centuries been reactionary.

Thinking that is reactionary always comes at a cost. It starts with an act of rejection, yet more often than not it remains committed to the logical framework that it apparently rejects. Reject the tyrant reason and elevate imagination, or reject irrealist imagination and restore sober rationality: both options assume that in human psychology there is a fundamental division, a dichotomy, between opposed powers. Instead of questioning the framework, most people simply declare allegiances or negate the formerly posited in order to affirm the formerly denied—which leaves the framework intact. The recent history of imagination cannot be understood without some clearer sense of framework questions—of the framework within which imagination and reason take their place and relate to one another—and how the framework came to be.

Broad generalizations about the past are tempting to make but hard to justify. Yet the temptation to oversimplification is raised to a higher power when it comes to imagination. In a history of matter theory or politics or cooking we may have to make significant mental adjustments for period and place, but we can ordinarily see that things, explanations, and practices separated by large stretches of time have recognizably common themes and family resemblances. These commonalities are sufficient to assure us that, despite all differences, we are still dealing with the same kind of thing. With imagination, it often seems that there is no agreement at all about its most basic phenomena and features. It is a long way, for instance, from Thomas Hobbes's (1588–1679) notion that images are vibrations in the nerves to the early nineteenth-century, post-Kantian claim that they are the creative product of the Absolute. Even whether the various words thinkers have used to name the phenomenon all have the same meaning is more than a little uncertain. At first glance, what Plato called *eikāsia*, what Aristotle named *phantasia*, what the Latin middle ages parsed as various forms of both *phantasia* and *imaginatio*, what we divide into imagination, fantasy, and creativity seem to be basically the same thing—but just a little investigation opens questions and even chasms. The more widely we cast our intellectual nets, encompassing more authors, more centuries, more disciplines and fields, the likelier that the diverse conceptions of imagination will simply bewilder us. We might easily conclude that today's confusions have their source in confusions that began long ago; we might come to agree with those who say that the past's understanding of psychological matters is hopeless. The best course, then, would be to leave the past behind and start over again, this time more scientifically.

This conclusion is fundamentally wrong, and the recommended new scientific course disastrously misdirected. The critical side of this book will show why.

****

Thirty years ago I began teaching in a philosophy program that educates students in (chiefly) Western philosophical traditions going back to the pre-Socratic philosophers of ancient Greece. I was by training and interests a philosopher and historian of the physical sciences who liked to keep an eye on larger contemporary questions, so I worried a little that the wide-ranging and backward-looking teaching the program demanded would impair the kinds of engagement my research required. Nevertheless, because I myself had had a decent liberal arts education, I understood its virtues. It had cultivated in me habits of thinking and inquiry suspicious of posing questions too narrowly or looking to a single time or place or discipline for definitive answers. I did not yet know to put it this way, but that education had provided me with the ability to think the modes, indeed the *matrix*, of natural and human existence in many different, concretely imaginative ways. Moreover, having recently become a parent, I began to feel a stronger sense of obligation to the future. Even though I might have preferred teaching courses geared to my special research concerns, I understood the need to cultivate and orient the minds and hearts of the next generation with regard to more basic things. And, after all was said and done, I loved thinking about classic questions and reading and thinking my way through the writings of great philosophers. So I threw myself into the task with enthusiasm.

This teaching put my research into new perspective. In my scholarly work I had been searching for connections between aesthetics and science, with the imagination an obvious point of focus. As I read and taught more widely and more historically I came to see that there had been a decisive break in the conception of imagination in the early modern period, that is, at some point in the seventeenth or eighteenth centuries, and thus well before the Romantic reaction. The prehistory of this break stretched back to ancient Greek antiquity. Aristotle (384–322 B.C.E.), who had attempted a careful explanation of imaginative power (*phantasia*) as part of a complex theory of human and animal psychology, went so far as to claim that for human beings there is no thinking without phantasms, or, as we would approximate, no mental activity without imagining. Because of Aristotle's long-lasting influence on philosophy and science, the claim had many important conceptual and historical consequences. Taken strictly, it meant that there could be no intelligent human activity—that is, any human activity that depends on thinking—without imagining. Without imagination there could be no complex pursuit of future goals, no moral or ethical action, no artistic or technical making, no asking meaningful questions, no scientific inquiry or knowledge, no intelligent mental activity whatsoever. Despite the old saw that no two philosophers agree about anything basic, something very close to Aristotle's notion was widely accepted for 2,000 years by Persian-, Arabic-, Hebrew-, and Latin-speaking philosophers. Yet ever since the seventeenth and eighteenth centuries, modern science has typically been understood as a rational activity that can do without imagination. The rational ban extended even to its innovative powers—though, to be sure, creativity was not an aspect of imagination that premodern philosophers, not even philosophers of art, had much acknowledged or valued.

So what happened? What decisively changed the attitude to imagination in the modern period? I thought these and other questions could be answered by perhaps 6 months of concerted historical and scholarly reading. I was laughably wrong.

I first learned how wrong from an unlikely source. If you think that something has gone awry with modern intellectual history you are very likely to make René Descartes (1596–1650) a chief villain. As the "father of modern rationalism" he is a very plausible villain when it comes to the decline of imagination. Early in his central philosophical work, the *Meditations*, he seems to abandon it as a source of truth (along with sensation and memory) and quickly advances instead by means of pure intellectual intuition to the certainty of the existence of the thinking ego and then to the being of God. To be human is to be *res cogitans*, thinking being; only accidentally do human beings imagine.

When I began looking into the writings of his youth, however, I found that the power of imagination was at the heart of his philosophizing, his mathematics, and his physics. He expressly called for its methodical use, and he deliberately and intensively cultivated his own imaginative talents. What we take to be one of his greatest accomplishments, analytic geometry—an invention that almost immediately produced an explosive development of rigorously analytic mathematics and science unlike anything that had been seen before—grew out of his practice of exact imagining. To us this may seem very strange, insofar as we are inclined to think of mathematics, especially mathematical analysis, as preeminently rational. Perhaps this was just another case in which a scientific advance was produced by something irrational or irrelevant that later development properly left behind. Yet even today, mathematicians and scientists working at the frontiers of their disciplines will be offended by the suggestion that their work is not imaginative. Are mathematics and sciences imaginative, or are they rational? By what right, and using what capacities, does one answer?

Let me not, however, give the misimpression that this book is simply about mathematical and scientific imagining and what has happened to it over the ages. Nor that it is just about old things, about what a few dead Western (and not-so-Western) people have *thought* imagination to be. It is about what imagination *is*. But my research has convinced me that we cannot really get at what imagination is unless we understand why we, today, think about it as we do, typically in ways that obscure its nature rather than illuminate it. Even expert philosophical and psychological research is affected by basic inadequacies in conceptualizing imagination and imaginative phenomena. This is part of our historical and philosophical heritage. Yet that does not mean that the only reason for pursuing the history of imagination is to clear out old ideas and theories littering our conceptual closet. If we have inherited from the past much that deadens our minds, that is not to say that there is no living inheritance from those who have thought before us. Our best current ideas often turn out to be reinventions of wheels that were far more perfect than the ones we manage to produce.

In pursuing my research I have learned to see imagining (my own and that of others) through the eyes of the past and have thereby come to understand many things about it. Here are just a few. Imagination is both familiar and elusive. It is not always easily distinguishable from sensation, memory, and intellect. It is intertwined with feeling and desire, and it can scarcely be understood without situating it in the entire "economy" of human cognitive and sensitive powers. It provides the element in which, as human beings, we live, even more fundamentally than fish live in water. Efforts to confine it to just a part of life (e.g., childhood, the arts, fantasy) fatally misconceive it. It is also fatal to conceive it as simply "having images in mind." If it is a source of creativity, that is because it is also the power of familiarity that provides the backgrounds against which our ordinary experience takes place. Backgrounds are elusive by their nature, however, because as soon as we turn our attention to them they lose their background character and become foreground. Backgrounds are, moreover, constantly changing: not just from epoch to epoch or person to person but even for a single person—over time, to be sure, but that time is sometimes just the passing of a moment. Against different backgrounds things appear to change—and sometimes it is not just a matter of appearance. If the relationship between object and background is changeable and often delicate, and if it is in this kind of situation that imagination excels, then perhaps it is inevitable that the more we try to bring it into sharp focus the more easily it slips away. Perhaps it is not surprising that not just today but for the past 2,000 years and more thinkers have almost always mistaken imagination for something else.

Almost always. What I have learned from reading and teaching the greatest[1] of the philosophers of imagination is that many of them—let us mention here only Plato, Aristotle, Descartes, and Kant—are largely exempt from this criticism. But this claim needs immediate qualification. What these figures have to teach about imagination is not simply what encyclopedias and other reference works or even monographs assert as their respective "doctrines."[2] They have in fact given us more than we have managed to see. It is as though we (a "we" that includes most commentators, historians, philosophers, and scientists over more than two millennia) have stared at their words with merely partial comprehension, then cobbled together some approximation, some simulacrum. Our accounts of what these originating thinkers wrote say at least as much about us as about them, as much about how we want imagination to be as what they thought it is. The situation is by nature ripe for confusion: about imagination, about what distinguishes it from rationality and other powers, about the nature of mind and soul, about ourselves. It should be no surprise that, as a result, studies of imagination almost always lose track of what it is—occasionally from the very first words.

Contemporary cognitive and neurobiological research has made relatively little progress in understanding imagination, especially when we compare it (for example) to research about vision and memory. One of the most elementary reasons is that the common scientific as well as popular and philosophical conceptions of imagination are dominated by an inappropriate model that misconceives imagining from the start. This misconceived model—which models imagining on the mental envisioning of an absent object[3]—is deeply rooted in the past. By itself, that fact alone justifies our looking backward, investigating historically. Even if it turns out that past philosophies of imagination are the nightmare from which we are trying to awake, it would be best to wake from those dreams with real awareness and face the consequences squarely.

But, as with most things, it is in the details that one finds both the devil and the divine. There is no doubt that the model of imagining as holding in mind a visual image is based in a theory of mind that arose centuries, even millennia, ago. But not all past thinkers have conceived things according to this model and its stereotypes. For example, if you look to what Plato, Aristotle, Descartes, and Kant wrote about imagining and try to place those conceptions within the context of their understandings of human psychology as a whole, and then locate that psychology within the

---

[1] In ultimate matters of the human spirit, the concept of greatness is essential. Although that is a subject for another day (as is also the discussion of spirit—I, too, have read Derrida!), one can make the case briefly. Greatness in philosophizing does not necessarily imply social power, morality, or even evident truth, but it does mean that someone has tried to think up to the level demanded by the phenomena in question and has to some ample and therefore inspiring extent succeeded.

[2] I do not mean to be simply dismissive of scholarly work. About every subject and author there are many good and even excellent studies, and even lesser ones have merits from which one can learn. Imagination is more difficult than most subjects, however, and in the indispensable authors it is necessary (as I will show) to understand what they say about imagination against the background of their highest philosophical ambitions. The best and most stimulating of the encyclopedic sources is without doubt Brann 1991.

[3] As will become clear, this kind of imagining is too limited and atypical to serve as a paradigm.

larger framework of their overall philosophical concerns, you find that they do not adhere to the stereotype. Even where they seem to accept or even ground it, the "acceptance" is hedged round with so many qualifications and caveats that the model turns into something else entirely.

What is perhaps even more surprising, given the usual portrayals of how much philosophers disagree, is the existence of remarkable parallels and continuities between what Plato, Aristotle, Descartes, Kant, and a small number of other thinkers have understood about the imagination—even if the later thinkers were not fully aware of the earlier traces they were following. To a surprising degree, these thinkers share a set of topics, themes, and orientations—a *conceptual topology*, to introduce a term that I shall develop more fully in the rest of the book. Despite many conspicuous differences between their accounts, these thinkers can be seen as working within and developing a common tradition. Yet basic features of this topology disappear in the works of their followers and even more in the secondhand accounts of later interpreters. To give just one example, albeit one that will be of core interest throughout this book: almost all the thinkers I will treat held a conception of imagination as a matrix or topographical power; that is, as involving *naturally* and *artificially* articulated fields, backgrounds, and foregrounds within which images emerge, are formed, and are determinately placed, and in the context of which the mind can engage in imaginative movement, in imaginative work and play. Such fields and grounds, such places of imagining, become routinized and familiarized in habits of mind, for good and for ill. These matrixes and topographies are also subject to being improved and reconfigured—more often for good than for ill, or so one hopes. Imagination is therefore the power human beings have of situating and reworking the appearances of things against, or among, different backgrounds, foregrounds, frameworks, and fields.

Within the conceptual topology of matrixes, human imagination comes to appear as more about making and remaking, contextualizing and recontextualizing appearances than about envisioning and fixing them in mind. The foregrounds and backgrounds of imaginative contextualization and recontextualization are, in turn, the element of thought's mobility, flexibility, and amplitude.[4] Without sensation there could be no such matrixes and topographies, and thus no starting place, for imagining; without the matrixes of imagination there could be no *effective* reason. They establish the fundamental characteristics of human imagining and mental activity; in particular they are the ground for both intelligence and creativity. But in the history of imagination, expert theorizing and subsequent popular opinion have almost always preferred models that overlook this. And the fact that critics and historians have almost always focused on theories and models means that they have been blind to the conceptual topology that underlies them and maintains unities where the critics and historians notice only differences. As a result, our theories and our practices of imagination have become arbitrary, unfocused, and placeless, and our conception of rationality has become ungrounded.[5]

---

[4]This sentence may be taken as the positive leitmotif of the book.

[5]This sentence may be taken as the critical leitmotif of the book. An example of unfocused, ungrounded reason is the notion that rationality can be purely *procedural*. There are no processes of mind and soul more rigorously and logically procedural than psychological compulsions.

There is at least one corollary that ought to be drawn here, right at the outset. Too exclusive an attention to any single power inevitably distorts our conception of it, and confusion about one power extends itself into confusions about others. If we, if our philosophical and psychological traditions, have repeatedly mistaken imagination, that means that we have inevitably made mistakes about sensation, memory, and reason as well. The most successful attempts to understand imagination have been those that do not isolate it or explain it as though it were a module, routine, or procedure separate from other human psychological powers.[6] Human imagination cannot be properly conceived apart from sense perception, from memory, and from rationality—nor even from pain, pleasure, aversion, and desire. Imagination is understood most clearly and amply when it is seen as integrating other human powers, as the matrix of the entire economy of the psyche. To use a metaphor that will gradually take on greater concreteness and urgency: imagination provides a place where the psychic powers *co*-operate in locating the possibilities and the faces that the world presents. Imagination is the human power that textures and contextualizes what we experience. It is the contextual and contextural matrix of experience.

The last sentences sound very much like definitions, albeit definitions with wordplay and new usages. At the beginning of Chap. 2, I will give a very complicated definition of imagination, the immediate point of which will be ironic: lengthy, accurate definitions rarely do much good for readers, especially at the outset of inquiry. If I were asked to give a quick-and-dirty definition, however, something that might help a reader begin to concentrate his or her attention, I would suggest this: imagination, or rather typically *human* imagination, is the cultivation and rational placing of images; it is reason's work with emergent images.

Definition invites questions, of course. For example: Does this quick-and-dirty definition intend to reduce imagination to a form of reason? (No.) Does it allow for some other kind of imagining than rational imagining? (Yes—there is animal imagination, though we must immediately remind ourselves that human beings are animals and that animal imagining may well observe some kinds of ratios and thus be, in a sense, rational.) How can imagination be defined as rationally placing images without explaining how we get images in the first place? (Perhaps animal imagination is *elemental, emergent* imaging; human beings would then have this and something more, or at least a different inflection of the elemental–emergent.)

Definitions settle very little, and the questions and contentions they raise are too complicated to resolve in parentheses. My parenthetical reflections are only suggestive anticipations of what is to come. The rest of this book, I hope, will provide ample evidence that these anticipations have real substance, and in surprising ways. If our histories of imagination have tended to obscure as well as illuminate the past, we might expect that there are certain traditions and conceptual topologies of imagination that have been lost from view, and in that sense they might be called *lost* or *concealed* or *occulted traditions*. It is even possible that some have been so well

[6]It is this conceptual isolation of psychological powers from one another, rather than the concept of psychological powers per se, that is the principal source of the disrepute of so-called *faculty psychology*.

concealed that they are scarcely evident even to people who work within them: the invisible heart of our most familiar conceptions of imagination.

Statements like these have more than a whiff of paradox, because they suggest that there are unconscious traditions full of gaps. But then how can they be traditions? It would seem that it is of the essence of tradition to be deliberately handed on from past to future. If we now try, by looking and arguing backward, to see the past as offering an unconscious and "gappy" tradition, wouldn't we be trying to create a tradition artificially? Wouldn't it be an act of dishonesty to treat a series of loosely connected historical episodes as a real tradition? Yet there is an important sense in which traditions by their nature have to be constituted by looking backwards, and in that sense *antichronologically*. The originator of a tradition is usually too busy doing his or her work to worry about establishing a tradition. It is only those who come later who feel the authority of the origin and become concerned about faithfulness to it. Without that active concern for faithfulness to the past, their work would be merely habitual, not traditional.

Whether we today still have a sensibility for such distinctions and concerns is doubtful. To use an old rhetorical trope: if we stand on the shoulders of those who came before us, we nevertheless tend to imagine that it is *we* who are the giants. We may be postromantics rather than romantics, but we still believe in the romantic myth that genius—our genius—has direct access to the truth of the world, with no more than accidental reference to the past.

But things are *never* that way. Everything we say, think, and do is adumbrated in what has come before us; and in the very languages we speak we are unconscious heirs to millennia of conceiving and speaking the world we inhabit. We live in matrixes of the past and the present that are largely products of the imagination of those who came before us. Insofar as we take over these matrixes, we reproduce their possibilities, at least until the tides of change swallow or transform them. This is another of the occulted grounds of imagination: that human imagination is communal as well as individual, that it is both creative and routine, that innovation is correlative to what is established. What I call conceptual topology is meant to capture the character of imaginative matrixes as both current and historical grounds and backgrounds of our living (and not just of our thinking). They are the places of our thinking, acting, working, and behaving. In this sense "conceptual topology" is no mere conceit but a concept that helps us articulate the peculiar character of imagining and its ground. Conceptual topologies are communal matrixes,[7] yet they can function only by investing the minds and hearts of individuals. It is by the marking, habituation, and regeneration of such matrixes that we set the course of our lives and hand over future possibilities to new generations.

This means that the stakes of imagination are very high. If we do not properly understand imagining, we fail to understand our world, our individuality and community, our knowing, our acting, ourselves. Such failure is no small thing.

[7]This is true even of the rare cases when a new matrix or topography is invented by an individual. It is almost invariably produced against the background of already existing topologies, and therefore even what is new in it is shareable or actually shared in action, work, and speaking.

If a thinker from a century, or four centuries, or two millennia ago provides us with resources to find our way in these difficult matters, then we must not be so proud of our real accomplishments that we end up neglecting what can supplement our shortcomings and perhaps lead to ampler satisfactions.

Despite the historical trappings—or, as I hope will gradually become evident, precisely by means of the historical investigation—the principal aim of this book is to clarify what imagination is by helping readers understand the phenomenon of imagining. I use the singular "phenomenon" deliberately: although I will be considering many different kinds of phenomena and many theories and claims about them, I am persuaded that there is an elemental, core phenomenon present in every more complicated form of imagining. Of course something elemental is not necessarily simple. There is sufficient structure in the core phenomenon of imagining to justify seeing it as complex: a complexity to which is due most of the historical varieties and elaborations of imagination that have contributed to its elusiveness and our confusions.

More than new hypotheses about imagination, we need to understand imagining as it is experienced, with all its territories and regions. How can we judge whether a hypothesis is a good match with the phenomena, whether it is a good map of the territory, if we do not have a serious and extensive familiarity with what the hypothesis is about? It is, of course, possible to argue that precisely having a decent hypothesis and background theory can make one's encounter with the relevant phenomena much more productive. To see what is in front of one's eyes, it helps to have a map. But in a territory as much traveled as imagination has been, over millennia rather than a few centuries, it is hard to avoid the well-traveled routes, which in some cases would be better named routines or even ruts—conceptual and theoretical ruts, as well as practical ones. Against one's better knowledge one finds oneself, over and over again, thinking, saying, and doing things not because they are most apt but because they are the things that people have thought, said, and done before. Even very ambitious theoretical works that have the stated intention of overthrowing the past are offenders.

Recognizing that one is in a rut, or at least a routine, does not guarantee that one knows how to get out of it, or whether one should try; after all, truth may in some sense involve routines, if not exactly ruts. Moreover, "overthrowing" the past in a dramatic act of rejection is unlikely to be successful. Philosophers and scientists are often motivated by the desire to be done with the falsities of the past, to turn their back on what has failed, to strike out in new and revolutionary ways. Revolutions, alas, have an inevitable tendency to bring us back to where we started. That is not a counsel of despair, however. There is something right, *well oriented*, in the will to see and do things differently, whenever we notice something wrong in how they were seen and done before.

Whenever people have made the persistent attempt to look at and to describe faithfully what they see and do, it is very likely that they got a great deal right. There is little doubt that *some* things they said are simply wrong and therefore ought to be rejected. But in most cases (beyond the simplest) it is not easy for those who come later to give a clear and full accounting of why something earlier is wrong and ought to be rejected. With most things, when we see some reason to think they are wrong we just turn away. It is like abandoning a residence that no longer works as a home. The residence is not totally worthless but rather inconvenient, unsuitable, badly situated. Often it is only after moving to a new place that we can see clearly what was wrong with the old one. If we are honest, we can also see that there were certain virtues in the old place that are missing from the new. We come around to a fuller, juster, more adequate sense of things. Perhaps, then, it is important to take the topos or theme of revolution more seriously than we ordinarily do. A revolution takes us completely around what we orbit. This gives us a chance to see things from not just one angle but from many, from *all*. Revolutionary knowing in the sense I am raising here aims at amplitude, at leaving nothing essential out. The attitude of rejection, on the other hand—call it rejectionism—is an offense against knowledge.

This is one reason to undertake the apparently unoriginal work of explaining what others have said and thought. Yet to explain imagination in Aristotle or Kant or modern linguistic philosophy is no mean undertaking, and surveying imagination in many lands, in many fields, over many centuries is in some senses more ambitious than to profess a new theory. It can also be more productive and even original, insofar as it shows us something of the ground or territory on which good, relevant theories can be built today. Philosophical historiography can set standards for whether new theories, contemporary or ancient, are ample enough to cover all the phenomena—and whether the supposedly new theories are so novel after all. The flip side of assuming that the past is irrelevant to the truth of things is antiquarianism—which is why so much philosophical and intellectual history gets written in abstraction from whether past accounts are true and makes a fetish of presenting thinkers in so detailed a historical context that the questions of our own age are muted to inaudibility.

This is not to dismiss, much less to condemn, love for what is old. It is rather to point out something that is often overlooked by the pride we take in what we consider our best knowledge: that in trying to know more about X, Y, and Z, we usually ignore or even forget A through W. In many circumstances this is harmless; but in some cases it squeezes the life and truth from our knowledge. This is as much the case in the natural and social sciences as it is in the humanities. It is a genuine paradox whose consequences we live every day: we have to ignore many things (which means "not know them") in order to know others. Often we act as though, eventually, we will be able to make up for our present ignorance, whether deliberate or accidental: we will one day know enough to act always in full knowledge of what we are doing. But that assumption transgresses a fundamental truth that is almost as old as philosophy, and certainly as old as the philosophizing of Socrates of Athens: no matter how much we know, we never have enough knowledge to know exactly what to do next, or how to do it. Knowledge is never enough.[8]

****

[8]By associating our best knowledge with pride I am not trying to diminish the modern sciences, which certainly embody some of the amplest and most accurate knowing human beings have achieved. It is instead to remark the ambition inherent in all claims to knowledge, and to intimate that the more strongly we assert claims, the less likely they are to be adequately supported by what we can show. More than knowledge—more even than the desire for knowledge—is at issue. In Chap. 4 we shall return to this theme, in the philosophically familiar form of the Socrates who knows that he does not know, and see that this Socratic trope in fact comes to a head where reason has to work with, place, and delimit images.

There will be plenty of history, theory, and description in the following pages, but all of it aims to clarify imagination, to bring to better focus what it is, what it is about, and what concepts most usefully and amply address and express its character. But none of that will mean anything if the reader does not also deepen his or her experience of imagining. Imagining is a practice. Neither *history*, nor *theory*, nor *phenomenal description* can be an end in itself. The most precious thing we can discover from past writers is hints of what they knew of the practices. From those we can begin to glimpse the full nature and scope of the actual imagining they tried to conceive in their theories.

Through years of teaching I have gradually learned that old texts from great minds can stir us out of inveterate ways of seeing, thinking, and speaking. The great minds have often founded schools, or had schools founded in their names, but no school really "contains" them. Unlike the followers who bind themselves to the conventions of a school or a master, the true masters' writings have a freshness always waiting to be rediscovered. But finding what is fresh requires working hard to follow the intersecting lines of their thinking, beyond the routines of schools and our own contemporary conventions. If a philosophical text is itself a net of words, a matrix, when it is most amply thought it is also most *concretely* conceived and imagined. The texts are, as it were, woven around the things they treat. And that saves them from being only text—or, rather, it opens up a way of understanding the nature of text and texture that will bring us closer to the experience of the things the thinkers thought.

It is possible, even certain, that thinkers other than those I have chosen to discuss in this book would have served as well. But if you are going to choose a handful of thinkers who most decisively shaped our conceptions of imagination, I do not see how you can avoid making four of them Plato, Aristotle, Descartes, and Kant. That they revealed far more about imagination than we ordinarily take them to have said makes them interesting as well as historically important—with the reminder that "interesting" means, etymologically, being among things, being in the very middle of what we are concerned with. My hope is that, by reactivating *their* thinking about imagination in a sufficiently ample way, *we* can gain a greater amplitude and freshness to our own thought. If the past cannot displace present thinking, it can nevertheless supplement and complement it—which means to supply what is lacking for the sake of bringing it nearer to completion. If sometimes my history-writing appears a little too detailed, that will be balanced by moments where it seems (especially to scholars) too sketchy, speculative, cavalier, or just mistaken. The only way to take another's thinking seriously is to try to think it for oneself, with all the perils that implies. In the face of a thinking that is both detailed and ample, one's own almost always falls short, even fails. Nevertheless, even one's failures can be suggestive if they are able to give indications of where the original thinking was headed as one's own gives out. And an awareness that a thinker was headed somewhere is itself already an act of *philosophical imagination*, a recognition that philosophizing is not just uttering and logically testing propositions but also orienting and placing our thoughts.

The theorizing I do here is almost all historically contextualized. There will be occasional moments where one or another historical concern comes to the fore, but my steady aim is to explain the conceptual topology of imagination as such. Here I will anticipate what that means by remarking that a conceptual topology is more basic than a hypothesis or even a theory. A topology can, and typically does, give rise to many different theories, even conflicting ones, in a single field of concern. In the first instance the field is an *interplay* between the basic topological concepts that mark the field of concern; the conceptual marking is the topology's way of placing phenomena in the field so that they appear as intrinsic to it, and with respect to which we locate the things of the field and our relationships to them. A phenomenal field with conceptual marking is a matrix or topography. The phenomena are *understood* when they are *properly* placed and *show themselves* as properly and *adequately* placed. At the core of every theoretical tradition, of any kind, there is such a matrix or topography; at the core of the different matrixes is the conceptual topology.

But that explains "conceptual topology of imagination" only as an objective genitive.[9] The aim of this study is not to apply extrinsic notions to imagination but to show the networking of concepts that are intrinsic to it. "Conceptual topology of imagination" is more essentially a subjective than an objective genitive. The human mind operates in its most fundamental senses by way of one or another conceptual topology, as the mind or its possessor finds, intensively grasps (*con–cipere*, according to etymology), and evolves the structures of a field of appearance. The very possibility of our having something like a marked field of concern is grounded in imagination. Thus there are likely no conceptual topologies without imagination, and no significant imagining without such topologies. If, then, the historical chapters show that there is in essence a common conceptual topology of images and imagination that has evolved from thinker to thinker, and if the features I gather from the history provide a large and ample sense of the phenomenal and conceptual scope that any minimally successful theory of imagination as appropriately topological must have, then I will consider that to be tradition and theory enough, for this book at least.

In the course of the exegesis of texts and what they are about I shall frequently undertake *phenomenal description* and *redescription* of acts of imagining, beginning with Chap. 2. If this book were an attempt at a structural account of imagination, say in the manner of twentieth-century, Husserlian phenomenology, it would have to be considerably longer. Perhaps in a more logical world I should have addressed this kind of phenomenological task first. But in a more logical world, philosophers and psychologists would long ago have already done such work. If, through the history, the reader sees how much has been forfeited by misconceiving imagination, this book will have served an important purpose. Yet even under the regime of Aristotle's motto, that there is no thinking without images,

[9]That is, it takes the grammatical object of the preposition "of" (imagination) as our object of study. But if we take imagination as the subject, as the *possessor* and *practitioner* of a conceptual topology (the subjective genitive), we view it as intrinsically occupying the conceptual topology that is proper to it.

imagination is something more and something other than a merely cognitive function. Treating it as though it *should* be cognitive is not only to misunderstand it but also to turn it into something other than what it is. If, by contrast, we recognize that imagination achieves its essential nature *as soon as it commences*, quite apart from our epistemological hopes and demands, and that it achieves this nature by the placement of appearances with respect to foregrounds and backgrounds, the phenomenon of imagining begins to take on a radically different character that has largely escaped almost everyone's philosophic and psychological ken. Chapter 3 will capitalize on these insights by developing the mutually reinforcing notions of matrix/topography and conceptual topology; it will show how they can help us conceive and recognize the pervasiveness of imagining in all kinds of theoretical, technical, artistic, and practical activities. This awareness, in turn, readies us for the ambitions of the more intensively historical parts of the investigation in the chapters that follow. As we shall see, the characteristics that a brief practice of actual imagining reveals are for the most part already identifiable in the classic philosophers of imagination. If we have overlooked them, the fault is ours.

The history in Chaps. 4, 5, 6, 7, and 8 will enable us to see that a basic conceptual topology of imagination has governed the *entire* history of imagination in Western thought, not excluding the present. Yet the line of thinking I trace and the full conceptual topology on which it draws are far richer and subtler than any of the specific theories that have been built upon it. If philosophy and psychology have, over the past century and more, been suspicious of older traditions, that suspicion has come at a high cost. It has led philosophers and psychologists to abandon traditional concepts and to ignore and overlook the conceptual topology to which they are still beholden. It has also deprived them of familiarity with basic facts and phenomena of imagining, facts and phenomena that were part of daily experience for those of past eras who were proficient in it. Philosophers and psychologists nevertheless (or perhaps one should say "therefore") have made unconscious use of and reference to imagination's topologizing, matrix power in the very attempt to suppress it. We might hope that the future science and philosophy toward which Chap. 9 points will cease ignoring these things and become more amply aware of what they are doing.

References

Brann, Eva T. H. 1991. *The world of the imagination: Sum and substance*. Savage, Md.: Rowman & Littlefield.

Castoriadis, Cornelius. 1997. *Fait et à faire. Les carrefours du labyrinthe, 5*. Paris: Seuil.

Chapter 2
Locating Emergent Appearance

There are two common, inveterate, even insidious misunderstandings of imagination. The more recent one identifies it with creativity. In a culture and civilization that not only prizes innovation theoretically but also rewards it practically, it is not surprising that the emergence of what is unprecedented would be valued highly. In such a culture, "imagination" is the answer to the question "what enables us to make something new?" Insofar as the answer is routine, it is more label than understanding. We do and make new things; whatever allows us to do this, we believe, must be the principle and origin of creativity. But suppose that what allows us to innovate is also the basis of routines and habits? What if the very power that allows us to innovate in desirable ways is exactly the same one that allows us to fall into ruts, and even to be *destructively* ingenious? Questions like these are unsettling, since culturally we have so much invested in imagination. We cannot easily respond to them because our culture has misplaced the resources required for an answer.

The second inveterate and insidious misunderstanding will probably not even strike the reader as problematic. For almost as long as anyone has thought about imagination in an organized way—in Western civilization for almost 2,500 years—the prototypical model of the imaginative act has been *visualizing an absent object.* "Imagine a friend," "imagine Jean-Paul," "imagine the Panthéon in Paris," "imagine Jean-Paul standing with a friend in front of the Paris Panthéon and counting the columns": not smells and textures and sounds come immediately to mind, but vision, with greater or lesser detail.

What I refer to as insidious is not even, in the first instance, the set of questions that rapidly come to mind as soon as you follow the injunction to visualize–imagine and then think about what is happening. Go ahead, imagine an oak tree; then ask yourself whether you are imagining it or remembering it. Try to remember an oak tree,[1] then try to imagine it. What is the difference? If you are remembering the

[1]Of course it does not matter that it is an oak tree. Other series of questions, too, are possible. For example, one might begin asking how *detailed* the remembering–imagining is, whether it includes leaf shape and bark texture, whether having an oak in mind differs from having a chestnut or a maple.

D.L. Sepper, *Understanding Imagination: The Reason of Images*, Studies in History and Philosophy of Science 33, DOI 10.1007/978-94-007-6507-8_2,

tree: are you remembering *the tree* or are you remembering an *event* of *seeing* the tree? (In the second case *your* involvement as *seer* moves into view along with the tree.) While holding it in mind, is "it" a tree or an image of a tree? To put the last in a slightly different way: if you are successfully imagining or remembering a tree, is the successfully imagined tree an imaginary tree, is it a real one that has been somehow translated "into" your mind, or have you set up some complex reference from a thought to a thing? Where, when all is said and done, *is* this tree or imagining? In your head? In a real space? In an imaginary space? Is this space (of whatever kind it is) in your mind, somewhere else, many places, or nowhere?

This only starts the questioning about an apparently very ordinary psychic event. In this and similar cases it is not immediately clear what the right answers are. What is more frustrating, the answers depend to a large extent on the theory that you hold about mind and imagination, and strong defenses can be provided for conflicting accounts. As with all so-called introspective techniques, there is doubt whether these psychic events are verifiable, repeatable, or even properly describable. No one else can confirm what has happened or verify the terms used to describe it. The very act of wondering about what is going on may well change the character of the event.

Now it may be that there are certain things and experiences—imagination perhaps being one of them—that cannot even be described without implicit or explicit theory. In the natural sciences this is commonplace with regard to things that lie far above or below thresholds of perception, like the distribution of galaxies in the universe or patterns of events at the quantum level. But what is really insidious about descriptions of imagining is that, with hardly a moment's reflection, one slips into and begins addressing psychological, epistemological, and methodological issues like those raised in the last two paragraphs, without asking whether, how, and why visualizing an absent object is truly representative of imagining. In the blink of an eye, one loses sight of one's aim and gets caught up in the well-developed but routinized conceptual machinery of formalized theorizing. Might holding an absent object in mind turn out to be too special a case that, in its specialism, leads us down unrepresentative paths? A question that one should ask very early on, whether there is a basic act of imagining and what it looks like, gets overtaken and eclipsed by other, more complicated concerns.

What *are* the truly representative phenomena of imagining? It is easy to adduce phenomena, one after another. Yet without some principles of selection and organization it would be impossible to say whether they were basic, or representative, or even, properly speaking, imaginative.

If one has to have a theory in order to describe imagination, it might be appropriate now for me to enunciate basic convictions. For instance: I might claim to know that human beings cannot think without imagining; that many other kinds of animals, too, have at least some basic imaginative powers; that in human beings imagination has as much to do with knowing (thus science) as it does with doing and making (thus human action and art) or with musing (as in daydreams); that it is the source of *familiarity* as well as of *creativity*; that it is radically individual but also social and linguistic; that it is both subjective and objective and thus really neither; that it is humanly elemental but, for all that, not simply reducible to elements from which images are sometimes thought to be constructed; that it is psychological but also worldly; that it is more evocative and incipient than representational or expressive; that it includes *having* images in a rather traditional sense, but only as a very special case; that it is more about something we are and do than about what we have and know; and, last but by far not least, that it is the practice of the psychological emergence, placement, and location of appearance.[2] I might sum up all these things in a definition, like this: imagination is *a (psychologically) evocative, anticipatory, abstractional–concretional activity that follows upon actual perception. It allows the imaginer to (1) dynamically (re)position herself and incipiently explore, place, vary, connect, and re–present appearances originating within a field of concern, (2) attend to and mark the field's potentials, and (3) exploit those potentials by projecting them to other fields (possibly new) in abstracted/concreted appearances.*

Are you, the reader, better off for "having" this definition, which you can memorize and repeat as a "teaching" or "doctrine"? Can you use it as a criterion for distinguishing imagination from other activities, or the elemental from the complex? The definition is long but not overly technical, as definitions go. Most of the terms are familiar to adult native speakers of English, but the whole is by no means self-evident. For example, why are there terms in parentheses? How can something be *both* abstractional *and* concretional, and what do those unusual forms of more familiar words mean exactly? Why are there parentheses in *(re)position* and a hyphen in *re–present*? Why contrast incipience and evocation with representation (or re–presentation)? The problem is not just that definitions use terms that need defining or explaining in their own right. It is even more that their meaning is about nuanced expression within a familiar, articulated setting. To put it more simply: you need to know your way around things, what you are talking about, and how to use basic and sophisticated terms about them. The significance of terms is known first and foremost to those who already work with them in a field where the terms are appropriately deployed. They are part of the *imaginarium* concerning the field, part of the imaginative repertory of those who occupy the field. Dictionaries and even encyclopedias are of limited help if you do not have this background experience yourself. They register appropriate uses in general without being able to supply more than a few indications about the fields in which they are used. More likely, usages of terms specific to fields of inquiry will not even make it into common reference works. Since it is ordinarily those *not* familiar with a relevant field who turn to dictionaries and encyclopedias for clarification, the entries, when they exist, often produce more confusion than satisfaction.

The history of imagination is littered with definitions, theories, and practices of every conceivable kind, for every conceivable purpose. At this historical moment what is more important than a new theory or even a comprehensive phenomenology is a clear conception of what the *basic* phenomena, conditions, and questions are that ought to be at the heart of any plausible study of the imagination. What must precede radical *comprehensiveness*, whether descriptive, theoretical, or some combination of the two, is something less ambitious—though not, for all that, easy

[2] Theseus' description in Shakespeare's *A Midsummer Night's Dream* (act 5, scene 1) captures this last point with perfect tone and emphasis: "And as imagination bodies forth/The forms of things unknown, the poet's pen/Turns them to shapes, and gives to airy nothing/A local habitation and a name."

or unambitious. I think that it is possible at this historical moment to aspire to a lesser but more manageable kind of comprehensiveness: an *indication* of what genuine *comprehensiveness* might look like in the study of imagination and a description of the *conceptual and phenomenal elements* that would be necessary as part of a comprehensive undertaking.

Before any real science is possible it is necessary to have a clear (though not necessarily scientific) grasp of the field one is trying to understand. This typically involves taking account not just of what the phenomena are but also of what those who have preceded us have taken them to be. If a breakthrough is necessary, it is also necessary to know the barrier that needs breaking through. If there is something malformed or lacking in what people have thought, then in order to measure the quality of a new or revolutionary idea it is necessary to assess it against the (in)adequacy of what it tries to overcome.

Too often we are satisfied, in our thumbnail sketches of history, with a parody of what people thought and did before the great innovation. This degradation in understanding history probably begins with the "revolutionary" generation. One thing at least is true: the immediately following generation (the generation of the *disciples*) no longer faces the same situation that the master did. She studied very carefully what was said and done by predecessors; disciples live in a postmagisterial world, in which the past is filtered and reconceived by way of the master's accomplishment. They have not lived with and seen all the things that were part of the master's revolutionary tour of the phenomena. By the second generation any living sense of how those who came before the master experienced and understood things will probably have died out.

With respect to imagination, we do not live in a postmagisterial era. There is no universally acknowledged theory holding the field, but only limited and often defective models. If we expect to make any progress, we need to know where we stand. We must look to what others have attempted and assess strengths and limitations. Whether these others are literally alive or dead is irrelevant. Of course we also need to look to the phenomena: to gain a sense of the totality of what is relevant, to sort the simple from the complex, and to cast a critical glance even at our own emerging conceptions in view of what appears. We need to be constantly attentive to how the phenomena speak to us today, and to recognize the degree to which our experience is shaped by past notions. The combination of (1) trying to imagine things ourselves with an eye to what is basic, (2) reconceiving basic notions and phenomena along with previous theories, and (3) through these efforts developing a sensibility for the elemental relationships between basic concepts and phenomena that encompass the field of imagining—this combination is the limited, but by no means simple, goal of the present work.

2.1 Some Practice of Imagining, and Thoughts About It

The default model of imagination considers examples like that of mentally picturing a tree or other object. Does it make any difference if we substitute a *scenario*—for instance, withdrawing money from a cash-dispensing machine? You walk up to the machine, remove the cash card from your purse or wallet, slide it in and out of the card reader, press the appropriate buttons or virtual buttons, take the delivered cash, put it into your purse or wallet, take the receipt once it prints, and walk away.

Does it make a difference that in this case, as opposed to the case of the tree, you tend to imagine yourself into the scene in a quite active way?[3] Being asked to imagine a tree likely induces us to picture it "objectively," without including ourselves. The case of the cash withdrawal makes observational passivity far more difficult, even if we are imagining someone else performing all the fine and gross motor activities and having all the expectations and intentions that were mentioned.[4] It is highly doubtful that the model of "holding a (in this case *moving*) picture privately in your mind" even begins to describe it, much less account for it.[5]

But why rest content with scenarios that ask you to *view* a familiar object or act, since there is more to imagining than eye can see? Let us change the imagining's kind and simplify. Imagine the smell of cinnamon. Let us repeat the first question that I brought up with the oak tree: are you remembering or imagining? It is scarcely credible, on the one hand, that if you have never experienced it before you will be able to evoke cinnamon's smell, even if it were described to you by someone with the narrative powers of Proust. Something like memory is involved—but can it be *pure* remembering, *without* imagining? What would that be? If the aroma you evoke is not *exactly* what you experienced before in a specific moment of cooking or dining, that would be an indication that something more, or other, than memory is at work.[6] If imagination is a special variety of memory, what would the special circumstances be that make that variety of remembering an act of imagination? On the other hand, some people regard remembering as the derivative action: that it is actually a special kind of imagining—imagining with a time stamp, so to speak. At the very least it seems that there are good grounds for differentiating the two, though exactly how is far from clear.

Let us move on to a slightly expanded exercise. Imagine the aroma of cinnamon mixed with that of nutmeg. (Stop reading for a minute—the next sentence will wait. *Really* imagine. If cinnamon and nutmeg are too elusive, try other smell combinations: apple and pear, strawberry and blueberry, or any other aromas, artificial or

[3]The imaginative cash withdrawal was described in terms of *your* doing it. Repeat the exercise by imagining that a friend does it, then just an anonymous "someone." Ask yourself where *you* are in the latter two situations. There are likely to be surprising differences in your overall experience of the imagined scenario. Questioning like this is pursued in Husserl 1980.

[4]This sentence puts the apparently objective description of the previous paragraph in a new light: no intentions were mentioned there, and fine and gross motor skills were no more than implicit. By calling specific variants to mind one can begin explicitly imagining the cash withdrawal in new ways.

[5]See Sect. 2.3, below, for a discussion of the Bergson–Deleuze–Agamben line of criticism of the inadequacies of the photographic image for understanding both static and mobile visual phenomena.

[6]Notice that this does not necessarily raise the epistemological issue of how you would *know* the aroma is or is not the same, nor whether the difference from remembering would be explicable in terms of a weakening or strengthening (of vividness).

natural, that work for you….)[7] Did you succeed in accomplishing this right off? Did you have to recall separately the aromas of cinnamon and nutmeg, then alternate "in your imagination" the aroma of cinnamon with the aroma of nutmeg? Do the individual trials undergo evolution or development, and does your ability to accomplish them gradually improve? After performing the trials several times, do you find that you can more quickly call to mind the two smells, individually or together? Can you evoke now the smell of nutmeg with a trace of cinnamon, or of cinnamon with a trace of nutmeg? Can you evoke varieties of these aromas that you are fairly sure you have never in fact encountered before?

Now shift focus a bit. As you try to conceive the smells, have you been *visualizing* cinnamon and nutmeg? If so, is it ground or whole? Have you been envisioning any texture, any colors? Has there been in this basically olfactory imagining an inkling of your taking deep breaths to savor the imagined aromas? In what space or place have you done this? Whatever you have imagined: are you able, when you try again, to eliminate some features and add others? Are you, in all this *imaginative work*, an impartial witness, or an active participant? Are you, in imagining, in any particular attitude or frame of mind? For example, is there a different attitude or frame when you are concentrating solely on the aromas, as opposed to a combination with, say, color and texture? Do you feel freer in one activity than another; do you sovereignly survey the play of aromas but struggle to conceive the colors? What kinds of pleasure or dissatisfaction do you feel? Can you imagine, and imaginatively vary, these feelings? Can you make yourself shift among these attitudes, frames, and feelings?

Last, and back to one of the old standby questions: does it seem to you that in all these imaginings you were exactly repeating what had already occurred to you in previous experiences? Even if it starts out as an exact memory, does the repetition that occurs in imaginative variation turn an original memory event into something different? Are you simply reassociating preexisting elements, or is something (possibly) unprecedented happening that makes talk of "preexisting elements" and "simple reassociation" beggar the reality of the event? How strongly does it have to "seem to you" that something is the case in imagining for you to have the confidence to pronounce it as *actual* or *true*? If you now go to the kitchen, combine some cinnamon and nutmeg, then discover that your imaginings do not match the real aroma, does that mean your imagining was a failure? Whether you imagined "correctly" or not, did your imagining prepare you in some way to appreciate better the fact you have now "ascertained" at the kitchen counter? There are many more questions than answers.

---

[7]Let me offer a general apology for this kind of authorial importunity. But remember: Books and articles about imagination lacking evidence of the author's acquaintance with the power are legion. One might even rank this as another insidious and inveterate misunderstanding of imagining: that it can be understood without doing. It is crucial to *experience* imagination *actively* in order to *understand* it. That you once did some relevant imagining—or dismissing claims because you recall having once previously thought about them—is not good enough.

Considering various imagined aromas by no means overturns the visual model of imagination, nor will it immediately be felt as a challenge to the primacy or representativeness of the visual model. If nothing else, habit is strong. On the other hand, there is a certain simplicity that attends imagining smells that is far more difficult to achieve with vision, and epistemological questions about their accuracy or exactness, though relevant, seem less urgent. One of the difficulties in researching imagination is that it is hard for philosophers and scientists to forbear prematurely asking epistemological questions, questions about the how and the why and the how accurate. They are premature because it is not at all clear that imagination should be assessed, primarily and for the most part, as though it were a form or anticipation of knowledge. This is clearer if we turn to fictional objects of imagining.

Suppose that I ask you to imagine as clearly as possible the hero of a favorite novel or play—say Hamlet. Then I ask you to tell me what color his eyes are, how many moles he has on his back, or whether he is right- or left-handed. Unless you have been asked such questions before, your imagining has doubtless not extended to all of them. (If it has, then think of a feature you have not previously imagined.) Once you have such a feature in mind, the situation can no longer be simply confused with remembering or prior perceiving; it seems to provide a purer example of imagining than any of my previous ones. You are not simply recalling something experienced, even if the result seems explicable as a "cut-and-paste" job (that is, as a "new" image produced by taking parts of several others you have experienced and recombining them). Recombination has traditionally been a favorite theory of imaginative novelty: the elements are old, only the arrangement is new. Perhaps that is what happens in imagining Hamlet this way.

It is hard to know, on the face of things, what this Hamlet-imagining proves. Still, it *does* certainly establish something that does not hold of sense perception or memory. In the presence of an actor playing Hamlet I may not have noticed his eye color before, but I can look again; if I am counting on memory and realize I did not notice the eye color, I can go back to the theater or look for photographs of the actor (granting, of course, that the actor cannot simply be identified with the character he plays). Even if there is sometimes, or even always, some remembering in the course of this imagining, it cannot be totally reduced to memory. More exactly: if memory presupposes an original that we more or less successfully, even if not perfectly, recall, imagination rarely has such a clear-cut standard to which it can appeal.

When I try to imagine Hamlet, "pure and simple," it is my conception of the character that is at issue. If I do not at first include eye color in my "inward image," I cannot expect any new information from just holding that first image unchanged in mind. This is not to say that I cannot reconceive the image of Hamlet I had a moment before, now with the desired or plausible iris pigmentation. In seeing and remembering, the goal is a more or less stable image corresponding to some real standard. But imagining counts on the situated emergence of images and their flexibility, and accordingly it takes given images as opportunities for variation, divergence, and situational change. Variation and divergence appear as enemies of accurate remembering, but they may well work to imagining's advantage. They are shortcomings only when we aim for a stable or standard object. Yet imagination can in fact supply itself with a standard: one act of imagining can serve as a reference point for successive ones, and imagined objects can often be stabilized *for the time being*. It could be that imaginative fictionalism and lability are in many circumstances virtues rather than vices. It is conceivable that precisely these characteristics allow us to adapt imagining, especially in its hypothetical modalities, to perceptual and cognitive uses. In view of these considerations, it could be that imagination's virtues are intrinsically transitional, that they help us along the way to something else (e.g., an accurate memory of a thing or event reconstructed by stages, or a progressively elaborated work of art, or a better-focused concept). Whether imagining, for all and any shortcomings in comparison to other, more complete acts of mind, has positive virtues of its own would be plausible but still insufficiently determined.

So where do we turn for answers? Imaginative tasks and reflective questions can be proliferated and complicated endlessly. If this book simply aimed to cultivate the reader's imaginative abilities it would need now to follow an appropriate pedagogical strategy and emphasize working on sequences of particular acts of imagining. Self-help books cultivating imagination often try to develop in the reader a specific sense or talent, often for artistic purposes. They proceed from elementary exercises to complex applications. Such approaches, however suggestive they may be, are too narrow for the purposes of eliciting very general features of imagining. Yet if wide experience of imagination in all its forms, developed in many ways and taken in many directions, is prerequisite for a truly ample understanding of what imagining is, what it can do, and what its relevant fields or matrixes are, then how can we avoid undertaking an ever-widening practical phenomenology of imagining—one that employs each and every aspect of our imaginative capacities?

A more manageable strategy, somewhat closer to the purposes of this book, would have us turn to a rigorous presentation of the leading phenomena of imagination, without the intention of practically developing our each and every imaginative capacity. That strategy might well eventuate in a phenomenological inquiry along lines pursued by Edmund Husserl (1859–1938) or Jean-Paul Sartre (1905–1980).[8] By declining to follow this route I do not intend to minimize or derogate from what phenomenology can achieve. Many of the questions I have already posed in this chapter can be enriched by reading even just a little Husserl. The formal phenomenology of imagination ought to be continued on every front, not least because its central aim—to differentiate and identify fundamental features of intentional acts of consciousness—is consonant with my concern to identify elements of imagining.

Yet, as much as other approaches and schools, phenomenologists have implicitly (and sometimes explicitly) taken visual imagining and the visual image as paradigmatic. Thus, against the fullness of their experience, they have tended to fall into

[8]Husserl apparently intended to produce a study of imagination, but his many reflections on the topic, beginning quite early in his philosophical career, were occasional and for the most part unsystematic. Some of his notes over several decades have been gathered by later editors into Husserl 1980. The kind of phenomenology of imagination most familiar to contemporary scholars is to be found in Sartre's two books on imagination, written in the middle and late 1930s, one historical (Sartre 1936), the other systematic (Sartre 1940). For a more recent phenomenological approach, see Casey 2000.

epistemology's reifying and objectifying tendencies—tendencies that cannot be overcome simply by suspending the question of the *existence* of the objects of consciousness. This is especially so for Sartre, for whom the question of the irrealizing power of imagination, its negative relation to an object, is overriding.[9]

The visual overemphasis of phenomenology is rectifiable, especially if its investigations into imagination were conducted more broadly. This would also be true for other limitations hitherto. For example, phenomenologists have not always adequately addressed questions of the simplicity and complexity of images and imagining. They typically begin with fairly ordinary objects and situations corresponding to real-world situations: imagining mourners at a funeral, an Oxford classroom where a philosophy professor holds a seminar, or dolphins cavorting in the sea. Whether and how far these can or should be "analyzed" or "decomposed" takes a back seat to phenomenologically more typical questions about noesis (the mental act) and its relationship to the noema or object. The act of imagining is invariably defined in contrast to sense perception, which implicitly reimports a natural attitude about the *unreality* of imagining that ought to have been suspended by the initial phenomenological "reduction."

Phenomenologists take imagining to be a conscious act, and as such it is one of many acts of consciousness. Consciousness is always intentional: it relates the mind act to an object. At a very high level of abstraction, this first-approximation description is not far removed from conventional empiricism.[10] On the other hand, phenomenologists do not shun the task of carefully distinguishing the ways in which the same object presents itself to different intentional acts. Husserl's famous consideration of the perception of a telephone emphasizes that although sides other than the one offered to sense are hidden, the phone is nevertheless perceived as having other sides, and that it is always possible (at least in principle) to turn it around. This contrasts with *imagining* a telephone: what is imagined is an appearance only from the point of view set in the imagining. An entirely new act of imagining would be required for something like an "other side" to appear, even if only as an intended rather than an actually perceived other side—perceived, that is, in imagination.[11]

[9] Sartre's irrealism does not, however, have to accompany phenomenological work on imagination per se. It is not, for example, particularly pronounced in Husserl or Merleau-Ponty, and Casey in fact takes Sartre to task for the one-sidedness that issues from it (see Casey 2000, 2–3). At his most irrealist, Sartre abandons phenomenology for ideology. There is more than a little irony in the fact that a professional philosopher who was also a novelist and dramatist produced a theory of imagination that could scarcely even begin to come to terms with works of literary imagination.

[10] As reflected, for example, in the claim of many twentieth-century philosophers that imagination is properly captured as an attitude to an object (in particular, a propositional attitude, for example "supposing," to a proposition P), or in the Lockean notion that the understanding takes an experienced idea and compares and contrasts it to others.

[11] These last sentences suggest that, even in very careful formulations of what differentiates imagining from perceiving, there is more than a hint of the conventional model of imagining as forming and holding a fixed view. Is the act of imagining a telephone intrinsically isolated from successive imaginative views conceived as variations on the original one? Is it not conceivable that the original act of imagining ordinarily or even always *intends* a subsequent amplification and proliferation of possibilities?

In essays and notes published posthumously, however, Husserl engages in far more complex exercises of imagination than this. In one, he imagines Christ raising the daughter of Jairus from the grave. Noting the mood that the scene conveys, he wonders what difference it makes to imagine oneself as a mourner rather than as a detached spectator of the scene, and further reflects that the scene includes Christ's compassion. The passage gives a finely nuanced analysis of the different kinds and aspects of situation and tone as consciousness shifts its activity from object–and–scene–imagining to imagining the subjective states of the different participating subjects.[12] Through wondering how imagination takes place, *takes up a position*, the object of imagination becomes less the scene than the resituated consciousness of the imaginer put in the different places of the mourner and the central actor. The imaginer's consciousness appears in a resituated and reconfigured form. The image as object turns out not to be simply separable from the act of the mind that forms and holds the object in mental view; the object–image bears within itself imagined consciousnesses that have imaginative points of view. Note what this phenomenon implies: that the imagining viewer and the imaginatively viewed are not easily and simply distinguishable into subject and object. Or, more strongly, that it is not that the imaginer simply constitutes or coconstitutes the object (as one might typically say in phenomenological analysis), but rather that the object itself is *an imagining in process*, imagining in development, that can even be conceived as incorporating its own imagining!

Is untangling these things an insuperable problem, for phenomenology or for any other approach? It is at any rate not easily done. Moreover, there is a deeper difficulty: that phenomenological investigation presupposes imaginative variation as its basic method. In the first instance that might look like an elementary problem of reflexivity or recursion, no more problematic than using consciousness to explore consciousness, and it is immediately reminiscent of the Husserlian theme that consciousness coconstitutes its objects. But this problem cannot be resolved by a theme. Before recursion can be invoked as a principle, its shifting positions and effects must be grasped. It appears, in the first instance, that in the case of the mourners at the funeral there are two kinds of imagining. The first is the "naïve" imagining that forms and holds a picturelike scene or object in consciousness; the second is the methodical imagining that takes the already-formed object and varies it. Whether and how they are the same imagining and whether one is more original than the other have to be established rather than merely postulated.[13]

Moreover, the introjection into the imagined scene of imagined characters with some kind of consciousness, even if that consciousness is a "borrowed" aspect of one's own, raises the thorny question of whether the methodological division of the

[12]See Husserl 1980, 464–477. This particular passage provides the jumping-off point for the profound analyses of Richir 2004; see esp. ch. 1 of Richir's introduction.

[13]Matters are further complicated if one adds to these "ordinary" imaginative functions the transcendental functions found in Kant: that the very having of a spatio-temporal imaginative field, whether for sensation, imagination, or memory, requires a prior, more primordial, unifying act of imagining.

basic phenomenological situation into the two poles of knower and object is adequate or even suitable for grasping imagination. If in first approximation it looks as though the imaginer is distinct from the image, a renewed engagement can make it appear as though the image or the imagined scene already implicitly contains lines of force for development, and thus that the imaginer in developing the image is following demands made by the image. Perhaps even Husserlian phenomenology has to be reconceived or abandoned if one aims to understand these types of phenomena.

2.2 Psychologism, Antipsychologism, and the Persistence of the Visual Model

Forming and holding a picture privately in one's mind—the visual model of imagination (and also memory), as I have named it—has been paradigmatic for over 2,000 years, since Greek antiquity. As we acquaint ourselves with key historical episodes in the history of imagination, we shall see better how this primacy came about, and that it was not inevitably central to the thought of some of the founders of imagination theory. For now, however, let us focus on recent assertions of the model that will help us understand why taking it for granted is mistaken.

Almost a century ago, the French philosopher Alain made a specific demand of people who believe that they can easily produce and hold in mind a well-formed visual image. Assuming that most Parisians and visitors to Paris would be familiar with the Panthéon (located prominently on the hill of St. Geneviève, on the Left Bank, not far from the Sorbonne), he asks them to form an image of it. Then he makes a simple further request: count the columns. "Not only can they not count them, they cannot even try. Now this operation is the simplest in the world as soon as they have the real Panthéon before their eyes. What then do they see, when they imagine the Panthéon? Do they see anything at all?"[14]

The point appears to be this: (most) people would say that they have before their mental eye a fully formed visual replica of what they are imagining, yet they would not have anything definite enough in mind to allow them to count columns. If they have any kind of image, it is not fixed but nebulous and fleeting—too fleeting to allow counting even to begin.[15] There are exceptions: for example, those who are gifted or cursed with eidetic imagery (photographic memory), who can tell you not only how many columns there are but describe each individually, down to the marks and streaks they bear from centuries of use, abuse, and exposure to the elements. The example of such people appears sufficient to answer positively the question

[14] Alain 1926, 338. The passage is part of the second of nine "Notes" appended to the second edition; titled "Sur les images," it serves as an appendix to book 1, chapter 3 ("Des Images et des Objets"). "Alain" was the pseudonym of Emile Chartier (1868–1951), philosopher who taught at the Lycée Henri–IV (opposite the Panthéon) from 1909.

[15] Shortly we will take up the rest of this passage from Alain. Note that the present example is another case in which the relationship between remembering and imagining is unclear.

whether human beings *can* have images in the "classic" sense of the visual model. But most of the rest of us have nothing like a mental photograph to consult, from which we could describe and count the object's features. As Alain says a few lines later, it is probably even false to think that we have *any* kind of *durable* image with definite features.

Already in the middle of the nineteenth century there was developing in Western scientific and philosophical communities a critique of psychological claims that relied on nonpublic methods like introspection. Alain's Panthéon example exhibits the core problem: asking a person to think, remember, imagine, perceive, etc., and then to observe and report what has gone on mentally is an inherently problematic "method." It is based on a conventional but perhaps quite false way of thinking and speaking about mind; it may well be the conventions, and not what happens, that decisively shape the descriptions. Alain is only one in a long series of people who have volatilized the conception of the fixity of the mind's seeing and of its objects. For instance, in the middle of the eighteenth century and in the late nineteenth century Hume and Nietzsche, respectively, launched devastating critiques of the givenness and fixity of the ego, the I: perhaps there is no *one* to do the introspecting that an ego is presumed to do. Even if there is, it is by no means clear that there is an "interior" where the introspector might look; and if there is such an interior, what is found there might not have sufficient stability to be located, much less investigated (this is the substance of Alain's criticism). And suppose, just for argument's sake, that at least a few people do have mental images of a stable kind: is it reasonable to expect them to have special insight into the nature of those images and how they come into existence? The worker who washes the car at the end of the manufacturing assembly line sees a finished product, but that does not mean he understands how it came about. The introspector at the end of the image assembly line is no better positioned to understand the imagining process in a scientifically justifiable way. Whether there are facts that one can acquire from asking people to introspect their mental workings needs to be verified and not just assumed.

The general tendency of this kind of criticism never coalesced into a full-blown philosophical or psychological movement, but it has since acquired a name: antipsychologism.[16] In its early phases, antipsychologism was directed at claims, whether tacit or express, that knowledge can be accounted for by the fact that it is the outcome of natural psychological processes—this claim is the basic contention of *psychologism*. "Three plus two equals five" should, according to antipsychologism, be true because of logic, because of the basic principles of number, because of the reliability of mathematical structures, and/or because of the derivation of theorems from axioms—not because it is the answer that the human thought process factually yields. If psychologism argues that we depend on what people actually think, antipsychologism counters that this is inadequate. We do not prove mathematical and scientific results by taking polls. Just as someone needs to make sure

[16]For the late nineteenth- and early twentieth-century psychologism/antipsychologism controversy in German thought, see Kusch 1995; for a broader historical survey, see Jacquette 2003.

that a calculator is constructed so it follows mathematical algorithms accurately, so do researchers need to ascertain that a knowledge claim adheres to rules, principles, or standards that are objective warrants for the claim.

This concern becomes especially urgent in the wake of evolutionary theory. According to natural selection, random changes in the organism lead over time to the emergence of new organs and actions and eventually to new species. The eye presumably evolved from light-sensitive cells in animal ancestors a very long time ago. We can be relatively sure, if the gene for that trait was passed on and the cells eventually further differentiated and organized into the eye, that light-sensitivity and vision conferred advantages of survival and reproduction. That does not mean that eyes, much less animal or even human consciousness, evolved in order to register things as they are in themselves. Helping us survive and reproduce is not equal to yielding truth. The naturalness of mental processes does not guarantee the scientific validity of the experiences they give us.

The logic of antipsychologism is ruthless and progressive. Indeed, the earliest antipsychologists themselves became the targets of later ones. For example, Gottlob Frege (1848–1925), one of the founders of modern predicate logic and a grandfather of the twentieth-century analytic movement in philosophy, was a severe critic of psychologism. So, too, was Husserl, whose early book on the foundations of arithmetic Frege actually criticized as too psychologistic. This criticism spurred Husserl to eliminate psychologism by developing a rigorous methodology for ascertaining and examining the characteristic intentions and structures of consciousness—the eventual result of which was the philosophical method and movement known as phenomenology. Yet both the analytic Frege and the phenomenological Husserl were labeled psychologistic by later criticism, because they still expected to gain insights into how the mind works in terms of its own states, structures, actions, and intentions.

As doubts increased about traditional accounts of psychological life and the concepts used in them, appeals to private acts of consciousness (like the inward beholding of a visual image) became ever more suspect. The behaviorist movement in psychology represents one of the severest forms of antipsychologism, whether militantly *denying* the existence of psychological acts and entities or moderately taking them as simply *irrelevant* to scientific explanation. Traditionally conceived mental acts like imagination were reconceived as sets of behaviors (including the utterance of sentences) and dispositions to behaviors that human beings display in appropriate circumstances. In Anglo-American philosophy there was concomitantly a movement to propositionalize imagination, or, more precisely, to conceive it as a particular attitude to propositions—for example, *supposing* or *pretending that* p, where p is the proposition stating what is imagined. Although the phenomenological movement resisted behaviorism and the reduction of consciousness to scientifically explainable material processes, and although its very method is the imaginative variation of consciousness, phenomenological discussions of imagination were also affected by antipsychologistic tendencies.

It is important to point out that if antipsychologism has been hostile to the notion of mental images and has tended to discredit anything resembling a traditional understanding of imagination, its effects on memory studies have been less thoroughgoing.

At first glance—mindful, for example, of Alain's criticisms—it might seem that memory would be subject to the same problems and criticisms, insofar as one needs to report what is going on "mentally." But memory studies can test what people say, write, or draw against documentary records (and that means recorded *in any way feasible*, in an era of ever more powerful recording technologies). Thus the report of an experimental subject can be reduced to propositions of the form "I remember that p"; the p–statement can be checked against the evidence. The resulting ability to distinguish the accurate and the inaccurate in remembering can be used to refine theories of, and experiments about, what goes on in the brain (for example, the sequence of neurological processes that occur when an event is witnessed, and then the corresponding sequence when the subject tries to recall it). Moreover, to a limited extent, this methodology opens up the possibility of a partial rehabilitation of the human being's introspective remembering experiences. With the help of more accurate theories that give a more definite account of what "going on in a mind" means, psychologists no longer need to rely solely, and in a completely uncontrolled way, on experimental subjects' descriptions (or misdescriptions) of what is going on in their minds.

Something similar may be happening in the study of imagination. To counter extreme forms of propositionalism and antipsychologistic claims that (visual) images do not exist, researchers like Roger Shepard and Stephen Kosslyn devised verifiable experiments that support the existence of traditionally conceived images. In these experiments researchers typically ask subjects to conceive a single object or situation as clearly as possible and then to perform imaginative variations, manipulations, and movements. For example, subjects familiarize themselves with a statue from photographs, in particular with how it appears from the front, the sides, and the rear. Then they are set a timed task: say, to start with the clearly recalled front view of the statue, and once they have done this to bring to mind as quickly as possible (say) the right-side view. They signal the researcher when the front view is mentally clear, and the timing begins; then they signal the researcher again to stop the clock when the right-side view is clear to them.

Many permutations and variations of the statue-rotation experiment, as well as other mental tasks involving viewing and moving in the space of imagination, have been tried with large numbers of experimental subjects. What the researchers have found is that (for example), beginning with the front view, it takes subjects twice as long to achieve a clear *rear* view as it does a clear *side* view. What the subjects typically report is that, in order to change views, they mentally rotate the imagined statue at a uniform velocity. The conclusion the researchers draw is that not only do people actually conceive a visual image, they perform operations on it in a way that reflects what would happen with real-world objects and situations. Thus the subjects are not reporting events of two discrete memory snapshots in succession but are remembering and viewing the object in spatial location where movements, both of the object and the observer, can take place. The second view of the object is achieved by imagining walking around the statue or imaginatively rotating it. It is not just having an image that counts, but also the imaginative location where it is placed, and the possibilities of variation that the emplacement permits. The researchers go on to argue that these results are not consistent with propositional theories of imagining that deny images or their relevance. These theories have no plausible explanation for why it takes twice as long to generate the propositional report "I can now clearly see the back of the statue" in comparison to "I can now clearly see the right side of the statue."[17]

The reader should not conclude from this very schematic account of decades of research that the matter has been settled. Behaviorists and propositionalists have undertaken a counterattack; and even if the weight of the evidence seems to favor the imagists, the matter will not be settled until more definitive results are achieved or researchers lose interest in the question. One thing that seems likely, however, is that, as psychological studies of imagination develop more sophisticated techniques, they will begin to acquire more interesting and theoretically productive results, just as memory studies have over the last half century.[18]

2.3 Limits of the Visual Model

There is nonetheless something very odd about this contemporary debate. The experiments performed by Shepard and Kosslyn do not require ultrasophisticated, high-technology equipment and techniques, nor are they based on any theory that has emerged only recently. They do, of course, require having reliable timing devices and ways to record promptly and accurately the signals given by the experimental subjects. But since it is *relative* rather than *absolute* times that are important, even devices as elementary as the water pipettes Galileo used to measure how long it took a ball to roll down an inclined plane 400 years ago would be sufficient. Why, then, did no one think to perform experiments like Shepard's pathbreaking ones of the 1960s much earlier? In the latter third of the nineteenth century psychology proudly advanced toward becoming experimental and scientific; why did no scientific psychologists perform experiments like these?

Perhaps in the last analysis there is no accounting for the fact that something did not happen. Yet notice that Alain's 1926 challenge to rememberers is implicitly an experiment: call to mind an image of the Panthéon (which you claim to know

[17]See Shepard and Cooper 1982 and Kosslyn 1994. One must not assume that simply reducing these rotations or circumambulations to a combination of memories (e.g., memories of the photographic representations of the different statue positions plus memories of rotating or walking around objects) explains much. The combination is not itself a memory—though whether imagination inevitably works with or combines memories is certainly open to discussion—and why the "rules of propositional combination" take precisely twice as long to follow in the circumstances would still be a mystery.

[18]Nigel Thomas has argued that, because of work like Shepard's and Kosslyn's, theories of imagination more or less aligned with traditional approaches were on the verge of a renascence in the early 1970s. But there intervened the rapidly evolving successes of the cognitive sciences (modeled on computation) and neurobiology, using the latest in high-technology brain and neuron imaging devices, and attention quickly shifted to mental functions, like vision and memory, that were technically, experimentally, and conceptually more tractable. See Thomas 1997.

perfectly well) and count the columns. Being a philosopher, and speaking chiefly to artists in the book where he gives the example, Alain can be excused for simply describing what happened when *he* tried to do something like this. Yet a clever scientific psychologist might easily have read Alain and said to himself: "Why not set up experiments to check how accurate people's remembering and imagining actually are? Besides asking them to imagine the Panthéon and count its columns, I can ask for other data as well. They can describe to me what seems to be going on as they try. Gradually I will be able to refine the experiments and come up with a list of better, even standardized, questions. I must be sure, besides counting, to ask them to do other things as well with the image they claim to be seeing, like viewing the Panthéon from different angles or positions." If someone had asked these things and gotten results of virtually any kind, it almost certainly would have spurred others to devise different and improved experiments—possibly very similar to Shepard's of the 1960s. Perhaps by now the issue might have been settled, or at least greatly advanced, and new phenomena might have been educed, different concepts developed, and new theories tried. This might have led to efforts to be less casual in our claims about psychological powers and to examine the traditions of philosophy and psychology more rigorously, so that a more accurate, if still necessarily imperfect, account of the workings of mind might have emerged. This, in turn, would likely have put us in a much more advanced position than we are today, from where it would have been possible to benefit more fruitfully and rapidly from techniques we enjoy today of monitoring and imaging regions of the brain and neurons. The most powerful antipsychologistic tendencies of psychologists and philosophers would not have been sufficient to stop a fruitful line of inquiry.

Yet the question of whether fixed images exist may be less important to the issue than the disputing sides think. On the one hand, without realizing it, by concentrating on whether these imaging events really exist in the mind, they have shifted the center of gravity of investigation away from the model of *holding* an image in mind toward one that highlights the active manipulation and transformation of images. It may turn out that the fixed (visual) image is something that imagining can indeed produce, but that such an image is only an aspect of what imagining is about, not the most central. Imagination may be able to form well-developed, fixed images, but that would be just the tip of the (imagined) iceberg. The imagination might more fundamentally be formative and reformative of perceptual appearances, and locative as well, that is, about positioning images in contexts. The center of gravity of imagination studies would then need to be shifted to the forming process and to imaginative placement; the finished image would be a derivative concern. Extreme antipsychologists might still be unwilling to yield ground. But even the friends of images are not fully aware that they are dealing with the distortive effects of more than 2,000 years of conventionalized tradition.

Already in his writings of the 1880s, the French philosopher Henri Bergson (1859–1941) tried to counter these distortions concerning the being and having of images. After rationalism and empiricism, but especially after Kant, Bergson argued that philosophy and psychology, whether idealist or realist, operated from fundamentally defective presuppositions about consciousness and its objects/contents. Idealist strains of thought portrayed consciousness as actively *unifying* experiences; realisms and empiricisms, more directly concerned with the fundamental units of perceptive experience and how we acquired them, either spelled out how the understanding manipulated them or reduced consciousness and its various levels to the sorting of these units accomplished by the natural organic powers of the human animal.[19] In either case, the tacit assumption was that perceptual units of some kind exist as such, and that the explanatory task was to get from these to their combinations/unifications, whether by a mechanical, an organic, or an idealist process.

Bergson commenced this line of criticism in his French doctoral thesis, *Essay on the Immediate Givens of Consciousness* (which served as the subtitle of the English translation, *Time and Free Will*). He extended it in the book that followed, *Matter and Memory*. As the introduction of the latter work argues, idealists treat the image as solely perceptual, realists and empiricists as purely substantial.

> These difficulties are due, for the most part, to the conception, now realistic, now idealistic, which philosophers have of matter. The aim of our first chapter is to show that realism and idealism both go too far, that it is a mistake to reduce matter to the perception which we have of it, a mistake also to make of it a thing able to produce in us perceptions, but in itself of another nature than they. Matter, in our view, is an aggregate of "images." And by "image" we mean a certain existence, which is more than that which the idealist calls a *representation*, but less than that which the realist calls a *thing*—an existence placed half-way between the "thing" and the "representation." (Bergson 1988 [1896], 9)
>
> In a word, we consider matter before the dissociation which idealism and realism have brought about between its existence and its appearance. (Bergson 1988 [1896], 10)

Gilles Deleuze, drawing on Bergson for studies of the cinematographic image, and Giorgio Agamben, appropriating both Bergson and Deleuze in his analyses of gesture, have argued that the photographic image (which is the contemporary prototype of the conventional model) is an abstraction from the living, developing, mobile world, not the simply given, technically produced equivalent of sense perception. The still photo, and by extension the perceived image, idea, or impression, is an artifact of theory. Although Bergson actually despised the cinematographic image, it was because he interpreted it as an illusion constructed according to the already-distorted still image of philosophical theory: project (say) 24 still images per second and get the appearance of motion. Deleuze retrospectively corrected Bergson by arguing for the primacy of the cinema–image over the still image; the latter is the greater abstraction, a cross section cut out of the mobile world. Agamben goes even further by assimilating images to gesture:

> It is necessary to extend Deleuze's argument and show how it relates to the status of the image in general within modernity. This implies, however, that the mythical rigidity of the

[19]The latter is basically a development of David Hume's contention, in a famous appendix to the *Treatise of Human Nature* (Hume 1739–1740), that he could find nothing corresponding to the I or ego beyond the experienced sequence of impressions and ideas. Bergson's early writings preceded the emergence of phenomenology, but his criticisms clearly apply to it as well, even if the ego is only a coconstitutor of experience.

> image has been broken and that here, properly speaking, there are no images but only gestures. Every image, in fact, is animated by an antinomic polarity: on the one hand, images are the reification and obliteration of a gesture (it is the *imago* as death mask or as symbol); on the other hand, they preserve the *dynamis* intact (as in Muybridge's snapshots [of animals and athletes in motion] or in any sports photograph). The former corresponds to the recollection seized by voluntary memory, while the latter corresponds to the image flashing in the epiphany of involuntary memory. And while the former lives in magical isolation, the latter always refers beyond itself to a whole of which it is a part. Even the *Mona Lisa*, even *Las Meninas* could be seen not as immovable and eternal forms, but as fragments of a gesture or as stills of a lost film only wherein they would regain their true meaning. (Agamben 2000, 54)[20]

That is, the "classic" or conventional image is like a cross section taken of a more robust, more fully dimensioned reality—or even of a more robust and fully dimensioned image (e.g., of a cinematographic image). What the image is depends in part on how it befalls us, how it is taken, what it is taken for: sometimes as solid, sometimes as fleeting; now as sufficient unto itself, now as pointing to what lies beyond it; under our control, or something we suffer. These and similar claims are, of course, not proof of a thesis. But they do provide sufficient motivation for wondering why the conventional model has been so convincing.

Let us, then, take a harder, quasi-experimental look at the visual model. Suppose we begin, as Alain did, with the Panthéon. For simplicity's sake, let us worry only about the façade. The chief presupposition we and Alain apply is that, if we claim that we can recall the façade of the Panthéon, we have previously seen it in person,[21] and that this previous experience has impressed in our psyches an image of the façade that we are able to call back to consciousness at will.

There is an implicit standard we are following here. The previous visual perception we had was (to use a term familiar from Descartes) relatively clear and distinct rather than obscure and confused. We might further specify the kinds of conditions that favor clear and distinct viewing: full daylight (though perhaps not bright, direct sunshine, which can dazzle the eye and produce deep shadows obscuring surface details), clear atmosphere, rested eyes, uncluttered mind, time to tarry over the viewing, etc. We could schematize, even quantify this by developing a questionnaire (with a scale from 1 to 10) that the observer could fill out as he stood in front of the monument. A few minutes, hours, and/or days later, we could place the former observer of the Panthéon in a quiet room, instruct him to recall the façade as clearly and distinctly as possible, and readminister the original questions. But to assure that he was not relying on propositional or conceptual rather than imagistic memory

[20] Translation slightly emended. For Deleuze's investigations of the image, one might begin with Deleuze 1986. Eadweard Muybridge (1830–1904), a British expatriate to the United States, invented techniques for the photographic capture of moving subjects (he produced, for instance, the photographic sequences of running horses that proved there are moments at which all four hooves are off the ground).

[21] One could easily alter this thought experiment to include the previous seeing of *images*, or contrariwise to restrict the kind of "seeing" we have done to the experience of photographic images of just the façade.

(for example, that he remembers having counted the portico columns),[22] we would have to add new questions that asked about features evident in original viewing but not directly pointed to by the first questionnaire.

Perhaps it makes perfect sense to say that there is such a thing as a memory image, but we have already pointed out several times that philosophical discussions and psychological experiments leave underexplored the differences between imagining and remembering. One reason is that as soon as memory enters the picture there is an epistemological shift. The example of viewing a fixed scene highlights what is stable and reliably given in an experience, and tests of memory look for whether memory is able to reproduce those stable and reliable givens. The visual image in the visual model is treated as a definitely determined, stable entity completely present at every moment. If it falls short of this, it is defective. This is a perfect situation for raising standard questions of epistemology, but it is doubtful that they help much in understanding the psychology of imagining. Perhaps the situation is not even entirely representative of perception and remembering, which, along the lines indicated in the Bergsonian tradition, is always part of a living situation. A very old philosophical tradition lays a heavy hand on imagining and twists it in a familiar direction. Put another way, the tradition turns the phenomenon to match the kinds of questions and terms the tradition prefers to raise.

Think again about imagining cinnamon and nutmeg, not as objects but as aromas. Some of the kinds of questions that a researcher might ask of experiments following the visual model can be adapted, but they give different results with rather different expectations. If I challenge you to count the columns of the Panthéon or to tell me how many moles there are on a friend's face and where precisely they are located, you likely will not be able to do this with any scientific, or perhaps even everyday, precision (though there are people who can do this quite exactly, and I suspect that virtually anyone can do *something* like this *occasionally* with respect to *some* particular thing he or she is very familiar with—even if the person will *always* be stumped by many questions we can pose if we the interrogators actually have the face or its photographic image in view or a battery of questions with verified answers at hand). Nor will you likely be able, by trying again and again, to come up with a version of the image that lets you count columns or moles.[23] But in the case of

[22] How many columns are there? The answer depends on where you stand and whether you are counting all the columns of the façade or only the frontmost ones. The Panthéon's portico has a front rank of six columns; but there is a second rank of six columns immediately behind the first, with the column at either end set outside the extreme columns of the first rank and thus, at a sufficient distance, appearing to belong to the first rank. Therefore one might plausibly say six *or* eight columns; and this is not to mention two additional ranks of columns behind the second. Few people would be able to recall the Panthéon with this accuracy of detail! Yet once one is aware of these complications and marks them—not necessarily in explicit propositions—one can more easily arrive at accurate remembering and reimagining of the building. Image and proposition are not an either/or but reinforce one another.

[23] I say "likely" because prodding by another or even ourselves sometimes helps us remember details we could not at first recall. Yet we also know from research that memories can be shaped and even induced in experimental subjects, to produce "false memories."

aroma appropriate questioning can easily lead to improved recall. My ability to evoke the aroma of cinnamon is perhaps at first very approximate, but repeated attempts improve the quality of the result; and if someone addresses to me questions appropriate to the phenomenon ("Do you notice the initial tang of the smell, and then the prolonged finish that makes it smoother?") I often can try again and recognize the sense of the inquiry.

As noted earlier, it is (nearly) pointless to ask me to remember a smell I have never experienced.[24] If I *have* experienced it, and am lucky enough to have a prompt memory, I can manage in an instant: there it is, the aroma of cinnamon. (In a nod to David Hume's distinction between the greater vividness of original impressions and the lesser one of recalled ideas, this is not always or even usually the fully redolent and robust aroma of really smelled cinnamon.) But if I am not one of the lucky ones with prompt memory, I may have to start from scratch. I will have to work at recalling it. It may happen slowly, step by step. Every few moments I will ask myself whether what I have managed to evoke is really the smell of cinnamon. Gradually it will articulate and define itself, however partially and fleetingly. But notice: it is not merely at the moment of full-blown success that I can say that I am forming images; *it is at the very moment of starting to try*. Even if I am in process of *remembering*, I am already *imagining* from the very start—imagining successfully, however effortful it is and however much I am falling short of actually remembering cinnamon's smell. This kind of imagining is part of the process of remembering and a prerequisite for it. At every moment in the work of remembering I am trying to summon back appearances; thus I *am* imagining, successfully imagining. I may be imagining the smell of cinnamon, or of nutmeg, or of one particular instance I had of smelling a quite extraordinary cinnamon aroma, or an artificial variant of cinnamon, or perhaps even some exquisite or bland variety of cinnamon that I have not actually smelled before. The smell may be (almost) as rich as direct perception itself, or it may have diminished notes and a more transient character. Successful perceiving and successful remembering require a specific result: but imagining is imagining, and whether it has hit on what epistemologists demand of perception or memory is not an index of whether one is doing it well.[25] The conclusion to be drawn is this: if we apply epistemological standards and questions to imagining, we end up trying to

[24]One must beware of categorically ruling out possibilities when the grounds for impossibility are not clear. If someone has experienced certain smells and noted characteristics that allow him to align them or place them in series, he might be able to imagine "positions" between or among them. See Sect. 3.1, below, for the case of the missing Humean blue.

[25]One might of course object that perhaps I am not managing to imagine *cinnamon* smell in most cases but only a simulacrum. Perhaps so: but is success in achieving an exact representation of something real or remembered the proper criterion of imagining? If to Alain's request that I recall the façade and portico of the Paris Panthéon I responded that I had succeeded but then proceeded to describe a building that exists nowhere, he (speaking colloquial English) might remark that I wasn't remembering the Panthéon at all but only imagining something else, something purely imaginary—an indication that in ordinary English usage (it works the same in French) we do not think that imagination has to have arrived at any particular real, fixed, or remembered object in order to be imagining.

force imagination into a mold that does not fit. Imagining can very well have approximate and emergent objects—one might say that this is typical of imagination—but it does not seem plausible to claim that it has fixed objects per se. It is a power that may have no proper objects, yet a power that lets appearances emerge and remodel themselves.

2.4 Elementary and Complex Imagining

In the continuation of the paragraph I quoted at the beginning of Sect. 2.2, the philosopher Alain presents a reflection that is typically omitted when the earlier passage is quoted.[26] Right after posing the question, "What then do they see, when they imagine the Panthéon? Do they see anything at all?" he writes:

> As for me, when I pose this question to myself, I cannot say that I see nothing that resembles the Panthéon. I form, it seems to me, the image of a column, of a capital, of a section of wall; but as I can in no way fix these images—as on the contrary the direct view, if one can say this, immediately places me back in the presence of objects that I have before my eyes—I cannot say anything about these images, except that it seems to me that I have perceived them for an instant. But as there is no lack around me of reflections, shadows, indeterminate contours that I perceive out of the corner of my eye and without thinking about it, it can well be that I take, from the memory of this chaos of a moment, the illusion of having evoked, in the time of a lightning flash, the absent parts of the monument that I name in myself. In all this, I ask only that one challenge oneself, and that one not describe [things] using discourse beyond what one has seen. (Alain 1926, 338)

Alain thus does not actually deny the existence of a mental or visual image in this situation of trying to recall the Panthéon and the number of its columns. What he denies is the existence of a *well-formed, stable* image whose columns are definite and countable. He is, in an important sense, making the point that I have made: imagination in activity is formative and reformative in a way that is often fleeting, almost constantly in process, and perhaps rarely successful by the standards of epistemology and its usual objects; it is nevertheless a mental activity with actual (in this case visual) presentations—flash presentations, as we might call them.[27] Alain does not go any further in trying to develop this insight, but he does acknowledge that there is something wrong with the way in which we ordinarily talk about these processes and activities. Perhaps, then, what we need more than antipsychologistic taboos is an effort to find ways to talk with more truth about images, imagining, and imagination.

It is important to acknowledge that there is nothing wrong with epistemology per se, when it addresses itself to its proper issues in proper contexts. Once we recognize that imagining is not in the first instance an epistemological situation—perhaps

[26] Sartre established the conventional truncated usage of Alain's passage about counting the columns of the Panthéon; see Sartre 2004 [1940], 38 and 88. For a more recent example, see Bouriau 2003, 10.

[27] Recall that Agamben 2000 refers to an "image flashing in the epiphany of involuntary memory"—see the last block quote in Sect. 2.3, above.

not even a cognitive, or perceptual, or memorative one—then we can begin to develop methods and concepts more appropriate to imaginative phenomena. It is also important to note that we must for the time being refrain from asserting that what I have said so far is only the rediscovery of an old truism, that there is reproductive imagination (as in memory) and productive imagination (as in creative art). That is looking at the phenomenon from a well-established tradition that draws a dark line of separation where there is really permeability. What the phenomena looked at anew demand is that we notice how the very act of trying to *reproduce* or *recall* an image requires incipient (productive?) and tentatively situated imagining, at least until we are satisfied that what we have produced is a good reproduction.

Alain's description of his own effort to picture the Panthéon suggests something more: that there is a sense in which the imagination has to actually construct a complex image out of simpler elements. We have to be careful here that we do not immediately turn suggestions into conclusions—in particular, that there might be imaginative "atoms" out of which the imagination constructs its appearances. There might be no atoms (indivisibles) in an absolute sense. What Alain's example intimates is that in trying to imagine a building our minds might work by combining elements at a level and with a background appropriate to the object being "constructed." One can also imagine that a stone mason would have somewhat different elements, and a somewhat different sequence, flash into his constructive consciousness as he tried to bring the Panthéon to mind, and that there would be a difference, too, in the ways that lithographers or art historians or mineralogists would approach the same task. It is, of course, possible that Alain was doing nothing more than suggesting the presence of some variety of associationism: that the mind works by linking and relinking the ideas it experiences to previously experienced ideas, in a manner contingent upon the individual's previous experience. But "association" is too general a concept to give us much purchase on the phenomena, even if we parse it into the traditional *contiguity*, *resemblance*, and *cause*.

The quotation from Agamben in Sect. 2.3, above, is useful because it calls our attention to another dimension of the phenomenon of forming images that is not captured by appeals to associationism. He says that every image "is animated by an antinomic polarity." At one pole the image is reified, becomes thinglike and independent of context; its involvement in an encompassing gesture or life situation is obliterated. This is the pole where philosophers and psychologists treat images as unit inputs or elements of cognition, apply to them epistemological standards of accuracy, and judge them according to whether they correspond to the truths they ought to serve. At the other pole, according to Agamben, images "preserve the *dynamis* intact." That is, the power or force alive in the gesture or action that was imaged is somehow contained in the image. He cites as examples photos made by Muybridge (e.g., of a galloping horse or a naked man in full stride) or a sports photographer. In these, in the fraction of a second it takes to produce the photographic image, something is preserved that shows not just the visible appearance of that moment but also the action it was part of: a winning kick headed toward the goal, a horse edging out the favorite by a nose, players piling on to celebrate a championship.

Agamben analogizes this to imagination's cousin, memory. The point of the analogy seems to be this: in voluntary memory we are striving to recall something that was determinate, and it frustrates us when we do not recover it in an intact, stable, standardized form; in involuntary memory things flash unbidden into consciousness, often with only tenuous connection to our current situation and what has preceded it in awareness. Yet this does not necessarily correspond to any comparable voluntary/involuntary distinction in imagination. Imagining can, of course, be voluntary or involuntary—think of the difference between an artist's planning a mural versus a poet's opium-induced hallucinations of Kublai Khan. Rather, what Agamben is claiming is that in every image, in the appearances of every imagining, there is evidence of both poles: of the reifying tendency and of the gesture- or activity-preserving tendency. But that is still too weak a way of putting things. Every image as image—as part of and related to gesture or action—is animated both by an impulse that tends to isolate it from the gesture and by another that presents it in the active, gestural relation. If this is true, then the conventional model of holding a visual image in mind does not so much emphasize a single pole as cut down the image at its root, by treating the image as an intrinsically psychological end or result in isolation both from what has inspired it and from every other image. A conventionalized, standardized image thus loses its ontological image character. Our philosophical and psychological theories objectify it by tearing it out of its originating contexture and resituating it in the abstract spaces of quasiscientific conception.[28]

We ought, however, to forbear being greedy for a theory that can explain imagination and its constructions and construals along such theoretical lines—although conjectures that help bring phenomena into focus and better articulate them ought to be welcome. What seems to be inarguable is that single and/or simple images can be "incorporated" into more dynamic or complex ones, that what we ordinarily experience as world objects are, compared to the image they present that can be recalled in the absence of the object, *complex*. To put an object together out of its various images is like building it out of parts, features, and aspects. To put it in a way that is about as simple as possible for the kinds of perceiving beings we are: as a prescribed task, imagining a single smell or a single hue of color or a single architectural capital is far easier and more spontaneously achieved than picturing the entire Panthéon or a five-course dinner, and imagining two smells or two colors

[28]This abstraction-and-resituation is, by the definition I gave in the introductory section of this chapter, the action of imagination. This is not the last time that a move to the scientific attitude will prove to be an act of imagination. Here and elsewhere in this book I take "context" in the first instance to refer to relations of things to one another, whether the things are texts or nontexts. "Contexture," by contrast, refers to the textures of a situation—which is to emphasize the qualitative characteristics of the situation, the qualities of the relations of things. The link between the two terms derives from the root metaphor: that of weaving. Texts weave words together; textures are the characteristics of both the pattern and the materials of the weave, characteristics that sometimes reflect a template according to which the weaving takes place. This dual character of text and context joined in contexture gives rise to a greater complexity than is usually betokened by the slogan "everything is text."

together or in quick succession is more difficult than imagining either alone. Whether imagining aromas outside a restaurant on the Place du Panthéon while contemplating Alain's questions is simply an extrapolated form of classic associationism is, however, a more challenging problem. Consideration of the nature of the image and its parts cannot, under pain of irrelevance, ignore the ontology of the image, especially the possibilities that it cannot be understood as purely objective or subjective and that to remain an image it must somehow retain an intrinsic relationship to its situation.

Perhaps this result, however provisional, seems too slight for the effort so far. Modern science, at least as it is often conceived, prefers ambitious theories. But before indulging explanatory ambition, one needs to become familiar with what it is that the explanation is about. Here, the "needs to" reflects, among other things, an *ethical* obligation. That is, one has to become familiar with, habituated to, the ethos[29] or place of the phenomena before one acquires the right to speak authoritatively about them.

2.5 Listening to Images

Before returning to what we should be looking for in visual imagining, and in order to provide ourselves with richer resources, let us consider yet another common form of imagining, that connected with hearing.

Hearing almost immediately presents us with a quandary. What heard things do we begin with? Spoken language? It is often said that, of the external senses, vision yields a *very* high percentage—typically 80–90 %—of the information that we acquire about the world. If we consider spoken language, however, we would have to lower that estimate and probably reconceive what we mean by "information"; then, turning to the written word, we would have to address the question whether it is primarily visual (in that we follow it with our eyes) or vocal (in that it is or can be a record of what is spoken, and in that reading often involves a significant degree of subvocalization). Language is in any case a complicated phenomenon that appears to involve far more than sensory images, whether visual or auditory. We started reflection about visual imagination with objects, like the Panthéon, that turned out to be too complex a starting point; speech would be an even more complicated starting point for considering audial imagination.[30] So let us look for greater simplicity.

Hearing appears to be less objectifying than vision. We can try to put ourselves into a frame of mind in which we watch the world as though it were a parade of lines and colors, but it is rare that we can refrain from assigning those features to objects populating the field of vision: trees, squirrels, clouds, buildings, cars, and so forth.

[29] In the first instance, the ancient Greek word *ēthos/ethos* means "accustomed place" or "ambience."

[30] Language as imaginative phenomenon is where we conclude this study, however.

We can, by contrast, close our eyes and just let sounds wash over us as sound (especially if no one is speaking to us). No doubt our inclination to identify things is still at work, but sounds appear to give us less information that immediately allows us to locate things spatially, to outline and configure them, and to identify them by kind.[31]

Suppose we have spent some time learning to filter out our penchant for identification so that we might simply listen to the soundscape. We can imagine performing an analogue of Alain's Panthéon experiment. We (the interrogators) will prepare a record—presumably a sound recording—of some stretch of sound and ask questions of people who have heard it. Of course this will not usually be experimentally productive, since it would typically mean nothing to the experimental subject if what we asked about were all the sounds she heard beginning at 10:05 a.m. and ending at 10:10 a.m. in the Place du Panthéon. For one thing, we need first to make sure we have called the subject's attention to the moment when we begin recording and signal the end as well. Yet this still would not be the sonic equivalent of visually examining an object for 5 minutes. It would instead be like having her watch a silent video a single time, or like displaying over a 5-minute period a somewhat random series of visual objects, each for just a moment or two. Unless our goal is to prove once and for all that human beings are inadequate sense-data recording machines, we need to come up with a different experiment.[32]

The Panthéon is a complex but familiar visual object. What might be a complex but familiar sonic object? How about a song, or, eliminating lyrics, a short instrumental piece that the subject knows well? That seems promising—but what sort of questions do we ask? Not how many demisemiquavers the song contains, what its time signature is, what key it is in—although many musicians could immediately identify the signature, those with perfect pitch the key, and a few all three. Yet if we demanded a quick answer to any of these we would be unreasonable. With the envisioned Panthéon we gave the person a chance to recall the building, then to *count* columns. These questions concerning sound are far more complex, relational matters. Perhaps asking about the number or kind of instruments playing would be a better analogue.

Since it seems natural to think of a musical piece or a melody as composed of notes, let us simplify further the question about counting demisemiquavers, which are better known as "eighth notes." So: how many eighth notes does the melody of the first sixteen measures of the piece we have chosen contain? Even most relatively untutored persons could be given a short lesson in identifying eighth notes sufficient for them to count the eighth notes in the relevant sixteen measures, not least because

[31] Letting sound wash over us can even occur with linguistic experience. Although it is almost impossible to focus on just the sounds being made by someone who is speaking our native tongue, if we hear a language we are totally unfamiliar with we can attend just to the sound. "Attending just to the sound," by the way, is a voluntary imaginative act that coconstitutes the perception.

[32] A distressing amount of psychological research (and related philosophizing) tacitly privileges the notion that our sense organs ought to be (or would be better if they were) accurate data-recording instruments. This is a peculiar, distortive presupposition, or rather prejudice.

they can hum or wordlessly sing it, aloud or to themselves. This ability to hum and sing, and even to improvise—though not necessarily in tune—is one of the most equally distributed talents in the world. It is also, whether done aloud or to oneself, one of the commonest ways in which human beings imagine. It allows one to perform certain tasks of remembering and imagining far more accurately than is possible with vision.

Sonic imaginative phenomena have been almost entirely ignored by researchers. Imagine how different the epistemology of imagination would look if we took sound as our typical model, or just as an occasional alternative. We can, and do, quite easily remember many complex sonic presentations, and we can verify this (if we are not tone deaf or completely hopeless as singers or hummers) to the satisfaction of the most behavioristic researcher by making appropriate sounds aloud—a behavior that our imagining/remembering a tune gives rise to. Of course any person can be faced with a question or task that exceeds his or her abilities. Some people can indeed count the portico columns of the imagined Panthéon, even if most cannot. Far more can count the notes in the melody of a favorite song. Counting the melody notes does not guarantee being able to tell the number of notes played by an accompanying bass instrument, however, or how many key changes there are. Yet, just as we have cases of people with photographic visual memory, we have that of Mozart (1756–1791), who after a single hearing of an unpublished choral piece in the Sistine Chapel (Allegri's *Miserere*) was able to write it down, note for note, with nearly perfect scoring of all the parts.

Is a melody a simpler entity with respect to hearing than the Panthéon is with respect to seeing? Suppose, instead of the Panthéon, we asked a person to study for a time a geometrical diagram, or a schematic elevation of a building (that is, the architect's simplified, quasigeometrical, face-on drawing that represents one of the sides of a building). The success rate of a person trying to reevoke that object and answering questions about it would likely be considerably higher than for the real, full-color, three-dimensional, street-surrounded Panthéon as object of recall—though putting a number to the columns might still require several concentrated attempts. Perhaps the complexity of this kind of reduced visual image has a certain parity with that of song and melody. Yet not even "Twinkle, Twinkle, Little Star" is *radically* simple, nor perhaps is the Panthéon so complex that it exceeds the capability of human beings to picture it by progressive imaginative reevocation.

Even if it is not immediately evident how to reckon comparative complexity between hearing and seeing, the fact is that in musical imagining the rate of successfully recalling a song is far higher than in visual imagination recalling a scene. Any human being who listens to music has an enormous stock of remembered songs and tunes. There are of course striking differences in the ability to recall lyrics, harmonies, key changes, and the like, but people typically have a fairly acute sense of melody and rhythm and of divergences from the performances, voices, and instrumentation they are familiar with. In the shower or on demand, people can also at least begin singing or humming a favorite tune. Success or failure is not determined by whether the sound has high production values. Moreover, people can easily improvise variations in rhythm or melody, or even make up a tune they have never heard before, with or without lyrics. Productively imagining sound is not just possible but easy—which should also allow for nicer distinctions between reproductive and productive imagining as well as between imagination and memory. But Alain did not ask his readers to imagine *Clair de lune*.

It is right, I think, to emphasize the positive role of behavior in imagining: imagination *is* an activity, and the behaviorist view captures that. Moreover, as I shall show in later chapters, for the founders of imagination it was almost always important in active imagining to mark, index, or even name some of the appearances, objects, aspects, and relations that occur in the course of imagining. Rather than blur the distinction between the conceptual and the imaginative, this will support the notion of their typical copresence in human imagining.[33] One learns how to imagine more powerfully as one learns how to name and conceive, although the original experience that leads to imagining does not need to be conceptual at all (or at least no more conceptual than the experience of a newborn infant, or of a philosopher trying to imagine a blend of nutmeg and cinnamon). In imagination one learns how to mark and distinguish fields of experience, at least relatively,[34] so that ultimately the activity of imagining is not sharply separable from the activity of *marking the imagined* and *conceiving the imaginable*. Our ability to *mark* increases our ability to produce, reproduce, and recall, and some degree of marking becomes second nature, no matter how limited our imaginative talents. But just because we mark an imagined field does not mean we have, always and instantly, the ability to name (to a researcher–questioner) everything there is in the imagining and in the imagined field.

Contrary to what behaviorists conceive, however, the preponderance of imagining is nonpublic. Any theory of imagining that insists on public behavior, including proposition–uttering, is interested only in the iceberg's tip, and, methodologically and systematically, it leads us away from the experience proper. A painter who is mulling over the shade of green to apply to a tree in a landscape may well engage in a great deal of publicly accessible motor and speech activity—making paint daubs of different greens on a spare piece of canvas or on a sketch of the main composition, and saying to herself, "Let's try forest green #102 instead of summer verdure #115"—but she will engage as well in a memorative and imaginative activity not directly accessible to anyone but herself. For example, as she runs through possibilities she might try to conceive what combining two tints will look like in the proportion 3:2. She *might* anticipate it exactly, but more likely she will have entertained one or several possible appearances that turn out to be not quite right.[35] Of course

[33]I would not positively say that the marking of an imaginative field of experience absolutely differentiates human beings from animals, but instead that this might be a fruitful dimension for inquiry into ways human beings are like and unlike other animals. An adequate phenomenological inquiry into this comparative question may, of course, be beyond the capability of merely human beings.

[34]This means, as I suggested before, that any image can be at least temporarily the standard in relation to which others will be differentiated, marked, or measured.

[35]I have no objection to saying that this is a behavior oriented to prediction or production, but that is to overlook the fully concrete phenomenon for the sake of capturing just one aspect of it.

there is no guarantee that anything at all she imagines will be perfectly right; or it may happen that in one instance she is on the mark for the color's hue though not for its brightness, and vice versa with another imagined combination. None of these private anticipations will be the publicly verifiable one; and although she can explain to someone afterward some of the things she thought in advance, propositionalizing does not capture the whole or even most of the parts. No matter how much she says before or after, there will always be a specific character—one really must say many specific characters—that will remain undescribed, and that perhaps are (in the current state of language, culture, and science) undescribable.[36]

2.6 Can Philosophers Sing?

The behaviorist and propositionalist inadequacy is probably most evident of all in singing or humming. To avoid the immediate issue of memory and/or imagining, let us push the scenario clearly to the imaginative side by making the tune we are humming or wordlessly singing an improvisation on the spot. As usual, it doesn't matter whether it is a miserable screeching of scarcely articulate tones or a masterpiece expertly performed. In one sense this scenario should suit a *behaviorist* just fine: the imagining is going on in a public, behavioral fashion. Neither the behaviorist nor the propositionalist is enchanted by the claim, however, that the singing imaginer might have first, or even simultaneously, intoned music inwardly.

Consider for a moment the kind of argument that the propositionalist uses about what is happening when someone says to him that she has been imagining a peaceful meadow shaded by live oak trees and cottonwoods with a brook quietly streaming through: "To account for your imagining," he says, "we do not need to assume that you have, 'in private consciousness,' the picture of the scene you have described. Imagining, properly speaking, is the disposition to produce statements of the kind you have just uttered: 'I am supposing myself to be lying in a meadow, the sun gleams through openings in the tree cover, I catch a whiff of the thick atmosphere of cottonwood blossoms, in the nearby widening of the brook the water silently forms eddies, then makes a burbling sound at a "falls" just 10 cm high, etc.' Your imagining is not just the utterance of that one compound statement, of course; it is the set of these statements and more, of everything you might ordinarily associate with being in such a scene."

[36]In an obvious sense I mean, for example, that color saturation was not nameable and thus not describable until quite recently. What was needed was a theory of color and a practice of color making that could break color down into different aspects capable of comparison and even measure. We have no reason to think that we have reached the point in history that has finally achieved the full theoretical and practical understanding of color. Comparing the state of color science 100 years ago to today makes it clear that there was progress to be made; we have every reason to expect that 100 years from now further, often unpredictable progress will have taken place. Doubtless we will retrospectively be able to identify antecedents of the new discoveries among today's artists, scientists, and critics who, in some partial way, will have anticipated the future developments. That means that some will have already gotten there ahead of science, if only in imagination.

The propositionalist could go on to make a similar argument about the medium of sound. "I am imagining a tune" is to be translated into a format like "I suppose that I am humming the simple melody–score of Y," where Y is a variation on "Happy Birthday to You," one of the further public manifestations of which is actually humming Y. From the perspective of the imagining and humming woman this probably seems like an impoverished understanding, more calculated for allowing the propositionalist to claim he is right than for unfolding the appearances in a manner convincing to the melody–hummer. After hearing the woman say that she first imagined in private consciousness the song she afterward hummed, and that she really had a sonic experience of it in private imagining, the philosophical observer might remark, "Your saying that you imagined the song first is one of those things that you are disposed to do; that is what imagining is, having the disposition to do and utter certain things. Also, in this case your utterances include actually 'uttering' the song, that is, humming or singing it."

The hummer might respond: "It sounds like you are more interested in preserving the appearance of truth of a theory than in describing the phenomenon. You are trying to establish a parallel between vision and sound, but it doesn't work. You are avoiding a more rigorous analysis that your approach requires. Here's the sort of thing you should say: People who believe in private visual imagining think that there is a sightlike experience going on in their minds even though there are no corresponding visible objects present. (Let's not right now discuss whether the "mirror neurons" that have excited neuropsychological speculation lately require us to complicate this picture.) What is really happening is that a set of related sentences and behavior potentials reach a triggering point, ready to burst into the public world. An example of this behavior, in visual imagining, is saying that you were supposing yourself to be seeing yourself lying by a brook, etc. If the person were, in addition, a painter, a corresponding imaginative behavior would be to pick up a brush and to paint someone lying by a brook. Musical imagining is analogous: when a singer claims that she is hearing a tune in her mind, what that really amounts to is that she is at a trigger point for making utterances like 'I am supposing myself to be humming a song that has the following melody and chord structure,' which melody and structure can be written down in musical notation. Or, with just a modicum of talent, she can begin humming or crooning it."

Our imaginative female hummer–singer continues: "But there are some real problems with this approach. The first might sound a bit like a debater's point, but it suits the style of your argument. You tell me that imagining really is nothing privately conscious but rather public utterance and behavior and the 'disposition' to such behavior. You say that I mistakenly assume that I have private experience before this public behavior, which means that you deny that there is such private experience—or at least think it is totally irrelevant. But I am not disposed to agree with such descriptions, and so my imagining includes (according to what *you* have said) the possible utterance of statements like 'I am privately imagining a jazz quartet accompanying my humming.' I think you can accept, as being within the scope of your definition and understanding, that imagining is the disposition to bodily and propositional behaviors, including, at least for people who are not philosophers, propositions (or would you call them metapropositions?) that say there was a preceding or simultaneous private imagining going on that mere propositions about imagining fail to describe. Nonphilosophers will also be disposed to behavior disagreeing with your denial of what they think about imagining. They are, for example, likely to say that you have no imagination.

"There is something less sarcastic to be said, however. All my statements and many of my behaviors that I currently and publicly utter and exhibit have a background or context that *is* invisible to you. Singing (forgive me for now allowing for the possibility of there being lyrics) brings it out very clearly. All you hear as I sing is the sounds produced by my vocal cords, lip and tongue positions, etc. What *I* 'hear' as I am singing—remember that I am not just an amateur philosopher but also a professional musician—is background music: a guitar riff, a drum beat, the accompaniment of a piano, with some songs even the swell of a symphony orchestra. And this is perhaps more strongly the case when I am at the same time actually singing, out loud, a song I know as opposed to one that I improvise, since with an existing song I am quite strongly aware of different covers and arrangements. You may be inclined to deny that I have such background *awareness*, even if you cannot deny that I utter statements about it. But what I say in response is that the more you learn about music—the more experienced you are in composing or performing, the more you have become acquainted with the timbre of different instruments, the character of different arrangements, the variety of performance techniques, the ways a talented singer or instrumentalist can use syncopation, melisma, and other common techniques—the more you know about such things, the more clearly this awareness is part of your experience of imagining: of *both* the completely private experience *and* the one that you can share by singing aloud.

"One thing you have overlooked—though it is hardly the only thing—is that there is no fixed referent in 'I suppose that X,' since the content of that X is precisely commensurate with the level of your experience of X-like things. In fact, it looks to me as though this is precisely where you, by casually referring to the totality of what I might say and do, commit two logical fallacies, begging the question and missing the point. What that totality is, is exactly what the question ought to be. Don't you have something more definite to say than that it is a function of my previous experience, my dispositions, the network of all associations that I have made (or that make me!)? Isn't it possible—necessary, even—that there is a finer-grained character, an intrinsic and natural structuring of experience, a significant portion of which is shared or easily shareable?

"You will, of course, be willing to grant that there is something very different going on with a tone-deaf person who says that he is imagining Beethoven's 'Ode to Joy' than there is with a conductor or a choir master who is reading through the score in preparation for a performance. I suppose that you can grant that, at least roughly speaking, the tone-deaf person is at least as far from displaying the musical imagination that I have as I am from the conductor, or as the conductor is from Beethoven. Perhaps you can grant this, as well: that even between the two extremes of the tone-deaf hummer and Beethoven there is some shared basis that has to do with matters like uniformities of human hearing, mathematical relationships between tones of the diatonic scale, the Western tradition of harmony, the acceptance over time of some tonal sequences as expected and others as exotic—not to mention performance traditions and memorable individual performances? But please do not confuse these things with the 'social construction of reality'—even though they create plenty of room for its possibilities.

"The problem I see is this: if you cannot grant all these things, I am afraid that it really does reveal more about the poverty of your conceptions than any problem with my musical abilities or my imagining—or with anyone else's, either."

2.7 Simple Imagining and Beyond

The imagined singer's words are polemical, but they are hardly unfair to the style of philosophizing at issue. Philosophers and other researchers easily fall into the habit of denying what does not fit their concepts and methods. Like a referee whose mere word determines that something is out of bounds, they order conceptual play to stop and direct everyone back to the playing field—or rather their preferred playing field. This often amounts to a kind of *local nihilism*: what is out of bounds simply *is* not; it does not exist, at least as far as their theory is concerned. The problem is that even today we do not have a clear sense of where the playing field of imagination is, much less its boundaries, so that it is premature, in fact presumptuous, to have someone telling us in advance what it is or is not. If harsh words and polemical attitudes risk heightening tensions and worsening conflict, they also sometimes serve the need to push back against existing prejudices. This is far from argumentative violence or terrorism. Sometimes pushing back simply clears the field of things that obstruct what is there to be seen.

Pushing back is not per se a nihilism or a counternihilism, as simply denying all behaviorist and propositionalist claims *would* be. Denial risks losing track of important lines of inquiry. For example, it is important to entertain the notion that one cannot grasp the specific densities and places of imagination or its most perfect forms without affirming its shared, public character. Doubtless a poet trying to complete a stanza might experience flitting rhythms, phrases, and rhymes analogous to the flitting, nascent picturing Alain described in his efforts to bring the Panthéon to mind. Putting some of the possibilities on a sheet of paper or on a computer screen is not only direct evidence of the imagining, it is usually necessary for stabilizing the imagination and for making progress toward the goal, the finished verse. Not for nothing did Schelling, one of the founding influences on the early Romantic movement in both Germany and England, argue that the completed work of art is the culmination and apotheosis of the historical work of imagining. The accomplished work, in turn, becomes a starting point for new thought and new imagining.[37]

[37] He made it the culmination of the externalization of thought itself. See in particular Schelling 1978 [1800], pt. 6.

As I have already said, however, we must beware of becoming prematurely overambitious. In philosophy's and psychology's attempts to conceptualize imagination, there has almost always been an astonishing carelessness about laying a firm foundation in basic events, phenomena, and acts. Even I have so far done hardly more than *mention* the question of what is basic in imagining and how we might identify it. Trying to recall the façade of the Panthéon, evoking the aroma of nutmeg, and humming a tune seem all to be fairly simple acts, but they are susceptible of further analysis and foundation that might identify a deeper stratum of imagining or of sense perception (as with Kant, for example, who claimed that imagination *in its transcendental use* synthesizes the unity of the manifold of sensibility—the space–time continuum, more or less—before, and as a basis for, any further, more particular acts of imagining). And before tackling the issue of works of art, it would be worthwhile to locate at least a few intermediate steps between the simplest and the most complex forms of imaging and imagining.

What this chapter has been implicitly proposing is a clearing of phenomenal and conceptual space for the sake of recommencing the study of imagination. "Clearing" does not mean abolishing or forgetting, and recommencing is by its nature starting *again* in the middle of a process where there have already been other beginnings. It might seem natural now to recommence with a full-blown study of imaginative phenomena attended by a highly articulated conceptual framework. But that would be to presume that we have already gotten a clear sense of the *field* of imagining, whereas all we have done is to recognize obstacles and offer a highly selective presentation of phenomena that help point out but certainly not exhaust things that have largely been overlooked.

The selection of phenomena has shown, for example, that our sense of what some simple kinds of imagining do varies according to its basic sense modality. Visual, olfactory, and sonic imagining display different kinds of features, a few of which we have only begun to elucidate. Many people regard as problematic, if not false, the long-traditional division of external senses into five types. Even so, we would still need, for relative comprehensiveness, to incorporate taste and touch, which are no less complex than the other three traditional kinds, each in its particular way, but even more neglected than they. We have not arrived at any clear criterion of simplicity, yet it seems evident that some kinds of imagining are more complicated than others and that greater or lesser simplicity is correlated with the cognitive success or failure, respectively, of imaginative acts. Imagining/remembering the Panthéon is relatively complex, and the traditional assumption that we accomplish it by summoning up or gradually developing a photographlike image stored in memory greatly distorts what typically happens. Yet we have much readier success in accomplishing another complex imaginative phenomenon, recalling a tune or song, and that makes typical criticisms of imagination based on visual epistemological standards seem shabby. On the other hand, imagining a yellow midway between orange and green seems like something that most sighted, noncolorblind people could accomplish, and that means that they would be able to imaginatively envision (at least) something like a flashed "patch of yellow." Imagination can be trained and improved, especially as we practice imagining a particular kind of thing or quality (the smell of cinnamon–nutmeg) or learn to recognize features or characteristics that a thing or quality has that can be varied in the imagination.

In some sense the work of imagination cannot begin at all if we do not have relevant previous experience, so that we cannot dismiss the possibility that imagination is considerably or even fundamentally dependent on memory. Yet we have also recognized imagination's ability to vary the features of a thing or to make appearance emerge in a way that cannot possibly depend on memory alone. We have seen that imagination sometimes works by a kind of recursion: we try to picture the Panthéon, we fail, then we pick up from the point where we failed and come closer to success. But it is partial rather than pure recursion, because often the input in the second phase involves far more (say, changing the context) than feeding back into the system the output of the first phase. Recognizing that we did not achieve what was wanted at first, we bring in *other* elements, *other* memories, *other* imaginative results, *other* situations that we recognize have been left out or were tacitly but unacknowledgedly present in the first attempt.

These are interesting, even tantalizing results, but where do they lead? Here I need to correct a misimpression that my representation of the current state of imagination studies may encourage: that the dead weight of older and newer philosophical traditions oppresses us, and that all we need to do to overcome it is to assert our freedom to experience things afresh, here and now. Imagine, and do it with all your might! Look, don't think! If that were the solution, then the best immediate course would be to resort to a wide-ranging, prephilosophical portrayal of imagining in all forms. That would be to presume that we already have an everyday, working conception of imagination at our disposal—for example, that we have at least a rough-and-ready understanding of its boundaries vis-à-vis sensation, memory, rationality, feeling, and desire. But that is precisely what we lack.

So where can we look for help, if even many experts are captivated by a false image of imagining, and this false image has roots deep in the past? After placing its *false* roots in the past, it may strike us as merely paradoxical to say that we need to inquire historically. Can truth arise from falsity?[38] There have been so many conflicting conceptions of imagination that turning to history might sound like a counsel of despair. Yet it seems to me that there are different ways of approaching the past, and that there are resources there that allow us to find conceptions and roots of conceptions that have been insufficiently explored—even in, or especially in, many of the thinkers who are considered to be figures and founders of mainstream traditions. I have of course implicitly indicated this already, for example with the question whether Alain's words might not have been taken differently and more amply than they were, and whether the kind of research Shepard and Kosslyn undertook beginning in the 1960s might not have been done half a century earlier.

[38]It should go without saying that the word *false* here is more likely to mean something like *misleading*, *partial*, or *insufficient*, rather than "assigned zero probability" or "designated F."

The alternative way I am recommending, once we begin traveling it, can open our eyes to elements and phenomena that have passed beneath notice almost from the inception of organized thought about imagination. One might even decide that these elements and phenomena ground a kind of concealed or even missing tradition, a tradition *manqué*, one with gaps and unexploited possibilities that later thinkers might have resurrected and filled out (and perhaps actually did, in some cases). The majority of the rest of this book is devoted, then, to researching the missing parts of this occluded-and-occulted tradition, not simply for the sake of antiquarian correctness but even more for the purpose of finding imagination and a way in and through it.

Yet we must delay the expressly historiographic mode for the space of one more chapter. Past thinking about imagination is endlessly fascinating, all the more so in those thinkers who decisively shaped Western conceptions. Without a guiding clue, it will be the easiest thing in the world to get lost in the labyrinth, and the deeper commonalities throughout history will be overwhelmed by conventional commonplaces as well as conspicuous but misleading differences. I have already said several times that the theme of position, location, or situation has been almost universally neglected—in that sense one cannot even say it is misunderstood—but that, once noticed, shows itself everywhere and requires a reradicalization of imagination.

I say *re–radicalization* for two reasons. One is that imagination is the psychological power that in Western philosophical and scientific tradition has most often been inverted or repudiated, from rationalist abolitions to romantic apotheoses. In this first sense I am suggesting that we need, once again, to overturn our conception of imagination, to *re*–radicalize it. Yet I mean the word in a more basic sense as well. To re–*radicalize* imagination is to locate again its roots.[39] We need to find the ground from which it emerges; if we are overturning anything in a useful way, it should be as a plow turns soil to make it productive. To speak a bit less figuratively: we must place imagination, we must find the location in which it exists and operates, and we must articulate what allows the place or places of imagination to be cultivated. To speak a little Greek, we need to develop our interest in imagination's *topos* or *topoi*, its place or places, into an account of the logic and structure of the place of imagination, into a *topology* of imagination.

The notion of placing or locating imagination bears another sense. As I will explain in the next chapter, a (conceptual) topology is related to the ancient Greek philosopher Aristotle's notion of topics. The term *topics*—which is also the title of a now little-read work traditionally grouped among the logical writings of Aristotle, collectively known as the *Organon*—refers to a set of interrelated concepts used in investigations that is largely definitive of the proper approach to whatever subject matter is in question; it has to be used by anyone who presumes to be talking about the subject in an intelligent way. In such a sense, if imagination is the subject matter, then there must be a set of topoi or concepts that define and differentiate imagination and its characteristics from everything else. One thing this book aims at, correspondingly, is to elucidate such a set of concepts for imagination.

[39] Latin *radix*, root, is of course the root of "radical."

There is another and deeper sense in which we need to speak of the place of imagination, however. If all this book aimed at were a set of topics of the kind just described, we could approach the thinkers whom we will be engaging opportunistically, seizing upon attention-grabbing things scattered here and there in their writings. That will not suffice. We have to dig deeply enough to see as an archaeologist would, whose principal interest was not curiosities but the way of life of those who made and used the things we find. Often enough that will require us to examine prehistories and orienting questions that may not at first seem of relevance. Yet without that kind of work, we will simply fail to see what is most deeply held in common by the thinkers we examine. As much as possible we need not just to understand a thinker's concepts but also to see them in place and at work in imagining, with all typical and proper objects, forms, motions, acts, and places.

And that leads to this final introductory observation. Only after we have sought the conceptual topology of imagination, and only after we have put its places/topoi into action, can we see the most decisive and fundamental sense in which imagination is a matter of place. Shakespeare's King Theseus was right: to put it in a formula, imagining is the very act of placement and location of transient things ("airy nothings"), of giving place, habitation, and name to things and their characters. Imagination is thus the placement and re–placement of thinking. It is active and gestural; it indicates beyond itself and its already-formed images as it reshapes their appearance.

At this point of the investigation, such claims are elusive at best. Their implications, even their basic meanings, remain to be seen. But the implications and meanings can be seen only by those who are willing to engage imagination where it is to be found. In the next chapter, then, we will continue to tease out some of the basics of imagination with an opportunistic mixture of phenomena, history, and analysis.

References[40]

Agamben, Giorgio. 2000. *Means without end: Notes on politics*. Trans. Vincenzo Binetti and Cesare Casarino. Minneapolis: University of Minnesota Press.

Alain [Chartier, Émile]. 1926. *Système des Beaux-Arts*, 2nd ed. Paris: Gallimard.

Bergson, Henri. 1988 [1896]. *Matter and memory*. Trans. N. M. Paul and W. S. Palmer. New York: Zone Books.

Bouriau, Christopher. 2003. *Qu'est-ce que l'imagination*. Paris: Vrin.

Casey, Edward. 2000. *Imagining: A phenomenological inquiry*, 2nd ed. Bloomington/Indianapolis: Indiana University Press.

Deleuze, Gilles. 1986. *Cinema I: The movement–image*. Trans. Hugh Tomlinson and Barbara Habberjam. Minneapolis: University of Minnesota Press.

Hume, David. 1739–1740. *A treatise of human nature: Being an attempt to introduce the experimental method of reasoning into moral subjects*. 3 vols. London.

[40] If there is a second, square-bracketed date, it indicates the year the work first appeared in its original language.

Husserl, Edmund. 1980. *Phantasie, Bildbewusstsein, Erinnerung. Zur Phänomenologie der anschaulichen Vergegenwärtigungen. Texte aus dem Nachlass (1898–1925).* Husserliana 23, ed. Eduard Marbach. The Hague: Martinus Nijhoff. English edition: Husserl, Edmund. 2005. *Phantasy, image, consciousness, and memory, 1898–1925.* Trans. John B. Brough. Edmund Husserl: Collected Works, 11. Dordrecht: Springer.

Jacquette, Dale (ed.). 2003. *Philosophy, psychology, and psychologism: Critical and historical readings on the psychological turn in philosophy*. Dordrecht: Kluwer.

Kosslyn, Stephen M. 1994. *Image and brain: The resolution of the imagery debate*. Cambridge: The MIT Press.

Kusch, Martin. 1995. *Psychologism: A case study in the sociology of philosophical knowledge*. London: Routledge.

Richir, Marc. 2004. *Phantasia, imagination, affectivité: Phénoménologie et anthropologie phénoménologique*. Grenoble: Millon.

Sartre, Jean-Paul. 1936. *L'Imagination*. Paris: Félix Alcan. English edition: Sartre, Jean-Paul. 1962. *The imagination: A psychological critique*. Trans. Forrest Williams. Ann Arbor: University of Michigan Press.

Sartre, Jean-Paul. 1940. *L'Imaginaire: Psychologie phénoménologique de l'imagination*. Paris: Gallimard. English edition: Sartre, Jean-Paul. 2004. *The imaginary: A phenomenological psychology of the imagination*. Trans. Jonathan Webber. London/New York: Routledge.

Schelling, F. W. J. 1978 [1800]. *System of transcendental idealism (1800)*. Trans. Peter Heath. Charlottesville: University of Virginia Press.

Shepard, Roger N., and Lynn A. Cooper. 1982. *Mental images and their transformations*. Cambridge: The MIT Press.

Thomas, Nigel J. T. 1997. A stimulus to the imagination: A review of "Questioning consciousness: The interplay of imagery, cognition and emotion in the human brain", by Ralph D. Ellis. *Psyche* 3, no. 4. http://www.theassc.org/files/assc/2287.pdf. Accessed 16 Dec. 2012.

Chapter 3
Locating Imagination: The Inceptive Field Productivity and Differential Topology of Imagining (Plus What It Means to Play a Game)

> *You speak to me of an invisible planetary system where electrons orbit around a nucleus. You explain this world to me with an image. So I recognize that you have arrived at poetry.* (*Camus 1942, 35*)

Chapter 2's examples and discussion have made clear that the generic definition of imagination as forming and holding an image in mind captures very little of what is involved in imagining. Even taking it as just a first approximation requires us to pay closer attention to the *formation* of the image—that is, the activity that takes us from not having to actually having an image—and the *manner* of holding the image in mind.

The philosopher Alain's description of what happens when he tries to imagine the façade of the Panthéon—the flitting, flashing quasiglimpses of visual fragments—might suggest, however, that there is little more to say about the uncertain phenomenon of formation. As for holding an image in mind, it might seem that, once there is some kind of well-formed image there, holding on to it explains itself or requires little more than short-term memory.

These are inadequate responses. If we rest satisfied with them we are likely to pass the nature of imagination right by. Even if imagination goes no further in formation than a flitting and flashing appearance, it is already something, and that something and its specific character need to be accounted for. As far as holding on to the image is concerned, it is not clear that it should be interpreted as memory.

For example, consider a worm—presumably without vision, hearing, or smell—that is able to turn away from a food source for a few seconds without, however, losing track of it. In terms of what is evolutionarily plausible, this short-term ability would seem to be a necessary antecedent for the emergence of something like memory. It is not immediately evident that it requires memory, however, especially insofar as almost any physically and physiologically plausible account of perception requires the at least momentary perseverance of awareness of an appearance. If conceptually all we need is some way of articulating the short-term persistence of sense images,

D.L. Sepper, *Understanding Imagination: The Reason of Images*, Studies in History and Philosophy of Science 33, DOI 10.1007/978-94-007-6507-8_3,

it might be better to explain image perseverance as perception prolonged rather than as memory. In fact, again considering evolutionary plausibility, appearance perseverance would seem to be prerequisite for memory—so that, at least from the perspective of evolutionary physiology and psychology, image–appearance and its elements, in some at least primitive form, must precede memory. And that would already go some way toward answering the old question whether memory is to be accounted a form of imagination, or vice versa.[1]

If we are concerned about developing, even just ad hoc, a vocabulary of nicely made distinctions, it is important to distinguish original appearance formation from image perseverance. It might be necessary, looking to the worm in search of food, to suggest that what it experiences in the first instance is not an image of food but an original appearance of what for the worm is food. Or is this a distinction without a difference? What speaks in favor of it is the temporal process: first there is the worm in proximity to food but not aware of it; next comes the moment of appearance (of the food to the worm, or however else we should formulate the statement); then comes (possibly) some kind of perseverance of the appearance even as the object or the worm retreats. It is traditional to call imagination the power of entertaining the appearance (of an object) in the object's absence, so there is a certain logic to calling this last phase an imagined image.

But then the question arises whether one can make an absolute distinction between original appearance and perseverant appearance/image. Either choice brings with it certain inconveniences and certain possible distortions of the phenomenon and an ambiguity in concepts. The issue might be resolved if there were an absolutely definite instant of appearance, an instant that had no buildup and no duration. If that were true, however, there would be no such thing as perseverant appearance (because at every passing moment previous appearances would be displaced by the instantly appearing presence of the moment), and therefore no such thing as an image. Moreover, it is not at all clear than any organism could have forms of sensation or perception of such absolute instantaneity. Since each instant–appearance would have to be immediately and peremptorily displaced by a new instant–appearance, whether the new could have any dependence on the old seems unlikely. Unless there were no time (other instants) between the two, there would also be an arithmetical infinity of other definable instants between them. Would each moment have a completed appearance that could be distinguished from all others, or would this infinitely rapid succession amount to a blur or even a blank? This may be speculative perceptual theory, but the alternatives suggest that something like a gradual but quick process of appearance formation and duration makes better sense of the temporal nature of what appears to animate beings.

In many circumstances it would be harmless to use "image" to denominate the appearance at any stage in the process, although it could be deleterious or at least

---

[1] I am not saying that this can be resolved by armchair helminthology, without actual investigations of invertebrate physiology. Only the phenomena of memory, articulated and correlated with neurological processes and locations, can give us real purchase on strategies for answering. But drawing evolutionary boundaries based in psychology is always difficult, especially in animals taxonomically remote from *homo sapiens sapiens*.

ambiguous if the term were inappropriately absolutized. This kind of distinction is at the basis, for example, of Kant's differentiating the transcendental and the empirical functions of imagination: the latter are largely reproductive and thus image-based and memorative, whereas the former are not at all about individual images but instead about the conditions and basic constitution of the entire field of appearance prerequisite for any particular perceptual, imaginative, or memorative image. We might further illustrate the need for carefully distinguishing images from other types of appearance by citing an argument about the inappropriateness of extending the optical sense of "image" to colloquial uses. Following the practice of optics, we say, for example, that mirrors and lenses produce images. Thus, when we are standing in front of our bathroom mirror, we say that we are seeing our image in the mirror. The claim has been made, however, that we should say instead that we see ourselves in or by means of the mirror—that is, we are seeing an object, not an image, nor even by means of an image.[2]

On the other hand, as soon as we draw out the circumstances of this kind of situation we almost inevitably feel the need to start talking about an image. We do this, for example, when there is some distortion induced by the mirror or even just the switching of right and left: it is not the object that is distorted but the look, or the appearance, or the image of the object. Something similar happens if we tack up on the mirror a photo of ourselves and begin comparing how we appear in the photo and in the mirror: we are comparing them as images. Our technical ingenuity only multiplies the possibilities. If there were invented a replacement for mirrors that consisted of a high-definition screen (with right-left switched as in a mirror), it would seem simply farfetched to insist that we were seeing ourselves by means of the device and not seeing an image constituted by pixels.

These different descriptions of what is happening allow us to make a valuable distinction. Sometimes one walks into an unfamiliar room that appears to be quite large, only to notice that what one is seeing at the other end of the room is a reflection of oneself: that is, one of the walls is mirrored. It makes sense, in the first moment, to say that we see people down there; in a second moment, that we see ourselves and our own motions mirrored; in a third moment, that we see the images of things (including ourselves) in the room. These distinctions are valuable. They are successive "takes" or "captures" of a phenomenon and may indicate some kind of shift in perspective on the objects. As such they are not nothing—and we should give them up only if there is an overriding, true reason.

3.1 Hume's Blue

The question whether sensation plus memory exhaustively accounts for all the distinctions we need to make about images and imagination arises in a classic and rather puzzling passage from David Hume's *Enquiry Concerning Human Understanding*

[2]For example, Hyman 2006, 124.

(Hume 1748, ch. 2, par. 8).[3] Hume had no qualms about reducing imagination to remembered sensation. For him our mental life originates in sensation, which gives us what he calls *impressions*; these impressions, when they are recalled to the mind, he calls *ideas*. They display the same appearances that the impressions did originally, only with less "force and vivacity." Hume's ideas are thus images, according to the long tradition of understanding images as the appearances of real-world objects without the presence of those objects to the senses.

In the second chapter of the *Enquiry* Hume mentions the one phenomenon he is aware of that contradicts his thesis that all ideas are remembered impressions. He asks the reader to suppose a person who has, over the course of 30 years, seen every shade of blue except one. If these were arrayed in a sequence from darkest to lightest, this person would, Hume claims, notice the gap where the unseen blue fits. The question quickly turns to whether something more positive might appear to the person as well:

> Now I ask, whether it be possible for him, from his own imagination, to supply this deficiency, and raise up to himself the idea of that particular shade, though it had never been conveyed to him by his senses? I believe there are few but will be of opinion that he can: and this may serve as a proof that the simple ideas are not always, in every instance, derived from the correspondent impressions; though this instance is so singular, that it is scarcely worth our observing, and does not merit that for it alone we should alter our general maxim.

Hume's discussions of imagination are extraordinarily subtle and rich, so if here I must sharply criticize him for incoherence that must not be taken as disparagement. He acknowledges the perception of the missing blue as an instance of imagination, thus not simply passive or receptive the way that an original impression on the senses would be. By dismissing the case as an anomaly not worth further reflection, he fumbled away a chance to break through to a deeper and more original stratum of imagination.

The very way in which he sets up the case of the missing blue is ingenious, though it clearly has a prehistory. Most notably, it is an adaptation of Isaac Newton's conception of the spectrum of colors produced by refracting a beam of sunlight (Newton 1704). In Newton's spectrum, each color has its distinctive place. Hume does not mention Newton—perhaps for good reason, since the continuity of the Newtonian spectrum can easily be turned against Hume's treatment of the case of the missing blue as an exception. Suppose, for example, that the experimental subject had considerably less than 30 years of experience, so that there were many more gaps in the array of blues. By a similar argument, wouldn't she be able to supply by her own imagination hues for each and every gap? That would make the imagination potentially far more active and productive than Hume says, and far less dependent on previous experience. The less experience the experimental subject had, the more

[3] The argument occurs even earlier, in the *Treatise of Human Nature* (Hume 1739–1740, bk. 1, pt. 1, sect. 1, par. 10). The essential identity of the two arguments shows that Hume found no reason to reconsider it in the interval between the works. Hume (1711–1776) was a leading participant in the Scottish enlightenment and the major proponent of modern skeptical empiricism.

room there would be for this production of new hues. Hume's missing-blue argument would of course hold for all other colors of the spectrum as well, and analogous arguments would apply to luminosity, color intensity, saturation, matteness, and the like.[4]

The argument could easily be extended to visible qualities not connected with color: for instance, we may have actually seen line segments and sticks and other things of an extraordinary number of specific lengths, but there will always be an infinite number of lengths that we have not experienced but can imagine. We can imagine these either by interpolation (as in the case of the missing blue) or by extrapolation (by imagining something longer or shorter than anything we have actually experienced, or, in the case of color, by imagining a shade of blue a little lighter or darker or more or less saturated than the ones at either end of a sequenced array). Similar arguments could be developed for the senses other than vision. Tone and music, for example: for any two tones that we distinguish, we can imagine and produce an intermediate tone by sharping the lower one or flatting the higher; for any volume of sound, we can imagine and produce a little more or a little less; for the timbre of a voice, we can imagine a timbre that is a little more or less "breathy," or "reedy," or "percussive," etc. More generally, for any quality of any sense, insofar as the quality admits of contrarieties (dark versus light, low versus high, smooth versus rough, etc.) we can easily conceive a similar ability to produce possibilities with a little more or less of the quality of interest. As we proliferate other examples of Hume's "singular instance" we begin to suspect that it is not the exception but the rule. Interpolating or extrapolating differentially resemblant instances, to give this phenomenon a name, does not depend on having actually experienced the exact color or tone or aroma or flavor or tactile quality before. One needs only enough experience to see gaps in experience. Imagination would thus not be reproductive only; it could actually produce a new idea, one that does not directly correspond to any previous impression.

Note that examples like the production of the new shade of blue (or, harking back to the previous chapter, an aroma midway between cinnamon and nutmeg) cannot be explained by the most traditional theory of productive imagination, that it works by dividing and recombining images that we have already acquired. Making the sequence or series that allows the missing quality to appear is, indeed, a synthesis of

---

[4]Hume appears to disagree with Newton's claim that the number of hues is limitless. If Newton were right, between any two shades of blue there would be an infinite number of intermediates. This does not invalidate the initial portion of Hume's argument; it requires only that he say we are capable of "supplying from our imaginations" *at least* one specific hue between two others. But, if Newton's claim were literally true, it would also virtually ensure that, no matter how many shades of blue we had encountered over 30 years, there would be an infinite number more that we had not. The exception to Hume's general rule would loom even larger, then, because between *every* two adjacent colors we could always add more, without limit. That is, the imagination would be infinitely more productive of new colors than actual experience. Similar conclusions might be drawn concerning other color qualities. But more recent physiological and physical considerations support the notion that between any two hues there can be only a finite number of discriminable intermediates; see, for example, Raman 1968, ch. 8.

phenomena based on some degree of preceding analysis and recognition (the analysis by which we recognize the relative lightness and darkness of colors, for example). But unlike the sphinx, obtained by grafting a human head onto a lion's body, the new phenomenon or appearance is not derived by a simply mechanical process. Although it sounds exaggerated to call imagination's supplying the missing blue *creativity*—the variation from the originals already given by experience is very minor—it is nevertheless an innovation.

Hume acknowledges that imagination's ability to produce previously unexperienced resemblant instances contradicts and thus threatens to undermine his thesis about the nature of human experience, and ultimately of human knowledge. Committed as he is to a strong version of empiricist epistemology—that what we know and think is radically dependent on what we have experienced—it is not surprising that he rejects its significance. Yet surely it is puzzling that a thinker of his acuity and his instinct for detecting inconsistency did not dedicate a little more reflection to the phenomenon. If a phenomenon appears to be at odds with a well-established result or an earnestly desired goal we expect at least some reasoning about why we have the right to set it aside—especially with a thinker who is so ready elsewhere to invoke the principle that a single exception disconfirms a rule. A philosopher more than anyone, even a skeptical one, is bound by standards of consistency that forbid him to remain indifferent to apparent contradictions, no matter how unlikely they seem. At least some plausible redescription of the exception seems to be in order. Since Hume in the *Enquiry* was trying to establish the truth of a very strong version of empiricism, we wonder even more about his almost flippant dismissal of merely imagined blue.

Of course a researcher cannot allow himself to be diverted from his goals by every apparent obstacle; otherwise his path would be constantly shifting. If every theory is born refuted, as some philosophers of science say, then even opponents of a theory probably ought to have the good grace to allow it time to counter problems and objections one by one. But this cuts two ways. Imagination's production of differentially resemblant instances may be at odds with radically empiricist epistemology, but it could well be a centrally important phenomenon for imagination studies and psychology.

3.2 From Resemblant Production to Schematized Activity in Fields

Let us try to tease further meaning from the production of differentially resemblant instances by reflecting a bit more on Hume's blue.

If someone simply posed the question, Can we actively imagine a shade of blue we have never experienced before, we would immediately face the problem of deciding how to verify our response. If asked whether we had ever experienced a pink and indigo zebra (presuming that we have had some encounters with zebras, with pink, and with indigo), we would probably answer "No" with confidence, both because we rely on personal memory and on our knowledge of zebras (viz., that they are white and black). Our answer might also be reinforced by imaginatively producing, through the recombination of elements we have experienced in other circumstances, a schematic zebra form with pink and indigo stripes. If we were asked on the other hand if we had ever seen a pink and *gray* zebra, and were first reminded that pink could include a white slightly tinged with pink and most blacks can as easily be called dark grays, we might pause, especially if our thoughts were augmented by the consideration that pinkish flesh might show through white fur and that the contrast of light with dark might make gray appear black (perhaps even with an indigo cast!). We might have to confess, then, that we were not sure about the pink and gray (or even the pink and indigo) zebra; and since most of us do not have eidetic memory images of our experience, we might have to confess further that we will never remember accurately enough to correct this uncertainty. Still, a few of us, our memories stirred by the reminder of pink flesh through white fur and the contrast of light fur with dark, might actually claim to remember seeing that very phenomenon.

Answering the question about our experience of differently colored zebras in effect works by progressive delimitation. We limit our focus to zebras and recall that we have seen a few—for those who live in North America, probably in zoos, and also in photographic and televised images—and further rack our brains to see whether we can recall the demanded indigo, pink, and gray. We might count all the more on the accuracy of our memory precisely insofar as a case of pink and indigo would be extraordinary and pink and gray counter to expectation. But if we are asked instead to think about all our experiences of blue (not just blue zebras), and all possible shadings of blue, and all the years of experience we have had of blue in nature, in social life, with pictures, magazines, and books, in museums, with crayons, paints, and other colorants, we would have to proliferate and expand the range of evidence virtually without limit. Unless we had perfect memories and some knowledge of how to organize all the data, we would be forced to confess that we cannot recall every shade we have encountered. Even if someone presented us with a shade we have factually never encountered before we might very well not know we had not.[5] Such weaknesses of experience and memory make Hume's claims about the missing blue even stranger and more puzzling. He might easily have made an argument for the improbability of imagining a new shade, along the lines of this paragraph: at even a very young age we have seen so many hues that determining whether we have seen a particular one is impossible. Instead, he used his orderly scheme for presenting colors to lend greater plausibility to the notion that there are exceptions to his empiricist rule. Whatever the oddities of Hume's treatment of the missing blue and the reasons for it, his idea of setting up an array

[5] I am leaving untouched for now the possibility that the "mental searches" of memory described in this section might be imagination-driven—that is, that we in some sense have to begin to imagine a possibility before we can interrogate our memories whether we have actually experienced it.

for organizing sense experience is ingenious and can be exploited further, for example to discuss whether it might lead to other devices for probing our experience, our memories, and our imagining power.[6]

So let us focus more closely on arrays. I mentioned earlier that Hume's array is an adaptation of the Newtonian color spectrum. This spectrum displays the so-called ROYGBIV colors (red, orange, yellow, green, blue, indigo, violet) stretched out in an elongated narrow oval. Newton gave many portrayals of it in his book, and for analytic and theoretical purposes he produced variants of it. For example, he produced a color circle by, in effect, bending the elongated spectrum until the red extreme and the violet extreme just touched. Hume's array, by contrast, is not continuous but discrete, like color-matching systems employing chips of color laid out in a series of minutely progressive steps (so that the blues, for example, range from greenish blues to blues with a violet tinge).[7] Both discrete and continuous arrays and displays have been standardized and are used for technical and scientific color matching and description. They can be either two- or three-dimensional (for instance, in the color sphere devised by the German Romantic painter Philipp Otto Runge ca. 1810). The two-dimensional standard color space of the CIE chromaticity diagram is an updated and modernized version of them.[8]

I want to ask three questions about such arrays: Are they products of imagination? Are they natural or artificial? Does the answer to either of the first two questions affect the answer to the other?

Perhaps the first thing to say is that, if we have needed so far to talk of sensation, perception, memory, and imagination, making such arrays requires us to talk of intellect, understanding, or reason as well, even if at this point it is unclear how we might precisely distinguish all these terms. One might be inclined to give a combined answer to the first two questions of the previous paragraph by saying that the array of colors is an intellectual artifact or model that systematizes our sense experience. We might say, then, that the array is a result of imagination, if imagination is the product of intellect working on accumulated sensation. This might be explicated further in a quite empirical way that is still open to novelty. We could say that the power of understanding (to use a term for intellect preferred by Locke and Hume) sorts our color experiences into gross resemblant classes (blue, green,

[6]An objective array heightens rather than eliminates our awareness of the limits of subjective memory. It does, however, allow us, for any specific hue we can recall, to place it in the series of colors, and to do this in a way that would typically match series produced by others.

[7]Is this the sequence that Hume had in mind? Or was he talking about all shades of a single hue varied by adding or taking away white light, that is, by progressing from a pale, slightly blue-tinged white to an intensely saturated blue? The principle that we can imaginatively produce an intermediate would not seem to be affected by which of these he intended.

[8]CIE stands for the Commission Internationale de l'Eclairage, in English the International Commission on Illumination. The original CIE chromaticity diagram was published in 1931. It is based on an averaging of the experience of very large numbers of observers with "normal" color vision. The space of color is usually conceived in three dimensions, but in the CIE diagram it is reduced to two-dimensional representation by the appropriate selection of a parameter for luminance (brightness) and then deriving two parametric equations involving the three color stimulus values (called "tristimulus values"). See Hardin 1988.

yellow, and other main colors), then refines the gross classes according to more specific qualities displayed by class members (e.g., the various shades of each of the main colors). Next, the understanding takes its experience of the lesser and the greater (recognized most definitively in the experience of mathematical quantity) and applies that to the ordering of the different colors according to their lightness and darkness. At the end we would have, really or virtually, an immense array of color images organized in patterned ways. It could be helpful in imagining color reproductively, but it obviously could also be used to help us imagine and produce missing blues, aquas, greens, chartreuses, etc. It is no mere "cut-and-paste" array. The old sensory experience serves as content that is organized according to an intelligent scheme productive of both old and new. Natural sensation is combined by art and understanding into an image-producing device.

In this framework it could equally well be argued that the entire process is natural. Just as in the case above, each of the individual experiences, even if experimentally contrived, occurs by a natural physical and physiological process (of light transmission and neural physiology), and the color arrays we make summarize the visible color relations that natural seeing displays. Blues naturally look darker than yellows, whitish blues are brighter than indigo, and so forth. Perhaps everything is natural–psychological up to the point of producing the *physical* array of artifacts, of color specimens in the form, say, of color tiles. That is artifice. Alternatively, one might want to draw the line separating the natural from the artificial somewhere after the initial unmethodical acquisition of the sense experiences and before the analysis according to concepts like blue and light/dark and their organization of experiences into orderly series. Thus the *imagined* array, not just the physical instantiation of it, would be on the side of the artifacts. What the understanding produces by acting on the input of the senses would count as artifactual. This likely would ultimately force us to say that all our *ideas* (processed and organized remnants of impressions, as opposed to first impressions) are artifactual, insofar as the understanding is involved in classifying/associating them and in calling ideas back to mind from memory in standardized forms.

These considerations are as much about the ontology of images and the imagination as they are about epistemology, since they are not yet concerned with truth and falsity. Whether any of the possibilities and tentative conclusions suggested in the previous paragraphs are true is uncertain. What begins to be clear, however, is that at least with human beings (that is, setting aside the issue of animal imagination) it might not be possible to talk about the functioning of imagination without taking into account an at least partial or occasional dependency on understanding, intellect, or reason. The production of color arrays, color circles, color solids, and the like yields images or representations that call upon and utilize what has been given sensorially (thus it is about having an appearance without the presence of the original object), but those arrays are organized rationally according to concepts of greater or lesser abstractness.[9]

[9]"Blue" would be minimally abstract, although it is certainly abstract by virtue of its equal applicability to teal, azure, cerulean, navy, etc. "Color" is yet more abstract, and "sensory data" more abstract than "color." "Sequence" is another abstraction, though not part of the series blue–color–sensory data.

This attempted boundary drawing with respect to some kind of "space" between sensation and intellection (for the time being within the presuppositional framework of early modern empiricism) makes the traditional decision to situate imagination as a power between sensation and reason concretely comprehensible. Yet it leaves quite vague what specifically distinguishes reason, imagination, and sensation. Is all reasoning in terms of, in the presence of, or with respect to images to be called imagining? In that case only purely conceptual reasoning, or pure conceptual reasoning with respect to concepts that have no attachment to anything sensory or imaginative, would be nonimaginative. Not even mathematics, at least insofar as it uses spaces, figures, and arrays—and perhaps even when it uses signs and symbols—would count as nonimaginative reasoning. These reflections also strongly motivate another traditional theme regarding imagination, that it *makes* images and idea complexes. What it makes is literally fictional—the word etymologically suggests being fashioned, formed, or molded in order to become a kind of show or display—but the fictional does not have to be false. A sculptor can produce a statue that resembles its subject to a tee and another that does not, yet we might judge that the less resemblant one nevertheless reveals something characteristic of the artist's subject that a more "accurate" representation might not (as with the art of caricature). A physicist can devise a figurative model of a theory (say a planetary model of the atom) that is quite literally false but that nevertheless, despite some clearly false implications, turns out to be theoretically and experimentally productive.

As the example of Hume's blue shows, imagination does not need to *re*produce a perception, whether attenuated or unattenuated. It deals in the similar rather than the identical. Insofar as it is related to reproduction it is often less about cloning an original than it is about presenting some sensible character or form of the original in a different medium—often simplified, with lesser concreteness than the original, or presented in more or at least differently complex combinations. The colors we see presented simultaneously in viewing a landscape are many and complex; so are the colors we see or imagine presented in color arrays or schemas, like Hume's blues or Newton's color circle. These latter are nonnatural presentations, in the sense that they do not display themselves in humanly unassisted nature. They are complex in that they bring together many different colors; they are conceptualized in that they present those colors in a configuration that aims at categorical comprehensiveness (all shades of blue or all possible color hues). All these factors suggest a strong, perhaps inevitable connection of understanding to imagination in human beings. And if this is not ground enough for further confusions, it takes only a little reflection to see that the more conceptually trained the imagination is, the more readily and variously it can assist the discriminatory power of sense perception. We would ordinarily expect that someone who has acquired a vast experience and knowledge of light and colors would concomitantly develop a more acute sense perception of them. What has been articulately imagined and conceived would direct (or provide parameters for) what is perceived.

Almost plain contrary to what conventional wisdom holds, our consideration of arrays shows that imagination is powerfully elicited and assisted by abstraction. What is imagined very concretely and systematically can, in turn, assist both further abstraction (a first-approximation array of hues might suggest new arrays for brightnesses or saturations) and subsequent concretion (when, for example, an inventor plans a series of color-coded products or an artist incorporates techniques and themes based on the abstract schemes).

If we associate the "power" of the imagination with the vividness and clarity of its presentations, we will see that, whatever its natural state in an individual (one might easily grant that different people have different native abilities to evoke different kinds of images), the imagination is made more capable when it is assisted by acute sensory powers, a prompt memory, and basic conceptual (including figurative) schemes. Indeed, conceptual schemes that lend themselves to figuration of prominent features of what is imaginable (like the continuous variation of hue presented by a spectrum or a color circle) might increase the promptness of imagining them and reinforce the adequacy of one's grasp of them *as interrelated*. It is easy to talk about "all possible hues," but without a schema to organize hues in a comprehensive presentation the reference of that phrase is vague. Schematized imagination might be to a certain degree possible for higher animals; it certainly has a great deal to do with the nature of human intelligence.

One might look back again to Alain's portrayal of image memory's inability to count the columns of the Panthéon. If we accept that images are attenuated sense impressions, and memory is measured by faithfulness to the standard of the original impression, it is still possible that the inability to count the columns is a problem more of remembering than of imagining. Descartes, who prided himself on the flexibility of his geometrical imagination, confessed that it was difficult for him to clearly imagine a polygon with many more sides than a pentagon.[10] Most people who have received elementary education in geometry would, I think, be able to imagine and count the sides of a square, a pentagon, or even a hexagon. With the universal standardization of road signs, they are also likely to be able to imagine very easily the octagonal shape of a stop sign. But a heptagon (seven-sided) is quite another thing.[11] They might also be able to imagine a Panthéon-like structure with a portico having six columns in the outermost rank in a way that allows for a count. A much broader portico would likely be beyond their capability of stabilizing the appearances sufficiently to count (since counting requires that we not lose our place in a series). But an architect asked to perform the same task, though endowed with an imagination no more agile than average, might be able to count the columns insofar as he is accustomed to mentally schematizing buildings according to their plans, elevations, and sections. His inveterate practice of moving quickly from real views to simplified elevations and plans and back could make it relatively easy for him to count columns using a combination of reproductive and productive imagination.

[10] See the conversation with Burman (Descartes 1964–1976, 5: 162–163). Descartes notes that we can "imagine a triangle, a pentagon, and similar things, not so however a chiliagon, etc."

[11] I have little doubt that, if a heptagonal roadsign were introduced and universally used, most people would acquire an ease in imagining heptagons. The imagination is trainable and extendable.

Medieval philosophers called the kind of contact we have with what we are presently, actively sensing *intuitive*; concepts applied immediately to these intuitive appearances were *first-intentional*; and concepts that applied to the concepts that applied to the intuitions were called *second-intentional*.[12] Second-intentional schemas applied to first intentions, just like first intentions applied to real things, provide an expansive and secure structure for both memory and imagination. Memory is always judged ultimately by accuracy. What we have rightly classified according to different first- and second-intentional schemes we are more likely to remember correctly, all the more so if there are many differentiating links in a familiar structure. Imagination, if it is less a matter of accuracy than of an emergent, quasisensory appearance to consciousness, might well be more flexible precisely insofar as it can follow the guidelines of different conceptual schemes or arrays, and the imagination might well be bolder where the schemes allow for a greater distance—a greater abstraction—from the original circumstances of concrete sense perception. If too many available possibilities might work to inhibit imagination, a schema helping to guide the re–presentation of possibilities, even if it were as simple as an organized array of shades of blue, might be useful, especially at moments when appearance was incipient—that is, at the moment when one begins to imagine something.

Thus, contrary to a cultural commonplace that conceives intellect and imagination as hostile powers, intellect and abstraction make imagination prompter and more agile. Imagination by its nature has an abstractive mode that can disentangle features of interest from a morass of complications and thereby assist the intellect, and intellect can indicate benchmarks in the imaginative phenomena and pathways that connect them to one another.

3.3 Imagination as a Release in/of/from the Conditions of Perception

If we are seeking the elements and basic phenomena of imagination we need to avoid getting lost in the details of complex acts of imagining, yet the basics of imagination need to be capable of development into more complex psychological phenomena. It is conceivable, for example, that one kind of brain phenomenon is responsible for the incipience of images or the elements of images, and another, at a higher level of processing, for the organization and combination of these image elements.[13] Still, this does not rule out some kind of unity of functions that would

---

[12] Thus a genus like rodent is second-intentional, because it applies immediately to various species concepts (rat, hamster, squirrel) and only through those to the instances of those animals. The species names, on the other hand, are first-intentional. On first and second intention, see Knudsen 1982, esp. 492–493.

[13] At this point of the discussion there is no justification for invoking the "faculty psychology" taboo. These functions do not have to be exercised by a single faculty or a single, discrete brain organ or module.

bear the mark of what was elemental in incipient appearance and thus justify the continued use of a term like "imagination" to describe them. That is, imagination might by its very nature, in higher animals at least, exist as a multileveled power based on elements or elemental functions of appearance. Speaking of different kinds or aspects—like pure sensory imagination, mathematical imagination, concretive imagination, and abstractive imagination—makes sense in terms of elements and functions. It might also account for some of the historical confusion over understanding imagination and its (dis)unities.

One conceptual need we have at the outset is a positive concept or principle that could substitute for the questionable premise that images are formed (and to be judged) according to whether they are rigorously faithful to an experiential original. We have already suggested that Alain's "phenomenology"—the flash appearances he witnessed in trying to imagine the Panthéon, and our own attempts to imagine visual, auditory, and aromatic images—highlight the incipience or onset of imagining as approximative, a process that approaches a more definite appearance by repeated efforts. This would be compatible both with a relative lack of organization in imagining when it is not strongly directed by intentional, cognitive purposes (as in dreams, daydreams, and hallucinations) and with the often slow emergence of images conformable to directive intentions (for example, when we are asked to imagine the aroma of cinnamon or to picture the Panthéon). The unguided kind of imagining does not seem measurable at all by cognitive standards of exactness and accuracy, and even the guided form cannot be measured solely by such standards. To be sure, if we did everything we could to imagine the aroma of cinnamon and then were given a jar of it freshly ground, we might be disappointed by what we had accomplished. But imagination "fails" here only in a relative sense, that is, if we are comparing it to something else. It is not clear at all that such comparison is the ultimate and authoritative one that reveals the definitive nature of imagination. It seems rather to be just another example of foisting cognitive standards on imagining.[14]

Being presented a jar of freshly ground cinnamon might also lead in a quite different direction and allow us to contest Hume's quantitative claim that the (imagined) ideas have less force and vivacity than sense. After taking a sniff, we might look at the person who hands it to us and say that it is inferior stuff not suited even to an immature palate. We connoisseurs, after all, have imagined a cinnamon purer not just than the one we were given but purer than any we have ever encountered. If we were also chemists, this imaginative insight might induce us to pursue a new line of research and lead eventually to an improved analysis of the active compounds in cinnamon or the distillation of purer aromatic essences. If we were chefs, it might spur us to imagine new, perhaps previously inconceivable recipes insofar as any cinnamon we have used till now would have been overwhelmed by the other ingredients. That is, success in imagining is not to be measured simply or intrinsically by conformity to a single empirical or theoretical standard, and the

[14]By now it should also be obvious that the intense attention to images we have been describing, though it clearly invokes what we habitually call reason or intellect, has a claim to be called imagination—in fact the distinctive kind of imagining that human beings do not share with other animals.

aim of imagining one thing is often to incorporate it into more complex imaginings and corresponding implementations. Intense attention to imagining can even drive out attention to sensation and focus on aspects of the presentations that are not pronounced in sensation—the quasimathematical imagining of shapes is an obvious example, and even the kinds of imaginative categories of wine appreciation developed by oenophiles.

In the twentieth century there were two notable attempts to conceptualize the flexible, labile, incipient aspect of imagining, one relatively well known and developed in the context of specifically poetic imagining, the other less well known but also more general in scope, and with an unrecognized connection to a classical conception of imagination. The first is found in the work of Gaston Bachelard (1884–1962), the second in the posthumously published writings of Walter Benjamin (1892–1940).

For Bachelard, the poetic image is variational. What this means is explained in the introduction to *The Poetics of Space* (first published, in French, in 1958). There he asks how a new and unique poetic image manages to communicate *transsubjectively* a meaning that can be understood and felt by a range of individuals who have widely different experiences and education. Because of the image's kind (poetic), its typical novelty, and the various preparation of the audience, the image does not evoke a *cognitive* standard that has been *accurately acquired*. He argues that only phenomenology, which he explains as a "consideration of the *onset of the image* in an individual consciousness," "can help us to restore the subjectivity of images and to measure their fullness, their strength, and their transsubjectivity." "The subjectivities and transsubjectivities," he says, "cannot be determined once and for all, for the poetic image is essentially *variational*, and not, as in the case of the concept, *constitutive*" (Bachelard 1994 [1958], xix; emphases in original).

This suggests among other things that, unlike the concept, which fixes a meaning and delimits the meaning's range, a poetic image sets off in people to whom it is communicated a set of variations. The concept is constitutive and limitative; the image is productive and proliferative. *The Poetics of Space* examines the variational possibilities connected with typical, or rather archetypal, spatial images; what it argues is that there is a network of symbolic and expressive connections that derives from a level of experience—of up-and-down, home, place, furnishings, and the like—that is universally shared by human beings. This network is not reducible to the factual associations of a person's experience as an individual human being, that is, as someone who has had a unique and unreproducible concatenation of life events, nor to a mere function of social construction or cultural assimilation. One can probably set aside most of the speculative features of Bachelard's analysis and still retain his basic notion that there is something about images—I see no reason to limit it to images produced and used in poetry and art[15]—that is intrinsically a variation or differentiation of the familiar (what I have called "differentially resemblant").

[15]Bachelard of course uses the term *poetry* broadly, in essence synonymous with ancient Greek *poiēsis*, "making." If every act of imagination is poetic in the described sense, then "poetic image" would not in any significant way differ from "image."

What this means appears more definitely from Walter Benjamin's posthumously published reflections about imagination from the mid-1920s (see Benjamin 2004, 280–282). If Bachelard is interested chiefly in what poetically communicative images convey, Benjamin looks instead to how images come about in the first place, what sets them going or moving as images rather than as percepts or concepts. Beginning with the very traditional thought that imagination is originated by and dependent on sensation, Benjamin introduces a twist: imagination commences not with the preservation of the perceptual original but with its *deformation* or *unforming*, its *Entstaltung*. This German word and the related verb *entstalten* imply a process that distorts or de–forms a thing.

Entstaltung is doubtless intended to put us in mind of the term *Gestalt*, which was central to the then-ascendant Austro-German psychological movement known as the Gestalt school. Reacting against empiricist theories that understood sense perception as building complexes out of sense data, Gestalt psychologists defended the notion that sense perception intrinsically and originally involves the recognition of patterns, forms, and configurations. What Benjamin appears to be suggesting against this background is that, even if perception is a matter of grasping a pattern or a form, we cannot talk of imagination until somehow that form is released from the precise circumstances of any given perception and the specific fixity of the appearance. This release is a release of an actual form into a loosened form potentiality that is open to realization in affine, variational, and differential possibilities. It is a release from perceptual conditions into perceptual and imaginative possibilities.[16]

Theories of imagination that conceive images as duplicates of originals distinguish between simple and complex ideas. The distinction is difficult to maintain, however. It cannot be that all simple ideas are originally received in and as a single totalized idea, that is, an idea completely occupying our consciousness. If that were the case, then our experience of red, blue, green, and the like, indeed of every shade of those colors, ought each and severally to have come to us originally in total form. For a moment at least our entire experiential field should have been a shade of blue and nothing else, at another time a shade of green, etc. If that is not what happens—and experiencing colors in this way requires extraordinary circumstances—that means that we rarely, perhaps even never, experience an idea of color as simple. Any particular shade is always part of a visual field in which other colors appear at the same time (not to mention the further admixture of qualities like brightness, luminance, saturation, etc., provided by vision and nonvisual qualities provided by other senses). So if we want to conceive red or a particular shade of red as a simple idea, we must have experienced it before, and we must have analyzed or detached or abstracted it from all other colors and all other sensible qualities with which it was intertwined. Simple ideas are thus experienced as simple not in the purity of

[16]If, for example, we have some kind of configurative impulse that allows us to perceive a triangle from perceptual cues—for instance, from three marked points not on a straight line—any Gestalt response has to have an intrinsic flexibility allowing it to follow these impulses in the most various circumstances. Thus Benjamin's reflection is an elaboration, not a critique, of the presuppositions of Gestalt theory.

perception untouched by "higher mind" but only *in* and *after* an act of higher mind. Quite apart from any strengths and weaknesses in the various attempts by empiricists and rationalists to make sense of this,[17] one can argue, as consequence, that color imagination proper cannot even *begin* until this red and all other colors we experience are loosened or detached—one might even say "abstracted"—from their original occurrence and repositioned among the possibilities upon which they seem to naturally open (which shortly we shall explicate in terms of a *field*). That is because the mind must have already taken control of the perceptual form as separable from perception. Thus, if in the perception of red we see a particular red, something must have already taken place in addition to make the perceived form subject to recognition and possible variation *as* red. If release and loosening from perception take place, then the possibilities of imagination in larger and more encompassing senses of mental activity can emerge.

3.4 The Repositioning of Imagination and the Problem of Reifying Consciousness

Later, when we examine Aristotle's basic definition of imagination as a motion that originates in sensation—a definition that is more than any other at the foundation of Western theories of imagination and yet rarely made explicit—we will see that Benjamin's notion amounts to a modern elaboration of it. A form is a form not in exactitude but within a range of variations on that form, as the locus where a forming/deforming/reforming power takes place. Perception in accordance with forms, and an imagination based on them, can work only if form is understood as intrinsically differential. What the first moment that separates perception from imagination requires is (to use the example of looking at a square plate) neither the perfect geometric square nor the most perfect possible ceramic square, but rather a differentially approximating shape or shaping tendency. We see natural and artificial things as square not insofar as they are perfectly square but insofar as they do not deviate far from squarish form. If perception works by drawing what is sensed toward normalized forms, imagination begins in the range of deviations surrounding the normal. Whether *perception* provides the standard for mature human experience, or *imagination*, or something else, is an open question.

But this suggests the need for a reassessment of what the prerequisites of imagination are, particularly what the ontology of images requires. There is a tendency, encouraged by developments going back to the seventeenth century, to treat ideas of all kinds as well-defined units in consciousness.[18] Images in particular have been

---

[17]Locke never provides a clear justification for calling an idea simple. Eighteenth-century rationalism, for example in the school of Wolff, simply accepted Locke's distinction of simple and complex. This was unfortunate, for rationalism, empiricism, psychology, and epistemology.

[18]I am expressing this thought in a way calculated to raise possible qualms. I do not agree with Sartre that the great originators of modern rationalism and empiricism all thought this way.

treated as discrete units of experience, and in certain radical versions of empiricism each image can be considered as in principle independent of all others. Whether a person associates two shades of blue with one another, then, depends chiefly on the particularized events of his life history. The resemblance between two ideas (to invoke one of the fundamental principles of association employed by this tradition) would then turn out to be not an intrinsic property of the pair but an imposed (or at least imputed) quality. Thus it is *conceivable* that someone might associate the redness–experience of red *mammals* with one another but might simply not notice the resemblance to one another of different red *flowers* (or might "naturally" articulate the redness experience in the two cases differently). This person would not have a "normal" experience of color.[19] Such a scenario is far more credible in the setting of epistemology, especially in the context of radical skepticism, than it is in the settings of anthropology, psychology, or physiology. In these latter settings it seems more natural to presume that, in seeing something red or blue, people are not simply experiencing a discrete unit of mental experience but rather are in the presence of particular determinations of the more generally determinable field of color vision. That is, as members of the species *homo sapiens sapiens*, people are born with vision that, if they are not afflicted with severe forms of color blindness, by its nature presents a field that can in whole or in part be determined to every possible color, and that in imagination these various possibilities (and the concepts pertaining to them) can be applied more or less indifferently to any and every kind of colorable object. Any particular experience of color is naturally, from the beginning, a potentiality of this field. The fact that a human being "associates" red with orange, with yellow, with green, with blue, with indigo, with violet, all under the rubric "color," is a function of the fact that each of these (and countless others) is a possible determination of the color field. (We shall forego adding that this might well derive from the physics of photons, the electrochemistry of different types of retinal cells, and other facts of physics and physiology.) Acts of perception focus on the particular, momentary determinations of the general determinability of this field. Imagination and memory can occur in animals that have the possibility of the field's being reactivated nonperceptually, without the perceptual presence of the object. In the first instance this encourages the expectation that perception, imagination, and memory will be interrelated by the appearance–possibilities they share, especially those that are close to one another in one or several qualities, like hue, brightness, saturation, intensity, shininess, and the like.

Understood in this way, the basic act of imagination takes what is perceived—for the moment we bracket the question of how liberally we should conceive the whatness of this "what"—detaches it from the intricate circumstances of perception, and in this abstractive and simplified detachment releases it to a new positioning with respect to differential possibilities. Imagination is an abstractive positioning for

[19]The word is in ironic quotation marks because, on the basis of radically empiricistic principles, it is not clear where the norm should be drawn from—though in immediate context I am invoking the reader's average, everyday experience.

differentially *inceptive* productivity and reproductivity. One thus has to beware of conceiving imagination in a way that attaches it too strongly to an original act (of perception) on the one hand or to a past or prospective *thing*.[20] Insofar as imagining is inceptive even before it can be representative, conceptive, or conceptual, past acts and their objects are important less for their specific attachments than for the range of the possibilities of appearance opened around and between what has been experienced. Inception is local: that is, it takes a specific place within an opened field of the possible. The possibilities are differential: that is, they vary features apparent in what has been experienced. This variability is an expression of the differential topology of imagination, of the emplaced differential logic of the topics of imagining. Here, the topics, the conceptualized positions in an appearance–field, are quite specific to the kind of imagining being done: for example, chromatic in the case of visual imagining, odorific in the case of aromatic imagining, humanlike in imagining drama. Previously experienced things are to imagination no more than templates and models to be modified in appropriate aspects. In fact, as we shall see later, a fundamental moment in the original event of imagination is its modeling character.

In this context it becomes evident that there is a powerful advantage to Benjamin's *Enstaltung*, because it clarifies (and to a certain degree resolves) an issue that has been only tacit till now—quite apart from the fact that it identifies the *fundamental* phenomenon of imagination! The powerful clarifying moment in *Entstaltung*, de–formation, is this: In the history of imagination theory there is ambiguity about whether the *fundamental object of imagining* is objects, or attributes and qualities of objects, because there was a similar ambiguity about perception. Empiricist theories of the seventeenth and eighteenth centuries usually left this ambiguity tacit; more recently the ambiguity has often been interpreted in a way that shifts the ambiguity to lower levels of perceptual function.[21] Benjamin's notion cuts through this ambiguity: whatever is susceptible of de–formation at any level, whether it is thinglike, or substancelike, or qualitylike, or otherwise variable in some way of likeness, is properly an object for imagining. The imagining begins as soon as an experience is loosened from its perceptual context.

When investigators present their experimental subjects with imaginative tasks, the tasks are usually couched in terms dealing with objects or object-like things: "Imagine a *sculpture* placed in frontal or side view, imagine the *Panthéon*, imagine a pink-and-indigo *zebra*, imagine a grassy *riverbank* on a summer's day!" This begs the question about what truly are the first or fundamental objects of imagination and

[20] Empiricism of any kind attaches imagination to a past act (e.g., Hume's original impression) or the thing the past act experiences. Sartre's phenomenological analysis, on the other hand, attaches it strongly to the present act of mind directed to a nonexistent intentional object. These are two extreme examples of philosophers' reifying and overdetermining images and thus misplacing imagination.

[21] For instance, in theories assuming that sense perception begins with sense data—say, a flash of color corresponding to each retinal receptor, like pixels of color—the natural, original object of vision seems to be a basic unit of the color quality that, alongside all the other data perceptions, is then synthesized into macroscopic experience.

what elements are most basic to the act of imagining. Most of the examples I offered in Chap. 2 focused on sense *qualities* (the smell of cinnamon) and emergent *virtual objects* (a song)—virtual (and emergent) in the sense that the object is processual and not properly present as a whole. Arguments from the perspective of evolutionary biology, in a third possibility, might emphasize the advantage to organisms of being able to imagine certain kinds of *events* (so that objects and qualities of objects might then be construed as derivatives of event imagination).

It is easier to raise such interpretative questions and possibilities, of course, than to resolve them. The nature and the inflection of the object, or objects, of imagination are, to be sure, fundamental matters that any credible theory of imagination must address. For the moment I want to emphasize that, once we have managed to imagine some*thing*—however literally or figuratively that "thing" be taken—it is easy to vary the imagining by holding certain characteristics constant and altering others (like the color of a zebra's striping, the hue of a pixel, the orientation of a sculpture, the emotional coloration of a dramatic scenario, or what happens next in the scenario). One might take a squirrel shape as given and imaginatively vary the coloration, or take the coloration pattern as fixed and project it onto different species of squirrels, other rodentia, or other mammals that do not ordinarily display the pattern.

This suggests once again that imagination has a variational complexity and engages in a multilevel inceptive modeling, even with supposedly stable objects. There is incipient imagining when we try to formulate an image in the first place, with the flashing of appearances in search of focus that Alain described; there is the stable holding in mind of the result of flash imagining if and when it reaches a relatively fully formed image[22]; there is the variation that we can apply to aspects of this fully formed image (e.g., holding one aspect fixed and varying the others); then there is the incorporation or the projection of these results into even more complex forms, as with works of art and engineering or with everyday practice.

If imagination does not begin in the incipience of the modeling of appearance, then it is not clear where imagination can begin. Nor would it be clear where we could postulate a stop to imagination without taking appearance modeling into account. Imagination can stop only where the incipient modeling of appearance stops. This of course is not to identify where it actually stops except in a formal or definitional sense. And that is not even to broach the more fundamental question of what it means for imagination to "stop" or "end." It is in the next chapter that we shall broach the question, in a more historical mode, and begin to witness the complexities of psychological life that it opens.

[22] "Relatively" is an essential rather than an approximative qualification whenever the medium of imaginative realization is different from the medium of its original. There is no degree of detail in a pencil sketch that fully realizes the pictured object's substance; and sometimes a few strokes manage to capture and highlight features of interest better than perfect ontological replication in the same fleshly matter could.

3.5 Fields

A claim that I repeat throughout this book is that the occluded-and-occulted tradition understands imagination as fundamentally about place and placement. In this and the previous chapter I have no more than implicitly begun to invoke that tradition by talking of *fields*. The express presentation of the occluded tradition will begin in the next chapter.

Although the term *field* has already proved itself to be suggestive, it is little more than a sound until what it designates comes more precisely into view. To begin with etymology: the word *field* derives from Middle and Old English *feld;* it is cognate with German *Feld* and Dutch *veld*. These all ultimately derive from the postulated Indo-European **pelt–* and its base **pele–*, **pla–*, with the meaning "flat and broad." In this way, *field* is etymologically kin to English *plane* and Latin *planus*, and to the ancient Greek and modern English words for the flat of the hand, *palamē* and *palm*.[23]

Many philosophers are skeptical of the usefulness of etymology, but in this case at least the modern word retains the force of its history. The first entry in dictionaries of the English language will define "field" something like this: a wide stretch of open land, a plain. Not counting subheadings, the dictionary that I have on my office desk gives 16 definitions for the term. Most include the notion of some sort of expanse, something spread out before a potential onlooker, whether on land or sea or air, whether literal or figurative. Although in the first instance "field" suggests the expanse itself, not infrequently the issue is the kind of thing that the field contains or what happens in the field (minefield, field of vision, field of play).

As used colloquially, the word almost invariably suggests a contrast with what surrounds it. If you were walking through a woods and suddenly came upon an open grassy space, you would in part experience it negatively as a place that was not filled with trees and thus open—even though the field as I have described it is not totally open (it is bounded and covered with grass). If the field you came upon were filled with soybeans, you would probably designate it in light of that fact: it would be a soybean field. The captain of a cargo ship coming upon waters filled with kelp, with sharks, or with contact mines would likely report them as a "field of seaweed," a "field of sharks," and a "minefield," respectively. In these cases there is both a contrast with the unencumbered sea surrounding the area and also attention to the salient kind of thing or phenomenon that occupies it.

There is also a mathematical usage of the word "field," which I emphasize because it will gradually take on thematic importance. In somewhat loose terms, a mathematical field is a set of elements upon which are defined two mathematical operations, with each operation combining two elements (regardless of the order, i.e., the operation is *commutative*) to yield a third element of the set, and with special rules assuring the consistent, systematic character of the two operations,

[23] See the etymologies of these words in the *Oxford English Dictionary*.

taken both individually and together.[24] An example: the rational numbers—numbers of the form *m/n* where *m* and *n* are (positive or negative) whole numbers and $n \neq 0$—constitute a field under the ordinary arithmetic operations addition and multiplication. To explicate: For any two rational numbers r_1 and r_2, $r_1 + r_2$ and $r_1 r_2$ are rational numbers as well, and the operations are commutative because the order of the operation does not affect the result: $r_1 + r_2 = r_2 + r_1$ and $r_1 r_2 = r_2 r_1$. Both addition and multiplication have an identity operator: for addition it is 0, since $r + 0$ is always equal to *r*, and similarly 1 is the identity operator for multiplication, since $1r$ always equals *r*. Furthermore, for any rational number *r*, there is another number (its inverse) that, when combined with it by the operation, yields the identity operator: for addition $-r$ is the inverse of *r* (since they add to 0, the identity operator for addition), and $1/r$ is the multiplicative inverse for *r* (they multiply to yield the multiplication identity operator 1; the only exception, which is covered by the formal definition, occurs when $r = 0$, since $1/r$ is undefined in that case).[25]

As is often the case with higher forms of mathematics, the immediate benefit you gain by redescribing in an abstract way something you already know in more concrete terms is a sense of alienation from your previous experience. Yet it is precisely the more general redescription that allows a mathematician to see and prove the existence of relations that do not depend on the specific circumstances of the familiar (the basic arithmetic of addition and multiplication, in this case—one might say that for the mathematician basic arithmetic undergoes a Benjaminian de–formation and at that moment opens into the imaginative universe of higher mathematics). To the mathematician these relations and the manner in which they are expressed appear just as real as ordinary numbers, addition, and multiplication do for the average person with a solid elementary school education.[26] And there almost always follows a further benefit that satisfies even the most pragmatically

[24]Here is a stricter definition: a field is "a set for which two operations, called *addition* and *multiplication*, are defined and have the properties: (i) the set is a *commutative group* with addition as the group operation; (ii) multiplication is commutative and the set, with the identity (0) of the additive group omitted, is a group with multiplication as the group operation; (iii) $a(b + c) = ab + ac$ for all *a*, *b*, and *c* in the set." S.v. "field" in James and James 1959. A group, in turn, is a set over a binary, associative—i.e., $(a + b) + c = a + (b + c)$—operation such that one of the elements in the set is an identity operator and, for each element of the set, there is an inverse element. It is possible to have noncommutative groups. It should be emphasized that the "addition" and "multiplication" of the group are not, in general, the addition and multiplication of ordinary arithmetic.

[25]This field of the rational numbers is infinite, but finite sets can be the domains of fields as well.

[26]Unpacking the implications of this sentence is key to understanding the nature of imagination. It is not just that professional mathematicians are well aware that their subject requires intense and subtle imaginative gifts that tend to be hidden from the rest of us (and even from many scientists who think of mathematics as something that is rationally-mechanically "applied" to other things). It is even more that imagination always has the dual character exhibited in mathematics: it is a way of conceiving abstractively what is more concrete, and it is also capable of taking on a more concrete character of its own. In the introduction to Chap. 1, I defined imagination (in part) as both abstractional and concretional; pointing to the imaginative character of mathematics is a first gesture toward explicating what that means. It goes almost without saying that "abstract" and "concrete" are, and thus ought to be grasped as, correlative, not absolute, terms.

minded people: that the more abstract mathematical conceptions and associated theorems allow us to see other categories of things in unexpected ways, and the resulting conceptual applications of the mathematical theory to other kinds of things often lead to new scientific results and practical applications.

In its most general, colloquial sense a field is a relatively open place contrasted with what surrounds it, with attention drawn to what does or might fill the field's openness. The mathematical definition of field concretely[27] develops some implications of colloquial "field," but as it were with reverse emphasis. It lets the specific internal structure and characteristics of that expanse or open place of the elements–plus–operations emerge into appearance. That is, the notion of mathematical field begins with a large number of individuals (abstract elements, often numbers) and then defines networks of relationship among them—a totalizing articulation of the place they occupy and the relations between them—by means of the operations. Any operation or sequence of operations performed on elements of the set produce other elements of the set. Of course the set as an abstract object is *de jure* conceived as given and complete—the set of all rational numbers, the set of all points on a line segment, the two-dimensional cartesian plane, the set of complex numbers with real coefficients—but *de facto* applying the operations generates new elements from old ones, at least in the sense that before the actual operational combination the mathematician probably has never focused on the specific individuals of the set that are the result of such operations.[28] In either case, the set thereby becomes less a congeries or heap and more a well-formed expanse of interrelated elements that spread out in increasingly articulated detail before the surveying gaze of the interested mathematician.[29] The field she surveys is not just the elements or the elements and operations but the resultant articulated "structure."

[27]This word is not intended ironically, even if to most people the definition seems anything but concrete!

[28]Kant's claim that "7+5=12" is a synthetic rather than an analytic truth rests on this distinction. That there is something unprecedented in the mathematician's experience is clearer when we add together very large, randomly selected numbers we have never dealt with before.

[29]What is at issue here is easy enough to conceive more concretely by thinking of how a child learns about fractions over time. Perhaps the younger child is introduced to them in terms of "pieces" (if a pie is divided into eight equal pieces and you are given three...); next she learns to form the mathematical representation using a stroke mark between two whole numbers (3/8) and is told that this is in effect a form of division; then she learns how to treat such representations as belonging to a set, the rational numbers, the elements of which she learns to add, subtract, multiply, and divide; and after achieving a certain mastery of these operations, she begins to understand fractions and all the arithmetic operations on them as a unified field of mathematical activity, learns alternative representations as equivalent (for example, decimal fractions), and grasps the set of fractions as, first, an extension of the concept of whole numbers and the division operation, and, second, a subset of the real numbers, which are not expressible as such fractions. Thus the student progressively acquires a sense of being at home in an ever-expanding field of numbers and operations, and fractions become part of the standard furnishings of her mind. That all this field-expansive knowledge is at least as much imaginative as it is conceptual is one of the themes of this book.

The mathematical notion of a field, being as specifically articulated as it is, provides some hints that will be helpful for understanding imagination and its occluded tradition. In the first instance the exactness of operations and the possibility of their almost limitless recursion and repetition is not of crucial importance to my metaphorical use of the term, nor do I wish to mandate that every time one sees the word "field" applied to some phenomena, one must imagine there to be strictly defined correlates of two operations on those phenomena.[30] In a looser sense, the usage suggests that there might be definable relations between elements or characteristics of appearances, and that there are ways of moving from one imaginable position to others, whether continuously or by steps.[31] For example, if we take our set to be colored illuminants, the set of all light sources, it would be very plausible to imagine *hue addition* to be defined as the hue achieved by projecting two illuminants onto the same portion of a screen. Alternatively but not equivalently (that is, being based on reflectance rather than direct illumination), one might take standardized pigments and define addition as the result of mixing two of them. Similarly one could, from a conception of the brightness of lights or pigments, define *brightness addition*. It is important here to realize that whether lights or pigments under the operations "hue addition" and "brightness addition" constitute a field is not settled by the fact that we use the word "addition" for both operations; they might as easily be called "operation 1" and "operation 2." The names of the operations are irrelevant: the sole issue is whether the operations are grouplike. Our ability to arrange standardized illuminants and pigments spatially in color disks or color solids is in fact largely based on mathematical group- and field-like behaviors of the properties of the illuminants and pigments. By extension—or by Wittgensteinian family resemblance—one can develop a fairly clear, though no longer mathematically distinct or rigorous, concept of field. Some virtual expanse or set must be surveyable and at least analyzable in part, and some of the discriminated elements must be capable of complex combination, relation, or variation in ways that articulate field structure.

3.6 Imaginative Topology and Topographies

The notion of *topics* has been part of rhetorical theory since Greek antiquity. The word means "having to do with place," from the Greek word for place, *topos*. Aristotle was one of the first to use the term in a technical way; and one of the works

[30] That is, I will not insist that the phenomena constitute a group in the strict sense, despite the fact that a mathematical field always implies two mathematical groups, the set over the "additive" operation and the same set, with the exclusion of addition's identity operator, over the "multiplication" operation. What matters for the analogy is that there are structured, operational relationships that can be analogized to fields and groups.

[31] In this sense, an even better mathematical model might be to replace group operations with functions. But that is a complication for another day.

that was included in the so-called *Organon*[32] goes by that title, *Topics*. Playing on its lexical meaning ("tool" or "instrument"), one might say that the *Organon* presents the conceptual tool kit needed by anyone interested in serious inquiry. The works of the *Organon* other than the *Topics* treat the basic kinds and categories of being and their expressibility in language; syllogistic logic and a corresponding theory of logical explanation according to causes; and argumentative fallacies that have the deceptive appearance of correctness. The *Topics* presents the art of dialectical reasoning and questioning based on commonplaces—that is, on commonly accepted concepts and principles that are the conventional "location" of discussions concerning any given subject matter. Aristotle does not define *topos* in the *Topics*, but in the *Rhetoric* he says that he calls "the same thing element and *topos*; for an element or a *topos* is a heading under which many enthymemes fall" (1403a18–19).[33] An enthymeme for Aristotle is a rhetorical form of syllogism, an argument form following the logic of likelihood or probable opinion. More specifically, Aristotle explores in the *Topics* the kinds of dialectical argument[34] that the basic forms and categories of being and logic make plausible, and thus these forms (and the argumentative strategies based on them) can be used to guide the formulation of propositions and questions in any organized inquiry.

There is, by extension, a more particularized conception of topics. For Aristotle the investigation of any subject matter requires that one bring along all the basic tools of logic and first philosophy (a.k.a. "metaphysics") as well as the results of any other relevant, superordinate investigation. For example, one brings the concepts and results not just of logic but also of the general theory of nature to studies of particular natural things, including animals, plants, or souls—since Aristotle understands soul as the natural principle distinguishing living things from the inanimate. But although these superordinate concepts may be necessary for the more specific investigation, they are not sufficient. Every subject needs to develop its own vocabulary, concepts, and schemas, and these are developed not by logical deduction from superordinate truths but rather inductively from experience and dialectically from the attempts of those who have, literally, tried to *come to terms* with the

[32]This is a title given by later commentators to a group of writings that treat of formal (scientific) and informal reasoning. It includes Aristotle's works *Categories*, *On Interpretation*, *Prior Analytics*, *Posterior Analytics*, *Topics*, and *On Sophistical Refutations*, often with the inclusion of the *Rhetoric*, not least because it develops themes of the *Topics*.

[33]Enthymemes are usually interpreted as arguments without fully articulated logical form, in particular without all premisses of the argument being explicit. The page–column–line numbers I use for Aristotle's works are Bekker numbers, a standard format of page marking indicated in the margins of nearly all modern editions of those works. A. I. Bekker was the editor of the Prussian Academy of Sciences nineteenth-century edition of Aristotle's writings. "1403a18–19" means lines 18 to 19 of the first column (a) of p. 1403.

[34]That is, where there is the possibility of uncertainty, either actually (for instance, when one is inquiring into what one does not yet know) or formally (when, no matter how certain one may be of one's own theory, there exist alternatives that need to be debated). Dialectic in Aristotle is the process by which we take different accounts given of a subject matter and argue out the logical consequences and conflicts.

subject matter. They are the conceptual forms of our specific experience of the field of inquiry we have undertaken to explore. With specialized studies that are confined to experts the relevant experience and formulations take place only for a select group; in matters that interest everyone, for example ethics, even the man or woman in the street can be, even needs to be, consulted. The topics that result provide a set of concepts and protoconcepts that anyone entering into the investigative field needs to know, because they are the means and instruments with which any sensible claims about the field have to be made.

Topology, then, would amount to a kind of "metatopical" investigation, a *logos* of the *topoi*—that is, an account of topics. A topology deriving from Aristotelian topics can be conceived as a form of self-consciousness about accounts of things insofar as they are developed from concepts and principles that bring those things to logical terms in basic ways. But such an understanding does not go far enough, especially when one takes into account a further Aristotelian inspiration. Terminologically this inspiration derives from a suggestion implicit in the mathematical use of "topology": that even before we develop theories to account for a subject matter, we need to have something more at our disposal than a well-defined set of elements and a well-developed language referring to and relating them. We need, in addition, a sense of the place or space or field of the phenomena, one that certainly is correlated with the terms we develop to speak about them but that is not simply and totally reducible to a terminology applied in the first instance to elements of a set. When a ship's captain looks out upon a sargasso sea, he is regarding a field, and not just a set of neatly isolable entities gathered into a set. The field can indeed be subjected to analysis into parts, but it is more than their collection or sum. The phenomenon of a sargasso sea is not comprehended by examining one or two specimens of Sargassum seaweed in an aquarium and then saying "millions," or even by assigning each plant the designation s_i, with i successively taking all whole number values from 0 to some very large n, and putting all the letters, separated by commas, between two curved braces $\{s_0, s_1, s_2, \ldots, s_{n-1}, s_n\}$ as representation of the set. It requires someone's firsthand experience of vast stretches of Sargassum seaweed in the ocean, and then the associated ability to imagine it. The field terminology and the subsequent symbolic representations emerge as a consequence of the experience of the phenomenal space or field. The resulting concepts and schemas are notional articulations of the space of the phenomena.

Put in this very generalized formulation, any conceptual articulation, any set of conceptual topics, could be called a topology just by considering the relational articulations against the background of the matrix they are embedded in. That certainly would serve many theoretical purposes and would in fact bear a strong affinity to Aristotle's conception of topics (as well as a weak theory of how one goes about producing "models" of phenomena). In the *Topics* he advises that one can frame questions and propositions in a specific inquiry by, for example, recalling that any substance can be characterized in terms of its attributes, and attributes can be divided into the essential, the proper, and the accidental; that a thing can be looked upon as an instance of a species, and as thereby having a relationship to a superordinate genus (in fact many superordinate genera); and that, since a species is the determination of a genus by differentiation, one can inquire into the subject matter by addressing differences and similarities that allow one to place related things into the differentiated network of species and genera. The inquirer does not know in advance precisely *how* these conceptual relations will apply to his subject matter, but *that* they will apply he does know. This means that he knows in advance that he will be able to place the subject of inquiry into a network of conceptual identities and distinctions and thus will be able to establish conceptually mediated relations to other things. The subject of inquiry, whether it is a kind or a thing, can be situated in a virtual space constituted by this conceptual network.[35]

This kind of topic networking matrix might all by itself justify using the term "conceptual topology." This quasi-Aristotelian topical usage would emphasize developing an inquiry according to relationships of substantiality and accidentality (substance, quantity, quality, passivity, activity, place, time, etc., called predicaments by the medieval theorists) or of species, genus, difference, sameness, and the like (predicables, in medieval parlance). A less Aristotelian but more general and abstract notion would use "conceptual topology" whenever there is merely *some* logical or mathematical relationship among terms, without giving any privilege to the substance–accidents logic of Aristotle. It could be formulated according to the canons of a more mathematical-symbolic logic or according to the structures of the most abstract varieties of mathematics.[36]

It is nevertheless also possible to conceive "conceptual topology" in more concretely developed ways. In studying a subject matter one often elicits a network of relationships that do not immediately fall into an already established logical pattern. When geometry was first conceived as a science, the reasoning about it abided by logical rules, yet the relationships that are treated in plane geometry had to be developed in terms of lines, angles, polygons, circles, etc. Those entities and structures were not exhibitable simply as, or in terms of, numbers and logical laws. In such a case, the relationships that hold between the various concepts one develops display a particularity not (yet) duplicated by any other actual subject matter. The representations of it are, for the time being at least, sui generis. It is, at the moment of initiation, a structure with one known instance. Over time the field and its structures may come to be seen as a "purely conceptual"[37] model. It might happen, as it often does in mathematics, that the structure turns out to be discoverable in

[35]This is not a reductive network, of course. A reductive network—for example, a biological one that claimed plants and animals were nothing more than devices to preserve a genotype, or a chemical one that said genetic expression is nothing other than the functioning of valence bonding—might nevertheless be conceived as a kind of variation on the network I have described, with the limitation that this type of reduction aims to grasp things not as a differentiation of the genus but rather as nothing but the genus.

[36]Logical and mathematical formulation does not imply that these things are beyond imagination, however, as should be clear already and shall become clearer as the book goes on.

[37]Here and in related locutions over the next few pages, these quotation marks are of the type known as ironic. My point is that the purely conceptual, the pure abstraction, is never absolutely pure. To put it differently, these pure rational phenomena have to be understood as formed in an imaginative field.

other fields of investigation or that it can be applied to many different kinds of thing and situation besides the original one. A similar argument can be made about fields and field structures discovered in nature, in culture, and in fantasy.[38]

This is the level at which one can speak of elucidating potentially or actually *isomorphic* and *near-isomorphic* structures—that is, structures having the same form, although in order to see the sameness one often has to first make separate abstractions from two or more actual situations that appear to be of quite different kinds. Any electrical circuit exhibits an electricity-flow structure that can be modeled by binary logic ("1" for "circuit on" and "0" for "circuit off"). Even if there were just one actual kind of electrical circuitry (say, copper wire), one could describe its structure using this binary model. If one also were the operator of a set of pipes and valves regulating the flow of liquid nitrogen, one might discover that the system is isomorphic under nitrogen flow to a copper-wire electrical circuit under electron flow, despite all the many differences between electricity and wires on the one hand and pipes and liquids on the other. The same diagram of lines and nodes can be used to represent wires and switches in the one case and pipes and valves in the other (ignoring or abstracting from, for example, the problems caused by the very low temperatures of liquid nitrogen or the shock potential of the electrical circuit). As long as a model is sufficiently abstract, one is inclined to say not that the model and the object share the same structure but that the model expresses the structure of the object.

Since human beings are finite and have to take relatively small steps in their theoretical innovations, almost every structure that is understood as a form capable of realization in many actual instances was originally conceived as the (at the time unique) structure of a particular subject matter, or, if it was devised "purely theoretically," without reference to any actual situation, as a "purely abstract" mathematical structure. Although our philosophical tradition tends to think of such abstractions as rational, they are actually imaginative (for the time being we can think of them as *rational imagining*, what reason does with images). Subsequently, once the conceptual expression of the form has taken on a sufficient consistency and familiarity, it can be used as a model for many different isomorphic phenomena.

A further usage of "conceptual topology," the importance of which will become more prominent in the course of this book, appeals to the notion of being isomorphic or same-structured but with a more dynamic notion of structure. A model for this usage is the mathematical field of topology. Historically topology appears as a generalization of geometry without a metric (that is, without a fixed measuring stick) and the objects in the space as plastic (that is, as capable of being manipulated and reshaped without actually cutting or breaking their fundamental shapes).[39]

---

[38]For example, Lévi-Strauss's analysis of the generative logic of kinship can be, and was, extended to the shapes of storytelling and then to social structures.

[39]The plasticity is of course closely related to the lack of a metric. If a measuring stick were constantly to stretch or contract unpredictably along its length, we could not rely on it for fixing distances between objects. Nevertheless, we could still use it to display certain properties of continuity and coherence, since the markings on the stick would maintain the same *order* with respect to one another even as the stick stretched or contracted. Topology studies precisely such matters and properties.

In *geometry*, congruence is an isomorphism that depends on the metric, on the measuring devices. Two triangles, for example, are congruent if they have three respective sides equal in measured length and the three respective measured angles equal. Traditionally one way of showing their congruence is to superimpose one triangle on the other to show that they match. Of course this means that one implicitly accepts that triangles are discrete objects within a two-dimensional plane, and that they are freely movable in the plane (can be imagined as moving) without any significant deformation—that is, all lengths and angles are unaffected by moving them.

More rigorous geometrical versions of superposition require exactly specifying the permissible movements or transformations (translations and rotations). Suppose that we have a cartesian plane with its mutually perpendicular *x*– and *y*–axes, and that we have in different parts of the plane two congruent triangles. One might define a set of permissible transformations that would superimpose triangle ABC onto triangle DEF by three successive movements: (1) sliding ABC parallel to the *x*–axis so that its midpoint (say, the center of gravity)[40] is directly above or below the midpoint of DEF, (2) sliding it parallel to the *y*–axis so that its midpoint is superimposed on the midpoint of DEF, and then (3) rotating ABC around that midpoint until the three sides are perfectly superimposed. These rules are not sufficient for establishing congruence, however, when the congruent triangles are mirror images of one another. We have to add one more transformation: (4) if necessary, one may start by rotating one of the triangles out of and back into the plane, with the axis being (for example) one of the triangle's sides. This is equivalent to picking up a triangle off a floor, turning it over, and setting it back down.[41]

Congruence is a *geometrical* equivalence of form based on exactly matching parts—exactly matching in measurement. *Topological* equivalences can be illustrated without worrying about angle and length measurements and allowing for various kinds of plastic deformation. For example, I could define the topological equivalence of all triangles if I defined transformations that allowed me to stretch or contract their sides (I thereby would also be changing the angles formed by those sides). It would be as though the sides of the triangle were made of rubber bands. I could similarly claim that all closed plane figures (not just straight-sided figures like triangles, rectangles, pentagons, etc., but figures with curved sides as well) are topologically equivalent if I defined transformations that allow me to curve, straighten, or kink and unkink sides (which would permit me, for example, to turn pointed angles into gentle curves, or gentle curves into sharp kinks).

[40]There are different kinds of midpoint that could be defined. I am assuming here that it is the midpoint determined by the intersection of the triangle's angle bisectors, which yields the so-called "center of gravity" of the triangle.

[41]An easily imaginable example: take an equilateral triangle (all sides equal), then from one of the vertices drop a perpendicular to the side opposite. This divides the equilateral triangle (which is also equiangular, with each angle 60°) into two congruent triangles, each with angles of 30°, 60°, and 90° and with corresponding sides equal in length. The only way to make them match point for point is to flip one of them over, by rotating or lifting it out of the plane, or "folding" the two halves of the triangle upon one another along the dropped perpendicular.

In three-dimensional space topologists often use elementary examples like these: a dinner plate is topologically equivalent to a wine goblet, and a cup with a handle is equivalent to a donut. The first can be shown by gradually deforming and reshaping the "notional matter" in the plate without producing breaks, cuts, or holes to produce the wine goblet. (Think of gradually massaging soft clay shaped as a plate into a bowl, then massaging some of the matter in the bowl toward the bottom and gradually elongating and reshaping it into base and stem—all without poking a hole in the clay or cutting out pieces.) Similarly one can flatten out the bowl part of a cup with handle, then gradually work that flattened clay toward and into the handle (which is a hole) to make the whole mass into a donut. Again, slightly more technically, one can say that if there are certain well-defined operations or transformations that can be performed on a mathematical object—operations that correspond to the intuitive idea of "massaging conceptual matter or clay"—then if one such object can be continuously transformed into another, the two are topologically equivalent.

On the one hand, the example of the relationship between congruence in plane or solid geometry and isomorphism in spatial topology illustrates the not insignificant consideration that "higher" or "more abstract" levels of mathematics do not abolish the use of imagination; they instead make those uses more rigorous and delimit them more precisely. This is not to dispute that from a formalist standpoint the imaginative model is merely a way station on a track headed toward a purely formulaic and rational presentation of the essence of the mathematical system in arbitrarily chosen symbols.[42] Even if a purer formalism is achieved, however, there is the question of whether mathematics learners and even mathematics researchers themselves do not still have to resort to various kinds of imaginative models and devices while actually "doing" mathematics. Or, more fundamentally, whether the very act of "emplacing" some more concretely presented form one has discovered into an abstract symbolic structure is not itself an act of imagining, with the symbols, rules, and sequencing employed in their combination being the matter of the field. Where there is matter, there is always possible variability of form and de–formation.

On the other hand, the geometric and topological examples also provide us with a different model of field, an analog or holistic one, as opposed to the more digitalized and discrete models corresponding to the algebraic/set-theoretical notion of a set of elements with "addition" and "multiplication" operations. One of the chief virtues of the "stretchable" topological model of topological space is that it is easily and directly imaginable. In any case, an imagined object can undergo modifications, variations, and alterations. The zebra can become pink and indigo, its tail can be lengthened or shortened, the proportion of leg length to overall body height can be altered, a horn can emerge in the middle of its skull to make it a zebracorn, etc.

[42] As I shall show later, however, it is highly problematic to assert without further argument (and an indispensable amount of historical investigation) that symbolic formulas are purely rational, without any admixture of or dependence on imagination. In fact the real genius of mathematics is that, in the long run, what was abstract becomes the element or field of mathematical imagining for a future generation.

If one of the inconveniences I mentioned earlier as associated with the set-theoretical conception of field is that it presumes the set is well defined in advance of constructing it, with a geometrical or topological conception one often need worry only about the immediate vicinity of the object of concern. We do not need to conceive of the whole of euclidean or cartesian space or the completeness and adequacy of transformation rules in order to rotate triangles, nor do we have to bother about the infinite reaches of topological space as we remold a plate into a goblet. All we need is just enough space to rotate and translate triangles or to draw out the plate–matter into an elongated stem. There are actually subfields of geometry and topology, called differential geometry and differential topology, respectively, that study what we can learn about more remote spaces surrounding an object (in particular, of a curve or curved surface) by examining features in the object's immediate vicinity. So, for example, given certain initial assumptions about the nature of space and of curves in the space, one can draw conclusions from variations in a small region of a curve about how the larger curve it is part of should look.

If a little bit of knowledge—that is, *insufficient* knowledge—can be dangerous, nevertheless it can also serve as a warrant that future research in the direction indicated by that little bit might be of assistance in developing a more articulate conception of what we are studying. The better articulation does not actually require that our hopes or surmises pan out as expected, since we learn even from failures when those failures are worked out in both concrete and abstracted detail. For the moment, the notion of imagination as structured something like a differential geometry or topology indicates to us a slightly more informed way to advance the insights of Bachelard and Benjamin. Because imagination begins in the incipience of appearance and reappearance, and because a fully developed image requires formation in considerable depth, that incipience of appearance would avoid mere chaos if it came with *directions of formation*. If we think of a formed image as capable of deformation, we can track the formative and deformative processes more particularly as we think of specific respects in which it can be deformed and reformed and conceive them as providing a tendency of change or a direction for differentiation. One might think here precisely of a fundamental insight of differential calculus: if you start at a point on a curve, and you draw lines from that point to many other points on the curve in the immediate vicinity of that point, you can approximate the line that is perfectly tangent to the curve at the point you started with. What is more, this process of approximation can tell you a great deal about the larger curve that all these points are part of, because the tangent to the curve is an indicator of the rate of change of curvature in that vicinity. It tells you where to look for the "next" points on the curve. To extend the analogy: the more we familiarize ourselves with relatively small reformations and deformations of something imagined, the more we are likely to find out about the larger place and processes of imagining itself.

A final point concerns occasionally useful distinctions involving the terms *topology* and *topography* (and their adjectival and adverbial forms). *Topology* is the most encompassing usage. As it is used to designate fields or spaces, it can reflect different levels of articulation. In the first, most general designative sense, *topology* refers to all fields and places that have a differentiated feel for a denizen or inhabitant (to be distinguished from an outside observer). A landscape may not have many landmarks, much less be highly conceptualized, but someone traversing, even for the first time, quickly acquires a feel for its formations and conformations. In this sense, topology simply indicates that there is a real or virtual field or space within which denizens recognize a "lay of the land." In more specialized usage *topology* refers to more highly articulated fields and spaces. *Conceptualized topology* (or *conceptual topology*) would thus be used of a field that has been elaborately marked out. A *topography* then could indicate a field or space that has been represented or written up according to the conceptual articulations of the conceptualized topology. Theories as we ordinarily understand them are topographies. A single conceptual topology can give rise to multiple topographies, i.e., multiple conceptualized representations. To summarize this using the example of music: The realm of music and sound is, for human beings, a basic topological place, a topology. At least a few features of it are universal, like higher and lower pitch, more rapid or slower beat, etc. As soon as people begin familiarizing themselves with it and marking it in more detail, it turns into a conceptual/conceptualized topology. Each culture, over broad geographical areas and considered over long swaths of time, cultivates its basic conceptual topology of musical experience. The more particularized ways of traversing the conceptual topology produce topographies. In this sense we can say that peoples East and West came to inhabit the common human topology of sound differently—in different conceptual topologies—and that each such conceptual topology has been diversified historically in many different topographies.

3.7 Placing the Topological Dynamics of Imagination

Before we turn to the historical reconstruction of the occluded tradition of the placement of imaginative appearance, it will be useful to broaden somewhat our experience and conception of common ways—and organized arrays—in which imagination presents itself.

One of the commonest forms of imagining in contemporary culture is game playing. After Wittgenstein's devastation of the notion that we can come up with a definition of "game" or even a listing of necessary or sufficient conditions for using the word, one can hardly claim that the examples and variations I will cite can stand for all. Nevertheless, I think that they will be suggestive and even comprehensively instructive.

Readers who have played games of solitaire (I am thinking specifically of varieties employing a standard, four-suit deck of fifty-two playing cards), whether using actual cards on a physical surface or virtual cards in a computer desktop window, know that in the first place you have to learn the basic rules, such as how you set up the starting array, which cards can be moved from one pile to another or to home, the use of the "buffer" where cards can be temporarily stored, etc. The first games you play usually involve little more than making moves almost randomly. In this way you start to acquire a sensibility for the game. The next stage involves learning basic tactics and recognizing how well a game is going. At this second stage you likely develop (or hear of) certain rules of thumb (for instance, "don't take home too many cards of a single suit prematurely" or "make sure that at least one seven can be moved"). As one becomes more experienced in the game it becomes possible to recognize a beginning strategy from a quick visual scan of the starting array and to anticipate what will happen several moves in advance. Even early on in this third stage one can begin to develop a style of play. By style I mean an approach not specified by the rules but that is identifiable to those who are expert players.[43]

At some point between the second and third stages, the game, in particular the array, takes on the character of a matrix, in which one is aware of the parts and the individual moves in relation to the whole game; the game begins to feel familiar. ("Matrix" suggests more than "field" that the array of things and features is segmented, discontinuous rather than continuous.) Before that point it makes a limited kind of sense to think that playing the game is chiefly a matter of following rules; beyond it, the rules slip into the background of awareness; they are called back into central focus only as needed. The game has become something more and other than rule–following.[44]

For example, pick up a copy of *Hoyle's Rules of Games* and actually read the rules for several card games, both those you know and those you don't. In the first instance what you find is that the games you know are hardly recognizable from the rules, and that with those you don't it is a struggle to grasp how play goes. The more complex a game, the more the rules merely distinguish illegitimate from legitimate moves and set parameters and limits to the game. With games involving elaborate or coordinated *physical* action this is even truer. The rules of tennis specify that to begin play the server toss and strike the ball, and although they limit where the server can stand they say nothing about the height of the toss, whether the ball is to be struck one-handed or two, overhand, underhand, or sidehand, gently or hard, flat or with spin. The rules indicate when a point shall be declared won, but they are totally silent about the manner of play that will lead to winning (apart from where the ball must fall and how many times it can bounce before it is struck). Even less do they say anything about how game play will evolve over the decades, as players become faster, stronger, and (because they benefit from past examples) more savvy. Rules set up the space of play and the basic moves and acts within it. The rest is left to prevailing practices (performance practices, perhaps a social form of imagination) and individual imagination.

[43]This paragraph is a simplified adaptation of Dreyfus's discussion of the stages of learning to drive a car, which he uses to illustrate levels of progress in advancing toward mastery (Dreyfus identifies seven stages in the first edition, six in the second). See Dreyfus 2008, ch. 2.

[44]This is not to say that one doesn't follow rules any more, much less violates them. Rather, they become second nature, to the point that one can attend to higher levels of structure because one no longer needs to focus on the basics. This is the most familiar experience in the world—which does not mean that it is sufficiently appreciated.

Imagination enters into every practice of a complex game. There is a remark of Wittgenstein's that says one cannot see something and imagine it at the same time, but playing a game appears to refute it.[45] Players imaginatively perceive the game actions and space; if they do not, they are doing something other than playing the game. "Objectively" speaking, the court is a grass or clay or concrete or asphaltic surface with chalked or painted lines and a fence made of netting. To a topographer plotting the earth's surface in this locale, all that counts for little; he charts indifferently the area within drawn boundaries and areas without. To a worm, what counts is not being struck by the ball, stepped on, or painted. But to players there is a world of difference between what is outside the markings and what is within.[46] An expert player preparing to volley at the net will usually see from the ball's speed and spin that she should not offer at it because it is headed out of bounds just a few inches beyond the end line, almost 40 ft behind her. She has developed such a refined "sense of the court" that almost every time she will be right.

Someone draws lines on the ground or floor, and people begin acting as though the drawn lines are real and constitute a distinct and highly structured place for the activities of a game. Players of soccer, football, volleyball, hockey, etc., develop an instinct for their position on a field or court, so that while performing at the highest speed and intensity they rarely take play out of bounds accidentally. They place their foot a hair's breadth on this side of a line and make the winning score; if they place it two hairs' breadths further on, their team loses. The best players rarely make a mistake in this.

Consider a basketball player at the very highest level of performance, say an outstanding NBA point guard.[47] When his team secures possession of the ball at the end of the court where they have been defending the basket, he is the player to whom the ball is usually thrown so that he can advance it by dribbling (repeatedly bouncing) the ball toward the basket at the other end of the court, where his team will try to score.[48] A point guard needs many skills. Since he has the ball in his control more often than any of his four teammates, he must be the most sure-handed of

[45] See Wittgenstein 1967, §621, pp. 109 and 109e: "Während ich einen Gegenstand sehe, kann ich ihn mir nicht vorstellen," "While I am looking at an object I cannot imagine it." The translation of *vorstellen* by "imagine" is not entirely unproblematic.

[46] Note that distinguishing various ways of experiencing the space complicates defending Wittgenstein's claim. Each agent perceives the place under a different set of abstract–and–concrete parameters.

[47] To accommodate those unfamiliar with the game, I will have to overdescribe. For those who hate sports and games of all kinds, I leave it to them to imaginatively construct an equivalent alternative.

[48] For our purposes here it is probably sufficient to point out that the court is 94 ft long and 50 ft wide, that, at both ends of the court's long axis, baskets (hoops with netting open at the bottom) are attached to the front of a vertically oriented board, with the hoop at a height of 10 ft from the floor, that at the beginning of every quarter each team is assigned a basket and scores points by "shooting" the ball so it falls through that basket (while the other team tries to prevent it), and that when the scoring chance for one team ends, because the team scores or loses possession of the ball, the other team moves (usually very quickly) toward the other basket to make its own scoring tries. Since in basketball walking or running while holding the ball is a rules violation, a player on offense has to move the ball either by bouncing it with just one hand ("dribbling") or by throwing it to a teammate.

players. He has to be a skilled dribbler, not just forward but also backward, between his legs, and behind his back. He is often one of the most agile players on the team (and therefore is ordinarily one of the shortest). He has to be able to control the ball with finesse, stop on a dime, change direction quickly, accelerate with explosiveness past defenders, feign movements and actions, keep track of the positioning and movement of his teammates and his opponents, and pass the ball quickly and accurately when he sees one of his men open, whether he lobs it in a looping arc over the heads of opponents or rifles it through a momentary gap in the array of defenders. Just as important as these natural and acquired physical skills, he must have a clear understanding of the team's designed plays and the likely variants that will develop because of the other team's defensive plan. Often he must improvise.

Coaches say that a good point guard has to be able to "see the floor." They don't mean that he has visual acuity of 20/20 with good peripheral vision, although those are undoubtedly desirable attributes. An even more important aspect of "seeing the court" is having a projective sense of what is happening and where everyone is, with emphasis more on what is about to happen than on what is currently visible. He not only sees the people in front of him, that is, takes in their current positions and the directed motions of the other nine players relative to the lines and circles marked on the floor, he also perceives the situation "in view of" schemes from the team playbook, the established habits and tendencies of his teammates and his opponents, and his highly developed talent for envisioning (forevisioning) the action. There is very little that is propositional about these actions and powers, there is little "supposing that" going on.[49] What the point guard does is certainly not purely rational, purely sensory, or purely a mix of the rational and the sensory. By their nature his activities require a highly articulated sense of the place and space of the game. The space of a point guard's game is not the space of the geometer, the physicist, or the engineer/architect; it is not the space of the plaza in front of city hall or that of one's living room. It is not even the space in the arena that, on the day following, will be an ice hockey rink or the orchestra pit for a concert, once the work crews have carted away the basketball floor and flooded the refrigerated subflooring area or replaced it with a stage. It is a space immediately present to his perception as memoratively and imaginatively saturated with an articulation by rules, court markings, playbook strategies, player talents, and the kinds of sudden opportunity that are constantly emerging in the course of the game's situation and that have to be seized as suddenly as they emerge. An analysis that neatly separates all this into categories of what is perceived, what is remembered, what is imagined, what is conceptualized, and what is desired is a philosopher's illusion.

---

[49] And even if one can say, for example, that the point guard is supposing that the power forward is about to make a spin move toward the basket (and so the pass will lead to a score if he is right or be intercepted if he is wrong in the supposition), that runs the risk of portraying (or parodying) the event as essentially cognitive and predictive when it is instead a dynamic situation of engaged activity. It is, moreover, quite simply wrong to say that the point guard is supposing that this side of a painted mark on the floor is in bounds and on or beyond it out of bounds, or that he is supposing that the players in differently colored uniforms are his opponents. Being in bounds or out of bounds is real, even if *imaginatively* real, as is also being an opponent—at least once you are in the game.

Games range from the very simple (tic–tac–toe) to the extraordinarily complex. They occupy or are played in physical space (even when that space is, in the first instance, a computer or television screen), and yet the physical space is not a place for the game unless it has been structured by the rules and possible moves. The place of the game is not simply superimposed on reality, it is an intentional rearticulation of real things and real space, and it is this intentional rearticulation that is perceived and felt by the players, and even to some significant degree by spectators. For the time of the game—and in fact *in* the time, the peculiar temporality, imposed by the game—it is the place that participants inhabit.

At this level of analysis it is not at all clear that it makes sense to talk of "objective" versus "subjective" space and place, and even referring to the reality of a game as "intersubjective" reflects more the effort to preserve the universal applicability of the subject–object dichotomy than genuinely trying to come to terms with the game space. What a complex game displays more than anything else is imaginative depth, density, intensity, and directionality. There are many different levels, many cross sections of the situation within which imagination takes place perceptively, memoratively, projectively, and even analytically.[50] At a very basic level the basketball players are *inhabiting* the place defined by markings on a floor and backboards and hoops elevated at the opposite ends of the demarcated space. They have an acute, elaborately drilled sensibility for their positions in the game space without having to pay explicit attention to their bodies ("Am I moving my right wrist properly as I take a shot?") or to the markings on and around the court of play. The point guard I have talked about is constantly anticipating the movements of his teammates. For him far more than other players, he is considering more than a living version of arrowed Xs and Os on a chart. He sees the emerging sudden turns and creative improvisations he knows his teammates are capable of and exploits the possibilities they offer. At the same time he has an acquired sensibility for the urgency of play, both as it is imposed by the course of the game (is it the middle of the first period or the last moments of the game?) and by the rules that require the offense to advance the ball to their end of the court within 8 s and to shoot within 24 s of taking possession of the ball. As we shall see, this is comparable to the many levels of the imaginative cross-sectioning of "ordinary reality" that occurs in both practical action and artistic making.

3.8 From Basketball Practice to the Biplanarity of Imagining

There is a simplification of playing basketball that, in its simplicity, should help refocus the concerns of this and the previous chapter. The simplification requires even less knowledge of basketball than the preceding, although perhaps rather more of child and adolescent psychology.

[50] Coaches will often tell players not to overthink a situation. But that is not the same as not thinking ever and at all. Once their experience of the game becomes habitual, it also becomes more easily *imaginable*—both in advance and in action.

Consider a solitary boy or girl, perhaps 12 years old, engaging in the activity known as shooting baskets on a schoolyard playground. Imagine, for the sake of concreteness, that the basketball backboard is fan-shaped and white, but also heavily weathered and showing a few dents (it is chipped, painted metal). The rim is a rusting orange set a couple of inches lower or higher than the standard height of 10 ft and a little bent from too many would-be star players grabbing on to it when they try to dunk the ball. The "netting" of rusted chain links has a few gaps, and the backboard is supported by a scarred aluminum pole with a three-foot offset (so that anyone rushing toward the basket will not immediately collide with the pole).

We observers are watching from a bench across the playground as the shooter does layups, one- and two-handed set shots, one- and two-handed jump shots, hook shots right- and left-handed. She rebounds or retrieves her missed shots and shoots again, and when close to the basket simply tries with one hand to catch–and–deflect the ball back toward the rim in a single motion with fingers spread wide. Every few minutes she stops, walks with the ball to a spot directly in front of the basket about 13 ft distant, studiedly bounces the ball five times with her right hand, five times with her left, five more with the right, then steadies the ball in front of her and pushes a shot toward the basket—something she sometimes does a single time, sometimes twice or even three times in a row. She will toss the ball away from herself and run hard to retrieve it. If the ball is approaching the edge of the paved playing surface she sometimes reaches far over to grab the ball, then reverses direction so suddenly that it seems she is trying to avoid an obstacle we cannot see; at other times she flings the ball backward over her head onto the court as her body goes flying in the opposite direction off the pavement. Occasionally she goes to the circle painted at the foul line, throws the ball straight up in the air, then jumps and slaps at it. She finishes a half hour of this kind of play by taking several long, looping shots, until finally one swishes (or rather clinks) through the netting, after which she prances around with both arms raised over her head and an exultant look on her face.

I have been describing actions whose significance is clear to those who know basketball and 12-year-olds, and doubtless it is sufficiently evident to anyone who has witnessed something similar. The girl is not simply practicing basketball, she is "playing" a game of basketball with teammates, opponents, and a crowd cheering (or booing) their performance. Perhaps it is the championship game of the National Collegiate Athletic Association Women's Basketball Tournament, and the shot that she followed with arm-raising exultation was a successful three-point shot with the clock running down to zero that won the championship for her team. A harmless fantasy that no sober observer will confuse with the actual play of the best women's collegiate basketball players!

Many kinds of condescension may be justifiable, but theoretical condescension is not. This is a most remarkable activity, one found nowhere else in the animal kingdom. Even if it were no more than a practice session, one could cite it as something that only human beings can do and that at least sometimes requires active imagining.[51] You can of course mindlessly practice jump shots, but if you are trying to use the practice to increase the height of your jump, with the aim of learning to avoid the outstretched hand of a taller defender trying to block the shot, it helps if you visualize such a defender as clearly as possible—sometimes as right-handed, sometimes as left-handed.

"Mindless" jump-shooting in fact usefully demonstrates a feature of the more complex game activities that it serves. The creation of the game and the play within it require the basic *instituting imagination* (including the institution of the rules and the field or court of play) of those who invented the game.[52] Players adapt themselves to the basic institution in learning to play the game, and while playing they practically imagine within various secondary institutions (like the team's playbook or the prevailing styles of play in the team's league). But then they can take another step, by which they (and their coaches) abstract elements and skills that can be developed by prescinding[53] from the game situation. They run various drills to practice some small feature, including shooting jump shots one after another from any and every point near the basket. That is, if one can say that the original institution of the game begins by taking a space and restructuring it for the purposes of the game, the creation of drills takes the game and its space, detaches a few features of it, and then creates an activity that, for the sake and the time of the drill, takes on an absolute character. The game imagines ordinary space and time as the place and time of play; practice presupposes the game and imagines one of its features apart from the game. In the course of the practice players often perform actions that, as part of an actual game, would be ruled out of bounds or contrary to the rules.

This is not the last time that we shall see this characteristic of imagination: that it begins by looking upon an original situation in a particular way, proceeds to leave behind the original situation while taking for granted certain aspects of the space and time of the original and of operations performed there, and then takes this new sense of a privileged space and time as a new original situation and restructures it once again, with the relationships of the first-order imagination being set aside or at least deemphasized for the sake of privileging a second order of imagining. What was initially taken as part of a larger field comes to be taken—imagined—for its own sake, and this part in its turn becomes the encompassing field for the partial activity. There is no reason in principle why this cannot be repeated indefinitely: a second partialization can be devised from the first partialization of the originally imagined whole, and a third from the second. One needs to note as well that this movement does not always have to be "away" from the original lived world, or always have to be partialized-abstracted. One can define within a field a subfield

[51] Kittens wrestling with one another or batting and chasing toys may be acquiring skills that will be useful in hunting, but they are not practicing, because practice has express intention toward the ultimate activity.

[52] I am freely adapting the notion of the instituting imagination from Castoriadis 1987 [1975].

[53] "Prescind" will take on a thematic role in Sect. 5.13. For the time being it can be considered a form of abstraction in which a part or feature of something is treated as though it existed apart from the whole.

that can be temporarily inhabited for its own sake, after which one fits it back into the original field. This creates a backward and forward movement, a "rolling" back and forth of imagined fields. The point of drills (second-order "games") in sports is to reincorporate what they teach into the first-order games. But sometimes they become games in themselves or come to be incorporated into new games that are related to but not simply part of the original one.[54] If partialization is a kind of abstraction, this contrary movement is a kind of concretion.

A grasp of these shifts of focus and plane that are characteristic of imagining, as we shall see in the next several chapters, are to be found as a constituting element in the occluded tradition, all the way back to the fourth century B.C.E. The psychological basis is the human ability to see something and its setting in terms of another thing and its setting. In *Descartes's Imagination* I showed how Descartes understood this as the *foundational* characteristic of imaginative consciousness. I named this characteristic *biplanarity*.[55] For example, one can be totally absorbed in the world as it is ordinarily experienced in sensation—although what this means is arguable—but we can also reconceive it by explicitly focusing on the world as presenting us with appearances, and when we do that we mentally accomplish a certain dissociation of world and world image, each of which henceforth (that is, for the time being) constitutes a field or plane. In the first instance we look upon the world through the medium of images, so that there is a kind of transparency– or template–character to the image field.[56] We can draw on an analogy with telescopes and microscopes, in which there is an optical plane for the object (the object plane), and another for the plane of the image (the image plane). Just as we can be totally absorbed in the world, we can focus upon the images in the image plane as though the image plane constituted a closed world (artists and mathematicians are particularly familiar with this phenomenon). A reversal of perspective is possible: rather than seeing the physical world through the template of images—for example, seeing it in terms of geometrical figures and solids in motion in Euclidean space—we can see what was originally the object plane as the medium in which is realized possible instances or exemplars of what was formerly conceived as an image plane. This is what happens, for example, when we take the laws of physics and their interactional possibilities as fundamental and then conceive manipulating things in the existing world as a special case of those laws. As we shall see in the next chapter, the ontological, epistemological, and psychological prototype for this conception was in fact and in principle invented by Plato.

The biplanarity of imagining produces a situation of *multiply differential (and often cross-sectioning) placement*. Consider: we can view the original field as

[54] An example of the former would be a passing drill that tests how many times two players, running up and down the court at full speed, can make legal passes to one another without dropping the ball; of the latter, the game of horse, in which competitors have to make shots identical to the ones their opponents have just made—often with fanciful conditions attached—in order to avoid incurring the letters H–O–R–S–E and becoming the (losing) "horse."

[55] See Sepper 1996, 49–58.

[56] That imagination has a template character is one of the principal conclusions Brann draws from the study undertaken in her magnificent compendium; see Brann 1991, 773–786.

autonomous; we can view it as object plane through the medium of the image field or plane; we can view the plane of images as autonomous and become absorbed in its possible forms and transformations; we can view the image plane through the medium of what was originally the object plane. Moreover, as the example of developing topology from geometry shows, one image plane (the geometric) can become an object plane for another image plane (the topological one); and then one can use the second image plane to view the original object plane, with or without the first image plane as an intermediary. One is limited only by the permutations. It is no accident that such imagining also implies an ability of the viewer to change points of view, thus also a certain mobility and detachment from the world and the other planes. In fact the point of view can be shifted to wholly within a plane, above it (to another plane), and between planes (in analogies, for example).[57] The practical situation of playing games suggests a further possibility: that the viewer can become absorbed into the complex of planes that the game implies almost as much as the player is, and the player can in some sense become a viewer. The player who knows the game best can become analytically and even aesthetically absorbed into the differential synthesis or unification of all these planes and spaces.[58] Coaches and educated nonplaying viewers can also be drawn into this complex experience, whether for the sake of strategy, teaching, critical evaluation, or aesthetic appreciation.

3.9 From the Biplanarity of Imagining to the Practice of Art

The work of imagination ranges from the simple evocation of a hummed tune, an enticing aroma, or a shade of blue, through the play and practice of children and adults, to the artistic works of chefs, musicians, painters, and poets and the theorizing work of scientists and mathematicians. How universal or even all-encompassing this imaginative work is remains to be seen. This and the preceding chapter, in an effort to clear the field of imagination of some obstructing prejudices and to reorient our attention, have concentrated on relatively "homely" and plain examples of imagining like humming and singing, or playing and practicing basketball. Even though they have included quite abstract moments, like the elaboration of mathematical concepts of field and topological space and the psychological concept of biplanar imagining, they have tried to avoid theoretical commitments that are not well illustrated in the examples.

[57] As I shall point out later in discussing Descartes, this framework provides a more exact understanding of what constitutes the ego. Descartes's ego is mobile; insofar as it is conceived as a fixed foundation, it is misconceived.

[58] I say "differential unification" because an inhabitable space is always differentiated according to myriad principles and is not the uniform, infinitely extendable space of Euclid or Descartes.

Until now I have largely avoided discussing art, except in passing. This has been deliberate, in large part because Romanticism too exclusively modeled imagination on the arts. The irony of the Romantic position is that, in trying to make imagination more fundamental and all-encompassing (the reasons for which need to be understood culturally and historically), it effectively distorted our understanding of imagination by conceiving it as creative pure and simple and setting it in opposition to rationality.[59] A conception of imagination as a differential field phenomenon, however, is likely to be more faithful to the process of making art than is hyperbolically creative imagination.

First, let us adapt some of our results in this chapter for artistic purposes. Hume's missing-blue argument can be applied to any sensory qualities and characteristics that admit of being placed in a series or array. As I noted in Sect. 3.2, Hume's sequence of blues was a step toward heuristic, technical, and scientific versions of color arrays in two and three dimensions that schematize practical and theoretical knowledge of color. Since they are historical inventions, it is obvious that modern color solids and circles have not been *essential* to artistic education and practice, always and everywhere. That is not to say that artists do not need devices like them that schematize or summarize knowledge of a material or an element of art-making; these devices can be both propositional (perhaps as rules of thumb) and array- or image-oriented. Over time artists acquire an immense amount of practical knowledge associated with their work and familiarity, or at least acquaintance, with different techniques and skills. When apprenticeship in an artist's studio was still commonplace in art education, the apprentices were gradually and sequentially introduced to the work of cleaning and preparing the studio, handling materials, preparing and deploying instruments, and executing tasks required by the genres of art and related techniques they needed to learn. Not every apprentice, or even every master, will be equally skilled in each aspect or phase—one will be a better draftsman, another superior at preparing pigments, a third at underpainting with gesso, a fourth at rendering colors opaque or translucent, etc.

This learning is not just a question of the ability to use materials and instruments, however. A much more sophisticated grasp of what is at issue can be developed from considering the situation of a contemporary artist who is commissioned to paint a Madonna and Child. With Internet access she could print color images of scores or even hundreds of paintings in the genre and tack them up on the walls of her studio. In doing the initial planning she might carefully take notes about each image. She would doubtless post subsets of the images in shifting series to study variations in one or another feature they shared (for instance, paintings in which the background was deeper or shallower, more or less filled with objects, featuring landscapes or interiors, executed in warmer or cooler colors, and so forth). In the course of this series making she might discover a previously unused blue (for the

[59] This sentence greatly oversimplifies the historical reality of Romanticism. One could begin to provide nuance beginning with Coleridge's differentiation of fancy and imagination; see chapter 13 of the first volume of Coleridge 1907 [1817], esp.1:202. But Coleridge's conceptions are one thing, the cultural commonplaces of the broader society's beliefs quite another. The popularly effective romanticisms of European cultures have been blunt intellectual instruments.

Virgin's cloak) or other types of gaps and spaces that would allow her to interpolate or extrapolate previously un- or under-developed possibilities. From one image series or several taken together she might find clues that would lead to an unprecedented composition, or at least to a novel treatment. Alternatively, she might take an image or set of images as whole, for instance as establishing parameters for a series of variations. Similarly for a sculptor commissioned to produce a heroic statue, or a musician hired to compose a song. The fact is that one of the commonest human traits is the ability to recognize, projectively and interpolatively, from a finite experience other possibilities of similar kind and to position them relative to one another. What makes this projection and interpolation possible in a productive way is that the existing phenomena activate a "sense" for the relevant field in which they are marked positions and for ways in which the field elements can be varied.

Not every feature of the processes of making works of art is susceptible to being sequenced in an explicit array, yet for every feature there will be characteristics subject to variation, the mastery of which is part of acquiring the art of painting. Although applying gesso to canvas to make it whiter and more uniform in shade and texture hardly suggests anything comparable to a color circle or a mixing chart, an artist will learn that he gets somewhat different results, at least occasionally important to the overall success of the work, by applying it with a brush that is wider or narrower, that has bristles longer, shorter, coarser, finer, or more or less even in length; or by using faster or slower brush strokes, a thicker or thinner gesso, more or fewer layers, and so forth. If there are different liquids that can be used as a solvent for the gesso or gesso–equivalent, the knowledge of their effects will perhaps be more discrete than continuous (diluted with water, a mineral spirit, albumen, etc.)—except when the different solvents can be mixed with one another at will. This would put us in a situation comparable to that of the person thinking about mixtures and intermediates with respect to the aromas of cinnamon and nutmeg. Every action of the artist and the appearances it produces can, like an individual experience of a hue, be subjected to some variation if it is seen as an instance in a field.

Any individual master artist will have wide experience; the cumulative knowledge and experience of a studio or a generational cohort will be much larger. The readier the networks and forms of communication and the more abundant the formalization and schematization of knowledge have become, the more accessible and effective this knowledge will be as a kind of publicly shared imagination.[60] Over time these networks and schemas will change, and even if there is general progress, some technical knowledge and practices will be lost. For example, with the introduction of commercially prepared oil-based pigments in the early nineteenth century, the general level of artists' skill in mixing pigments from scratch declined. But this was compensated by greater consistency in the product and the introduction of new and more brilliant hues producible only by industrial techniques; and the portability of tubes of color made painting outdoors or outside the studio more feasible.[61] Every time a new invention, expedient, or device is created, it takes its place in the

[60] A publicly shared imagination or system of images is known as an *imaginary* (French *imaginaire*).

[61] See, for example, Townsend 1993, 41.

field of practices and undergoes adaptive variations that are likely to lead, eventually, to new imaginative fields or subfields.[62]

By the same token, one can establish a scale of notes to mark out the tones produced by singing; one can develop, further, a notation that expresses the complex simultaneous and successive making of multiple sounds on one, then several staffs; and thus the making of music that was once done only spontaneously can be expressed in a repeatable, symbolic format. The format is abstract, yet it also has concreteness and is imaginable. At some point in human evolution or human history someone recognized the melodiousness of voice; that person or someone else noticed the repeatability and variation of notes; someone remarked the octave, someone the fifth; another noticed that marked tones could be arrayed between and beyond these. The octave divided into 12 semitones was divided further; if Westerners do not typically do this, some other cultures do, and these articulations are at least possible even in the tonal fields of those to whom quarter tones are unfamiliar. They may sound strange, but they are implicit (if remotely) in the topological or topographic field, and the common way of marking the field can already suggest alternatives and divergences. At a yet more abstract level—but not so abstract that just about anyone who is not tone-deaf can distinguish—we can hear and have a sensibility for major and minor keys, and for the differences and expressive possibilities of compositions structured by key changes. At another level, more abstract but still accessible to perception, especially for experts, we might hear the expressive differences between scales, so that a composition in C major transposed to D major sounds different. More abstract and complex would be the ability to hear the expressive possibilities, both abstract and concrete, between rondeaux, rondelles, and canzoni. Even more radically, one might take ordered arrays of the entire diatonic scale to serve as the structural principle of compositions, as with the twelve-tone method.

Almost any sighted human being can recognize the scale of hues that Hume remarked; it requires training in notation to look at a set of staffs with graphic marks and simultaneously hear the music, but many people can do this. They have not acquired just a greater quantity of discrete ideas and their associations. They have cultivated new fields of imagination as such, as whole fields; they have learned to mark out special positions in the field; they have come to recognize significant and repeatable relations between the marked positions and learned to isolate (or section out) subfields; and sometimes they learn how to relate the various fields to one another in a new entity or new field, as the person who discovered how to use key changes in a single composition did, and as composers who learned how to exploit the techniques of twelve-tone music in a harmonic setting did. This innovative interplay of fields is not limited to the arts: as we shall see, Descartes learned to relate the newly developing field of algebra to the established field of geometry and thus invented analytic geometry.

[62]One day an artist incorporates a plastic object into a collage. A second learns how to mold plastic to acquire greater control over the pieces added to the collage, a third starts molding plastic to serve as the ground of the entire collage, and finally someone begins producing large molded pieces as the whole work.

It seems to me that, if the secret of human creativity is associated with imagination, as many people think, the creative imagination, whether it is practical, artistic/technical, or scientific/cognitive, is rarely a matter of a flash of genius, or at least not a flash that comes out of nowhere. The beginning practitioner must acquaint herself with all the elements of the practice, art, or science, the elements of the fields in which those elements operate and interact. A painter learns to synthesize the experience of colors into a color circle or a color sphere, to explore the mixing and luminosity of different pigments, to work with the surface effects of different pigment-bearing media like oil and tempera. Then she must work out a comprehensive practice of achieving diverse effects of spatial articulation by drawing different kinds of lines, by using different densities of shadow, by adjusting color and its opacity, by varying the texture and layering of the different paints and other media. Acquiring mastery in an art and preparing creative inspiration is in large part a matter of learning, and occasionally discovering, the different fields constituted by the literal or virtual spaces in which relevant differences take place and recognizing ways of interrelating and correlating the fields and the effects. The more expressly developed is the artist's sense of these topographies, the greater the possibility there is of genuine mastery in her art. A similar argument can be made for how a research scientist becomes familiar with experimental equipment and techniques, practices from different scientific and mathematical disciplines, theories from different fields, etc., and finally projects them into an experiment or a theoretical possibility that has not been marked before. In fact the artist and the scientist learn not simply to "sit back and imagine," they learn to imagine as they work, to fill out the space with the new possibilities they encounter, and to mark specific locations as rich in possibilities. Every realized possibility, every sketch drawn, every variation made present, can become a point of reference, a new mark in the field from which, by manipulation and differentiation, new possibilities of the field might be discovered.

The field, note well, is not something merely "psychological" and "subjective," nor is it simply "objective." Rather, it is a situation, an emplacement where objects are experienced, and the sense of the emplacement, of the field that is recognized, can be experienced by others as well, at least if they have the talent, the time, and the diligence to come to know it. The field is a place that gives rise to more possibilities than any single individual can exploit, and thus it creates the possibility of developing different topographies and different styles of inhabiting and traversing the field and relating it to others. The fact that the technique of varying small planes of color developed by Cézanne in his late work can be mimicked but never perfectly reproduced is evidence both of the real basis of these imaginative fields and the unique ways in which they can be inhabited.[63]

[63]These considerations suggest a path to understanding even feeling, emotion, and passion as differential responses to fields of experience—perhaps these psychological phenomena themselves can be seen as a topographical field or virtual space. Today an attempt to do this might begin with a work like Damasio 2003, especially if supplemented by a direct encounter with Spinoza and Descartes on affect. Spinoza, inspired by but radicalizing Descartes, defines passions/emotions as "the affections of the body by which the body's power of activity is increased or diminished, assisted or checked, together with the ideas of these affections." But previously he defines the ideas of bodily affections as the work of imagination. Thus emotions turn out to be the activity-increasing or activity-diminishing affections of the body that accompany imagination.

All these examples help make evident that there is a basic relationship between tradition and innovation, between routine and creativity. Not every variation is truly innovative, nor is what is routine necessarily unimaginative. One cannot look at a work of art, determine the quantity and intensity of innovations with respect to each possible field or cross section of a field that enters into it, and designate as "most creative" or "most a work of genius" the one that has the highest score. A work of art, to be successful, has to achieve some significant degree of unity. Most features of a work of genius, from the underpainting to the disposition of figures to the harmonization of colors, will not be precedent-shattering variations or innovations. In many respects it is precisely the fact that the artist has a secure repertory of standard techniques and materials to call upon in fairly conventional ways that allows him not just to practice his art but also to reach rare moments of beauty and sublimity. Thus I am not suggesting a metric for determining whether a piece of art is good or great, but rather making clear that the actions of the artist, both standard and innovative, are typically explicable as field variations on exempla. They are interpolations in or extrapolations from an organized sensibility, a sensibility for the multifarious relationships of standard techniques and expedients to the appearances and the expressiveness those appearances can produce. This is to conceive the process of artistic production as a networked series of an enormous number of moments of imaginative reproduction, variation, and innovation—the elaborate work of placing incipient appearances.

3.10 Transition: Reversing the Occlusion and Occultation of Tradition

We have already come rather far in beginning to reassess what imagination is. We have already glanced at a few authors and episodes in the history of imagination, some as warnings, others as encouragement. For the next five chapters we will move into a more emphatically historical mode. The aim is not to provide a survey. I have claimed in this chapter that there is a deep but largely inapparent conceptual topology that has governed and continues to govern our experience of and theorizing about imagination. What I have been able to say positively about imagination till now is a result of my having discovered that topology's history. Following the thread that leads through it has clarified for me not just our mistakes and omissions but also resources that can lead us beyond them. Exploiting them will allow us to recognize and cultivate the true breadth, depth, and importance of the fundamental questions concerning imagination.

In Chap. 1, I noted an irony of the modern sciences, a deep paradox built into how they are structured, is that they most fruitfully commence when they discover what they can safely set aside and ignore. The astronomical revolution of the sixteenth century required setting aside a good part of the evidence of the senses, for example the part that saw the stargazer as motionless. Ancient optics had been predicated on explanations of how the eye could see things by means of light; seventeenth- and eighteenth-century optics set aside any deeper investigation of the seeing eye in order to investigate, and to theorize more intensively about, the light and the paths it takes. Until the late eighteenth century, chemists tried to explain the sensory qualities of the materials they worked with; the revolution introduced by Lavoisier and others required setting this aside and focusing instead on quantification and the relations of matter to matter. In the laboratory, experiments are abstracted from their natural settings and their circumstances are simplified as much as possible; everything extraneous—or apparently extraneous—is stripped away for the sake of analyzable results. When reliable results are achieved, they are then projected back as explanations of natural occurrences and projected ahead into future experiments. Thus do the sciences construct the differential fields of theory and experiment, by setting things aside and learning how to creatively ignore what obstructs progress.

There is, of course, the expectation that ultimately much of what is set aside will be recuperated, and often enough this happens. The tendency of knowing is, by its very nature, universalizing, and the more fundamental the things being investigated the stronger this tendency grows. All living things are cellular, so eventually microbiologists hope to explain something about every living thing, and perhaps everything about every living thing. Everything material is matter and energy, matter–energy, so physics discovers truths about everything that is matter–energy and hopes eventually to explain everything about matter and energy and, if everything is matter–energy, to explain everything about everything. There is common to all these expectations the logic of analysis: we can take any thing or phenomenon and break it down into parts; some of those parts can be set aside as we focus on others; later, we will extend our investigations to those things we set aside and unify them with what we have already discovered.

The paradox fully emerges when we reemphasize that progress in knowledge requires deliberately neglecting things. In order to know, we must selectively and judiciously ignore. Our best science is built on deliberate and systematic ignorance. I do not say this in order to unsettle our confidence in the sciences. It is important to emphasize, as we reemphasize the deliberate ignorance of the sciences, that we really have no choice. It is a consequence of our finitude: that we are here rather than there, now rather than then, that our attention has been drawn in one direction rather than in others, that the state of science is precisely what it has historically come to be.

There is no simple way of compensating for one-sidedness. We sometimes try to invoke "everyday life" or the "lifeworld" in order to remind ourselves that our various one-sidednesses are part of something larger. But it is the easiest thing in the world to show that one does not eliminate one-sidedness and partiality simply by summing many together. Just a little travel will make clear that each society takes many things for granted, and if we traverse time and space through wide reading we recognize that although we always share commonalities with people far removed from us, many things about their worlds are alien to ours.[64] Especially in response

[64] If we think of the mortality rate in the past, the rigidity of social structure, and the overwhelming proportion of people engaged in agriculture, we can begin to see how different the life world would have been in the past. One does not have to be a professional anthropologist to recognize this—though it helps!

to the more ambitious claims of scientists, we may hear voices reminding us that they leave out passion, emotion, feeling, faith, imagination, and a host of other things. But these, too, can end up being little more than partialisms. If the problem we are addressing is one-sidedness, then substituting partialism for partialism is no answer. Quite apart from the tradition of defining the human being as a *rational* animal or *wisely knowing* human being (*homo sapiens*), one can also make the pragmatic rejoinder to critics of science that we will gain little by setting aside knowledge in favor of any of these other things. We need knowledge of many different kinds, depths, and extents in everyday life, and we almost always find that even a little more knowledge is better than less.[65]

The deep philosophical past of the West is to us both familiar and strange. It is near to us in the sense that it shaped the prehistory of our intellectual concerns and methods; it is distant in that we often feel that we had to liberate ourselves from its various provincialisms. This past is our heritage, but it is also a "different country." Our present is sometimes visible in it, although as if reflected in a distorting mirror.

History, and even more narrowly the history of concepts and theories, cannot immediately answer our most pressing questions. Studying the past can nevertheless provide a dimension of depth and resonance to our world. This happens most effectively when we try to find the common things beyond the strangeness—which, when it succeeds, can often make the strange more familiar. Achieving this requires that we reactivate thoughts, many of which have became ossified—sedimented, in the language of Husserl—in the course of centuries and millennia.

It is useful to remind ourselves of a meaning that resonates in the words *historia* and *histor*: witness. Witness is testimony of what one has seen, giving an account of what the world has shown. The *histores*, and the *historiae* they narrate, are, within the bounds of this book, the people who have witnessed the place of imagination, traversed it and felt out its character, and given it names and marked out its forms. Our goal is to take up the resources they have left us and to rethink them, singly and together, to the point where we might discover the possibilities of an occluded-occulted tradition we have foregone and lost track of.

But summarizing too broadly, too much in advance, is trivializing. Actually investigating something familiarizes us with its peculiarities and its place; a summary attempted without that familiarity is precisely a reification of knowledge apart from place, utopianism. The historical narrative would thereby be deprived of both narrativity and historicity! It is not self-defeating, however, to remark in advance that the concerns of this and the last chapter will constantly be reinforced and extended by the historical sensibility we acquire as we find well-articulated alternatives to our present conceptions and theories. We will see that the things we think and say about imagination and images is almost always derivative of what was better thought out in the past.

[65] Considering the different approaches and one-sidednesses of the sciences leads quite naturally to understanding what science does, fundamentally, as establishing rigorously constituted imaginative planes and spaces. This is not the last occasion on which I shall point to this fact, which could easily be taken up in the philosophy of science.

The imagination we will find in Plato, Aristotle, Descartes, and Kant (and a few others) will have mobility and multidimensionality. The notion of imagining as based on and accomplished through placement (in and through sensation, reason, and other psychological powers) will be recurrent, a leitmotif. We will see that even when their theories differ, they for the most part share a conceptual topology in which the individual theories all take their place and get their sense. These thinkers will repeatedly suggest and sometimes outright argue that images and imagining are the true element of human being—although an element we can imagine even transcending, to the point that we can think about and conceive the very limits of imagining. And they will face, though not always straightforwardly, an insistent question that addresses imagination's heart and that has become the unacknowledged background to all our thinking, our science, our practice: whether and to what degree we must consider language, *logos*, and logic as themselves fundamentally imaginative forms. By and large they will say that imagination is inextricably woven together with them.

This hardly means that the thinkers I treat in the following chapters speak with one voice. As should be evident from this chapter, sharing a conceptual topology does not mean articulating identical or even consistent theories. It does mean, however, that each thinker has theories that illuminate the same field and articulate it in resemblant, if not isomorphic, ways. Despite the limitations of each, there is an amplitude to their investigations and their concerns that puts into question the adequacy of our own. Moreover, by the end of our historical investigation we will begin to see that, despite their differences, these thinkers have worked within a common topology of imagination—of imagination and reason—that still prevails. Although it has also been home to the distorted and one-sided theories that litter our past and present, in its ampler forms this conceptual topology still offers the potential of a future, productive development.

It is not possible, however, to recognize the common topology of these thinkers by taking a cursory glance at their writings, as most surveys do. It would remain largely invisible if all we tried were to cherry-pick just the "relevant" sections of their writings. Each of our major authors has passages and works where imagination is featured. It is not so much in these individual passages, however, as instead in the manifold filiations of their thought, in tacit indications and connections, that their understanding of it is revealed. This is a virtue of their thinking rather than a defect. If imagination could be grasped by summarizing a few pages of their writings we would have long since arrived at a better and fuller understanding, both of these thinkers' "doctrines" and of imagination itself. That is not how great philosophical thought works. The greatest is measured, if it can be measured at all, by an articulated amplitude that strives to leave nothing important out.

The inconvenience this poses to both me as author and you as reader is not small. It means that the limits of the presentation cannot be narrowly drawn. It is often necessary to look beyond imagination, and not just to other psychological powers like sensation and reason—this part is indispensable in any case, no matter how much it goes against the still prevalent antipsychologistic temper of our philosophical age—but to epistemology, to ethics, to physics and philosophy of nature, to metaphysics and ontology. This is all the more necessary insofar as the antipsychologistic temper has encouraged interpretations that slight the psychological element of our authors' thought—an element that they regarded, quite precisely, as having epistemological, ethical, natural, and ontological consequences and roots. As a result, my presentations will argue the need for basic revisions in how we interpret not just the imagination theory and psychology of these thinkers but even their work as a whole. If, however, most histories of philosophy of the past century and longer have neglected the element and even the fact of psychology, it should not be surprising that a re–placement of imagination in a re–placed psychology might require a major shift in our conception of our philosophical history. By the time we have reached the final chapter of this study, we will see that, if it is not true that our illustrious predecessors anticipated everything about imagination and its reasons, they nevertheless knew a great deal more about it than we have managed to conceive.

So let us begin the work of rediscovery.

References[66]

Bachelard, Gaston. 1994 [1958]. *The poetics of space*. Trans. Maria Jolas. Boston: Beacon Press.

Benjamin, Walter. 2004. *Selected writings*, vol. 1: 1913–1926. Trans. Marcus Bullock and Michael W. Jennings. Cambridge: Harvard University Press.

Brann, Eva T. H. 1991. *The world of the imagination: Sum and substance*. Savage, Md.: Rowman & Littlefield.

Camus, Albert. 1942. *Le Mythe de Sisyphe*. Paris: Gallimard.

Castoriadis, Cornelius. 1987 [1975]. *The imaginary institution of society*. Trans. Kathleen Blamey. Cambridge: The MIT Press.

Coleridge, Samuel Taylor. 1907 [1817]. *Biographia literaria*. Ed. J. Shawcross, 2 vols. Oxford: Clarendon Press.

Damasio, Antonio. 2003. *Looking for Spinoza: Joy, sorrow, and the feeling brain*. Orlando: Harcourt.

Descartes, René. 1964–1976. *Oeuvres*, 2nd ed. Ed. Charles Adam, Paul Tannery, et al. 11 vols. Paris: Vrin.

Dreyfus, Hubert. 2008. *On the internet*, 2nd ed. London: Routledge.

Hardin, C. L. 1988. *Color for philosophers: Unweaving the rainbow*, 2nd ed. Indianapolis: Hackett.

Hume, David. 1739–1740. *A treatise of human nature: Being an attempt to introduce the experimental method of reasoning into moral subjects*. 3 vols. London.

Hume, David. 1748. *Philosophical essays concerning human understanding*; 1758 edition retitled *An enquiry concerning human understanding*. London.

Hyman, John. 2006. *The objective eye: Color, form, and reality in the theory of art*. Chicago: University of Chicago Press.

James, Glenn, and Robert C. James (eds.). 1959. *Mathematics dictionary*. Princeton: Van Nostrand.

Knudsen, Christian. 1982. Intentions and impositions. In *The Cambridge history of later medieval philosophy*, ed. Norman Kretzmann, 479–495. Cambridge: Cambridge University Press.

Newton, Isaac. 1704. *Opticks*. London.

[66] If there is a second, square-bracketed date, it indicates the year the work first appeared in its original language.

Raman, C. V. 1968. *The physiology of vision*. Bangalore: Indian Academy of Science.
Sepper, Dennis L. 1996. *Descartes's imagination: Proportion, images, and the activity of thinking*. Berkeley/Los Angeles: University of California Press.
Townsend, Joyce. 1993. *Turner's painting techniques. Exhibition catalog*. London: Tate Gallery.
Wittgenstein, Ludwig. 1967. *Zettel*. Trans. G. E. M. Anscombe and G. H. von Wright. Berkeley/Los Angeles: University of California Press.

Chapter 4
Plato and the Ontological Placement of Images

Histories of imagination, and of psychology in general, usually begin in earnest with Aristotle (384–322 B.C.E.). His *On the Soul—Peri psuchēs* in the original Greek, *De anima* in the Latin rendering—is the founding work of scientific and philosophical approaches to psychology. In it imagination (*phantasia* in Greek) is defined and located with respect to other powers in a way that has figured ever since in psychological tradition. *On the Soul* provided the conceptual template, the thought schema, for psychology, a template that even today has consequence and effect.

Aristotle was a student of Plato (ca. 428–347 B.C.E.); he entered the Academy in Athens at about the age of 18 and remained there until shortly after Plato's death. It is not surprising that Aristotle's theory of imagination was influenced by his master's discussions. What precisely he learned at the Academy is a matter of conjecture, however, not least because Plato wrote in dialogue form. Because the dialogues, fictional reports of conversations, rarely attempt to present straightforward doctrinal truth, arriving at their "teachings" requires no little interpretative work. Imagination and images come up in the dialogues not infrequently, though often not in the vocabulary that became standard later on. If they do not present a single, unified doctrine, they nevertheless lay down an understanding of images and imagination as enabling human beings to "see" the ultimate intelligible things, known as ideas or forms, through the "shadows" they "cast" in the things of the visible world.

Even before Plato began to develop a framework and a vocabulary for these phenomena, before there was any explicit concept of imagination or a word for it, there had emerged in early Greek thought a concept—perhaps one should say a protoconcept—of "image" and a context for conceiving it, in what we might call the emerging *problematics* of imagination. Imagination understood as a human power or activity arose in Greek thought as part of the effort to understand images, what they are, and where they take effect. This effort occurred within the context of trying to understand how human beings experience the things of the world. The basic concepts

D.L. Sepper, *Understanding Imagination: The Reason of Images*, Studies in History and Philosophy of Science 33, DOI 10.1007/978-94-007-6507-8_4,

and vocabulary for such matters had to be gradually, and not always consistently, worked out as a part of theories about nature, natural things, and natural events.

Section 4.1 will sketch a history of how a theory of images was first developed. Initially, the terms "image" and "imagination" will not be thematic. They will be implicit and foreshadowed in the background, before they emerge as topics in their own right. The account will also highlight theorists' imaginative use of metaphors and analogies—at first almost haphazard, then more systematic—for which I use the term "thought–scheme," although "image–scheme" is as appropriate. One thing we shall see almost from the beginning is the progress of abstractness in these schemes. Whether this progress (and abstraction as the process that achieves it) implies that images eventually give way to purely rational ideas will become more explicitly a theme as our investigation advances.

4.1 Pre-Platonic Philosophy and the Emergence of the Image–Bearer

The question of the image preceded the question of imagination in the history of Greek philosophy, and the image itself emerged as a philosophical concept from reflection on what *carried* or *bore* appearance from a thing to a perceiver. The image was first conceived as a result of the appearance–bearer's action. The appearance–bearer was understood as flowing or detaching itself from the thing whose image it carried. The appearance–bearer, in its turn, derived from early Greek philosophizing about physical bodies and physical actions (where *ta phusika* are "the things of nature"). The being or ontology of the appearance–bearer was understood within the context of early philosophers' conceptions of nature and its operations. The nature-based ontology of the appearance–bearer preceded both the ontology of the image and the psychologization of the process of imagination.

When around 600 B.C.E. Thales of Miletus said (as we are told)[1] that everything is water, he appears to have wanted to indicate that the most basic element in nature was not just inert material but something capable of taking on different states and appearances. It seems likely that he had in mind what we call the phases of water: liquid, solid, or vapor. He certainly noticed that the bodies of all living things, though usually solid and dry to the touch, contain liquids, which suggested that liquid animated them. The different things and their appearances in nature would thus be due to the transformations and combinations of a fundamental liquid, water,

[1]None of Thales' writings survives; what he and other early philosophers wrote and thought is recorded at second hand in the doxographic tradition, that is, in the accounts of later writers reporting about them—usually centuries later. For no philosopher preceding Plato do we have any fully intact works. The authoritative collection of the fragments reporting their words and stories about them can be found in Diels and Kranz 1974 [1903]; a usable selection with English translation and commentary is Kirk and Raven 1957.

in different modes. The ordinary water we encounter and drink every day was only one form of this fundamental water, and not its purest.

This *physiological* materialism—an account (*logos*) of matter as basic to nature (*phusis*)—differs from the somewhat later atomism. Thales and the other early physiologues (*phusiologoi*) of Greece did not typically assert the existence of elementary units or particles of the primary substances they postulated. There is no evidence that Thales thought there were ultimate particles of water. Water and the other postulated primary substances like air and fire were conceived as continuous rather than particulate.

There is a subtle but important difference to which this gives rise. In atomism, what we experience, macroscopically as we say, is due to the actions of microscopic, indivisible particles that combine and separate. What we see, the *phainomena*, conceal to some extent these true and real microscopic actions. Thales and the other Ionians, by contrast, conceived matter as a continuum, and its changes had as much an *ideal* and *phenomenal* as a *material* character. Water is as much the variety of its appearances as it is a common, unitary, underlying thing. Thales apparently did not name or explain how water takes on its variety of appearances, however.

When Anaximenes, also from Miletus, designated air rather than water as the primal element, he was probably trying to outdo his predecessor while following the same basic thought–scheme.[2] He argued that air produced the variety of phenomena by becoming either denser or rarer, processes he called *condensation* and *rarefaction*.[3] Perhaps it seemed more plausible to him that the prime element would be more airlike than waterlike, since phenomenologically atmospheric air surrounds us but is ordinarily invisible, colorless, and odorless and offers no resistance to motion through it. One can imagine air, more easily than water, extending indefinitely beyond the earth—and this *is* imagining, especially at a time when the experience of flight was limited to birds and insects. The thought–scheme thus gained greater elaboration and definition and became more easily universalized. If the processes of condensation and rarefaction were hardly more than metaphors on the way to analogy, they nevertheless suggested the possibility of further application and development. Anaximenes' theory better corresponded than Thales' to the universal dynamism of nature that is reflected in all the physiologues' theories.

Not even the ancient atomists conceived their "elements" (the "indivisibles," *atomoi*) passively. They introduced them in the first instance as the invisible causes of all the variety of visible things and events. Atoms moved, turned, oriented themselves to one another, joined, separated, etc. By contrast, we today, unless we are well schooled in the dynamism of the quantum realm, are inclined to picture elements as more or less static structures consisting of protons, neutrons, and electrons, and

[2]Tradition sees Anaximenes either as a younger contemporary of Thales or as belonging to the next generation.

[3]Condensation and rarefaction do not have necessarily atomistic implications. If matter is continuous rather than divided into units, rarefaction would be like stretching out a volume of matter in three dimensions, condensation the opposite.

molecules as static arrays of atoms. Perhaps this suggests something like a typical social, cultural, or civilizational imaginary.[4]

If Thales made water the primal element, and Anaximenes air, earth had no takers—perhaps because it is so static. Anaximander, another resident of Miletus who is traditionally placed chronologically between his fellow townsmen, chose the more expressly abstract course of designating *the unlimited* (*apeiron*) as that from which everything else came and to which it returned. He is the earliest Greek philosopher whose words are preserved, in a sentence quoted by Simplicius of Cilicia (ca. 490–560 C.E.), a mathematician and commentator who flourished more than a 1,000 years later. Simplicius embedded the quotation in a description of Anaximander's understanding of the unlimited:

> ...some other unlimited nature, from which come into being all the heavens and the worlds in them. And the source of coming–to–be for existing things is that into which destruction, too, happens, "according to necessity; for they pay penalty and retribution to each other for their injustice according to the assessment of Time," as he describes it in these rather poetical terms.[5]

The words given in quotation marks are believed to be Anaximander's. The clause can be seen as making fully explicit what was tacit in his predecessors: what is has to have a nature flexible enough to give rise to *all* appearing being, and what appears does so by determinations added to or taken away from whatever already is. The logic underlying all determination is that determination is external with respect to what is, and determinations, whether they are added or removed, transform what already is into something else. What underlies the constant re– and de–determination exists as the limitlessly determinable yet intrinsically undetermined. It is an unlimited whole that is constantly, but just temporarily, delimited this way or that. The emergence of anything arrives at the expense of something else, which is re– or de–determined out of its previously determined existence. Change occurs by the re–determining of what is already present; this gives rise to something new. The new displaces the old, and in that sense the new does injustice to the old by doing away with it, by destroying it. The fundamental principle of order in this process is time. As process it is governed by the justice of giving and taking away. More clearly than his predecessors, Anaximander develops the Ionian thought–scheme to portray the cosmos in its *totality* as constantly, intrinsically *dynamic*, and the dynamism as governed (using the analogy of justice) by something like the necessity of law.

The conceptual movement from Thales' water hypothesis, to Anaximenes' conception of expanding and contracting air, to Anaximander's process of justice-governed determination and redetermination of the otherwise undetermined "stuff" of nature illustrates how an explanatory thought–scheme develops. The development moves in the direction of greater amplitude of explanation, greater

[4]"Imaginary" used in this sense indicates a way—even a system—of imagining (including a large store of interrelated images) that is characteristic either of an author or of a social group. It is also a name for a kind of conceptual topology.

[5]Quoted from Kirk and Raven 1957, 117, substituting "unlimited" for *apeiron*.

detail of explication, and increasing abstraction of basic terms. But the fact of abstraction should not mislead us. Clearly the scheme is strongly imaginative to begin with, but it does not through abstraction lose its imaginative character. The governing images became less simple, more dynamic, and more subtle and speculative; through use they became more familiar, and precisely as such they appear to be more concepts than images.

Describing the thought–scheme in this way immediately begs for some articulation of the difference between imagination and abstraction, since Anaximander in particular appears to provide an abstract conception of the Ionian thought–scheme precisely by using the metaphor of political justice, extended or projected from the city to the cosmos and to some degree fused with cosmic order. The basic components of the scheme are placed against the largest conceivable background, and both the scheme and the background get more expressly developed. A measure of the success of a scheme—in fact of its existence as scheme rather than simply as hypothesis—is its potential for fruitful development. The more universal the claim implicit in the scheme, the more that later thinkers will find gaps and incompleteness in its previous forms. A recognized gap is a gap that quickly gets filled, in one form or another: Thales' theory as it has come down to us scarcely raises the issue of what happens to water to make it appear different in different circumstances, whereas Anaximenes' theory makes the processes of condensation and rarefaction basic to the nature of the fundamental element and invokes large principles (Friendship and Strife) to explain why the processes happen. Then Anaximander relativizes the need for naming the element specifically (it becomes the unlimited) and conceives the principles, according to an analogy with the political realm, as united in a system that requires what is taken away to be subsequently compensated for—a principle of conservation of cosmic justice.

What tends to be overlooked by claims that explanation becomes progressively more abstract over time is that greater abstractness imposes the need for a greater and more diverse flexibility in the concrete deployment of abstractions. This ongoing development is not simply deductive, nor is it primarily inductive or even abductive. It is, however, extremely common, and by most understandings of imagination must count as imaginative.

Another Ionian, Heraclitus of Ephesus (ca. 535–475 B.C.E.), took the scheme in a new direction by making a link to a realm that had not originally been part of the Ionian thought–scheme. At first glance, however, it looks as though his philosophy is a throwback to Thales and Anaximenes, since he proposed fire as the primary element. Several hundred fragments of his writing are preserved, so despite their fragmentary character taken individually it is possible to acquire a much more specific sense of his style of thinking than for his predecessors. He seems to have had in mind as much fire's symbolic value—the flickering flame, always changing but always the same, maintaining itself by consuming what it burns—as its physical character. Like Anaximander with his notion of the unlimited, he also made more explicit than Thales and Anaximenes that the primacy of the primary element has more to do with its intrinsic way of acting on and changing things and appearances than with its specific kind or external principles that affect it.

More fraught with the possibility of future development was Heraclitus' conception of human experience. Heraclitus argued (at considerable length in the existing fragments) that the lawful ways of nature can be recognized and known by human beings who are sufficiently attentive to what is happening around them. He in fact took this thought a step further, in a direction that deeply influenced later philosophy and science, by declaring that this power of recognizing and knowing is shared by all human beings. Human beings are not just one kind of natural thing among countless other kinds but also cosmically oriented beings: beings aware of the cosmos as a wholeness ordered and organized by the cosmos's governing principle: *logos*.

Previously, the Ionian thought–scheme had implicitly assumed that human beings—or at least some human beings, notably Thales, Anaximenes, and Anaximander—were capable of recognizing and speaking deep truths about nature. Heraclitus made this assumption thematic. In the later doxographic tradition we are told that he held a hereditary political office in Ephesus, but that disgust with the ignorance and venality of people led him to abdicate and abandon the life of the *polis*. Whatever the truth of this story, it appears to conform roughly to what Heraclitus' fragments say about the typical ways of human being. They say that people prefer their own private ways of understanding, but that it is possible for them to abandon those conceptions in favor of the *logos*.[6] The *logos* is not materially visible but governs and patterns all visible changes and is the same for everyone. In this sense Heraclitus was the first to expressly set the question of human existence, knowledge, and nature in the context of the ways and means of physical nature. He strongly argued that human beings can see through the striking surface appearances of things to underlying and overarching unities and patterns.

What he meant is not, in the first instance, very complicated. An example is the very simple fragment "The way up is the same as the way down."[7] If there is a path that winds up a hill, it is also a path that winds down the hill. If we consider that many experiential encounters involve things that we judge only according to the narrow limits of our immediate concerns, and that the very same things show themselves differently to others in different settings, we perhaps get an inkling that there are all sorts of unities, and aspects to unities, to which we are ordinarily oblivious, because we see them only from our private perspectives. Still, it hardly takes any great exertion of conception or imagination to see that, whether we take the path up or down, it is the same path—and every person can naturally become aware of myriad such unities. There is a patterning principle or principles at the foundation of both physical nature and human experience. In human beings the principles need to be developed into a fuller awareness—though most people stop short of the common *logos* and prefer staying with their own. At least that is what Heraclitus says. As we shall see momentarily, without this conceptual topography (another name for

[6] *Logos* eventually came to have as part of its ordinary meanings "mathematical ratio" and "reason," but in Heraclitus' era it still had the primary sense of "speech," "words combined into an account." Not until very late in antiquity could it be used to mean "individual word."

[7] Diels and Kranz 1974 [1903], B60.

a thought scheme, emphasizing its extensiveness), Plato's philosophizing would scarcely have been possible.

It was commonplace in later antiquity to consider Heraclitus one-sidedly as the philosopher of constantly changing reality, like the flickering flame. In this form he was contrasted to the slightly later Parmenides (born ca. 510 B.C.E.), who in his philosophical poem "On Nature" characterized being as incapable of change; if it changed, being would have to move from what is to what is not, and vice versa; thus being would turn into nonbeing, nonbeing into being.[8] Yet Heraclitus' understanding of *logos*, which governs everything that is and is known and that therefore is the principle according to which what is can truthfully communicate itself to human beings, is not a world apart from another basic Parmenidean contention, that there is a common character to being and knowing.[9] Both Heraclitus and Parmenides thought that the vast majority of human beings talk about and even see things other than they are. For Heraclitus, human beings prefer their own, their "private" *logos*. For Parmenides, human beings constantly posit a divergence between being and thinking by trying to traverse "the impossible way," the way of not–being. The way of not–being treats things as constantly coming into being and passing out of being. That is, it assumes that being comes from and returns to nothingness, that being is constantly being created and annihilated. This way of presenting and describing things collapses as soon as one starts speaking and thinking in a way commensurate with being—which, by the same token, turns from the way that treats nonbeing as though it exists.

Parmenides appears to represent a decisive turn in ancient Greek philosophy, away from the accounts of nature that the Ionians had nurtured and toward one of the most radical of rationalisms that Western philosophy—a tradition that has had no shortage of rationalisms—has witnessed. It is so radical, especially in its common interpretations, that it has always had about it an air of absurd rigor. Whether Parmenides himself was guilty of absurdities is arguable, especially if one takes into account the second part of "On Nature." In the second part he gives an extended example of how we might explain the appearances of change in nature without violating his prohibition against following the way of nonbeing. It is far more fragmentary than the first part, however, so it may well be that we lack the key for interpreting it. The Italian philosopher Emanuele Severino, who has gone as far as anyone since

---

[8] It is important to resist the facile interpretation that Parmenides exaggerated the significance of a merely apparent contradiction of terms, since that is to underestimate from the outset the status both of terms or words and of contradiction. The verbal contradiction needs to be thought through rather than swept aside.

[9] "Knowing and being are the same" is a common translation of Fragment 3 of "On Nature." Kirk and Raven 1957 renders it (in continuity with Fragment 2, but noted as independent), "the same thing can be thought as can be," but they also remark that very literally construing the syntax would produce "the same thing exists for thinking and for being." Two alternatives that sound more alien in English are "knowing and being are with respect to the same" and "the same is for knowing and for being." Notice, then, that the most literal rendering and other plausible alternatives to the simpler and more familiar "knowing and being are the same" do not assert any simple identity between knowing and being. The more-literal translations are more easily assimilated to Heraclitus' position.

Greek antiquity toward developing a rigorously consequent account of Parmenidean thought, argues that it is not incompatible with positions that are ordinarily believed to conflict with it: in particular, he argues that a multiplicity of beings is possible for Parmenides, and that the appearance of change is neither a refutation of Parmenides nor contradictory of being—which by its nature must be and cannot not be.[10]

Many accounts of the history of philosophy take Parmenides as a turning point that, to use the rhyming German expression, establishes a radical split between *Sein und Schein*, being and appearance. If so it would be an important turning point as well for the possibilities of a well-developed theory of images and imagination. On the other hand, it is possible that the *Sein/Schein* split is an (over)interpretation of a more subtle point of Parmenides' claims. One might put it this way: with respect to what is, there is no right to treat it as though it is not—perhaps not even as though one thing is less or more than another with respect to being (and so the lesser might be discounted and the greater exalted). One of the problems of making sense of the far more fragmentary second part of "On Nature" is that, given the typical interpretation of Parmenides as *opposed* to change and multiplicity, his proposed style of explanation of natural events seems to *accept* both. But if his physical theory ascribes being equally to the various elements of the theory of the second part, then a good deal of the paradox is lifted. Treating being as superior to appearance would then amount to a fundamental violation of his central tenet. Anyone who discounts appearance as less real than something else is trying to mix nonbeing with being. What appears, *is*. It is not opposed to being, or even reduced in its being status because it is supposedly an effect of a cause (causes are usually ascribed *greater* reality than effects, even if only implicitly: this, too, amounts to nihilism). Appearance, even the appearance of change, must not be presented by feckless human beings as though it were simply unreal, but attributed its proper being instead.

The words of Parmenides had a profound influence on the history of what eventually came to be called philosophy. Atomists like Democritus (ca. 460–370 B.C.E.) and Epicurus (341–270 B.C.E.) accepted the radically contradictory character of being and nonbeing. They presented their atoms as what really is, the fundamental, unchangeable beings without which there is nothing; and their empty space corresponded to the nothing—the no–thing, so to speak—which is not any *being* at all but "is" still necessary for the things composed of atoms to disassemble and re–form by motion of the atoms. That this is a compromise of Parmenidean principles or even self-contradictory is not the point, but rather that the atomists felt

[10] At least one aspect of Parmenides' position can be easily interpreted as consistent with common sense: "being" is something that is not subject to degree. What is, is, and to treat it otherwise is the foundation of nihilism. So, for example, an extreme reductionism that denies reality to appearances insofar as they are "really" something else is nihilistic, *insofar* as it denies being to appearance. This greatly expands the ranks of the traditions of nihilism! Severino argues that Western thought has been and continues to be thoroughgoingly nihilistic precisely because everyone agrees that change is real and requires some kind of annihilation—of form, quality, orientation, position, or the like—and the emergence of something else that did not exist beforehand. See especially Severino 1982, 19–61, an essay that first appeared in 1964.

sufficiently bound by Parmenides' stricture that they organized the fundamental pattern of their thought in a manner that conceded the stricture's basic correctness. The compromise effectively "reduced" appearance to the positions and behaviors of atoms. Atoms were being; the space in which they moved was a no–thing that was necessary for the atoms to take up different positions. The boldness of atomism was precisely that it affirmed the stark dichotomy of being and nonbeing (while not, strictly speaking, granting being to the no–thing of space) in a way that accommodated the constant appearance of change.

The sophistic movement, by contrast, was affected more by the *logic* or way of reasoning of the Parmenidean dichotomy. Parmenides' followers—best known were Zeno (born ca. 490 B.C.E.) and Melissus (born ca. 480 B.C.E.)—supported and developed their master's thought by exploiting the logic of *reduction to absurdity*. The ordinary appearances and events that seem to militate against Parmenides' affirmation of the unicity of being were shown to be contradictory. Achilles appears to overtake the tortoise when it is given a head start in a footrace, but this is impossible, since when Achilles reaches the tortoise's starting point the tortoise will have already moved ahead, and similarly for Achilles' catching up to that point. Achilles would have to go on forever and ever, always trying to make up the new distance that had opened up between them. What appears and what reason shows stand in contradiction: paradox.

Insofar as the sophistic movement was predicated on a mastery of logic and rhetoric, which allowed experts in sophistic to weaken strong arguments and strengthen weak arguments, it developed and exploited the argumentative power of the logic of contradiction, which to this day is the foundation of Western logic. If Parmenides in particular called attention to the problem of the ontological or metaphysical falsity we incur when we illegitimately talk about what is as though it were not, those who developed the art of argumentation gradually created a logic that appeared to be autonomous, that is, to have and retain its validity apart from any particular application of its terms to any conceivable world. The logic of speech thereby becomes autonomous. Some *representations* of things in speech are simply impossible by virtue of the very (logical) *structure* of speech representation, and this appears to mean that logical truth (and all logical truths) transcend being, appearance, and representation. Thinking, regarded as speechlike, is not the same as being but superior to it, and logical thinking consigns certain lesser forms of thinking (like the perception of change) to nothingness.[11]

If this looks like one of the ultimate forms of ancient rationalism, it is nevertheless important to see that the *force* of the argumentation depends on seeing impossibility in a given situation, that is, in seeing a logical schema and its violation in the thought–image of a situation. A logical schema taken in this way is a representation or image of an aspect of the situation. "Achilles can overtake the tortoise" is evidenced by sense experience, "It is not the case that Achilles can overtake the tortoise"

---

[11] Taken in this strong sense, this ancient rationalism claiming descent from Parmenides violates his basic stricture, because it understands mind as sovereign over being rather than coordinated with it.

by the schematic analysis of the situation according to an indefinite division of time and the race course. The force of the impossibility comes from the recognition of the incompatibility of the two different representations of the situation. Later, logic-chopping forms of sophistic rationalism focused more on the linguistic or rhetorical forms of contradiction than on the exhibition/showing of conflicts between appearance and logical schema.

Both Heraclitus' *logos* and Parmenides' being–knowing unity prescribe how human beings might see things aright and speak more truthfully than they ordinarily do. They left obscure, however, how the benighted might advance to this level of thinking, seeing, and speaking. Although these philosophers were not concerned with images and imagination in any specific sense, it is not a misrepresentation to say that they saw the typical problem of human experience as false portrayal or representation: representation of the world in a way that cannot stand up to more stable aspects of the world, its parts, and its processes. Although false accounts, false *logoi*, are the major part of the problem, *logoi* are not intrinsically false, since a true account of things is possible for one who takes due care. The physiologues of Miletus had placed natural things and events at the heart of their thinking, and the accounts they gave had those things and events as their direct objects; Heraclitus and Parmenides changed the focus, the direct object of philosophizing, to the ways of thinking about things and the ways of speaking in accordance with that thinking. They thus created a new level or layer to reality. Henceforth, naïveté about speaking and thinking would no longer be acceptable or even possible. This is the true moment of philosophy's birth. It is the birth, in particular, of the theme of relations between various levels of being and knowing, which was crucial to Platonic thought and became a basic element in all later theories of imagination.

It may seem like an abuse of language to assert that, in ancient Greek philosophy, questions of images and imagination first arose as a question of the nature and function of *logoi*, accounts of things. Certainly it required further conceptual development and refinement to progress to *images* as such. The image per se became a matter of thematic concern a little later, precisely as some thinkers tried to conceive how the things of the world communicate themselves and their presence to the senses and sense organs. In the first instance this is a development of the physiological schema of understanding: how natural things give rise to the variety of appearances.

The leading example is the image-bearing particle introduced by Empedocles of Agrigentum (ca. 495–435 B.C.E.) and his followers. Empedocles proposed that the four basic kinds of matter—earth, water, air, and fire—combined and separated in accordance with opposing principles that, like Anaximenes, he called Friendship and Strife. The mythical or figurative language should not conceal to us the intention to give an intelligible account of the cosmos and of how human beings experience it. Matter that, to begin with, comes in four basic forms is combined and separated by virtue of attractions and repulsions. But Empedocles also thought that the kinds of matter in things were strongly associated with the qualities that we perceive to be in them. This led him to a theory stating in general that we perceive by virtue of material particles that detach themselves from things and travel to our sense organs. The name he gave this kind of particle was *aporroē* (plural *aporroai*), effluence or effluvium.[12] These effluences are physical but also very tiny and rarefied; moreover, they are not just tiny chunks of matter but bearers of the configuration of the whole thing they are part of. That is, they are tiny representatives of the whole thing, and because of this they can convey its appearance to the sense organ.[13]

Empedocles' effluences are a physical answer to a question that has both a physical and a psychic aspect: how do we see, or otherwise sense, and ultimately understand things as they are? The primary things to be seen and understood are the things of the natural world. Tiny components of this realm carry the representation or appearance or semblance—we would say the image—of the object from which they come to the sense organs of the perceiver. Physical reality is not just a static or inert physical arrangement, it is a physically dynamic environment in which objects are constantly emitting images of themselves. It is these that enable a human being (and presumably animals as well) to experience things in the world as they are.

Physical/natural things are involved in processes by which they communicate themselves integrally. In sensation the terminal point of the process is a sensitive being, like the human being to whom the things show or reveal or otherwise indicate themselves. The process by which they show themselves does not, however, directly show itself in turn. To recognize and understand that process, one must have the key—in Empedocles' case, the theory of the four kinds of matter subject to the conflicting forces of love and strife, and of the effluences that bear the thing's image. This or something like it counts as one of the earliest schemas for *how* human beings and animals can sense and perceive the things of the world. His predecessors had recognized that this perception took place but had not offered a physical theory of how.

But that is not yet all the way to a thought–scheme that has a place for theories of imagination. Without any clear distinction of imagining from other acts of mind, it is probably anachronistic to treat the effluences simply as an element of a basic schema for imagining. It is, in itself and as part of Empedocles' matter theory, a forerunner or adumbrator of questions that would *lead to* a later, full-blown conception of images and imaging processes (including imagination).

The crucial contribution of Empedocles' theory to the future of imagination is the explicit emergence of an *imaging event*. In the first instance, as we have just seen, the image is presented as the result of the physics of the world, a natural event undergirded by natural elements. Although one has to be careful about simply asserting inevitabilities, it seems to me nevertheless true to say that it was virtually

---

[12] Another image-bearing particle, the *eidōlon*, was introduced a century and a half later, in the atomism of Epicurus. In Homer's *Iliad* the word is used of the soul of Patrokolos when he appears to Achilles in book 23. Similarly, in book 4 of the *Odyssey* it is used of the dream figure of Iphthimē when she appears to her sister Penelopē. The *eidōlon* is not the person but a phantom–double of the person. We will see the term play an important role in Plato's *Sophist* (Sect. 4.4, below).

[13] Empedocles conceived macroscopic things as proportioned mixtures of the elements. The effluences were conceived not as indivisibles (i.e., atoms) but as tiny replicas that bear in themselves the same proportions of the elements as the macroscopic thing that emits them.

inevitable—if there was going to be continued reflection on the kinds of questions to which Empedocles' effluences were a response—that the physical and psychological consequences of image– or semblance–recognition would be developed further. The effluence is the bearer of the *eidos*, the typical look or appearance of the thing. It is thus an image–bearer and image–producer, before *imagination* was distinguished from other sensitive and perceptive powers. Such a distinction occurred in the course of the following hundred years or so, as a schema for theories of imagination formed and then solidified itself, especially in the work of the greatest of the ancient Greek philosophers, Plato and Aristotle. Perhaps their particular elaborations of the image process were not inevitable. That it happened was nevertheless decisive for the future of imagination.

4.2 Image–Bearers, Figures, and Images in Plato's *Meno*

Unlike earlier thinkers, Plato's works are by and large preserved. Accounting for what he thought is accordingly far less dependent on speculation that extrapolates from hearsay or fragmentary records than for the pre-Platonic thinkers. But since most of his writing is in dialogue form, presenting what he thought about any particular subject has its own special difficulties. In a dialogue we are witnesses to a conversation. Often we will not know with certainty what the author thinks about the credibility of any particular statement or passage, or what functions a statement has in context. In addition, the Platonic dialogues often have something of the nature of drama about them, so the conflict of personages and situation works alongside conflicts of argument. The more fully the dialogue participants are delineated, the more the author can take advantage of the particulars of character and shifts in focus to advance or modulate our understanding of what is said and what it is about. Just as in conversations of everyday life, progress can be highly contingent. What is said can be due to the mood of the moment or even to happenstance rather than to logical necessity. And sometimes the Platonic dialogues can surprise us by what is not said when we think it should have been.

The governing principle of the following interpretations of Plato is goal-oriented: to gain a sense of the role that Plato played in developing the original schemas and topologies for understanding imagination. There is little doubt that Plato decisively shaped the schemas, and even more that he was the first to offer a theory simultaneously comprehensive of both the being of images and the psychology of imagining. The "thumbnail Plato" of basic reference works, of course, is someone who denigrated images as at best faintly resemblant of reality, who claimed that the realm of matter and sense was deficient in reality, that what truly exists is ideal, and that only the philosopher can access true reality. That this thumbnail version is at best a pale image of the reality of Plato will emerge as we look more closely at how images and imagination figure in the dialogues.

A place to begin is Plato's brief treatment of Empedoclean effluences in the dialogue *Meno*, in which his favorite protagonist, Socrates, engages in discussion with

Meno, a visitor to Athens from the remote Greek province Thessaly. The nominal subject of the dialogue is how virtue is acquired, in particular whether it can be taught. Socrates insists on an orderly inquiry, so before trying to answer whether virtue is teachable he suggests understanding what virtue is. But Meno has a problem: whenever he is asked to give an account of virtue or anything else he instead provides a listing of things of that kind. To show what it means to understand a *kind as such*, Socrates suggests considering the question of what color is. Meno recognizes that listing colors is not a satisfactory answer. Socrates knows that Meno is a follower of the sophist Gorgias, and that Gorgias taught an Empedoclean-style theory of effluences, so he proposes an answer like one Gorgias might give (*Meno*, 76C–E).[14] Objects emit effluences. The effluences enter (or are blocked from entering) the organs of sensation because of their shapes and the shapes of channels that conduct the effluences to the various sense organs. Touch particles will pass into the organ of touch, but not into the organ of hearing or sight, and particles of sound and vision each pass into their respective organs but not into the organ of the other or into the organ of touch. These various particles convey, from the thing to us, the qualities we perceive. Thus, the effluent particles that are of the right shape to be admitted to the eye give rise to colors. As Plato's Socrates summarizes the notion, to Meno's enthusiastic agreement, "color is an effluence from shapes which fits the sight and is perceived" (76D). Whether this is an exact rendering of Empedocles' theory is not so important as that Socrates' hypothesis is a very plausible development of the imaginative schema originated by Empedocles.

Although he has produced a credible account of the Gorgian-Empedoclean hypothesis, Socrates is not as enthusiastic as Meno. Socrates grants that it is superior to a mere listing of colors. Meno, he says, likes the idea because he has heard it before from the mouth of Gorgias, and because it is simple. You can define what each sense perceives in a similar way—sensible quality X of object Y is produced by an effluence from Y that is commensurate with the pores of the organs dedicated to hearing–touch–taste–smell (pick one), and it is perceived as a quality of hearing–touch–taste–smell (pick the same one again). But that simple adaptability of the definition to each and every sense suggests that the definition says very little, or perhaps nothing at all. We might elaborate: we do not see or have any particular evidence for effluent particles, much less for different particles for each kind of sensible quality, nor do we see or understand the channels in the organs, nor do we have any understanding about what the particles do once they get into the organs or how all the various qualities they would bear are produced and differentiated. The reason we talk about particles, effluences, and channels in the first place is that (1) we see, hear, touch, etc., things, and (2) we think that these phenomena must be associated with a physical process. But our "knowledge" and "experience" of that process is entirely hypothetical. That lack of evidence does not, however, prevent

[14]Citations of the *Meno* and other dialogues of Plato will use the standard Stephanus page and section references. Translations will be drawn from Plato 1997 for the *Meno* and the *Sophist*, and Plato 1968 for the *Republic*. I will occasionally make slight emendations.

people like Empedocles, Gorgias, and Meno from hypothesizing it as true, or further developing the effluence schema, for instance along the lines of Socrates' interpretation.[15]

Socrates says that he really prefers a different way of defining things: determining what the common element is in all the instances of that kind of thing, then to define the thing in question *to be* that common element or *to be closely associated* with it. The whole discussion of color began because Meno asked him how virtue is acquired; Socrates responded that they should first define virtue before determining how it is acquired; Meno then tried to explain what virtue is by offering assorted examples—the virtue of old men, of women, of children, of leaders, etc.—to which Socrates responded by urging him to think of what all these have in common and suggesting he use shape and color as analogies. Socrates wants not lists but an account of what they all have in common that makes them shapes or colors. He even says that he would accept as answer something that was always connected with color, for example that it appears to be in the surface of objects. Socrates thus thinks that the discrimination and correlation of appearances is a legitimate path to knowledge. But Meno prefers the physically complex, hypothetical way of explaining instead, which requires him to talk about a whole series of things (particles, pores, sense-organ channels, etc.) he knows little about—not even that they exist.

Coming as we do at the opposite end of a tradition, we may have some difficulty understanding Meno's almost childish inability to adapt himself to Socrates' request.[16] It is worth taking a moment to conceive it adequately. For the moment we can express the situation in terms of abstraction, concreteness, and analogy.

Meno has no difficulty identifying things that fall under familiar terms like shape, color, and virtue. Thus he can use these terms as appropriate labels. As soon as you ask him *why* "color" is an appropriate label for the things he calls white or black he is nonplussed. If we call "color" an abstraction, at least in comparison to the more concrete terms "white" or "red," it looks as though his preference is to deal with the concrete and to exemplify the abstract by its concrete forms.[17] He has the same inclination in the dialogue with "virtue" and individual virtues, and with "shape" and individual shapes.

It would be untrue to say simply that Meno is incapable of exercising abstraction. Besides the fact that he has no trouble noticing similarities and differences between

[15]This observation holds just as much of light waves or photons as it does of effluences. That is not to criticize light waves or photons, but to point out that the *structure* of attempted explanation is similar. Centuries of research, observation and experiment, and theoretical differentiation have in fact made the contemporary understanding of physical processes, vision, and the like much more strongly supported than the effluence theory ever was or could have been.

[16]A resurrected Socrates might similarly object, for example, to those who want to account for thinking in terms of ion cascades across synaptic gaps in neural networks, without first talking about what makes thinking distinctive or about what all kinds of thinking have in common.

[17]One obvious difference between "color" and "red" is that at some level the only way to define red is to give samples. Concreteness and abstraction are of course relative. "Fruit" is more concrete than "food," while bananas, apples, and mangoes are more concrete than "fruit."

objects sufficient to identify and discriminate between them, his "natural element" (so to speak) is arguments about things. He shows a penchant for recalling what others have said about things and seems eager to add whatever Socrates has to say about virtue and its teachability to his memory bank. As has often been remarked, it is undoubtedly indicative of Meno's character that his name is a slightly distorted form of the Greek stem *mne–* used in words having to do with memory. But the way he remembers arguments is mechanical, more like association than understanding. Rather than attend to what Socrates asks him and look at things from the question's perspective, he searches his memory for some at least tenuously related account that he has heard before. Meno is like the clever student who always has something to say because he can easily remember what he has heard or read before. Whether he really *understands* is another matter.

The relationship between memory and understanding is of course one of the principal themes of the *Meno*, if for no other reason than the slaveboy's geometry lesson halfway through the dialogue. The geometry lesson is quite literally a demonstration of the use of images in the approach to understanding. It is perhaps the earliest preserved discussion at length of one of the crucial questions about imagination: what is its role in relation to the human powers that lead to knowledge?

Just before the geometry lesson begins, Socrates is forced to change tack by Meno's persistent inability to inquire methodically. Meno appears to have learned nothing from the preceding thought exercises. As soon as Socrates gives him free rein, he repeats the question that started the dialogue, whether virtue is teachable, etc., nearly verbatim. When Socrates demurs, Meno says that, just as others had warned would happen, Socrates has put him in a state of perplexity: like a torpedo fish, Socrates has stunned him. Socrates points out that they are now exchanging images of one another, but that the image is defective in an important point: presumably a torpedo fish stuns others but not itself, whereas if he produces perplexity in others it is only because he himself is perplexed.

Meno raises an objection that threatens to stymie any further discussion: Why search for anything at all? You want me to search for something I don't know, but if I don't know it how do you expect me to recognize it when it appears? And if I know something well enough that I would recognize it if it turned up, why would I search for it? Socrates counters by saying that Meno is using a debater's trick, a ploy to maneuver himself out of difficulty.

It is precisely at this moment that Socrates changes approach. It must seem to him that Meno has tired of the demand that he think for himself about the questions Socrates asks. The clever student wants to know what will be on the test so that he can prepare to regurgitate; Meno wants to hear answers that sound good and that he might use in the future. But Socrates, besides being a master of all the tricks of argumentation, is also a master practical psychologist and sees through Meno's ruse. So he begins speaking vaguely of a secret knowledge that priests and priestesses have shared with him—and, as quick as that, Meno is hooked, he simply has to know the secret. With due allowance for all of Meno's other limitations, from that point onward he remains genuinely engaged in inquiry, to the very end of the dialogue.

Socrates' professed secret is this: knowledge is a form of memory, knowledge is something that we recall. When we come to know something, we are really being reminded of it; everything we can know is, in some sense, already in our soul. This means that everything we know ultimately comes from a previous life. As he summarizes the moral of the story, once the intervening geometry lesson is over: "if the truth about reality is always in our soul, the soul would be immortal so that you should always confidently try to seek out and recollect what you do not know at present—that is, what you do not recollect" (86A–B). Whether readers are simply to accept the myth, the lesson, and this summary as true is doubtful, given how Socrates then concludes the summary:

> I do not insist that my argument is right in all other respects, but I would contend at all costs both in word and deed as far as I could that we will be better men, braver and less idle, if we believe that one must search for the things one does not know, rather than if we believe that it is not possible to find out what we do not know and that we must not look for it. (86B–C)

And immediately they resume their quest for virtue and how it is acquired, using better method and without debater's tricks.

What is the "proof" of the secret, priestly myth that precedes this conclusion? Socrates asks Meno to produce one of his servants. Meno picks a boy who has been in his household since birth, and who (as we find out a little later) has never studied geometry, though he clearly knows some arithmetic and simple figures like lines and squares. Socrates draws a divided square. He then asks question after question, each time waiting for the boy's answer. Socrates verifies that the boy understands that squares have equal sides, that lines like EF and GH (see Fig. 4.1) joining the midpoints of opposite sides are the same length as the square's sides, and that if the sides of the square ABCD are each two units long the area will be four. He then poses the question to be resolved: what is the length of the line that, when used as the side of a new square, will produce an area twice that of the original? In other words, given a square like ABCD or EMHD, how do you find a square with double the area? The boy answers swiftly: "Obviously, Socrates, it [=the side] will be twice the length" (82E).

Socrates pauses to direct a comment to Meno. "You see, Meno, that I am not teaching the boy anything, but all I do is question him. And now he thinks he knows the length of the line on which an eight foot figure [i.e., an eight-square-unit figure; the original square is four square units, two by two] is based. Do you agree?" Meno does. Socrates then continues the questioning, and proceeds to use and re–use, to re–evoke and modify, the figures he draws. He draws a square that is two by two, then doubles the length of one of the sides to make a four-by-four square that he superimposes on the two-by-two (in Fig. 4.1, if EMHD were the original two unit by two unit square, then ABCD would be the superimposed four-by-four square). When he asks how large the new square is, the boy easily sees that it is four times as large, sixteen square units.

Socrates urges him to try again. The boy suggests taking a length halfway between two and four, three. Socrates produces a new figure (Fig. 4.2) and has the boy attend to its various dimensions. What the boy has proposed, in essence is that

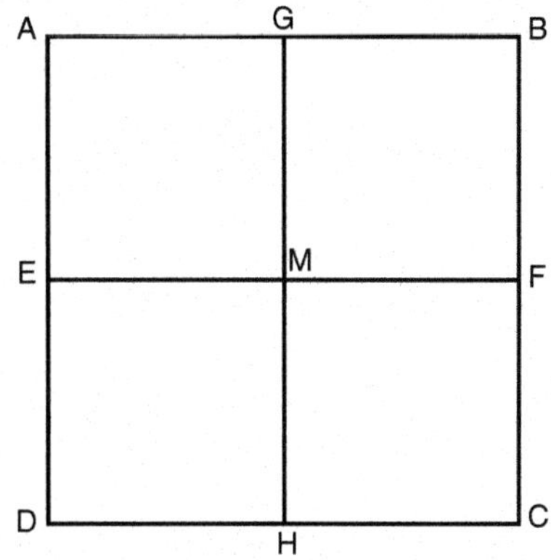

Fig. 4.1 The first figure for the slaveboy's square-doubling attempt in the *Meno*

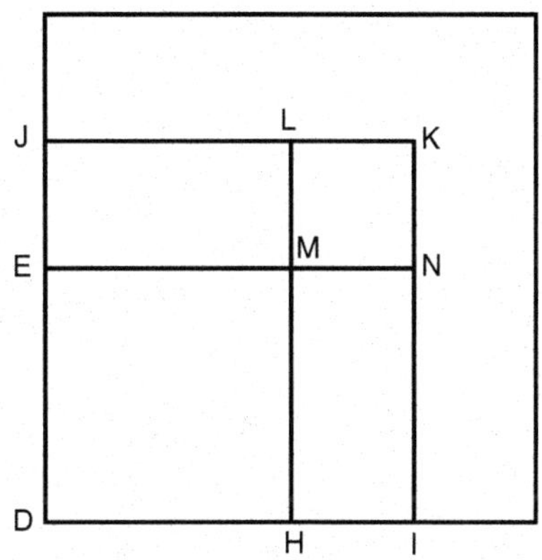

Fig. 4.2 The second figure for the slaveboy's attempt at doubling a square

the square JKID is the solution to the problem. The square JKID, with side three units long, is presented as divided into four parts: square EMHD (the original two-by-two square), square LKNM, and two rectangles, JLME and MNIH. The area of the original square EMHD is four, LKNM is one, and JLME and MNIH (being two-by-one rectangles) are each two. So the total area of the square with side three is nine, which is too large by one square unit. At this point the boy announces that he no longer has any idea how to proceed. Socrates points out the similarity to Meno's reaction when *he* did not know how to proceed further in the inquiry about virtue. The boy's perplexity is real, and now he truly wants to know the answer.

Thus the boy is well prepared for the swift denouement that follows. Socrates starts again with the original square, EMHD, adds equal squares to it above and at one side (Fig. 4.3), and then encloses the "step" space of the resulting figure (Fig. 4.4), in effect adding yet another equal square to turn the stairstep figure into a large, four by four square. That is, he has in effect created Fig. 4.1 once more. He proceeds to draw the four lines EG, GF, FH, and HE (Fig. 4.5) and tells the boy that these lines are called diagonals (of the four smaller squares). The boy sees that the

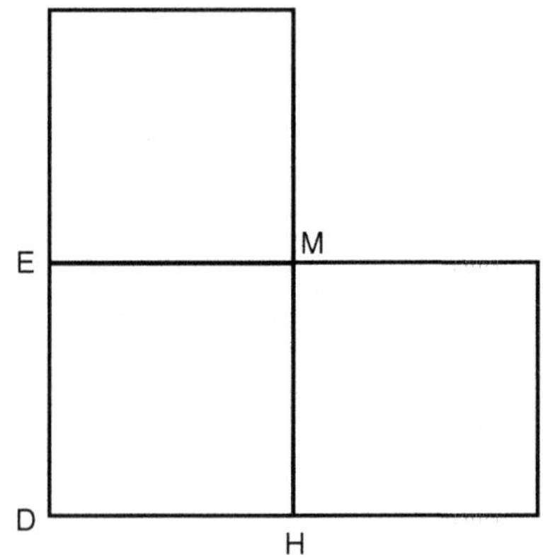

Fig. 4.3 First stage of the final figure for the slaveboy's attempt

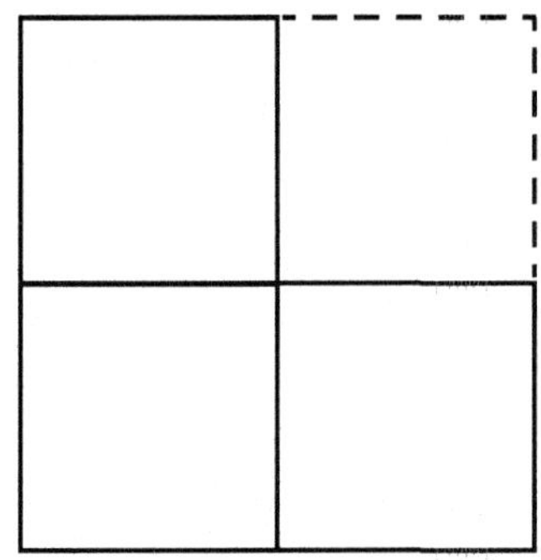

Fig. 4.4 Second stage of the final figure for the slaveboy's attempt

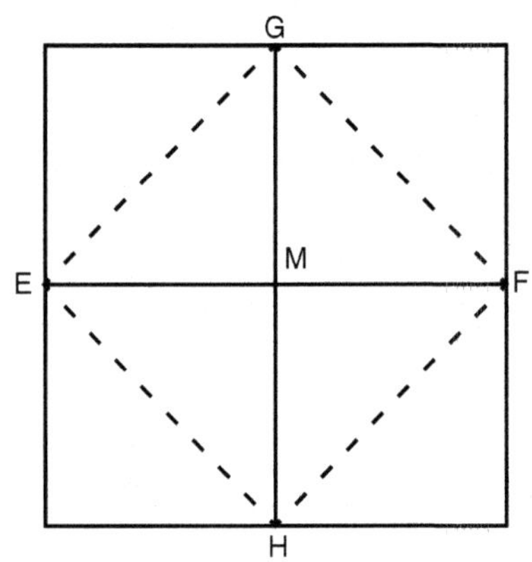

Fig. 4.5 Last stage of the final figure, leading to the slaveboy's answer

triangular spaces EGM, GFM, FHM, and HEM are equal to one another; that each is one half the area of one of the four squares that make up the large square; and, finally, that the total area of the square created by the diagonals, EGFH, therefore has to be twice that of the original square EMDH. The problem has been solved. The desired line, the one that is the side of the square twice the area of the original square, is the diagonal of the original square.

4.3 The Use and Abuse of Images

As a proof of the theses that all knowledge is recollective and that therefore the soul must be immortal and have preexisted this life, the slaveboy's geometry lesson leaves much to be desired—and Socrates' concluding concession that he would not insist the argument is right in all respects seems to acknowledge this. Even as a geometric proof according to the standards of Plato's day, it would need greater punctiliousness. For example, although it is easy enough to prove that the quadrilateral figure formed by the four diagonals in Fig. 4.5 is a perfect square, it is only assumed to be so by the three interlocutors.

As an object lesson in the role of images and figuration in acquiring and perfecting mathematical knowledge, however, the lesson is spectacular. The development of image and imagination in the *Meno* is tacit but rich. In several important senses, Plato is showing how the inquiring mind works and establishing a place for images in its functioning. Image and imagination are not incidental to the *Meno* inquiry, they are essential.

If Meno has a problem with memory, it is because he is slavishly dependent on recall. Recall is a basic mechanism of memory but has little cognitive value without establishing the relevance of what is recalled. It does not necessarily reactivate or repeat past understanding, much less give birth to new understanding. Meno's way of recall is largely verbal and associative. He has great difficulty proceeding from an initial use of terms to a new perspective on them. His access to things thus largely stops with labeling, at least when that labeling appears to be successful. What Socrates' *anamnēsis*, remembering or recollection, aims at, by contrast, is to traverse what has been grasped earlier, and entrusted to words, in new ways and contexts.

Recollection requires, first of all, that the inquirer amplify his sense of the placement of what he is calling to mind. Meno wants verbal answers before he has acquainted himself with the things and contexts they are about. He recalls texts and speeches (*logoi*) but does not appreciate the textures of thinking all around an object of inquiry; his understanding of words puts a remembered word or text next to another remembered word or text, but this falls short of becoming a context. Thus he proves to be blind, deaf, and unfeeling not just toward the textures of things but also to the feel of their contexts, places, and environments. The feel for these contexts or places can be called, by analogy, contextures. Without a sensibility for texture and contexture, Meno often falls dumb—that is, he is incapable of expressing in words the sense of the places and the contexts of things.

The "doctrine" of imagination of the *Meno* is most directly evident in the slaveboy's geometry lesson. Whether, and to what degree, geometrical figures are properly called images is a question that must be carefully answered. (But it is our problem, not Plato's.) Nevertheless, at first glance they seem to qualify as a kind of simplified image, especially if we take seriously the criterion that Alain emphasized in wondering whether and how we image Paris's Panthéon. If there is anything at all for which we can adequately summon a visual image, things like line segments, circles, triangles, and rectangles seem to be leading candidates. We may be embarrassed by the question of the Panthéon's color or the number of columns at the front plane of the façade, but not by the number of sides of a square, its color, or the angles at the vertices. Most people would probably contend that they can, for a moment at least, clearly conceive the shape of a square—even if they do not really require having a square image in mind to tell us how many sides it has. That they do succeed in this is further evidenced by contrast with geometrical objects that most people would not claim to be able to conceive clearly—for example, to imagine very clearly a thirteen-sided polygon.[18]

Socrates does not ask the slaveboy to conceive squares mentally, of course; he draws them. Let us assume for discussion's sake that he draws them in the dirt or sand, though doing it on paper or on a display screen would serve equally well. First question: is drawing a square an act of imagination, at least as far as *our* use of the word is concerned? One conventionalized account of how drawing happens would say that first Socrates mentally conceives or imagines a square privately, then in a second, public act he draws it in the sand according to the mental pattern. In a third act, he recognizes or verifies that the sand drawing conforms to the mental figure (or not, if he draws very badly). A perhaps equally conventionalized class of criticisms of this kind of description would say that it is extravagant in invoking so many separate mental acts. A behaviorist would try to reduce the extravagance by speaking chiefly of behaviors, dispositions, and propositions uttered; other critics might speak in terms of brain–region inputs and outputs and associated motor activity.

Unfortunately, almost any alternative description of the process of drawing figures will be saturated just as heavily with *our* concerns and preconceptions as with those of Socrates and Plato. If all that were left to do is recite verbatim the text of the dialogue, we would be following the example of Meno, who knows how to speak the words he has heard but not how to explain them. Plato's Socrates may not have concerns and concepts identical to our own, but he clearly believes that something new in one's experience can emerge by thinking recollectively about what one has already experienced. If there is something to focus on, and that thing can be worked and developed, one can gain familiarity with the "territory" it occupies and learn how to survey its major features.

---

[18] The point is relative rather than absolute: if I claim I can imagine an equilateral triangle, a square, a pentagon, a hexagon, and an octagon quite clearly, but disclaim the ability with a heptagon, a nonagon, a decagon, etc., it means that I am not simply making extravagant claims but also implying that I have a way of discriminating success from failure.

Platonic imagination in the slaveboy's lesson begins with Socrates' action of drawing with a purpose. The slaveboy would not be able to answer any questions at all without looking at and talking about a figure (so much for any notion that he has *direct* or *immediate* access to so-called Platonic ideas!). Of course Socrates' overriding purpose has nothing in particular to do with geometry, but rather with demonstrating the truth of the claim that knowledge is recollective. Socrates has also already made the claim that all things are akin. "As the whole of nature is akin, and the soul has learned everything, nothing prevents a man, after recalling one thing only—a process human beings call learning—discovering everything else for himself, if he is brave and does not tire of the search, for searching and learning are, as a whole, recollection" (81C–D). One thing naturally leads to another—at least as long as you are resolutely in the mood or mode of inquiry. Once you draw a figure and begin asking questions about it, you will learn unexpected things. Of course mathematics as a formal, propositional undertaking exhibits this character as well. Propositions already established will in turn help establish others as well, with no limit to the number of new true propositions that can be discovered.

Whatever status the act of drawing and what precedes it might have, Socrates counts, in the first instance, on a commonplace human ability that is at the root of Platonic imagination: we can look at one thing and, through it, see another. A literalist who comes upon the inquiring trio of Meno, Socrates, and the boy might wonder why they are fussing over ridges in the sand. There are indeed ridges in the sand. They are there because Socrates conceived the need for illustration, and he has asked the boy to look at the markings in the sand in order to see a (geometric) square. I put parentheses around "geometric" because, according to Meno, the boy does not know geometry as such. He knows how to apply the word "square" to certain things he sees, he may realize that the sides of the square are equal, he may dimly recognize that if the angles are not of the right kind the figure is not a square: but he does not know much about what a square is, about its internal relations, or about its relations to other things. In his knowledge of geometry, he is like—is an image of?—Meno with respect to Meno's knowledge of virtue. Meno, however, has on many occasions spoken fine-sounding words about virtue (see 80B). It is not likely that the boy has ever given lectures about squares.

There is something paradoxical about Plato's imagination. It counts on an odd power human beings have: they can look at something and, at a glance, take or conceive it as something else. Ridges in sand are taken to be line segments that, seen together, constitute a square. There is a medium, a substrate—the sand, as representative of the geometric plane—which is capable of almost limitless, significant formation. This is the root of Platonic imagination, its ground and basis, perhaps its ground bass. This pun has a serious intention. In a musical composition, the ground bass is a repeating sequence of notes in the bass register, above which melodies and harmonies come and go and play off of. The interest of the composition is almost always in the other voices and registers, but the ground bass is present throughout, providing the entire composition with a foundation and unity. Without it, the relationship of the voices to one another would be weakened.

It is important to notice that taking something as other than it first appears brings with it certain corollaries. The first is that this looks like an act of falsification. *Here* are ridges in the sand, chalk on a board, glowing pixels on a screen—but *there* is Socrates calling it "square." One can heighten the paradox by adding more content to the scenario. In a different context, warfare, Socrates might produce ridges in the sand and call them the encampment of the Athenian army that has to be defended against Spartan attack. Of course if we dismiss this kind of act as simple falsehood—ridges in sand are *not* a military encampment—we deprive ourselves of one of the chief instruments we have for conceiving our world, taking possession of it, and inhabiting it. We also have to wonder whether we can make any statement at all, or have any intelligent conception, without this ability. Even when we say "There are ridges in the sand that we will now take to represent a geometric square," our attempt at precision will not quash every hint of comparative thinking, allusion, metaphor, and analogy. Is a ridge in sand really a ridge, if the paradigm case is a mountain ridge? The very attempt at precision presupposes that, in the first instance, we take or conceive things in a partial way, under a certain aspect. There is no guarantee that any number of precise specifications will change that. Exactness is rarely as exact as we hope. Imagination may help us arrive at exactness, but exactness is not a fundamental desideratum of imagining. Focused attention, however, is such a desideratum.

The second corollary is closely related to what the first has suggested: that there is a kind of motion in imagining. The mind does not rest in what is before it, but by a movement "joins" one grasp of a thing to another. Perhaps imagination's specific character is to be quite explicit about there being a difference between levels or frameworks. I know that this is sand, but I take it as being part of another level, as part of geometry; I take it as belonging to a different frame, that of the geometrical plane, even though it is "really" just a patch of dirt in the agora. In saying the movement "joins" the grasp of a thing to another, I put scare quotes around *joins* because the act should not in the first instance be interpreted as a form of voluntary or even involuntary association. Perhaps a better term—at least more consistent with Plato—is that the mind is directed toward something beyond the immediate object. Of course the question-and-answer method of Socrates strongly encourages the mind's being carried in the direction of the current tendency of inquiry. But Socrates is surely right that a question alone does not force or even provide an answer: it is in the first instance an invitation to think or see for oneself the object against a background the previous discussion has prepared, to re–collect it thus.

Thus asking the slaveboy to see a square in the ridges of sand is only the first act of Platonic imagination in the *Meno*. The second act, or rather the second register, is to act upon and treat this figure in the sand as though it were a square in "geometric space." The boy does not yet have any clear sense of such a realm, but nevertheless he is able to take his first, halting steps in that space. He will not turn into a full-fledged geometer overnight; in his duties as slave he may never think geometrically again. Nevertheless, he is now closer to being a geometer than when the day began.

Till now I have kept silent about a small but significant misrepresentation I have made of the geometry lesson. The squares and other figures are described in the

Greek text, but Socrates never uses an economical symbolic designation of points, lines, or figures (that is, there are no As, Bs, Cs or their Greek alphabetical equivalents). Although Socrates' approach is "mathematically equivalent" to using figures labeled with letters (which is already an ancient practice), it is perhaps not imaginatively equivalent.

Let me explain briefly what I mean. Anyone who has undertaken much mathematics knows, tacitly at least, that there are different styles of mathematical thinking. To some degree the styles correspond to major divisions and subdivisions of the field (for example, geometric and algebraic/analytic), yet even within a division or subdiscipline one can find different styles. Geometric thinking tends to take place in and thus to favor a continuum; arithmetic and algebra, bound to the manipulation and functional processing of individually specified quantities "plugged into" equations, tend to favor the discrete. This does not mean that there can be no crossover—quite the opposite! The geometric continuum can be divided by points, lines, and planes; an equation can be conceived as describing a continuous process (especially if there is a time factor) or simultaneity. One mathematician may be inclined to look at his subject matter in search of processes, another in search of structures. Styles can be communicated and shared, so they are not irreducibly subjective. They can also be complex, composed of multiple style strands. An experienced mathematician can often recognize in another a distinctive style that is melded from different approaches and schools of thought.

Styles can be mimicked in part but never simply reproduced whole. They doubtless have some natural basis in the human capacities for thought, imagination, and perception, but they must also be historical. A "euclidean" or a "cartesian" style of doing geometry is distinctive, and even if we sometimes misidentify a moment of origination (perhaps Euclid and Descartes simply built on the practice of a contemporary we have neglected) there *is* such an origin. Think of style, then, as the way of engagement with the field. The slaveboy cannot be said to have a style of doing geometry, since he has hardly done any geometry at all, and at first he is little more than a witness to what Socrates presents to him and at most a judge of the aspects of the figures to which Socrates' questions draw attention. A style is exhibited chiefly in how one juxtaposes or moves from theme to theme, item to item, object to object, question to question.

Whether Socrates has a style of geometrizing is a more complex question.[19] The style of his questions and his way of proceeding suggest that he could approach

[19]There are further questions, such as whether each person has a unique style of doing things (not generally, I would say), or whether style is flexible and occasional (almost certainly—especially among those who are the greatest masters of style). Deeper reflection on these matters might fruitfully commence with a reconsideration of Heidegger's understanding of *attunement* (*Stimmung*, often subjectivistically translated as "mood") as a mode of *being found in place* (*Befindlichkeit*, ordinarily subjectivistically translated as "state of mind" or "disposition"). See section 29 of Heidegger 1927. To my knowledge the most sophisticated conception of styles and their role in (scientific) experience and knowing is J. W. von Goethe's theory of *Vorstellungsarten*, ways of (re)presenting things, which he developed while writing histories of different sciences he actually practiced himself; see Fink 1991, 115–125.

geometry problems in a more technical way than he does here, but that he is accommodating himself to the geometrical naïveté of the boy. He is counting on the boy having had enough experience to notice regularities in shapes and to use names for a few basic kinds—with a 3-year-old this conversation would probably not be feasible.[20] If he would resume the questioning again with the same boy on another day, he could count on what they had talked about in the presence of Meno and build toward a more rigorous approach. A master can adopt different styles to different purposes at different times.

Socrates himself suggests this, when he points out to Meno at the end of the demonstration that we cannot say that the boy at that moment fully understands the square-doubling problem and its solution. We can nevertheless be confident that he has followed closely enough and displayed sufficient interest that, if he chooses to try again tomorrow, he will be better able to negotiate it than he did today, even though he may forget many details of today's proof. This in fact seems to be the upshot of the geometry lesson as Socrates' "demonstration" of the recollection theory of knowledge. The path to knowledge is not just about seeing truths but also about the character of the inquirer and of his way of negotiating the path (this last is another, more articulate name for the style of a method). Whether, for Plato, truth is univocal, whether the path to it is unique, and whether there is just one kind of inquiring character are all arguable—in fact, I would say, all doubtful. At any rate, the way to truth and the aspects of truth that are most salient will vary according to one's previous inquiries and experience.

Nevertheless, the example of the slaveboy allows some tentative conclusions about how imagination is involved in the pursuit of (geometrical) knowledge. Though some may believe that in a pure mathematics of ideal relations one achieves knowing by intellect alone, the slaveboy needs to think from and about figures. In the first place, he needs to be able to see ridges in the sand as standing for not just ridges, not just this particular square here that is not anywhere else, but as "the" square two units on a side. He has to be able to see it as capable of being modified (by new lines), as decomposable (into four smaller squares each one unit on a side), as constructible (out of those same four smaller squares); now the square in the sand is a two-unit square, but in beginning another proof it could serve as a one-unit square, or a four-unit square, or even just a square of some kind, with arbitrary dimension. He has to focus sometimes on one of the sides of the square, sometimes on all the sides, sometimes on the area the sides enclose. He has to see line lengths and areas as equal or unequal to one another, and if unequal he has to conceive the proportion. He has to see each line as divisible, and any point of the division as a possible source of new lines. He has to look upon the sides of the original square as extendable, and those extensions as being joined to produce new squares—or triangles—and see the relationship between the wholes and the parts. To make this geometrizing productive, he has to learn how to proceed from one thing to another in an orderly way (for instance, to gather together

[20] This is not to say that no geometrical conversation at all would be possible, however.

the four central triangles of Fig. 4.5 one by one until he sees them as forming a square that has the desired area).

If imagining is having a single image clearly before the mind then all this activity is a sequence of discrete imaginings, and this sequentializing activity would probably need a new name (perhaps imagination$_2$, or hyperimagination). If, on the other hand, imagining intrinsically involves incipiently seeing something as more or other than what is immediately presented, then the "mental motion" leading up to the registering of a single image, as well as the "mental motion" leading away from it to the next one, is also part of ordinary imagining, even "part" of the image. Since the *Meno* does not treat imagination and images as the primary focus of its inquiry, Plato says nothing directly about this.

Yet before we leave the dialogue behind, it is important to notice at least one other point. The geometry lesson with its explicit deployment and consideration of image–figures is itself an example of inquiry, and it is presented by Socrates as a kind of representation or image of the process of recollection. More precisely, it is presented not just as an example of recollection but as a demonstration of it—the distinction residing in the fact that the lesson *is* the thing in question (recollection) and at the same time *shows* what it is, shows its imitable and imageable character in an imitable way. The geometry lesson as a whole thus takes on the character of a Platonic image: something that we can see but that participates in and stands for, and allows us to see, something else as well. It has its own look, and also the look of something else. The lesson itself also occurs in the form of a *logos*, an account in words; and that *logos* images certain truths of geometry. Which raises yet another question: are all *logoi* images? Does language, do words, work by way of imaging?

4.4 Speech as Image, Reason as Imaginative, and the Platonic Ontology of Imaging

In two dialogues usually attributed to his mature literary production, Plato directly addresses images and imagination. The dialogue *Sophist*, which one scholar has called "the drama of original and image,"[21] concentrates on the character of images;

[21] Rosen 1983. As so often happens with Plato, details and basic circumstances of the dialogue suggest or reinforce themes that occur in it. The leader of the dialogue is from Elea, the city of Parmenides, who claimed that being and thinking are one; and although the stranger from Elea seems to be a practitioner of Parmenidean reasoning, he also grants that they must "kill" the doctrine of "father Parmenides" in its most literal and radical form. One of the consequences of this parricide is that images can be ascribed reality. Other details also strongly corroborate a more than incidental importance of the theme "image and original": there are two men present named Socrates (one a young man, the other the old philosopher); Socrates the elder tells the Stranger that when he was young he was present at a dialogue with the old Parmenides; and when Socrates the elder urges the Stranger to ask questions of the young Theaetetus, the latter says if he needs help he will call on his friend Socrates the younger, since (emphasizing how alike they are) "he's my age and exercises with me and he's used to sharing lots of tasks with me."

and the *Republic*, one of the two longest of the dialogues and ostensibly about justice, presents a framework that places images with respect to other kinds of thing and imagination with respect to other powers of soul. Even if careful interpretation cannot achieve from these dialogues a fully explicit Platonic theory of images and imagination, it is nevertheless possible to see that the dialogues articulate the basic phenomena and concepts of imagination in a way that was decisive for the future.

In the *Sophist*, Socrates is present, but the inquiry is led by a foreigner from Elea, the home city of Parmenides. The Stranger, as he is called, hopes to arrive at an understanding of what a sophist is through the application of *diairesis*—the method of division, which at every step of inquiry classes the thing being pursued as belonging to a category or to its opposite.[22] A puzzle that the group constantly faces in the course of the inquiry is that the sophist and the philosopher are similar and thus hard to distinguish.

The Stranger introduces the method of division by asking for a definition of "fisherman"—or, more precisely, "angler"; they are practicing the method with "something trivial and [will] try to use it as a pattern for the more important issue" (218D). Beginning with the question of whether the angler has an art or not, they narrow their focus step by step and conclude the following: the angler is a man with an art; the art is not mimetic or productive but acquisitive; the acquisition is a form of coercion rather than a voluntary exchange; it is a coercion that is secret and hidden (called hunting) rather than open (fighting); it is a hunting of living rather than lifeless things; a hunting of swimming animals rather than of land animals; of animals that swim in the water (the kind of hunting called fishing) rather than of those that swim in the air (birds); the kind of fishing that strikes rather than encloses; the kind of striking that takes place in daylight, using sharp hooked objects (barbs), rather than at night; and, finally, not the kind of striking with barbed instruments that takes place from above (called "spearing") but rather what occurs from below by means of a hook ("angling").[23] Definition proper is then just the result of gathering in order all the terms that were positively identified with the thing or activity in question. *Diairesis* produces definition, and it breaks definition into a (long) series of steps, each of which asks a binary, either–or question that appears to have a definite and clearly correct answer. Not least important: the definition is an image–*logos* of the thing defined.

The inquiry into the sophist then takes a shortcut: the Stranger remarks that the sophist seems to be a kind of hunter of land animals, so they begin the pursuit of the

[22] The method is not always applied in a precise way. In particular, it is not always clear that the division of the field into two is exhaustive. But in essence it is a binary method based on dichotomous contraries, and thus it has an at least superficial relationship to the binary logic on which computer algorithms are based.

[23] The divisions leading to the definition begin at 219A and end at 221C. The last step is more complicated than one might expect: the kind of day–fishing in question is "done with a hook, not to just any part of the fish's body but always to the prey's head and mouth, and pulls it upward from below with rods or reeds" (220E–221A). This suggests that, at least at the end, they are rushing to judgment. But just at the end?

sophist by borrowing the binary scheme they used for the angler, from the point where it splits hunters into hunters of swimming animals and hunters of land animals; they make new divisions as needed. The result is that sophists are hunters of the tame land animal called "man." They return to yet other points of dichotomy in the angler division, and correspondingly the Stranger leads the group to a total of six distinct definitions of what a sophist is.[24] This leads the Stranger's principal interlocutor, Theaetetus, to exclaim that "the sophist has appeared in lots of different ways. So I'm confused about what expression or assertion could convey the truth about what he really is" (231B–C); to which the Stranger responds that confusion is the right reaction. Even the sophist they are hunting down would be confused. The sophist seems able to escape being pinned down by any of the accounts or *logoi* given of him.

What the Stranger does then is to drop the example of angling in order to discuss this new result: that the sophist evades definition and seems to be many different things. Getting Theaetetus to agree that no one can know everything—although the sophist seems to have something to say about everything and knows how to contradict everything that someone might say—they conclude that the sophist has "appeared as having a kind of belief–knowledge about everything, but [does] not [have] the truth" (233C). To "exhibit [the sophist] more clearly" the Stranger offers what he calls a new pattern or paradigm (*paradeigma* in Greek). Immediately he produces new confusion in Theaetetus by mentioning those who claim they can, by a single kind of expertise (*technē*), make and do everything. (The Stranger is implicitly calling attention to precisely the point where their original divisions had begun, with the division of arts into the productive-mimetic and the acquisitive.) If earlier Theaetetus had balked at the notion of someone's knowing everything, now he wonders what it would mean to make everything and suggests the Stranger is making a joke or playing a game. The Stranger responds by asking the rhetorical question whether there is a game that involves more expertise and charm than the kind (*eidos*) that is imitative or mimetic. The person who plays this game says "he can make all things by means of a single kind of expertise." "By being expert at drawing he produces imitations that have the same names as beings….[W]hen he shows his drawings from far away he'll be able to fool the more mindless young children into thinking he can actually produce anything he wants to" (234B). Similarly there is an art of words "that someone can use to trick young people when they stand even

[24]They provide a summary of their six attempts as follows (231D–E): the sophist is (1) "a hired hunter of rich young men," (2) "a wholesaler of learning about the soul," (3) "a retailer of the same things," (4) "a seller of his own learning," (5) "an athlete in verbal combat, distinguished by his expertise in debating," and (6) someone who "cleanses the soul of beliefs that interfere with learning." The last definition corresponds to the longest discussion, which borrows least from the example of the angler; it begins with the isolation of a kind of art (which is the point where the angler definition *began*) that they had not previously considered, the art of dividing or discriminating things into different kinds—like combing, carding, and sifting. The Stranger does not call attention to the sudden appearance of a tripartition of the original starting point, nor to the fact that this new kind of art would include the method of division itself!

farther away from the truth about thing." He would put "words in their ears, and by showing them spoken images (*eidōla legomena*) of everything" he would "make them believe that the words are true and that the person who's speaking to them is the wisest person there is" (234C). The sophist is, therefore, "a cheat who imitates beings" (234E).

The Stranger then suggests that they undertake a fresh application of the method of division to understand the art or expertise of mimesis. Formerly it had been included under the art of *production*, which was opposed to the art of *acquisition*. Now he gives mimetic art an alternative name, the image-making art, *eidōlopoiikēn technēn* (235B), the art of making *eidōla*. This is a diminutive of *eidos*: form, kind, typical appearance (from which meaning Plato extrapolated his theory of ideas). He divides this image-making art into two kinds (*eidē*), only the first of which (he says) he can make out immediately.[25] To give just the name: the first kind is the making of *eikōnes* (singular *eikōn*), in an art he calls *technē eikastikē*. He calls the second kind the making of *phantasmata* (singular *phantasma*), by means of the phantastic art, *technē phantastikē*. Thus the *icon* is contrasted with the *phantasm* (or simulacrum), the eikastic or icon-making art with the phantastic or phantasm-making art.[26]

The icon is a true likeness. The Stranger points out that the icon is produced when the imitation follows "the proportions of length, breadth, and depth of his model, and also by keeping to the appropriate colors of its parts" (235D–E). The phantasm, by contrast, in one way or another diverges from the proportions of the original. The example used by the Stranger is monumental sculptures of gods and heroes that have (for instance) disproportionately large heads: they violate the proportion in the original object, even if it is for the reason that otherwise the heads would *appear* disproportionately small to the typical viewer at ground level. This example means, of course, that the Stranger perfectly understands that viewing involves perspective.

Theaetetus and the Stranger are too quick to moralize against the proportion-altering phantasm production as a kind of deception, however, because they do not

---

[25]The Stranger described the sophist at 234C as having the art of making spoken *eidōla*, so the question now is which of the two subkinds the sophist immerses himself in. Note that although the method of division is an improvement over Meno's listing of instances to understand a kind or *eidos*, it is perhaps not all that much of an improvement. One begins with something grasped in an approximate way—an angler or a sophist—and then tries to determine into which of two contrary classes the thing fits. The classes are themselves understood only approximatively and so add to the ambiguities.

[26]"Icon" and "phantasm" have the virtue of being cognate to the original Greek words. "Icon" is a relatively unproblematic rendering. "Phantasm" has a more complex history; after Aristotle it became the generic term for what English names "image." "Phantom" is a possible cognate rendering that would suggest more strongly the inadequacy of the *phantasma* to that which it images—it is, for example, used in Allan Bloom's translation of the *Republic*—but it has its own problematic associations. Below I will use "phantastic art" for *technē phantastikē*; for *phantasma* I will use either "phantasm" or "simulacrum," a Latin word that conveys the pejorative sense of a likeness's falling short of what it tries to resemble.

adequately discuss the nature (rather than just the kinds) of mimetic production.[27] The first issue is that the distortion of proportion in the case they discuss, monumental statuary, has a solid reason. The distortion, which in a quite literal sense is a disproportion, nevertheless is introduced to make the overall impression more similar to the original than it would if the proportions were strictly iconic. An iconic image of the original would look less like the original than the phantasm does. Distortion can sometimes serve truth, and accuracy can be misleading. This suggests that a deeper investigation of images than they offer is needed.

The second issue is how far the iconic production of images can be taken. Unless the icon preserves the same proportions as the original in *every* respect, it will diverge from the original. Thus it will be a phantasm instead. A two-dimensional image of a three-dimensional object, say a photo or painting representing a statue, may exactly preserve the original's proportions of height and width and colors, but it distorts the third dimension. Even a three-dimensional hologram would distort the original statue's proportions in certain respects—you can pass your hand through it—and if the hologram is not the same size as the original the proportions of the thing to its surroundings are altered.[28] A life-sized statue of marble and an otherwise identical statue of bronze would each diverge from the original living, *fleshly* human body that both portrayed in different ways, textures, and colors.

Color, the one nonmathematical quality the Stranger mentions in describing the iconic image, produces its own problems. If you reduce one spatial dimension by half, in order to maintain the original proportions you would have to reduce all other dimensions by the same fraction. But what does it mean to reduce the color dimension by half? Presumably we would expect the colors used to be identical, whether we were dealing with a statue twice life size or half.[29] The color in flesh is produced by melanin (as we know), and flesh has a certain degree of depth and even translucency; but a pigment used to paint a marble statue would be made of some other inorganic or organic material and be opaque. If we consider all such inevitable divergences in proportions between original and *eidōlon*, we realize that there is something defective in the Stranger's distinction between *eidōla* that are icons and those that are phantasms/simulacra, and thus also between the eikastic art and the phantastic art. In a strict sense, the only way to eikastically mimic an original is to

[27] This shortcoming is partially reversed near the end of the dialogue, beginning approximately at 263D, as an extension of the discussion of the true and the false.

[28] Technically this might not violate the stricture if one distinguished between external and internal proportions, with only the latter counting as proportions of the thing. But then there would be internal problems to be reckoned with: for example, if a sculpture reduces an object's height, width, and depth by half, the surface area is reduced by three quarters, and the volume (and weight) by seven eighths. Some strictly determinable proportionality holds for each characteristic, but not the same proportion for each and every one.

[29] There might, however, be certain perceptual color effects produced by having a larger or smaller expanse to work with, so an artist might have to vary the colors somewhat in a smaller statue to produce the same effects as in a larger one. That would, of course, mean that the image was a simulacrum rather than an icon.

produce an exact duplicate, a clone. But even here there would be a difference, because one would be the original, the other the artifact. Furthermore, since the Stranger treats speeches (*logoi*) as *eidōla*, and since speeches cannot have height and width and depth and color, how is it possible at all for words to be an icon? Mustn't they all be phantasms/simulacra?

To summarize and refocus: a perfect icon preserves all proportions of the original, so eikastic art is primarily oriented to re–presenting the original's *total state*, whereas in the simulacrum there are changes to the original's proportions in order to preserve the *appearance* of the original that would be distorted by accurately preserving the proportions. The phantastic art is thus more concerned with the resemblant appearance of the image to the thing than with the thing itself. This way of formulating the distinction allows a sharpening of the question. If an icon appears less like the original because eikastic art refuses to change any proportions, is that a defect or a superiority? Wouldn't an accurate reproduction in that respect be misleading, that is, false? And is the Stranger overlooking the possibility that perspective, viewing from a standpoint, might exclude the possibility of perfectly maintaining all proportions? Put radically: is there not a *being of appearance*, to which one should also strive to maintain a proportion?

Such problems and questions, individually or taken together, do not amount to a "refutation" of the Stranger's distinction. A refutation in real argumentation does not necessarily work the way it does in pure logic, with the definitive rejection of what has shown itself to be untrue. The problem, to give a name to it, might be called the imprecision of truth and untruth. This imprecision derives from the fact that we typically judge something, formulate a true proposition about it, by looking at it against a backdrop that discussion has put in place. This is precisely the issue that the apparent precision of the method of division raises but that makes the search for the sophist's nature more elusive. Is the sophist a hunter or not? That question demands a simple yes or no answer, but we are judging an impression of something in its typical field of operation (the sophist in the agora) against the background of something else in its own field (the hunter on land). The method appeals to analogy and metaphor.

In the schematic world of binary logic, truth and falsity are sharply delimited and mutually exclusive in a perfectly dichotomous way, but elsewhere it is rarely so. In fact, except in its proper, perfectly abstracted realm, logic is better at calling attention to real and imaginative places where there is an issue to be considered further, rather than at deciding matters definitively. Often one can say of a statement that it is true, *as far as it goes*; and of a false statement that *nevertheless there is still something to be said for its basic idea*. The scientific hypothesis of Nils Bohr that the electron–proton system in the hydrogen atom is like the moon–earth system was in some senses true but in others false. If a friend we have failed calls us an Iago or a Judas it may be unfair, but there is nevertheless something in the charge's tendency that, despite our being offended, we cannot entirely deny. As long as the clarity of what we say and think is less than perfect—whatever "perfect" might mean here—this situation holds. Improving clarity depends crucially on living with the distinctions for a while, in order to make them familiar and place them accurately. To make explicit for the first time the connection with our Chap. 3 discussions: propositions in Plato cannot typically be answered with a logical yes or no; instead, a proposition is an invitation to affirm or deny that the topology of what is used as an image in its field corresponds well to the topology of the original object of concern in its field.

The distinction between icon and simulacrum is thus not as clear as it first seemed. Proportion in the discussion of the *Sophist* is a mathematical notion.[30] Taken as holding between real things (including artifacts) and their images, it cannot be strictly and perfectly maintained. Yet it seems plausible that some revision or modification of the underlying notion might make it viable for distinguishing between good and bad images. Maintaining proportion, in some shape or form, seems to tend in the right direction, as far as it goes. The real difficulty is determining how far that is. It is not just a question of static being but also of dynamic appearance. It is when one reaches a point like this that we need to remind ourselves that the Platonic dialogues rarely have a strictly dogmatic purpose. They are in the first instance invitations to think about things in an ample way, as well as to think about how we think about them, with the goal of discovering clues and concepts that go some way toward meeting the phenomena. In the last analysis it is up to us to make our images and word–images well, thus to see the advantages and disadvantages of envisioning things through our definitions and our prototypical examples.[31]

Later in the *Sophist* the Stranger does something more with these clues and concepts about imagination by returning to the eikastic and phantastic arts at the end, in the context of discussing the true and the false in speech and thought (and, of course, all still within the context of trying to understand what a sophist is). Things are more complicated than earlier, however, precisely because all the themes, including those that were only tacit before, are now explicitly woven together. The Stranger gives in these passages the earliest extant definition of imagination in Greek thought, a definition that brings together things that had hitherto (in the dialogue and in the philosophical tradition) been kept distinct. It is also a definition that Plato's student

[30] In the 235D discussion of the icon as maintaining proportions the word is not *logos* or *analogia* but *summetria*, which means due proportion in the sense of having a common *measure*.

[31] This seems to be one of the key lessons of the *Sophist*. The method of division works well enough for defining things that are already clearly apprehended, but for objects that are more vaguely conceived it can be much harder to know what distinctions to make. Moreover, often enough the interlocutors discover that a division that seemed clear-cut is not. In the initial division of arts into the productive and the acquisitive, for example, *mimēsis* (later renamed *eidōlon*– or image–making) is included as a part of production, but later it becomes either a third kind of art coordinate with production and acquisition or superordinate to production. It is worth recalling here that, in the middle of the dialogue, the interlocutors see that the most abstract divisions of all (like being and not–being, motion and rest, sameness and difference) *interparticipate* with one another in complex ways that have to be determined by special and insistent inquiry. This interparticipation is a way both of seeming and being.

Aristotle expressly took issue with in *On the Soul*: that what appears in imagination is a mixture of sensation and opinion. It is a *summixis* of *aisthēsis* and *doxa*.[32]

At 264C the Stranger reminds Theaetetus that earlier they did not know to which of the two varieties of image making, eikastic or phantastic, the art of the sophist belongs, because they became sidetracked by the question of whether there can be any falsity at all. The problem was that something's being false seemed to imply a kind of nonbeing in the false, in violation of Parmenides' injunction. They solved the problem by resorting to "parricide": they had to kill the strict doctrine of father Parmenides, although a weakened version would be retained. To say that something was not beautiful, for example, was not to assert nonbeing in the nonbeautiful thing but to ascribe to the thing something *different* from the beautiful. Thus what is beautiful, what is not beautiful, and the beings to which beauty and nonbeauty are attributed all involve being or existence. More generally, appearance involves being, existence, and difference. The mind looks to the thing and finds beauty or some other quality in it similar to other things, and finds also that it is different from things that are not beautiful (and, contrariwise, of a thing that is not beautiful, the mind recognizes that other things have beauty, but this thing does not). And a thing that is beautiful in one respect (say, in graceful arrangement and proportion of body) may be different from beauty in another (say, in color). Instead of a strictly "Parmenidean" cosmos of unitary and identical being, then, things are arranged so that they are mixtures or blends of one and many, of sameness and difference, of lasting and transient, even of being and nonbeing—not to mention the mixtures of more particular attributes like good, beautiful, true, just, and even more mundane qualities like the white and the nonwhite or the proportional and the nonproportional. The examination of things requires exemplars that have to be compared to other exemplars. Examined in this way, a way that is memorative and imaginative, the absolute conceptual differences and dichotomies that the interlocutors sought earlier in the dialogue cannot be sustained.

These conclusions have decisive consequences for the understanding of images and their relationship to things, and by extension of the sophist, insofar as he is portrayed as a maker of images in speech, and thus also for understanding how subtle the differences can be between a sophist and a philosopher, who in many respects resemble one another. Things themselves, not just the images or appearances of those things, are subject to similar uncertainties. Things have proportions and participate in qualities. Although what is, is—as Parmenides insists—the being of what is, when it is assayed or assessed in particular respects, has a more and a less about it and its various appearances. A political action may be just but less just than another action and more so than a third; it may fall short of having all the ideal characteristics or proportions of an ideally just action but still deserve to be called

[32] 264A–B. Technically, it is a definition of appearance rather than imagination: "So since there is true and false speech, and…thinking appeared to be the soul's dialogue with itself, opinion the conclusion of thinking, and what we call appearing [*phainetai*] the mixing together of sense perception and opinion, it follows that since these are all akin to speech, some of them must sometimes be false."

just. A man's nose may be slightly crooked, but we can still call him handsome (he is handsome in general, but not in this particular). Perhaps, in a statue we make of him, we even feel justified in slightly "correcting" the imperfection, leaving only enough so that viewers immediately recognize who it is.

It should perhaps be less surprising, then, that our accounts (*logoi*) as well as our other forms of representation are subject to similar kinds of mixtures. If we make a very fine oil portrait of the handsome man with the crooked nose, it will present only some of the (visible) characteristics of the living human being. We will not be overscrupulous in rendering every blemish or pore, for example. Moreover, some of the ways in which the image falls short of the original can be due to limitations in the *medium* we use and the *techniques* we apply. The translucency of the skin in his nose may be impossible to reproduce using any available paint, and the strokes of the brush may leave the texture of the flesh obscure. The third dimension of space will only be suggested, and the aspect of time's passage will be entirely missing.[33]

What this suggests about images, then, is multiple. First, an image, insofar as it is, is something and not nothing. But its being is not brute and isolated being. As image, it is an image of something: thus it stands in relation to that something, and in a far closer relation to that something than most things do. The image is an image precisely insofar as it is a portrayal of something else. As a portrayal it is not a clone of the original—but not even clones are identical to what they clone. All images are imperfect in comparison to the original, both because nothing other than the original thing itself can "contain" its whole being and because the similarity of image to original holds only up to a certain point, to a certain degree, and only in a few respects. This is in part due to the nature of derivation (the derived thing is always in a kind of subordination to the original) and also to the characteristics of the medium in which the image or representation is realized.

But there is more. The image's being is not exhausted by being in a medium (a photograph as paper and ink), not even with the addition of its representing an original thing (the man with the crooked nose). An image has qualities due to its medium. A camera aperture can distort dimensions in a photo, and too much or too little light can affect the color. The pixelation by which a digital camera produces an image will introduce bizarre artifacts in extreme light conditions. Every image has relationships of more and less not just with respect to the original but also with respect to the medium, and thus with respect to other images made in that medium. A particular photographic image will have properties and qualities that another photographic image made in different circumstances will share, others that it will not—and sometimes the present image is made the way it is precisely by following or avoiding what happened with another image. Moreover, the photo's representational value can be altered in manifold ways by including or excluding other objects in the field of the photo, by altering the depth of field, or by changing the

[33]This is not to say that there is no temporal aspect possible in the "static" medium of painting. Consider the dynamism of painting of the European Baroque, or, more subtly, the ravages of time or the weight of the impending future apparent in portraits by Goya.

background, middle ground, or foreground. The plane of the medium and the manner of presenting it encompass a constellation of axes along which variations can be introduced.

An image turns out to be a far more complicated thing than it at first seems: complicated, and also complicating, because it adds so many relationships to that of "simply" being. Simple being is not so simple, of course. Being things, too, quite apart from any attempt to represent or image them, have about them the more and the less, the similar and the different, with respect to other things and in particular aspects. They participate in many forms that might not be at first apparent and in ways that are in tension with other forms of participation. Perhaps they even participate in various degrees of the representable and the unrepresentable. This has come a very long way from Empedocles' image-bearing particles and the *Meno*'s question of whether and how such effluences bear the image of the physical thing from which they flow. The *Sophist* is relatively indifferent to such questions. Rather than physics or a psychology of imagination, its focal interest is the *being of images*. This is because even if in the last analysis our concern is the being or existence of things, images too have a certain being, and the being of things shows itself precisely through how the things *show* themselves: through the *self-imaging* that we call their appearances.

In our treatment of the *Republic* we shall further consider the ontology of images. The only additional matter to detain us here is the alteration the Stranger makes to the method of division at the end of the *Sophist*. What the pursuit of the question of the relationship of image to original has revealed is that things cannot always be put simply into a single dichotomous classification in an unambiguous way. Pursuing the sophist and his nature reveals that there are many different and sometimes conflicting appearances that lead to placing him differently in the network of divisions (for instance, he was a retailer of the opinions of others, then a wholesaler of them, then a retailer of his own opinions). The method of division can be useless or, worse, misleading if the person doing the division is not attentive to context and circumstance or if he absolutizes a feature that turns out to be relative. At the end of the dialogue the Stranger therefore produces a revised format for division. The original method had taken a thing, asked whether it was A or non–A, put it in the appropriate half, then divided that half into B and non–B and asked which of those it was, all the while ignoring the half of the A/non–A branch that was not followed. The new method in principle follows out the successive division of *all* paths in the network, even those irrelevant to the object of interest. It thus turns into a pursuit of the way in which the world comprehensively presents all the variety of things, rather than simply of a single path through the network that ignores all other possible paths.

Understandably the Stranger and his dialogue partners do not carry out this more comprehensive division very far. What the method gains in amplitude it begins to lose in unwieldiness. But the context in which the revised method comes up is itself revealing: it returns to where all the divisions started, the arts, by attempting the accurate division of arts into the eikastic and the phantastic. It turns out that, because the gods make things that more or less resemble ideal forms (presumably they are responsible in some way for the production of the handsome man with the crooked nose), there must be a *divine* art of imitation. Part of it is eikastic, part phantastic; and similarly for human beings. Thus the pursuit of the question of human imitation and the kind and character of human-made images cannot be adequately undertaken without taking into account the non– or extra–human as well. To determine similarity comprehensively, one must first discriminate all paths along which two things might be similar. It makes sense to distinguish between less and more accurate representations of an original in a medium, but even the best (most eikastic) representation will, in certain respects, be imperfect (and thus phantastic as well). Moreover, if the gods themselves imitate not only in perfect proportion but also with divergences from perfect proportion, it becomes much clearer that there is nothing illegitimate per se in the phantastic as opposed to the eikastic. If the gods make simulacra, can phantastic, simulacral image making be simply and always wrong?

4.5 The Multilevel Look of Things in the *Republic*

What does all this imply? Some answers are provided by the second-longest and perhaps most famous of the Platonic dialogues, *Republic*, which is said to be a dialogue about justice. As it turns out, that representation is misleading, not least because it is just as much and as fundamentally about the ontology of imaging, and, in a different turn for Plato's thinking, the psychology of imaging. As such, the lessons to be drawn from it are quite different from what the conventional "Platonist" wisdom says. Not coincidentally, even if Plato intended Platonist "doctrines" about images and imagination and they turn out to be wrong, he nevertheless established a way of thinking about imagining that has pervaded Western, and not just Western, approaches ever since. The *Republic* is one of a handful of truly indispensable works in the history of Western thinking about images and their being, about imagination and its place in mind and soul.

If you search the *Republic* for passages in which image–words are discussed systematically, you will get the impression that images and imagination do not count for much. If, on the other hand, you notice how many images Plato and Plato's Socrates use, if you examine their functions, and if you compare those with what the characters say about images and image making, you begin to get the feeling that there is much more to imagination than first meets the eye.

To put it as simply as possible: at the very heart of Plato's thinking is the notion that the cosmos is by its nature a place and process of imaging. It is by tracing out this cosmic process that human beings come to truth. The most revealing sign of imaging's centrality is, of course, Plato's use of the terms *idea* and *eidos*. "Idea," a simple transliteration of the Greek word into the Roman alphabet, is the feminine form of a noun that also has a masculine form, *eidos*. The latter provides the stem in *eidōlon*, a term we have seen used as the generic "image" of "image making" in the *Sophist*. Plato used *idea* and *eidos* as more or less interchangeable, with a certain preference for the former.

Idea, *eidos* indicates the look of something, the thing's ordinary or conventional appearance that we notice and recognize whenever the thing shows. The phenomenon is so basic that even so simple an explanation seems too complicated. What allows us—or for that matter what allows an infant at an early stage of development—to see a cat as a cat? It is the *idea* or *eidos*, the look or shape or form or ordinary appearance that cats have. We may not be able to describe it in so many words, but sure enough we are able, when we see a cat we have not seen before, to grasp immediately that it is a cat. What is more, we can do exactly the same thing for dogs, squirrels, pigeons, horses. Even if very early in its development a human baby cannot discriminate all these from one another, it quickly acquires the ability to accomplish this with very few mistakes, and over time the ability becomes only surer and more sophisticated. Even without the ability to articulate the distinctions in words, the child can tell a cat from a dog, a Persian from a Siamese, a domesticated shorthair from an ocelot, a Bengal from a saber-toothed tiger. In some sense human beings never stop progressing along this path of both common and specialized knowledge, whether it is about cats, sports trading cards, automobile fuel-injection systems, or subatomic particles. It culminates in the effortless familiarity with the things in their world possessed by experts and consummate amateurs (that is, those who love the knowledge for its own sake).

It is tempting to conceive this commonplace ability in an overintellectualized form. If it is intellectual, it is intellect at work in everyday life. It pervades human experience, language, and rationality. Although, as is often said, we cannot get into the subjective state of mind of dogs, cats, bats, and bears, it does not seem very plausible from their behaviors that they have anything approaching the general scope and depth of the human ability to see a thing not just as A, but also as B, and thereby to put A and B into relationship with one another. For human beings the discrimination of thing from thing and kind from kind easily lends itself to systematicity, and the scope and depth of discriminations already made enters into, or even simply colors, the ongoing, active discrimination of things from one another. It is not just anthropomorphism that makes us doubt that, when a dog sees another dog, it registers the dog equivalent of "Airedale," or when it sees a cat, it comes into possession of the same type of *felis felis* cat–look, the cat–*eidos*, that a human does, or a prey– or enemy–look—much less a proposition, express or in terms of behavior, that relates the two.

These may be hard matters to decide definitively. But it seems likely that, even on the threshold of language acquisition, a human infant has a surer grasp of the *eidē* or looks of a wider variety of things than does a chimp that has had hand–signaling or symbol–indicating or even vocalizing language taught it for years.[34] We should, on the one hand, avoid immediately identifying this ability with reason or intellect; we may not, on the other, want to sharply distinguish this "perceptual categorizing" from intellectual concepts. Such perceptual abilities seem to be an

[34] These matters cannot be settled from the philosopher's reading and viewing chair, of course, but from that chair it can be easier to hit upon relevant questions and concepts with dispassion than from other, more partisan places.

essential prerequisite for subsequently making articulate the abstract distinctions that are the pride of rational human being. But something analogous to them seems to be part of the psychic armamentarium of many animals.

As we have already seen for Plato, issues like these are inextricably linked with the mimetic power, the eikastic and phantastic powers, of human beings—powers that we encompass with the term "imagination." Not only the sophist and the philosopher but the ordinary human being are constantly using language to portray things of the world, not just their color or their height, breadth, and depth (a very basic sense of imagining), but their being in all other respects as well—even their being, pure and simple. Human beings as such take the looks of things and render them in many ways and respects.

Plato, or perhaps Socrates, took the Greek term for the look of a thing and raised it to a higher power. The Ideas, also called Forms, and usually capitalized in English translation to mark their status,[35] are an extension of the ordinary looks of things that ordinary human beings ordinarily notice. Although there is reason to assert that, for Plato, the Ideas are apprehended by a "higher" power than that of sense perception or imagination, we shall see momentarily that it would be a mistake to think that Plato's intellect functions autonomously, independent of, and without any important relation to images and imagination. The fact that the Western tradition has constructed and reinforced a conception of radically autonomous intellect or rationality—to no small degree encouraged in this by interpreting Plato—does not mean that this is how Plato understood things.

The entire Platonic corpus is predicated on educating the human ability to appreciate the looks of things. The *Republic* goes as far as any in providing a comprehensive scheme for understanding the implications of this look–taking and –making. It articulates the cosmic structuring powers that produce images and the human powers implied therein. It is less important—and not just for those interested in imagination—that Plato developed out of these concerns a theory of ideas that was subsequently further elaborated, and parodied (not least in thumbnail sketches of Plato's philosophy), than that this theory tried to make comprehensive sense of the manifold interconnections of things and the human experience of them. Aristotle's criticism to the contrary notwithstanding—that his predecessors, Plato included, had failed to give an accurate account of the nature and kinds of thinking and perceiving (see esp. *On the Soul*, 427a17–b17)[36]—it is only on the basis of something like the *Republic*'s account of the different levels of apprehendable things and the human capacities required to grasp them that Aristotle's psychology became possible. Aristotle claimed that there is no thinking without images, yet this would have made no sense without the Platonic background[37] to support its plausibility.

---

[35] A practice that extends to the individual Ideas, for instance the Good, the Beautiful, the True: a practice that I shall in general *not* follow after this paragraph.

[36] Aristotle's writings are cited here using the Bekker page–column–line numbers.

[37] By "Platonic background" I do not mean the "definition" of imagination as "sense perception with opinion," but rather the conceptual topology of an ontologically grounded human psychology against which such a definition makes sense.

That the *Republic* is about justice is true, as far as it goes. Considering the ground that the dialogue traverses, however—from the heaven of ideas to Hades and back—that does not go very far at all. Although our present interest is motivated by a concern for images and imagination, it is still necessary to make certain global claims about the dialogue. One is that in the first of the ten books the dialogue partners seek to *define* justice, but that the definitions given beggar its reality and possibilities, so that beginning in the second book the dialogue takes a sharp turn that forces Socrates to *portray* justice ever more comprehensively and in ever more comprehensive context—to make an ever more detailed *image* of justice—right up to the end of the dialogue.

In the first book a question comes up: what good does wealth serve? The person to whom Socrates addresses it—Cephalus, a rich man, a resident foreigner in Athens, and the father of one of the group of young men who have brought Socrates to Cephalus' house—says that wealth allows him to compensate for wrongs he has done. This leads to a first definition of justice, giving each his due and telling the truth—though the second part is quickly dropped and never *expressly* returns. Socrates quickly comes up with an example showing that the "giving each his due" part fails: if a friend who lent you a knife demands it back in order to commit violence, both friendship and justice appear to demand that you withhold it from him.

After Cephalus takes his leave to offer sacrifices to the gods, his son Polemarchus proposes a new definition: that justice is doing good to one's friends and harm to one's enemies. It is formulated to respond to part of the objection to the first definition, and, just as much as Cephalus' attempt, it reveals something about the concerns of the person proposing it. The young Polemarchus spends his days running with other young men he likes and doing all the things they think are good for them, whereas the ageing, soon-to-die Cephalus seems to be worried about what will happen in the afterlife. Polemarchus' definition falls short, precisely insofar as it is hard to know who your real friends and real enemies are. When Socrates argues further that it makes no sense to do harm to anyone at all—because harm makes the person worse, and do you really want to live among people who have been made worse?—the young sophist Thrasymachus denounces him as an idealistic fool and puts forward a new claim: that justice is whatever serves the strong, that might makes right. After a long analysis that occupies nearly the entire second half of book I, Thrasymachus, seeing that he is about to lose the argument, declares himself indifferent to the impending conclusion. Socrates seems to take this as a challenge and in short order turns loss into humiliation. Thrasymachus, supposedly an expert in making arguments, is forced to admit that he really understands nothing at all of his argument; and he blushes. Having thus "tamed" the savage theory of Thrasymachus, Socrates declares in quick summary that justice makes the human being good by giving good order to the soul, which in turn produces good actions. But he also concedes that their discussion (like other Socratic dialogues) has shown that even after much questioning and discussion they do not really understand what justice is, and he gets ready to leave.

But the young men won't put up with this. They insist that, this time at least, Socrates will have to stand and deliver: he will not be allowed to confess ignorance and then run off. He will have to stay and tell them about justice. They explain why this is important. Although parents and other adults praise justice, they value it more for what it leads to (especially good reputation) than for its own sake. They advise being just because it is profitable. What is worse, merely *seeming* just appears to be good enough, whether or not one is truly just. This is obviously an ethical version of the traditional problem of being versus appearance, and thus a question for which the categories of the *Sophist* come into play. It is not simply that a just act may appear unjust, and vice versa, but that one can, with deliberate expertise, "stage" words and events in ways that make something appear as what it is not (like the phantastic art of producing simulacra the *Sophist* described).

But things are even more complicated than the young men realize. What they understand is that their parents speak about justice in a way that presents it as different from what the children see with their own eyes. What glimmers in their consciousness is this: if citizens are concerned about reputation—not what they are, but how they seem to others—they will need something like the sophist's art in order to make whatever leads to wealth, power, or pleasure appear just. You thereby end up trying to deceive people by making things appear other than they are. What the young men clearly understand is that, given these confusions of reality and appearance, it may be impossible to know whether it is justice that motivates words and actions. This leads them to make an extraordinary, and perhaps unreasonable or even impossible, demand of Socrates. They want him to prove not only that justice is preferable to injustice, but that it would be so even if the world were arranged topsy-turvy, where real acts of justice would be considered unjust and disvalued or even punished by society, and acts of injustice would appear just and be rewarded. That is, in the world they hypothesize[38]—one that they will have to portray and imagine more concretely in what follows—the just always has the look of injustice, and vice versa.

If it is clear enough that the question of eikastic and phantastic images and imitations is intertwined with their demand, the problem of images and image making now arises almost immediately from another direction. Perhaps, Socrates suggests, the problem they face is that, if they are looking for justice in the individual human being, in the human being's soul, this is a "small" place where it is hard to find and make out what it is. Suppose, he goes on, there were a message written in tiny script somewhere, so small that it was almost impossible to read, but that there were another place where it was written very large. Obviously, if they wanted to understand the message, they would read it first in large format. So, he explains, if they looked for and found justice in a larger place, like a city, they could in turn use that knowledge to see it much more easily in the human soul. The young men agree, and accordingly, in multiple waves over the next several books, Socrates and his friends set out to portray a *city in words* (that is, in *logoi*), where they hope to find the image

[38]The significance of such hypothesizing, which is both an intellectual and an imaginative act, will become evident in book VII, in the last phase of explaining the nature of the good by using the sun, a geometrical line, and a cave allegory as images of the good.

of justice writ large.[39] Portraying things in words is, for Socrates as much as for the Eleatic Stranger, mimetic, and either mimetic in the proportionally accurate eikastic way or in the proportion-distorting phantastic way; or, since we found reason to question whether this distinction is as clear as it at first seems, portraying things in words raises questions of accuracy and proportion just as much as portraying things in paint or marble or bronze.

4.6 The Paradoxes of Imaging

One of the many ironies noted of the *Republic* is that, in the course of agreeing on what and how children should be taught, the young friends of Socrates agree to ethical and epistemological standards and rules governing artistic portrayal that, if applied to the *Republic* itself, would require its being banned from their ideal city. For example, at 392D Socrates points out the difference between narrative and mimesis in giving accounts of what people do and say. Basically the distinction is that between producing third-person indirect narrative (where we say "A said that X," with X expressing the content, though not necessarily the exact words, of what A said) and the direct rendering of the person's words (where we say, "A said, 'X,'" with X being an exact quotation). The objection to direct quotation is moral: that it requires the person doing the quoting to "play the part" of, imitate, the person speaking. Given that the character of some people is problematic, even evil, this means that the person who uses direct quotation is imitating a morally problematic person and thus perhaps becoming more like that person. Indirect quotation, by contrast, establishes a certain distance between words and character. We might say

[39] A "*polis* in words" or "in speech" is how they refer to the city they are designing at several points in the dialogue, especially when the question arises of whether such a city could ever be realized. It is Socrates who calls the phases of the argument "waves." After sketching out the structure and education system of their ideal city, they have to revise the plan when they address the status of women and children in the city (the second wave), and again when they ask to what degree the city must be based on knowledge rather than opinion (which leads to the third wave, in which the kings must be philosophers). But there are, implicitly, smaller waves as well: for instance, in the second book they conceive first an idyllic community of herders and farmers, which is rejected by these urban youth as too unexciting. That leads very quickly to an "overheated" city, very much like Athens, with the eager pursuit of international commerce to feed refined appetites for commodities and pleasures. That leads in turn to what later they call the actual first wave, a city divided into those who produce things, those who police the citizenry, and those who govern. This is a city that, at least according to the end of the third book, is based on a "noble" lie: on the claim that roles in life must be assigned to citizens according to their genetic natures (according to whether bronze, silver, or gold flows in their blood, making them eligible for, respectively, productive occupations, the military/police force, or leadership/guardianship). Unfortunately, whether they notice or not, the very problem that led them to demand that Socrates justify justice returns in the "ideal" city: the triumph of what might be merely apparent justice over real justice. The rest of the dialogue is proof that Plato's Socrates does not fail to notice the irony of this development, and gives the lie to interpretations that he advocates the tyranny of knowledge or pseudoknowledge.

that it objectivizes the person's words, whereas, by subjectivizing a person's words, direct quotation makes the moral danger more intimate.[40] After a fairly lengthy discussion of the different characters of those who imitate others, the group acknowledges that it is appropriate for a good man to imitate other good men but not bad ones (396C). Even though the *mixed* style of a little bit of mimesis added to some indirect narrative is judged to be pleasing, especially "to boys and their teachers, and to the great mob, too" (397D), they decide that, in the city they are designing, it will be best to allow only poetry that is nonimitative narration.

Plato's dialogues, insofar as they all predominantly employ the technique of direct mimetic rendering, violate this stricture. Plato presents the dialogues as though they were being spoken, here and now. This would be Plato's problem, of course, not Socrates', since the former is the one who "wrote down" the dialogues; the participants in the dialogue have no reason to note this, though readers of the printed dialogue do. Socrates does, however, occasionally himself use mimetic rendering in the *Republic*, for example in book X (618D), where a herald in the afterlife explains the process of reincarnation to the assembled souls. Presumably such a character is morally unproblematic and so not a danger. Of course Socrates and his friends do not live in the city they are designing in words, so they are not bound by its rules. Yet the irony goes a little deeper, if for no other reason than that the opening lines of the *Republic* establish that the work as a whole is a mimetic performance by Socrates: it begins with Socrates saying, "I went down to the Piraeus yesterday with Glaucon, son of Ariston, to pray to the goddess," and continues throughout as Socrates' mimesis of the entire conversation through the ten books of the dialogue, with occasional description of the circumstances of the conversation—the mixed style, as they call it, though with a preponderance of the forbidden direct rather than indirect narrative.

We must avoid dismissing this as just an oddity, not least because Socrates most intimately connects imitation with the question of justice. The ultimate reason they disallow the imitative style and the mixed style is that they made a decision very early in the construction of the city in words that quickly became the fundamental principle underlying the city's structure. When they were still at the level of constructing a primitive community of herders and farmers in book II, Socrates asked his audience whether (a) each person ought to produce and secure all basic needs or (b) productive activity ought to be specialized. That is, should I make my shoes and clothes, build and fix my shelter, raise and prepare my food, etc., or should I do just the single one of these activities that I am best and most productive at and then share or trade for the other necessities with fellow citizens, who likewise specialize their activities? The young men opt for specialization, without any reflection other

[40]This is a traditional moral objection to acting in stage plays. One way to call into question the distinction between the kinds of narrative would be to show that it does not really reduce the risk, since even direct narration invokes a certain distance from the original and indirect narrative involves a degree of mimesis. This is in any case a problem intrinsic to images and imaging: how close does the appearance of an image of something bring us to the real thing the image brings to mind?

than that it seems more efficient.[41] Over and over again they refer back to this specialization of activity, each person keeping to his own, ever more narrowly defined work. The specific consideration when they discuss poetic mimesis is whether one person can adequately imitate all other people, or only those similar to him. When Socrates reminds them that each person has been trained to keep to and practice just one activity, they settle on the latter as the correct answer, and that negatively settles the fate of the purely mimetic style and the mixed style, too. He points out explicitly that the guardians, those who are in charge of and make decisions for the city,

> must give up all other crafts and very precisely be craftsmen of the city's freedom and practice nothing other than what tends to it—they also mustn't do or imitate anything else. And if they do imitate, they must imitate what's appropriate to them from childhood: men who are courageous, moderate, holy, free, and everything of the sort; and what is slavish, or anything else shameful, they must neither do nor be clever at imitating, so that they won't get a taste for the being from its imitation. (395B–C)

Once you have admitted any imitation at all into the city, however, you introduce a train of consequences. As the education of those who govern continues, for example, it becomes clear that they must know both justice and injustice in order to distinguish the two.[42] However they acquire their knowledge of injustice—one is tempted to say that precisely here fictional representations have an important role to play—the rulers need to be able to make accurate word images of such things.

When, in book IV, Socrates and his friends have just finished the first wave of the construction of the city in speech, they look for justice along with the three other cardinal virtues, wisdom, courage, and moderation. The method Socrates introduces then is rather odd: he suggests finding the other three virtues first, so that justice will be "what is left," as though that were an unambiguously defined notion (see 428A). When, after identifying the other three, their first attempts to figure out what is left founder, Socrates so to speak slaps himself on the forehead and says in embarrassed astonishment that all along it has been there staring them in the face: that justice is precisely what they settled upon as the first organizing principle of the community,

[41] Of course it *is* more efficient to have each person do what he or she is best at. But the ethical and political problem they fail to consider is that if we specialize too much we may disproportion ourselves by developing only one talent, to the neglect of others essential to good, just human being. The problem had already emerged in book I, when Socrates asked whether shepherds look after their own interests or the interests of their sheep. Socrates said they must do both; they have to acquire two arts, not just one. But in the second book they promptly forget this, and it is fateful (one might even say *fatal*) for their undertaking. It is the consequences of this unanalyzed step, which Socrates implicitly criticizes throughout, that has misled some commentators to portray Plato and his master as advocates of totalitarianism.

[42] At 409B–E Socrates discusses the need for judges to acquire knowledge of badness in others; at 484C–485B he says that the true guardians must have both knowledge of things and experience of them, and that the lover of wisdom prefers holding on even to knowledge of contemptible things. This is an extension of a theme that receives its first lengthy development beginning at 437A, that knowledge extends to contraries or opposites: that is, one knows the good and the bad, the just and the unjust, by the same standard. And that theme is ultimately subsumed in the thesis that what is, as well as what is known, "rolls around between opposites."

that people should specialize their activities and do just one thing. Thus justice is doing your own work and not meddling in the work of others (432D–433A). Reading retrospectively, it was justice itself that dictated banishing imitation from the city, except in the restricted sense that you were allowed to imitate what you are like.

It certainly is astonishing that they so accidentally stumbled upon the nature of justice with virtually the very first constructive step they took in book II. Astonishing—and deeply false, since almost every further step they take after this astonishing discovery at the end of book IV undermines the adequacy of this conception of justice's nature, although no one in the dialogue ever expressly points this out. Perhaps they ought to have recognized the problem already in the earlier books when they insisted that each person should do only one activity. It is well and good that the shoemaker not try to bake bread and the baker to fix shoes, but if a fire breaks out in the baker's shop and threatens the whole block, do they all simply keep working at their jobs, because it is the job of *firemen* to put out fires, not theirs? No one protested that guardians, if they are to be good guardians, need to assign the tasks to others in a reasonable way, which would require their becoming acquainted with jobs other than their own; moreover, that in contravention of the definition of justice the guardians must meddle in absolutely everything in this city. Later, in book VII, when they come to the highest education and training of the guardians, no one objects to the fact that the guardians need to acquire expertise in many different subject matters, nor that they will spend 15 years after their education doing an apprenticeship in all the low- and mid-level jobs of administration and policing that the city requires. When in book IV they define *moderation* as each person harmonizing with all others—Socrates explicitly uses the model of each person singing his part in the chorale of society—no one points out that this can be done well only if each person is aware of and responsive to the parts sung by others. In book V (462C), when Socrates compares the best governed city to the human being who, when his finger is wounded, finds that the entire community "is aware of the fact, and all of it is in pain as a whole along with the afflicted part," no one points out that this suggests *all* people of *every* class must have similar cognitive and affective awareness. Nor, when at 469C they begin to discuss the enslavement of enemies captured in war and distinguish between Greeks (who will be treated as friends) and barbarians, does anyone remark that they are conceiving human beings as part of larger communities beyond their particular city, and that the principle of such an enlarged sense of belonging cannot be that of specialized activity.

One could go on identifying similar contradictions and tensions, almost indefinitely. I will add just two more, from the final book, book X. One, from 619B–E, nearly at the conclusion of the dialogue (which ends with 621D), undermines the thesis that Plato was the original, approving architect of totalitarianism. In the afterlife, a soul preparing for reincarnation that came from an orderly city—presumably like the "ideal" one they constructed in words—makes the worst possible choice for his next life, a life of tyranny, which will result (after he dies in that life) not just in a 1,000 years of disciplining punishment but in eternal perdition. The soul makes this mistake because it "participated in virtue by habit, without philosophy." If in the "best" city only the guardian class learns how to philosophize, the city will make most of its citizens unfit for choosing a new life. Many will *ultimately* come to a bad end, in Hades, because of their inability to discriminate the just from the unjust. Clearly the gods' notion of justice—and Socrates describes the judges in the afterlife as *true judges* (thus truth makes another surreptitious return in defining justice after being forgotten in book I)—is in conflict with the notion of justice in the "best" city they have described in words.

The second episode from book X that I want to emphasize occurs at the beginning. It illuminates further the problem produced by the conventionalization of virtue and the good, a conventionalization that culminates in inadequate education about justice and the good. It quite precisely involves the question of imitation. Socrates and his dialogue partners attempt to make the most decisive possible distinction between philosophy and art, in particular between philosophy and poetry. The judgment that they make is, not surprisingly, in philosophy's favor, yet the terms in which it is made, the different kinds of image making of philosophy and poetry, undermine the legitimacy of the argument both philosophically and poetically.

The basic strategy of the argument is to determine the "distance" of poetry and of philosophy from truth. Poetry is described as imitating real-world objects (like the heroes of the Trojan war). Real-world objects are, unfortunately, highly unstable. Craftsmen, at least the best, operate differently. A bed maker does not merely imitate an existing bed, he realizes in matter a pattern or form that has more durability than any real bed. The philosopher looks more deeply yet into the fixed and unchanging, and what he produces, *logoi*, are in accordance with even more durable patterns, ones that deserve to be called eternal—and thus the human being who attends to such things is as close to the eternal as it is possible for a human being to be.[43] Socrates determines that the poet is much further from reality than the philosopher, because his portrayals are of things several "levels" below that with which the philosopher is concerned.

In a moment we shall come to the basis for the pseudomathematical theory of levels that supports this calculation. For now it is important to notice that what the philosopher does and what the poet does are both considered a kind of mimesis, image making. One thing Socrates does not consider is whether he has characterized poetry rightly by calling it a mimesis of real things. Although one can probably argue that most of the characters and events of Greek epic, lyric, and drama were regarded as having really existed—and thus that, compared to later fictional art, there is a dearth of merely imagined or made-up characters—it seems obtuse to characterize the works of Homer, Hesiod, Sappho, Aeschylus, and Sophocles as merely mimetic. Poets could argue that even if they occasionally abstract truths from particulars (something that the philosopher almost always tries to do), truth, if

[43]This prepares the way for the second of the three major episodes of book X: between (1) the competition for truth of philosophy and poetry and (3) the story of what happens in the afterlife (the myth of Er) comes (2) the "proof" that the human soul is likely to be immortal because it can, through philosophizing, consort with what is eternal.

it is to mean anything at all, cannot mean only abstract eternal truths. The truth that the poet deals with is more concrete than that of the philosophers—a truth of local habitation and local naming. Moreover, it is rare that a poet simply imitates an already existing person. Poetic characters can be a mixture, drawn partly from life and partly from possibilities of human being that may have been realized nowhere but that can still perform a function just as clarifying as philosophy's—and, for the average or even the highly educated reader, even *more* clarifying, because the imagined people are concretely presented in circumstances familiar to the audience rather than among the abstractions of philosophy.

It is in fact easy, in a mixed group of philosophy and literature professors, to descend into turf-defending vituperation when discussing book X. But there is a deeper criticism to be made of the conclusion that philosophy triumphs over poetry. It is a deepening of the earlier criticism that the *Republic* violates the rules that it lays down for poets. The dialogues of Plato are more strictly mimetic pieces than even the most traditional of epics or lyrics (dramas are, of course, strictly mimetic). A Platonic dialogue is a "picture" of several real people engaged in a conversation and all its circumstances: for instance, an account of what happened to Socrates and his young friend Glaucon as they were walking together back to town after having seen Athens' officials and priests welcome the new goddess Bendis, when the two were accosted by a larger group including Polemarchus and Adeimantus (the brother of Glaucon, and also of a fellow named Plato who does not make a direct appearance in the dialogue), and all of them then proceeded to Polemarchus' house, where his father Cephalus was about to make sacrifice to the gods, etc., in a parade of (irrelevant?) concrete detail.

There is more. In the book X argument about the deficiencies of poetry we find an argument reminiscent of the *Sophist*. In the latter dialogue the Eleatic Stranger characterized the sophist as claiming to possess the art of imitating everything. The Stranger then provided an image–analogy for what this meant: the visual artist's ability *to portray everything visible*. In book X of the *Republic* Socrates presents poetry's claim to portray the world as the equivalent of holding up a mirror to the world (just as the painter does, though he uses the medium of paint). The mirroring that poetry engages in is a mimesis that reproduces in another place, in words rather than on the surface of the mirror, the look of all the things in the world. It is image making on a grand scale, just like the image making of the sophist. In the *Sophist* the Stranger frequently called attention to the difficulty they were having in distinguishing the sophist from the philosopher. By contrast, Socrates and his friends in the *Republic* accept that the philosopher's art is much different from the poet's (and the sophist's). But they are as ready as sophists and poets to make images in words of everything and anything, both within and without the world (e.g., the realm of the ideas and the myth of Er). Their narratives are filled with seemingly irrelevant touches about real and fictional things, for example in their description of caves and the afterlife.

Plato was certainly not oblivious to this. The *Republic* is not just a portrayal of a conversation. With its characters, actions, and discussion it attempts to construct the image of a world that is very much like that of Athens ca. 400 B.C.E. Its personages are at least as concrete as those in a Greek drama, and if what those personages talk about is often more abstract than what is talked about in drama, we nevertheless see their personalities and characters through their speeches. There are proud characters who are brought down by their pride (Thrasymachus, for example), and persons of modest station who play small but significant roles in the drama (the slave who stops Glaucon and Socrates in the first scene). The young men who follow Socrates have different kinds and degrees of intelligence, spirit, sensitivity, and education. Many of them have been alienated from their parents and from what their city teaches them to value and so are in search of something more worthy of pursuit. We hear of other real and fictional characters who engage in remarkable exploits (like the shepherd of book I who finds the miraculous ring of Gyges, which allows its holder to become invisible and thereafter to seduce the queen and overthrow the king), those who live in a strange world unlike Athens (the denizens of the cave of book VII who are chained to benches and stare at a wall that seems to be their whole world), and others who even get to experience heaven and hell (Er in the concluding myth of Er, who being left for dead on a battlefield is allowed by the gods to see the afterlife so he can bring back an account of it to the world of the living). And poets would probably be inclined to urge upon Plato and Plato's usual defenders that the moral and even philosophical lessons of the dialogues are all the clearer for being presented in vivid, poetic form. Thus maybe the philosopher *is* a kind of poet—though not necessarily a good one—writing in unpoetic genres. At the very least, the principles Socrates and his friends use to differentiate philosophers from poets are neither clear nor convincing.[44]

4.7 The Ontology of Images and the Psychology of Scenario–Imagining

We are ready to turn to the heart of Plato's conception of images, imaging, and imagining. The heart of those concerns is to be found in the discussion of the good, which comes in the middle of an account of the personality and training of the philosophical character—the character of the person who will rule the ideal city. This discussion of the good presents a framework according to which the West has ever since conceived, understood, and misunderstood imagination.

Book V (at 449C) begins the second wave of city design with a question about the status of women and children—the so-called community of women and children.

[44] At 484C–D Socrates parenthetically mentions painters, "looking off…toward what is truest, and ever referring to it and contemplating it as precisely as possible." So much for the notion that poets and artists must be several times further away from the truth than philosophers! By this comment (and similar ones throughout) I am not ridiculing Plato for inconsistencies but pointing out how this master philosopher–artist constantly challenges us to subtle reading and thinking. An inconsistency may seem like a grave philosophical sin; but, as Aristotle knew, it is more fundamentally an invitation to think about different respects in which the conflicting statements might agree.

Women and children are not the property of the men but of the city as a whole; more exactly, men, women, and children are all property of the city. Later, book V also starts the third wave with these words:

> "Unless," [Socrates] said, "the philosophers rule as kings or those now called kings and chiefs genuinely and adequately philosophize, and political power and philosophy coincide in the same place, while the many natures now making their way to either apart from the other are by necessity excluded, there is no rest from ills for the cities, my dear Glaucon, nor I think for human kind, nor will the regime we have now described in speech ever come forth from nature, insofar as possible, and see the light of the sun. This is what for so long was causing my hesitation to speak: seeing how very paradoxical it would be to say. For it is hard to see that in no other city would there be private or public happiness." (473C–E)

From this point late in book V up to the middle of book VI Socrates discusses the traits of character of those who are capable of being philosophers. Those traits are based in love, a love that wants every kind and part of what it loves. The philosopher loves learning of all kinds and "is willing to taste every kind of learning with gusto, and…approaches learning with delight, and is insatiable" (475C). Of course this avidity seems to violate the principles of moderation and justice set for the city. There is no limit to the philosopher's desire for knowing, and he wants to "meddle" in every kind of knowing rather than stick with just one kind that is proper to himself. Moreover, such a person so loves truth and hates lies that it becomes quite unclear how he would respond to the "noble lies" told to maintain social harmony, or to the initial "nondecision" in book I that eliminates truthtelling from the definition of justice.

Book V concludes with a prolonged discussion of knowledge, ignorance, and opinion (which lies between the two), in which it is argued that what is opinable participates in both being and nonbeing. Opinable things "roll around somewhere between not–being and being" (479D) and are not "addressed by these names [like big, little, light, heavy] any more than by the opposites of these names" (479B). This is, of course, reminiscent of the Eleatic Stranger's revision of the Parmenidean doctrine in the *Sophist*, though without the fundamental qualification he made: that although being and not–being are opposed, things nevertheless participate in both in various, *differentiated* ways. The philosophical character among human beings is the one who has a sensibility for the difference between the opinable and the knowable, and thus between the problematic "understanding" of the unstable and the lasting understanding of what does not change. This establishes the claim of the philosopher to rule. The first half of book VI discusses other characteristics of the philosophic natures beyond their love of all learning. They have no taste for falsehood, they are moderate because their prime concern is the pleasures of the soul rather than of the body, they are great-souled and recognize that human life is nothing great, they do not fear death and thus are courageous, they are not attracted to vices and thus are just, they are tame and measured spirits rather than wild ones, they are quick in learning, they have good memories, and they are musical and graceful (that is, filled with the gifts of the Muses).

After this multiplication of good qualities Adeimantus raises a problem. He and others similar to him often have trouble knowing how to answer Socrates' questions and wonder whether they are not being slowly led in the wrong direction. Lots of people who engage in philosophy look quite strange, even vicious, and they end up useless to the city. Socrates agrees, and then says something very pertinent to our immediate concern: "the question you are asking needs an answer given through an image." To which Adeimantus responds, probably with affectionate irony: "And you, in particular, I suppose, aren't used to speaking through images." The word they use is *eikōn*: if it means what it means in the *Sophist*, it indicates a veridical image that maintains the proportions of the original.

The image Socrates offers as a response to Adeimantus portrays the city as resembling a ship on which the owner, taller and stronger than anyone else on the ship, knows little about piloting it and is nearsighted and a bit deaf to boot. He is pestered by the sailors, each of whom has his own notions of sailing and wants the owner to turn command over to him. They conspire to do away with those who get in their way; eventually they "enchain" the owner, take over the ship, and feast on its supplies. They praise as the true pilot the one who knows how to seize command and do not believe that you can both take command and also spend time acquiring all the arts needed to pilot the ship. In the meantime the true pilot studies the seasons, the stars and the heavens, the wind and the weather—and is dismissed by the others as a useless stargazer.

For our purposes, perhaps even more important than the allegory itself is what Socrates says about images before he tells the story of the ship owner, in response to Adeimantus' friendly taunt.

> "All right," I said. "Are you making fun of me after having involved me in an argument so hard to prove? At all events, listen to the image so you may see still more how greedy I am for images. So hard is the condition suffered by the most decent men with respect to the cities that there is no single other condition like it, but I must make my image and apology on their behalf by bringing it together from many sources—as the painters paint goatstags and such things by making mixtures." (487E–488A)

And when he is done with presenting the complex, artificial image of the ship owner's peril he tells Adeimantus, who obviously grasps the meaning of the image, to "teach the image to that man who wonders at the philosophers' not being honored in the cities, and try to persuade him that it would be far more to be wondered at if they were honored" (489A–B).

Note that the images we are talking about here are not like the "snapshot" images of empiricist psychology; they are complex and dynamic. We are not asked merely to imagine a ship, or a ship owner, or a sailor, but all of them together, interacting in a developing scene—one might thus call the kind of image Socrates and his friends are talking about a *scenario*. The scenario–image is woven together out of elements that all show their typical figures, the *eidos* of each, but taken together what the whole scenario exhibits is an *eidos* that is far more than the sum of its parts. The scenario is deployed in a particular way that suits the purposes of the narrator; but as a scenario, the image also has a certain lability or flexibility, in that the deployment can be modulated, varied, extended, or even made to take alternative paths of development, more or less plausible. Moreover, the use of the image is not finished with the conception of the scenario in action and its possibilities of development.

For, after all, Socrates introduces this and other scenario–images in order to induce his audience to see something more or even something else. He expects them to recognize a form, a structure, a sameness that can be seen elsewhere—in this case, in the government of cities.[45] This is the kind of case in which the imaginary *logos* turns out to be at the same time a *logos* about something else: a *logos* in a scenario that applies, to some degree point for point, to another scenario in another field of concern. This means it is, literally and figuratively, *analogia*, analogy—which in the first instance in Greek means an extended proportion (one of the meanings of *logos* is "fraction" or "simple proportion")[46]—in which we say that, as A is to B in circumstances or field M, so is C to D in circumstances or field N.

We can think of both *logos* and *analogia* in relation to conception and projection. In the first instance it seems that the most basic kind of *logos*, a simple proposition relating terms A and B, is a representation of a recognition that is more or less passively received. One must not, however, put entirely out of mind that uttering "A is B" or "A is doing B" adds an element of positive activity and projection to the receptive one. If reception is always joined to at least a certain degree of activity of the receptive mind, whether affirming/positing or negating, it would be easier to understand the imaginative moment as involving a shift in the balance between reaction and activity.

Suppose I see, in late spring, that a tree in my neighborhood is dropping its leaves. The degree to which I even take cognizance of this depends on circumstances. If the tree comes into view while I am driving home from work elated or depressed by the day's events, it will be one of innumerable things in my field of vision that are at best marginally registered. If I am on a walk through the neighborhood I am more likely to notice it with attention, perhaps to the point that I say something about it later to my wife. If I have recently been pruning my own trees it

[45]One might argue that Socrates' use of scenario–images is itself developmental, moving from metaphor and allegory to analogy (for the audience at least), from an inexplicit but felt similarity to a more clarified and articulate elaboration.

[46]The connection between *logos* used to mean speech on the one hand and proportion on the other is more intimate than first appears. The basis of speech is a relation between things that is expressed in statements like "S is P": to predicate one thing (the predicate, P) of another thing (the subject, S) is to express this relation. The mathematical *logos* or ratio is conceived similarly. A ratio is a proportion between two (whole) numbers, two line segments, two surface areas, or two other like things. Whereas we think of the ratio of a to b (where a and b are any numbers) as just another number, ancient Greeks conceived it as essentially a *relation* between two things of the same kind. If a is a line length and b a line length, "a:b" stands for the *logos* or proportion of those two lengths. Two areas, c and d, can similarly be related in the *logos* "c:d." But a line length cannot be put directly into proportion with an area except by way of an extended proportion or *analogia*. If line a is half the length of line b and area c half the area d, we can express this in the *analogia* $a:b::c:d$; and by rules of manipulating proportions one can say that $a:c::b:d$. We mimic this algebraically by saying that if $a/b = c/d$ then we can multiply each side of the equation by the fraction b/c to get $a/c = b/d$. The difference is that we think of all these fractions as being the same kind of thing, numbers, so that no reconciliation of kinds is necessary. For the Greek understanding, however, a line length compared to an area is not a kind but a relation between (different) kinds; it has no absolute value, but can only be reexpressed by other, analogical relations between the kinds.

might set off sympathetic concern, or selfish concern insofar as it could be the first visible sign of a tree virus, or impartial, arborological interest in what might be causing the fall. In all these cases there has been a shift from receptivity to wonder and active concern. The shift will become more pronounced if a few days later I notice other trees with a similar problem and decide to investigate. In none of these cases is it quite right to affirm that the image or scenario–image has been simply translated into a proposition, a set of propositions, or even a set of propositions with attitudes. In all cases (except the scarcely registered seeing–while–driving) there is a placement of the scene into a context with concerns and a marking or inflecting of the context and concerns with terms, statements, and questions. The image takes up a place among others in a field of possible and variable relationships, and what I say and think about it marks ways I have of reiterated or renewed access to the field.

The point, at the moment, is not whether these all correspond to our contemporary sense of what imagining is, but rather that all of these elements are at work in the Socratic/Platonic notion of making an image by using many *eikōnes*. There is, as the first moment of imagining, the deployment by the imaginer of something that was implicit in noticing something, noticing that A is B or A is doing B. The things of the world show themselves to us by giving us their looks, both static (the bushy tail of a squirrel) and dynamic (the typical ways in which a squirrel flicks its tail). Things and activities are recognized not as isolated from the parts of the world they occupy, except perhaps when we maximally deploy our abstracting and objectivizing capacity (to derive the notion of the squirrel species or to focus on a single squirrel in abstraction from every external relationship). That means that already in seeing squirrels—at least as part of the world of *doxa*, opinion,[47] where things roll around, constantly shifting between being and nonbeing (479D)—part of our recognizing their look is already to grasp, as much in anticipation as in articulated form, some of the immediately possible modifications and variations of squirrel appearances. The squirrel sitting under the red oak holding an acorn is one that we recognize as capable of nibbling the acorn in the next moment, of putting it down and digging a hole for it, of hopping with it to the base of the tree or running with it up to the top branches, of discarding it—or of screeching at a philosophical type who, caught up in wonder at possibilities of the scene, is blocking access to his acorn hiding places. In an important sense all these thoughts lead to others that are related, are akin. It is not necessary, at least in the first or even the second instance in accounting for this work of the human mind, to invoke a knowledge of squirrel *essences*.

The squirrel scene is relatively simple, and yet it is capable of almost limitless variation and complication, and it can be intertwined with other scenes (the red–oak–scene, a curious–cat–scene, a German–shepherd–scene), limited only, as we say, by one's imagination. Very little, and according to some theories virtually none, of this would be part of a newborn infant's experience, yet many aspects of it will in short order become part of that infant's world and indissolubly entwined in the

[47] Or rather of *seeming*: as Arendt was fond of pointing out, *doxa* belongs in the realm of the *dokei moi*, "it seems to me" (Arendt 1978, 77).

infant's future experiences. The scene is part of my grasp of the squirrel's being present. A major and indispensable part of the scene is itself present: its being located in a place, its being in the midst of a few crucial items, its being engaged in at least one of its postures or possibilities of action. Also part of this scene is the ambiguity of its boundaries, in part due to my abilities (e.g., to look right or left or up or down, to focus on the tail or the teeth), in part due to what the squirrel can immediately do to change its posture or activity, in part because the scene of the squirrel is already entwined with other scenes and things I have at least marginally noticed (e.g., holding an acorn that dropped from the oak, sitting at the edge of the tree's shadow) that themselves are capable of imaginative alteration and new placement. This is the peculiar presence of things subject to change: we can always be surprised by what happens next, but we are never surprised that something *relevant* happens next, because that is already implicated in the presence of the thing, and we already anticipatively recognize aspects of things that might happen and the new "places" where new aspects might lead. So the development of the scene into a scenario (a complex of scenes) is inevitable and expected, though the exact course of development is not, insofar as it is guided by multiple interests and possible emplacements. We must recall, however, that in the first instance an "interest" is less a subjective stance than a way of "being between" or "being among" the things of the world.

Part of the presence of the thing with its typifying looks is its dynamic situation. Presence is already projective beyond the limits of the present. In the first instance the look of a thing leads us to treat it not as forever fixed but as active and reactive in a field that includes it. If we call the thing an object, we can indicate the relevant field or fields in which it is involved as the thing's objective field. Human beings grasp things as having characteristics or being in basic ways, and as acting in basic ways, and this natural progression, this "ringing of changes" on what has presented itself, brings into view a stage, a field of play, a typified place or field opened up by the initial interest. Once this stage is seen, the interest and concerns can shift to the stage itself, to the scene as a whole: not just to relations of things within the field, but to the relations that the field itself enables and that offer points of focus for comparison to other fields and the things that occupy them.

Imagining may ordinarily begin as oriented to objects and their actions, but it does not end there. Not all imagining is object-oriented. In one sense this should be obvious, if we reflect for a moment on the finite openness of human experience. Human beings (and probably many animals as well) are aware of things as situated. Any animal like the human which can take the look of a thing in at least partial detachment from "immediate" consciousness can perceive and operate in the light of that situatedness's developmental possibilities. Any animal, like the human, which can further detach the look by deploying a particular feature in mediate awareness has a capacity essential for imagining some of the possibilities of development. By incorporating multiple aspects in covariation, this capacity might be extended and complicated to the point of, say, presenting a story about shipboard quarrels of sailors and ship owners over who should captain the ship. The tendencies revealed in such imagining might—if the capabilities of the animal could grasp a much more complex look of the scenario itself, detach it, and further project it—be used to picture things even remoter from ordinary experience (for example, power relations in cities) or even to present an aspect of the world that is ordinarily invisible, perhaps a world that is invisible to the eye per se.

This is what book X's myth of Er does. It is a more radically developed and concretized scenario than that of the shipboard quarrel, but it is nevertheless the same in kind. If the shipboard quarrel is intended to reveal to the dialogue partners something that they at least implicitly know about what happens in cities, so that they see it as an analogy, the myth of Er situates its characters and events in a world that they do not know, and that nobody knows (except Er—but he is fictional). That nobody knows it and that it is a fiction do not mean that it is unintelligible. It is a kind of possible world—some such scenarios might in fact be likelihoods, though unexperienceable ones (because unliveable, except in fantasy)—with the kind of complexity, depth, and intertwinement that makes the world we live in a world. It is a world of sufficient complexity that we can imagine ourselves as being in it or something like it. That means not simply that we can imagine ourselves as physically situated in a space–time with gods, souls, rewards, and punishments, but as acting and reacting in it from the basis of our acquaintance with our ordinary situations in our everyday world. On the basis of our being able to detach ourselves to some degree from the particularities of situation, we adapt ourselves to the projectively or fictionally developed features of the imagined world.

4.8 The Grand Image–Sequence of the *Republic*: From the Good Itself to the Dialectical Education of the Philosopher

Is a *theory* of image–scenarios present as such—explicitly, and not just as a kind of necessary presupposition or plausible extrapolation—in the *Republic* itself?

The answer must be no if by "explicit" is meant a set of statements flagged as such by Plato or Plato's Socrates. If "theory" means propositionally expressed doctrines requiring allegiance because they have withstood the dissolvent power of dialectic, then there are scarcely any at all in Plato's writings. But if a theory is a way of looking upon things, highlighting some, and attending to as many of them together as is possible, then there certainly is a theory of imaging and imagining in the *Republic*.[48] It is revealed by the grand image sequence that stretches from the middle of book VI to the end of book VII, and includes two of the most familiar

[48]The Greek word *theōria* seems to have arisen within such a network of concerns. In the classical period and long thereafter *theōria* was conceived as a kind of intelligible viewing. Any propositional network of the kind we typically call theory thus presupposes, from the Greek perspective, a field or fields opened by the concerned consideration, and the consideration is itself the *aboriginal theory*. The notion of conceptual topology enables us to gain a more articulate and focused grasp of the backgrounds and foregrounds that are intrinsic not just to ancient theory but also to theory in the modern sense of the term.

image episodes in the Platonic corpus, the analogy of the divided line and the allegory of the cave. What I wrote of squirrels and analogy in concluding the previous section was in effect a commentary in advance on this extended sequence.

Commentators have noted that the *Republic* is marked by multiple ascending and descending movements.[49] That is certainly true of the sequence of books VI and VII. Two of those movements are largely responsible for the conventional interpretation that, for Plato, images are the least real kind of thing and imagination the least reliable and most misleading of human apprehensive powers. The interpretation goes something like this: On the divided line, imagination is assigned to the lowest part of the line corresponding to the human power of seeing shadows and reflections[50]; and in the allegory of the cave the prisoners who are chained there take the shadows projected on the cave wall as the real world. On the divided line, reason as understanding (*dianoia* at first, later rebaptized *noēsis*, at 533E) is placed at the top; it apprehends the *ideas* or ultimate *forms* of things. In the cave allegory, this level is achieved by first being freed from the chains that keep one staring at the wall of shadows, then looking around the cave, and finally ascending out into the open where one experiences the real world and sees the overarching heavens and the all-illuminating sun. Unfortunately for the philosophers raised in the ideal city, however, they must descend again. They must return to the cave to govern it in accordance with the ideal truths discovered outside. They are obligated by justice to return to the cave and govern the city. Their ascent to the ideas was made possible by the philosophical education with which the best city, the city in words, provided them. The city demands in return that those who have benefited from this educational curriculum employ their abilities for the good of the city. As they grow old, the philosophers will be allowed to devote ever more of their time to philosophizing, though they will still need to offer the city occasional services.[51]

It does not take much of an ironic sensibility to notice that Socrates thereby creates a city that makes the world safe for philosophers, since they get to rule it! It requires a more subtle irony to notice that Socrates' explanation of the education undermines the foundations on which the city in words was previously constructed,

---

[49] For a discussion of the significance of these movements, see Wood 1987.

[50] Socrates initially says nothing about how the line should be oriented, although near the end of book VI he and Glaucon refer to the section with the forms as "above," though without any discussion of why things should be pictured this way. In the imaginary of Western thinking, imagination *must* be lowest, and the forms highest: it is the configuration commensurate with the elevated dignity of reason.

[51] The education outlined in books II and III occupied eligible young men (and ultimately young women) until the beginning of military service, around age 20. The higher education that is described in the last half of book VII takes place in six stages, with 2 years devoted to each of the lesser studies (arithmetic, plane geometry, solid geometry, astronomy, and harmony) and 5 years to the culminating dialectics. That takes them to age 35, when they begin 15 years of community service in administrative and policing responsibilities. At age 50 those who have proved themselves most worthy become true guardians of the city. From that point onward their governing responsibilities are no longer continuous; when their wisdom and decision-making ability is not needed, they spend their time as they wish—presumably in the contemplation of the very highest things, the ideas or forms.

in particular the principle that each person should keep as much as possible to one kind of work and activity and cultivate only knowledge and images that are appropriate to it. For example, Socrates indicates that the higher studies undertaken are useful not just to the governors but also to the defenders and to the craftsmen and producers of the city. That is, the inferior classes would perform their jobs better by knowing things that are not included in the education (as described in books II and III) to which they are limited. By the very terms according to which justice in this city has been defined, that means that the city is unjust: it denies them an education that would let them practice their trades as well as possible. But it should also have long since begun to dawn on the reader that the city is organized in a way that potentially leads to great harm for its people.[52]

One of the most pervasive of historical ironies is that although throughout books VI and VII there is a cascade of images (that is, scenario–images), these two books are typically believed to offer one of the most fundamental portrayals of Platonic idealism, and of idealism pure and simple. It is also taken as evidence of Platonic rationalism: that apprehending these transcendent ideas is the highest activity and capacity of human beings, and that this apprehending is done by pure intellection or rationality. The irony here is that, however many passages there may be in the Platonic corpus tending toward such conclusions, only when removed from the dialectical and imaginative economy in which they function do they support this distorting, short-circuited portrayal of Plato's thought. At the very least, if there are such transcendent ideas, they are portrayed by Plato in a way that shows them always working by a process of involvement in mediums other than the ideal. The only way for human beings to approach the ideal is through recognizing the multiple levels of being in which the ideal *images* itself.

Socrates, invoking the difficulty of the task his young friends have set him, begins to explain the curriculum of studies that will turn the philosopher candidate, who possesses all the capacities that a philosopher needs, into an actual philosopher. The previous account of education has been defective; it has failed to "come to the end of the greatest and most fitting study" (504D). The young men

> have many times heard that the idea of the good is the greatest study and that it's by availing oneself of it along with just things and the rest that they become useful and beneficial. And now you know pretty certainly that I'm going to say this and, besides this, that we don't have sufficient knowledge of it. And, if we don't know it and should have ever so much knowledge of the rest without this, you know that it's no profit to us, just as there would be none in possessing something in the absence of the good.[53] (505A)

---

[52] See 619B–D, discussed in Sect. 4.6, above. I reiterate: those who take the city in words as Plato's definitive vision of the best kind of city need to think more about irony as a virtue and about whether Plato understands philosophizing as they do.

[53] Thus, once again, a "just" or "good" city that denies any of its citizens the best possible knowledge of the just and the good would be in truth an unjust and evil city. I hope the reader has already drawn a further conclusion: that Plato plays these ironic games precisely because he is letting Socrates carry out the early demand of the young men, that he prove the superiority of justice to injustice even in a city where people think that what is unjust is just and vice versa. The city in words they have devised is precisely that city.

Most people think the good is pleasure, while the more refined think it is prudence, but neither group is able to give an account of why and how. Yet

> this [good] is what every soul pursues and for the sake of which it does everything. The soul divines that it is something but is at a loss about it and unable to get a sufficient grasp of just what it is, or to have a stable trust [*pistis*] such as it has about the rest. And because this is so, the soul loses any profit there might have been in the rest. Will we say that even those best men in the city, into whose hands we put everything, must be thus in the dark about a thing of this kind and importance? (505D–506A)

Thus it appears that the good is not quite like any other *thing*, and that guardians, auxiliaries, and craftsmen–producers alike will get no true good from life if they are devoid of knowledge about it. It is a knowing that is no mere specialization. Anyone denied access to it will suffer harm.

Not for the first time in the dialogue the young men pose a question that postpones completion of an impending task. Adeimantus asks whether the good is knowledge or pleasure or something else, to which Socrates replies ironically that Adeimantus will not be satisfied with anyone's opinion (*doxa*) about the good—except for Socrates'! In words that resonate with the concerns of the entire dialogue, Adeimantus says that it does not seem just for Socrates to talk about other people's opinions (*dogmata*) but not about his own, when he has spent so much time "dealing with these things"[54] (506B). Socrates points out that opinions without knowledge are shameful and ugly, or at best blind; perhaps it would be better to hear "bright and fair" things from others.

At this moment Glaucon (whose name means "bright" or "shining") breaks in to say, as at the beginning of book II, that they will not let go of him now, he will have to stand and deliver. He adds: "it will satisfy us even if you go through the good just as you went through justice, moderation and the rest" (506D). (Socrates had shown them justice by contrasting it with moderation, wisdom, and courage.) Socrates insists that he does not have the power to give an account of the good itself; he asks for indulgence but also makes a concession. If they will permit him to set aside what the good itself is, he will be "willing to tell what looks like a child of the good and most similar to it," or, expanding the metaphors because of Glaucon's reply that at some other time Socrates can "pay what's due on the father's narrative," he will now give them the interest and the child rather than the principal and the father (506E–507A). By proliferating metaphors and images Socrates is communicating something about what he has already been doing and intends to do further. He reminds them that they have frequently in the past agreed that there are many things of every kind (beautiful things, good things, etc.) that they distinguish both in being and in speech, and that they "assert that there is a beautiful itself, a good itself, and so on for all the things that we then set down as many. Now, again, we refer them to one idea of each as though the idea were one; and we address it as that which really is.…And, moreover, we say that the former [i.e., the many things called by the

[54]Or "occupied with these things." The word is a past participle used as a noun, *pragmateuomenon*. It implies not just thinking about these things but also dealing with them in all aspects of living.

idea's name] are seen but not intellected, while the ideas are intellected but not seen" (507B). Then he proceeds to explain, well into book VII (at least through the allegory of the cave) everything—including the ideas—as the offspring of the good; and his mode of explanation is the production of a series of images, each of which illustrates the others. He offers a cascade—one is tempted almost to say a riot—of images, each generated from what precedes.

First, Socrates presents the good as figured by the sun, which both gives rise to all (living) things and produces the light that shows them for what they are. At 509D Socrates proceeds to draw out consequences from this image according to the analogy between (a) how the sun rules over the flourishing of physical things and their visibility and (b) how the good rules over the realms of beings of all kinds and their fundamental way of appearing (their truth). To illustrate what he is saying, he introduces a line segment (without at first mentioning an orientation) that he divides into two parts, according to a proportion that is not specified; then he divides each of the two parts again, using the same proportion as for the first division. He proceeds to clarify what he is after by explaining that each segment represents a different level or realm of beings, and that to each of these levels there corresponds a power of the human soul that allows us to know the things of that level. The first cut of the line, into two parts, separates the intelligible realm from the visible realm. The second set of cuts divides the *intelligible part* into a realm of the ideas and a realm of things belonging to "geometry and reckoning and such subjects" (510C; one can conveniently call these "mathematical things"); in the *visible part* there are, first, the things of the natural world and the artifacts made by human beings, and, second, the images (*eikōnes*) that derive from the natural and artificial objects as their shadows, reflections, and the like. He gives names to the human powers that apprehend and work with the things of these different levels: corresponding to the forms of things in the intelligible realm is *noēsis* (intellection); to the mathematical things *dianoia* (discursive thinking); to the things of the natural world *pistis* (trust); and to the images *eikāsia* (image–perception).[55]

The divided line further articulates and distinguishes, using a mathematical image, what is produced–and–shown by the good. The image of the sun portrayed

[55]Bloom translates *pistis* as "trust," but translators typically use "belief" instead. That too much intellectualizes the relationship to the things and artifacts of the world unless we qualify it as "belief that things are as they show themselves." "Trust" is the first English equivalent given in the standard Greek–English dictionary of Liddell–Scott, and it nicely expresses our basic relationship with the things of the world: we trust that they will behave in the way that such things do. For example, I trust that the chair that appears before me will support my weight and not collapse, or prove to be a phantom, when I sit on it. As for *eikāsia*: in the *Sophist* the *technē eikastikē* is the art of icon making, the making of images (*eikōnes*) proportional to the originals, and could be rendered as imagination in that icon-making sense; here in the *Republic*, it is nature that produces shadows and images in proportion to the original object, and we have the power (called *eikāsia*) to see those things—shadows, reflections, painted images—not simply as realities but as realities imaging other realities from which they derive. It is a perceiving that allows us to see both the thing that the image is (a shadow, a reflection, a painting) and its reference to something that exists elsewhere, in another format or plane.

the good as giver of life/being and giver of light/perceivability; the line expands on this duality, enumerating four basic kinds of thing the good produces and four basic human capacities for grasping them. Perceiving is mobile and complex; seeing something on one level is to begin to see its proportional relationship to things on other levels, either directly (the shadow as cast by a tree) or mediately (the shadow as bearing mathematical and even ideal relationships). The complexity and mobility of perception is correlative to the complexity and mobility of the expressions of the good, which are signified by the entire cascade of images presented in book VI; it is an intrinsic part of what Socrates is trying to convey about the nature of the good and the being and possibility it originates. This is where book VI concludes.

At the beginning of book VII, Socrates goes on to illustrate the significance of the line with a complex, dynamic scenario–image known as the allegory of the cave. People are chained in a cave to a long bench; they cannot turn their heads but always look toward the cave wall, where images are constantly appearing. It is these images that are reality, as far as they are concerned. They cannot see one another, or their own bodies, except as shadows projected on the wall. When they speak, their voices echo, so the sounds they hear seem to come from the wall. In the allegory it turns out that the shadows on the wall come from a large projector fire set behind them; unidentified "image makers" carry objects, cut-outs, or stereotypes to cast the shadows, and these images and the names that the image makers call out constitute the "reality" that the chained inhabitants "know."

The allegory goes on to describe what happens when one of the chained bench-sitters is freed, by an otherwise unidentified person. The first thing the freedman does is look around, but he is immediately blinded by the bright projector fire. His instinct is to turn back toward the wall to avoid pain, but with the urging of the unidentified liberator the freedman's eyes eventually become accommodated to the firelight. He comes to understand that what until now he had taken as real was a projected image made by others. Then the liberator begins drawing the freedman upward, along a steep path past the fire and toward the mouth of the cave. When the freedman reaches the mouth and comes into the open, he is blinded once more, this time by the light of day. Once again, with encouragement, persistence, and the accommodation of vision he will gradually be able to discern things. First he will look toward dimly lit places, like shadows, images, and reflections; then attend to the flora, the fauna, the clouds, and the sky. Finally, he will be able to glance at the sun itself. He will conclude that the things he had taken for reality in the cave were very pale shadows of the better and more real things he has found in the open world.

Socrates makes clear to his friends that such a person would not envy the life of those in the cave. If offered the choice of living as a slave who gained rewards by cleverly figuring out the empirical sequences of images cast on the cave wall or instead as someone who had discovered the outer world, he would choose the latter. Socrates immediately adds: if such a person returned to the cave he would appear to the others to be in a laughable state, especially while his eyes were readjusting to the darkness. The strange stories he would tell—that what they see is a reality that reflects other, more basic realities—might incline them to do away with him.

Socrates concludes with the following interpretation of the cascade of images in books VI and VII:

> "Well, then, my dear Glaucon," [Socrates] said, "this image as a whole must be connected with what was said before. Liken the domain revealed through sight to the prison home, and the light of the fire in it to the sun's power; and, in applying the going up and the seeing of what's above to the soul's journey up to the intelligible place, you'll not mistake my expectation, since you desire to hear it. A god doubtless knows if it happens to be true. At all events, this is the way the phenomena look to me: in the knowable the last thing to be seen, and that with considerable effort, is the idea of the good; but once seen, it must be concluded that this is in fact the cause of all that is right and fair in everything—in the visible it gave birth to light and its sovereign; in the intelligible, itself sovereign, it provided truth and intelligence—and that the man who is going to act prudently in private or in public must see it." (517A–C)

The irony of the long account (*logos*) of the good is that, although the good is the most intelligible thing, Socrates has to present it to his friends exclusively in terms of images: images that first attempt to portray the nature or most important aspects of the good, then images to illustrate the previous images, and finally word–images (*logoi*) to recast the significance of the visual images. In fact the visual images are always conveyed by words, further compounding the involvement of logos and image. He concludes by telling them they must connect the images as a whole to the images that came before. Someone might suggest that, with a more learned or philosophical group, he might have used a more rational approach. But Socrates almost always uses stories and humble images, so that is unlikely. Perhaps Plato, especially in the private lessons at the Academy, provided a more rational account (perhaps an esoteric account revealed only to his closest students). But there are reasons to doubt even this, and at any rate there is no direct evidence for the claim of esotericism—though it does flatter a millennia-old idea of the privilege of philosophers.[56]

A closer look at the divided line can help elucidate the real role of images and their relationship to rationality. First, we will consider the images and the image-perceiving power it portrays; second, the significance of the proportional division of the line; and, finally, the dynamics of reality and mind that the line reveals.

Socrates gives two examples of the *eikōnes* that are on the fourth part of the line: reflections in shiny surfaces, and shadows. Reflections and shadows are *of* physical objects, which are placed on the adjoining, third part of the line. Although the physics of reflections and shadows was perhaps still somewhat uncertain, Plato and his Socrates understand this production of images to be regular and proportionate to the dimensions of the original object. Taller trees cast longer shadows than shorter ones, whatever the position of the light source. If you know the height of a control object and the length of its shadow and at the same time measure the shadow of a second object, you can easily calculate the height of that second object. The color and clarity of an object reflected in water or polished silver may be less bright than the object is when perceived by direct viewing, but the qualities in the reflection stand in

[56] Because the more abstract an account is, the more intellectually sophisticated the audience must be. We will see in a moment, however, that the philosopher's education sketched out in book VII does not move to ever greater abstractness but rather toward an ever more comprehensive concreteness.

determinate proportion to those of the original. Just as in the *Sophist*, good imaging is vitally related to preserving proportions.

That concerns the production of shadows and reflections in nature. How do human beings *apprehend* images? "Through vision" is not a satisfactory answer. There are many animals that can "see" something, but only a few can take that appearance as an image. Mirrors are used by psychologists and animal ethologists to show the difference. Primates and elephants are capable of using mirror images in ways that humans do; for instance, elephants will notice a mark researchers apply to their foreheads and rub at it. Cats, on the other hand, will claw at the "cat on the other side of the mirror" until they lose interest; they do not seem to grasp that what they see is themselves. So, beyond vision, human beings (and any other animals capable of perceiving images) must be able to see not just that something appears in the mirror but also that the appearance *images* an original thing. This awareness of the relation to an image is evidence that the mind is "in two places" or planes at the same time: in the plane of physical things and in the plane of images. One has in mind the original thing while looking upon, or through, the image. The image is, no doubt, a reduced or aspectual being in comparison to the original, but that does not mean that it has no being at all. In terms of the discussion of the *Sophist*, it is like the physical object a mixture of being and nonbeing—one presumes with a higher proportion of nonbeing than the original.

It is the proportionalities established in the divided line that emphasize (and also mirror or image) both the ontological and the psychological proportionalities in imaging. It is difficult to know precisely how to interpret the proportion used to divide the line, not only because no specific proportion is named but also because we do not even know which division is larger than the others. Socrates, as noted before, says that the line should be divided in a certain proportion, and then the two parts divided again in that same proportion. Interpretatively it seems to be important to know that some proportion holds, but it is not crucial to know it exactly. Socrates does not point out a mathematical consequence of the method of division he uses: that the interior two parts will be equal, *no matter what proportion is chosen*. Since mathematics played an important role in Plato's Academy, which was probably the world's leading center of mathematical research in the fourth century B.C.E., it seems unlikely that Plato would have been unaware of this. There is not merely a proportion between the two middle parts of the divided line but an equality, and that strongly suggests that there must be some essential equality of the things they stand for—the mathematical things and the physical things—with respect to both their being and their being known.[57]

[57]The next section will address what this equality might mean. In much of the older literature it is interpreted as something Plato must unfortunately have overlooked—"unfortunately" because the equality subverts the symbolic representation of increasing reality and clarity (of vision and understanding) as one moves from the visible to the intelligible realm. For a brief (chiefly negative) discussion of these claims, see Pomeroy 1971. Pomeroy believes that the division of the line is according to the Golden Section; that would intriguingly relate the length of the whole line to all the parts. If the proportion were golden, the whole length of the line would be to its longer part as the longer part is to the shorter; put arithmetically, if A and B are the lengths of the parts after the first division, with A larger than B, $(A+B)/A=A/B$. There is no specific evidence in the dialogue to justify the correctness of this specific interpretation, however.

Whatever the significance of the equality, Socrates emphasizes that definite proportions hold between the different parts of the line. Proportionality in one direction (for example, from thing to shadow of the thing) implies proportionality in the other direction (from shadow of the thing to thing—the inverse proportion). Moreover, Socrates points out that physical things can be represented by mathematicals (physical things have a geometrical shape, for example), and that the mathematicals can be represented by things and by drawings (like Socrates' tracings in the sand in the *Meno*). The proportions in the line stand for the fact that one can move from the visible to the intelligible parts of the line and back again with some sureness. Thus one can think of a physical thing and draw a representation–image of its physical cross section (from the third to the fourth section of the line)[58]; one can use the drawn figure as a representative of a mathematical figure and, from one's knowledge of mathematical relations, arrive at conclusions about the drawing (from the fourth section to the second and back again), and then apply the conclusions from the drawing to the physical object (from the fourth to the third). Moreover, when Socrates explains the correlations in the line he also explicitly remarks that one will use the mathematicals as representing what is on the first section of the line, ideas:

> Consider also how the intelligible section should be cut….[I]n one part of it a soul, using as images the things that were previously imitated, is compelled to investigate on the basis of hypotheses and makes its way not to a beginning but to an end; while in the other part it makes its way to a beginning that is free from hypotheses; starting out from hypothesis and without the images used in the other part,[59] by means of ideas themselves it makes its inquiry through them. (510B)

The line, far from being static, is thus to be interpreted dynamically, insofar as the mind moves from the things of one part of the line to the others and back. The power of *eikāsia*, image–perception, is most fully itself not when, like a cat or a very young infant, we stare at the thing in a mirror unaware that it is us, but when we see the image and at the same time see through or by means of it the thing it images. As I remarked earlier, this means that our mind is in two places, on two levels, at the same time. Of course Socrates himself suggests through the allegory of the cave that most people take things simply at face value: the slaves chained to their seats take images as fully real. Nevertheless, at least some of them, and perhaps all, possess a capacity for seeing the relationship of the images to their causes. But first they must be released from their chains and turn around. Those who never escape the cave may nevertheless have at least a nagging suspicion that what shows itself on the wall is not all there is. Those who have seen the process of image production but have not left the cave will likely either join the image makers, or become political radicals who try to overthrow the image makers, or turn cynical. Those who have left the cave and returned will know that there are several levels of real being beyond that of the images—but by the same token they may be less successful denizens of the cave.

---

[58] The ideas are the first section, the mathematicals the second, physical objects the third, images the fourth.

[59] Whether this qualification implies that at the highest level reasoning transcends all imaging will be discussed below.

4.9 Singing and Hearing the *logos*

The allegory of the cave is a complex image, a scenario–image, implementing the scheme of the mathematical image of the line, which itself embodies a dynamic interpretative scenario. The cave is also, more expressly, a scenario–image of the (political) life of a city. One question, not unimportant but that we cannot address here, is whether it is an image of all cities, of all cities except the ideal city, or only of this ideal city, the one designed in speech. What at any rate becomes evident from the cascade of images of books VI and VII is that the images closer to the good are simpler yet more generative than those further from the good. Any thing that images another, whether the imaging moves closer to or further away from the good, images in the "matter" of its appropriate level. Geometrical figures imaging physical things are adapted to (Euclidean) two- or three-dimensional space rather than to the physics of earth, air, fire, and water (not to mention hydrogen, oxygen, lithium, and four fundamental forces). The adaptation to the different material substrates produces certain material differences (and distortions). The overall narrative itself—the narrative cascade of images of the good—makes the point that each earlier image elaborates itself in the different element of the image that follows. In general it looks as though this process is attended by a progressive adaptation and systematization that is at least as much about the articulations of the previous image that are possible in the substrate of the imaging matter as it is about the things being imaged. The line does not have any explicit politics, but the cave does, and in fact it is an image articulated in order to reflect the proportions of the line as they might be expressed in the political realm—even though caves are no more political than lines.

If every metaphor quickly starts to limp, the same is true for images and analogies. At some point they begin to fail to convey information about the original or, worse, begin to distort it. We are back at the problem of the icon and the simulacrum. Just as at the end of the *Sophist* it appears that the gods themselves make images, not always perfectly iconic, it now begins to dawn in the *Republic* that the good itself images itself in ways that are only partially iconic: some proportions hold in the process, but others do not. Every icon is a simulacrum in certain respects, and every simulacrum has the appearance of being, in some respect, an icon. Wherever there is appearance, the work of making similarities and distinctions in and through images never ends.

This is not, however, to fall into the usual interpretation of the divided line, the allegory of the cave, and the rest of the cascade of images, as though they prove the inferiority of the material realm to the spiritual realm or the nonbeing or minimal being of matter and, worse, of images. The fact is that there are ontological and cognitive relationships between all parts of the line, thus between all parts of "reality" and all cognitive powers. Certainly *part* of this is expressed in an old standby of Platonism and neo-Platonism: everything that is gets its being ultimately from the most intelligible of things, the ideas. Although both political life and geological caves seem far from the good itself, the goodness, being, and value they have comes to them, invests them, insofar as they reflect or participate in the forming power of the intelligible. Through the dimness of the realms of the changeable and the illusory—that is, the realms of physical things and their images—one can discern the traces of the ideas.

Even this partial rehabilitation of matter and images is still too bound up with the desire to correct conventional interpretations of Plato, however. The cascade of images and the individual images themselves (even that of the cave, if taken in the right way) suggest something rather different and more radical. When Socrates disclaims the ability to explain the good directly yet is willing to take the path of image explication, the images he uses to justify this alternative way present the good as productive: it is like parents whose children resemble them, like financial principal put to work in order to earn interest. He expressly compares the good to the sun because the sun both gives rise to things (in their being) and illuminates them (to let them be revealed and known). The good, he says, is beyond being, which makes sense if it is what gives rise to all beings but cannot be properly understood simply as one among them. It may mean further that the good is responsible for everything that is possible as well as everything that is. One could argue that this is a near corollary of being productive, since one would not expect that what is productive produces *everything* that *might* be produced. Leibniz, for example, says that God creates the best possible world, not every possible world, though each possible world is conceivable precisely as *possibly* real. In this sense, the source of the good, or rather of the best, would be much richer than, thus beyond, being.

The process by which the good of the *Republic* is productive should be called *ontological imaging*. This makes a certain sense even in more conventional interpretations of Plato. But unlike them, the *Republic* goes a step back beyond the forms, or rather tries to conceive, through the process of ontological imaging, how the good can be the unifying source of everything real and possible. Socrates mentions that he has talked before about the idea of the good, and since the idea of the good would appear simply to be one of the forms, thus on the first part of the divided line, we are strongly tempted to understand it that way. In a limited sense that is right—the good as the idea of all ideas, so to speak—but that would also be to forget that the entire line and its interrelations, and not just its first part or the extreme of the first part, images the good (or, rather, images the sun, which is itself already an image of the good). The good produces images "all the way down" (to speak according to an up–down orientation). The cosmos is and works as an imaging machine. Thus to take images in Plato's understanding of them as ontologically deficient is not only not justified by the *Republic*, it would contradict, deeply and fundamentally, what the narrative shows and says, what it images in speech.

There is one last point I wish to make to illustrate this claim and thus to revise and deepen our understanding of the proper relationship between Platonic imaging and intellection. The point arises from examining the curriculum for educating the philosopher, which Socrates draws up in the second half of book VII, after he has explained the allegory of the cave and the image cascade of books VI and VII. The philosopher candidates will first study number; we can call it arithmetic, but we should probably think of it more as number theory than as learning how to add, subtract, multiply, divide, etc. (though doubtless the study does not exclude these things). Second comes plane geometry, the science of two-dimensional figures.

Next comes stereometry (three-dimensional or solid geometry)—though for a few moments Socrates and his friends make the misstep of leaping over it, directly from plane geometry to the fourth study, astronomy. Astronomy is the study of the disposition and motion of all the bodies in the physical universe, the study of the good ordering of things in the *cosmos*. The fifth study is one that perhaps seems to us oddest or most out of place: *harmonia*, harmony. In a sense it does break the series, or rather point to the fact that from numbers and the two geometries one can proceed in different directions regarding motion. Socrates says that

> motion presents itself not in one form but several, as I suppose. Perhaps whoever is wise will be able to tell them all, but those that are evident even to us are two....In addition to astronomy...there is its antistrophe....It is probable...that as the eyes are fixed on astronomy, so the ears are fixed on harmonic movement, and these two kinds of knowledge are in a way akin, as the Pythagoreans say and we, Glaucon, agree....[Those who study harmony in the proper way] do the same thing the astronomers do. They seek the numbers in these heard accords and don't rise to problems, to the consideration of which numbers are concordant and which not, and why in each case....And I suppose...that if the inquiry into all the things we have gone through arrives at their community and relationship with one another, and draws conclusions as to how they are akin to one another, then the concern with them contributes something to what we want, and is not a labor without profit, but otherwise it is. (530C–531D)

From one perspective it looks as though Socrates might easily have introduced harmony immediately after arithmetic, as harmony is about the numbers governing accords. If initially skipping over solid geometry was a misstep in the middle of the curriculum, perhaps harmony should have come earlier in a natural progression. Yet from another perspective the introduction of harmony after astronomy is justified because harmony, as the harmony of the cosmos, presupposes spatial motion, which in turn presupposes the geometry of space; and of course the numbers discovered in harmony must first be studied in the arithmetic of units and measures. In that sense, harmony integrates all four disciplines that precede it. After harmony comes the final stage of the philosopher's education, dialectic. Each of the previous five disciplines is assumed to take two years; dialectic will take five. After it is done those who have completed the curriculum, aged 35, will be "called back into the cave" to do service work for the city for 15 years, until the age of 50.

The philosopher candidates will learn *true* dialectic, something different from the argumentative dialectic of the sophists. In describing it Socrates continues to use the imagery of poetry and song that he began when he referred to harmony as the antistrophe of astronomy. Glaucon remarks that determining the work of the preceding five disciplines and their community and relationship is a very big job. Socrates responds:

> "Do you mean the prelude or what?" I said. "Or don't we know that all of this [the five preceding disciplines] is a prelude to the song itself which must be learned? For surely it's not your opinion that the men who are clever at these things are dialecticians."
>
> "No, by Zeus," he said, "with the exception of a very few whom I have encountered."
>
> "But," I said, "was it ever your opinion that men who are unable to give an account and receive one will ever know anything of what we say they must know?"
>
> "To this question too," he said, "the answer is no."

> "Glaucon," I said, "isn't this at last the song itself that dialectic performs? It is in the realm of the intelligible, but it is imitated by the power of sight. We said that sight at last tries to look at the animals themselves and at stars themselves and then finally at the sun itself. So, also, when a man tries by discussion—by means of argument without the use of any of the senses—to attain to each thing itself that *is* and doesn't give up before he grasps by intellection itself that which is good itself, he comes to the very end of the intelligible realm just as that other man was then at the end of the visible." (531D–532B)

A curious thing about this passage is the sudden shift in its last paragraph from the imagery of music and song back to the imagery of sight. Perhaps one should respond with an intellectual shoulder shrug: even before Socrates and Plato there was a strong tendency to privilege vision as the best sensory analogue to knowing, a tendency present even in the origin of many of the Greek words for knowing. But this shift makes all the more surprising the evocation through harmony of the need to accommodate the sense of hearing. Can one analogically hear, perhaps even feel, the presence of the good and its working? Is hearing closer to dialectic than is vision? Are the dialectical accounts that the true philosophers give a kind of song? Earlier, in book IV, singing was used (in explaining the virtue moderation) as the image–model of the complex unity of the city, where each performs his part in harmony with everyone else. Extending that image analogically suggests that the good itself is what unifies and harmonizes—by means of multileveled, ontological imaging—everything in the cosmos, not excluding human beings per se or even the dialectical human beings who recognize and express this unity. The problem with the visual model of knowing is that it puts the seer at a distance from the seen object; the advantage of the musical model is that it presents the dialectical knower himself/herself/itself as part of the harmonization. A singer in a choral group has a part to sing, but she is a better singer of her own part the more clearly she hears the other voices and grasps the harmonization. This represents an engaged participation rather than the isolation of an envisioned object over against the envisioning subject.

Whether or not Plato's or Socrates' conception of seeing is engaged or distant is certainly arguable. For the ancient Greeks, the analogy of seeing to knowing did not necessarily imply a subject–object *dichotomy*. There is no isolation of the subject or the object, nor any implicit postulation of a view from nowhere, the kind of "objectivity" that puts the subject in a nonperspectival state of knowing. The Greek viewer shares the same place as other viewers in the world of the object. Moreover, the Greek visual model by no means excludes the model of hearing. Heraclitus' appeal to the *logos* as what humans hold in common, for instance, privileges the modality of *hearing* meaning or *hearing–and–speaking* meaning rather than seeing it. As we shall see momentarily, however, Socrates implies in this discussion that there is a kind of mismatch between intellectual seeing and saying, and raises the question whether seeing through the medium of an image—even a *logos*–image—is, in this special case, less desirable than seeing it direct.

In the book VII exchange with Glaucon we have quoted from, Socrates affirms that the final stage of the philosopher's journey is the true dialectic rather than the dialectic of the sophist. All the preceding effort expended in the journey he calls the "activity of the arts," beginning with "the release from the bonds and the turning around from the shadows to the image–stereotypes [*eidōla*][60] and the light," through the climb out of the cave, to the struggle there to look at everything from the "divine appearances [*phantasmata theia*] in water and at shadows of beings and not merely, as before, the shadows of image–stereotypes [*eidōlōn*] cast by a light that, when judged in comparison with the sun, also is like a shadow." All this effort "has the power to release and leads what is best in the soul up to the contemplation of what is best in the things that *are*, just as previously what is clearest in the body was led to the contemplation of what is brightest in the region of the bodily and the visible" (532C–D). When Glaucon asks for a fuller account of the character of dialectic and its forms, Socrates responds with these words:

> "You will no longer be able to follow, my dear Glaucon," I said, "although there wouldn't be any lack of eagerness on my part. But you would no longer be seeing an image of what we are saying, but rather the truth itself, at least as it looks to me. Whether it is really so or not can no longer be properly insisted on. But that there is some such thing to see must be insisted on. Isn't it so?"
>
> "Of course."
>
> "And, also, that the power of dialectic alone could reveal it to a man experienced in the things we just went through, while it is in no other way possible?"
>
> "Yes," he said, "it's proper to insist on that too." (533A)

Socrates reaffirms the benefit to be drawn by going through all the stages of this ascent that aims to "grasp with respect to everything—about each several thing itself—what each is," and he notes that, in comparison with what this dialectic aims at, everything below can no longer be called knowledge but rather an intermediate between opinion and knowledge (533D).

This is an astonishing result: Mathematics and even the knowledge of forms—the intelligible things of the divided line—seem to be reduced to kinds of image, and knowledge of them to kinds of higher opinion. The distinction between knowing and opining, grasping intellectually and presenting in an image, is thereby preserved at this highest, dialectical level at the price of introducing uncertainty into all the knowledge represented on the divided line, which is turned into a mixture of opinion and knowing. And this is because all that preceding opinion–knowledge is considered as an image, an image of the good. So on the verge of the last stage of ascent, maintaining the distinction between image and original relativizes everything else. Dialectic aims to see the truth rather than seeing images—or so it *seems*!

The logic is impeccable insofar as it is governed by the logic of the proportional participation holding between the different realms or planes of experience. Dialectic

[60] The context makes it clear that Socrates has in mind the stereotypes that the image makers of the cave hold in front of the projector light to cast shadows on the wall. Since we have already traversed the path from the cave to outside the cave, we understand that the image makers' *eidōla* are shaped in imitation of things outside the cave, though not necessarily as accurate representations; they are little ideas, little *eidē*, used to project shadows. Some of the shadows may well maintain proportions not just to the stereotypes/*eidōla* but also to the originals, so they may be either icons or simulacra, in the *Sophist*'s terms. My choice of "imaging stereotypes" for the stereotypes/*eidōla* is intended to hold open these possibilities and to avoid the derealizing connotations of other translations.

is a song (532A) and a journey (532B, 533C) that opens and explores a new realm. At the beginning of the journey the realm can scarcely be glimpsed, but at the end it turns out to have its own topology. We are reminded, perhaps, of how in the *Sophist* the principles of dichotomous distinction broke down when the questioning turned to the interparticipation of being, nonbeing, sameness, and difference. Whether we can maintain the dichotomy between seeing images and seeing realities at the highest of highest levels, the good itself, becomes problematic as well.

To pursue this much further would require further amendments to the interpretation of the divided line—how significant it is hard to tell.[61] Perhaps it is not important to insist on a more accurate description of what they have already considered, says Socrates. "Then it will be acceptable," he says,

> just as before, to call the first part [of the line] knowledge, the second thought, the third trust, and the fourth image–perception; and the latter two taken together, opinion, and the former two, intellection. And opinion has to do with coming into being and intellection with being; and as being is to coming into being, so is intellection to opinion; and as intellection is to opinion, so is knowledge to trust and thought to image–perception. But as for the proportion between the things over which these are set and the division into two parts of each—the opinable and the intelligible—let's let that go, Glaucon, so as not to run afoul of arguments many times longer than those that have been gone through. (533E–534A)

The man "who grasps the reason [*logos*, the proportional reason] for the being of each thing" will be called dialectical. But to the extent that he is not able to give an account of something, to himself or another, he will be denied "intelligence with respect to it." The oddity of this conclusion is that it contravenes—or oddly confirms!—what they have just said (at the end of the block quote immediately above). The man who grasps the true proportion of each thing is the truly dialectical person; but they have just agreed to abandon further research into the proportion that governs the line and so have abandoned the way of true dialectic on the very threshold of the realm of the good, which only the true dialectician knows.

Socrates concludes by bringing up the *logos*–account of the good itself.[62] He asks whether we can deny that someone "knows the good itself, or any other good," who has the ability "to separate out the idea of the good from all other things and distinguish it in the argument," testing it with regard to being rather than opinion. "And if he somehow lays hold of some image–stereotype [*eidōlou*] of it, you will say that he does so by opinion and not knowledge, and that, taken in by dreams and slumbering out his present life, before waking up here he goes to Hades and falls finally asleep there?" (534B–C). After quickly affirming the truth of this, they immediately return to the changes they will make in their city in speech in order to implement the philosophical education, from arithmetic to dialectic.

---

[61] In the first instance we might have to make the good itself a new "level," rather than the source and motive power, of the line. But would that not amount to destroying the very logic of the image–cascade of the good?

[62] He does not, however, bring up beauty, although he set it in parallel to the good at 531C. That evocation and its immediate suppression are, perhaps, a reminder that there are questions of the nature of appearance that are being left tacit.

This long summary of the philosophical curriculum *seems* to settle the issue of its goal, and of the goal of philosophy pure and simple. When all is said and done—but that moment never comes in the discussion!—knowledge must displace opinion, and the recognition of ideas in themselves will displace images that are surrogates for understanding. But even though Socrates and his interlocutors go no further, there are still two questions to pose, lest we overinterpret ourselves into too strong an anti-image rationalism. The first is whether, after all the concessions, the distinction between knowledge and opinion can stand at the highest level when it has been relativized at all others. The second, which begins the response to the first, is whether Socrates' uncertain description of a situation where image and opinion appear to be overcome, even transcended, does not hint at a truer state of affairs.

4.10 Forming an Equable Icon of the Cosmos

In commencing the discussion of the good in books VI and VII Socrates denied that he could give a direct account of it, and in the course of the argument he progressively narrowed what can count as knowledge and expanded what counts as opinion based on making images. Over and over in the dialogue the participants declare themselves satisfied with results they arrive at, only to decide later—sometimes almost immediately—that there is a better account of things. What seemed to be true absolutely comes to be relativized.

Thus it is hard to know what to make of Socrates' words at 533A, where he says that in the best account of dialectic and its forms we would see "truth itself." The tradition of rationalist idealism would read the words as clearly pointing to the purely intelligible idea of the forms, or the form of the good. Yet the concluding passage may have remarkably different consequences if we take it literally. Dialectic is contrasted with all kinds of knowledge discussed earlier; they are all mixtures of knowledge and opinion. When Socrates tells Glaucon that they must put aside worrying about the proportion to be used for dividing the intelligible from the opinable, he is acknowledging that they cannot resolve the matter with the tools, visual images, and *logos*–images they have used to this point. Still, one needs to emphasize that, if there is any (nonzero) proportion at all, the ratio of opinion to knowledge will be exactly representable in a geometric image. Opinion is, by its very nature, proportionate to knowledge, thus it is an image of knowledge and shares in its being. Moreover, one cannot avoid looking at things from the opposite perspective: the good itself deigns to image itself in fields other than the field of ideas; thus it images itself even in opinion, *doxa*, in the "it appears to me" of *dokei moi*. So there must thus be some iconic and knowable good in *doxa*, and not just a simulacrum of good that is "merely" opinable. Perhaps Socrates' dismissal of the image–stereotype that some people make for themselves and others, and are satisfied with, is not a dismissal of image making per se, but rather of the kind of image making that distorts things so that they look right only from the very limited perspective of cave denizens.

The traditional image of rationality places us on the level of the object of reason in an unmediated, face-to-face view. That this conception of truth is itself a metaphor based on an image–scenario does not appear to trouble easy rationalisms. Socrates' accounts of rising to the ultimate knowledge always maintain a distance between idea and gaze, and the gazer never takes up permanent residence among ideas but quickly descends from them. Nor do easy rationalisms reckon with the distance that Socrates (and presumably Plato) sets between transcendent vision and subsequent accounts of that vision. We need to be constantly reminded that Socrates treats accounts, *logoi*, as scenario–*images*. That must not be dismissed, by a presumably more logically-knowing generation, as a quirk of the Platonic/Socratic conception that has been subsequently corrected by our more rigorous approaches. For Socrates the paradigm of a reliable *logos*–account is one that stands up in the face of all the scenarios of the real and the possible, that is, mindful (mind–full) of the kinds of things the account is supposedly about. The account without the network of the matters to which it applies would be vacuous; without some person's witnessing–and–accounting, there would be no seeing or true account.

Socrates prides himself on a very special knowledge in particular, his knowing that he does not know. He typically invokes this claim not before but after—especially after—having gone through careful argumentation about whatever has been the subject matter of a dialogue. No matter how thoroughly one has considered things, no matter how far one's vision has stretched, there are always more tests to be applied, more situations to be distinguished, new questions to bring to the account. Nor should it be surprising that a frequent practice in the dialogues is reminding (re–minding) oneself of what one has already gone through. The entire *Republic*, we pointed out earlier, is a (presumably) exact recital of what happened on the previous day. The object of the repetition is not to recall the words but to regain the state of mind and seeing that one achieved previously, with the prospect that one can then understand better, taking yesterday's engaged account as part of today's object of consideration to produce a further, deeper, and ampler engagement.[63]

With the sequence of the image cascade of books VI and VII and the partition of the divided line, Plato's Socrates gives an articulated account that is acutely sensitive to the multiplicity of standpoints of human practice and cognition. It also, not coincidentally, constitutes the earliest extant elaboration of the phenomenon of imaging between multiple fields or planes and the corresponding mental mobility in and among them that I described in Chap. 3 as a basic characteristic of imaging and image perceiving. This field– or plane–awareness is *elementally* and *fundamentally* present in Socrates' explanation of *eikāsia*. It is the ability to take the phenomena of

[63] This approach is not subject to the objection that it fetishizes knowledge as presence. It is quite the opposite: a recognition that we cannot know beyond what we have engaged (and not just seen), that what we bring to mind escapes and becomes *impresent* almost as quickly as it appears, and that however deep our engagement with things, there are always further aspects to be considered, some very remote, some near but inapparent because they have been beneath notice. When considered Platonically rather than Platonistically, "presence" is not total illumination but chiaroscuro, light–dark, with atmospheric perspective (the blurring of things at a distance) and an uncertain horizon.

shadows and reflections as a natural group, to understand that they have special (physical and mathematical) features because of the kind of phenomena they are (e.g., reflections by means of light in water or in mirrors), and by the same token to see through them to something else they put us in mind of on another level of being (e.g., a reflection in the mirror while I am shaving puts me in mind of my children—they are playing in the bathtub behind me). This pattern of imaging and image taking also holds beyond the "lowest," material levels of the world. We learn from the divided line that we can look at the sand at our feet (on the third, physical level) and take tracings made in it as drawings (presumably on the fourth level of images) in order to think about a square (on the second, mathematical level); and, holding this entire mobile practice of imaging and image taking in mind, we can see it as representing our ability to recognize unifying intelligibilities (on the first, ideal level) manifested in exemplars of any type.[64] Perhaps we can view a painting of an event in a political revolution as a representation of a new standard of justice (arguably the painting is on the fourth or icon level, the event takes place on the third, physical level, although it is present now only in memory, and the concept of justice in view might be on the first level, that of the forms/ideas).

Even at the moment when he makes his final attempt at a radical distinction between knowledge and image- or stereotype-making opinion, Socrates' words offer an alternative to this dichotomy. The problem with the dichotomy is twofold. (1) *Eidōla* cobbled together by lazy dreamers are unlikely to be well proportioned to the things outside the cave. Of course, as I have argued already apropos of the *Sophist*, no image can be proportional in every respect to the original it images, unless the image *is* the original. One does not, however, evade the question of proportion by contrasting seeing *eidōla* to purely rational viewing. It is precisely in offering the possibility of such a rational view that Socrates refers explicitly to the *idea*, that is, the typical look, of the good. So instead of a simple gaze holding a simple object with no intermediates, what we see is the *look* of the good, not the good itself; moreover, in order to claim knowledge we have to give an account (*logos*) of it. That places two things between our gaze and the object! And that is not even to reiterate that, in the dialogues, accounts are images. The problem of iconic appearance versus simulacral appearance once again cannot be avoided, and precisely at the moment when it seems to be overcome.

(2) If there is a difference between the thing and its look, and also between the look and the account we give of it, the philosopher's dialectical work is not done, although for purposes of this-or-that Platonic dialogue it may be over. In the *Meno* Socrates used the geometry lesson with Meno's slave and the attempt to define virtue to assert one of the few things he claims to know: that we will be better for continuing to search whenever the search has not reached the truth. In the dialogues recounting his trial, imprisonment, and death, the *Apology*, the *Crito*, and the *Phaedo*, a constant theme is that (the threat of) death will not deter him from inquiry and from questioning his fellows. In fact in the *Apology* he portrays his vision of the

[64] This is a fairly precise description of what happens in the *Meno*.

afterlife as a place where he will ask the others there all the same questions he posed in Athens! In the *Phaedo*, the dialogue that quite literally ends with his death, he continues asking about the possible immortality of the soul to the last moment—but every time his friends think he has proved it, he immediately pokes a hole in the argument by showing that something has been left out of account or left uncertain. They are downcast at the repeated failures, but he maintains his equanimity throughout. He even, in the middle of the dialogue, encourages them to understand what his real legacy is: never to despair of trying to give accounts of things, never to give themselves over to misology, the hatred of *logoi*, the hatred of making word–icons, word–images, of things. If he is their role model, then the only way to be faithful to him is to keep inquiry and imaging–accounts going until they are satisfactory in every respect—which may be never.

In this perspective, we need to remark that in book VII Socrates portrays those who hold on to an *eidōlon*, an image–stereotype of the good as lazy: they rest content with it, it puts them to sleep. He contrasts this behavior with the vigorous activity of the person who uses intelligence: he separates out the idea of the good from all other things and distinguishes it in the argument or account, he goes through every test of being and "comes through all this with the argument still on its feet" (534C). In this energetic fashion one arrives at an account that maintains, as far as one is able, a right proportion to the things that need accounting for. One is, in fact, trying to produce an icon, not a simulacrum, of the idea of the good, a perfect image with no difference in proportion from the original. But as we have argued, this goal appears to be, strictly speaking, impossible. Perhaps tomorrow the inquirer will find something that does not match today's account or detect a false proportion in what seemed accurate yesterday. So the inquiry will continue, and if necessary begin all over again. There is no rest for the intelligent human being who wants the good. The presence of laziness, not the use of an image or *eidōlon*, is the source of the problem.

Indeed, there seems to be something fundamentally defective about translating into Plato what is really a more Aristotelian trope: that truth ultimately brings us to a finality that is a kind of rest. In Aristotle human excellence or virtue is activity, and the highest and noblest, the purest activity is in accordance with what is highest in us, our knowing power. The full realization of that power is portrayed, in the final book X of the *Nicomachean Ethics*, as the ultimate (human) happiness. Its activity is *theōria*, traditionally translated "contemplation." He describes it as the most godlike of activities, and says it is (like) thought thinking itself. All other human activities tend to wear us out, but the contemplation of accomplished truth, our knowledge of the ultimate cosmic things, is the most restful of all. In this conception the distance between thinker, thought, and object is annulled; we become, as far as is in our nature, precisely our thought. Our thinking becomes, as far as is possible, its object, and thus we become, as far as is possible, cosmic, godlike, and wholly one. In the long tradition of trying to reconcile the Platonic with the Aristotelian philosophies, the Platonic-Socratic conception of the culmination of human being has been assimilated to the dream of thought thinking itself. That has been to overlook that Plato's Socrates always maintains modest tact when it comes to what is ultimate, and that even when his words appear to say something like Aristotle does, his practice, as well as a more careful interpretation of his words, is quite different.

4.11 The Perfect Image of the Cosmos as the Goal of Dialectic

If the good itself naturally and necessarily gives rise to all realities and possibilities, then perhaps when we come close to it we both nearly touch it and at the same time move away from it emulatively. That is, in the most direct apprehension of the good we can muster, we would immediately begin imaging it ontologically, *in ourselves* (we would be transformed into something more like it by our viewing it) and *through ourselves* (we would begin producing further images of it in other media, places, and levels of being with which we communicate). To apprehend the good itself, to apprehend the beautiful itself, to apprehend truth itself without such effects would amount to a falsification of them. There is at such moments a distinction that our understanding must make, separating image from original, and yet it turns out to be a distinction that is more notional than real. The original is imaging, thus it is also imagelike. The original and the image tend to fuse. But this is a special fusion that is by no means confusion: the image does not become the original. The articulation of the cosmos into all its levels and aspects remains, and remains active. In this special fusion we do not become confused about what or where we are; rather, we see and engage in it in a way commensurate with how far and deeply we have seen, heard, and engaged.

As with almost everything that appears in the dialogues, one has to wonder whether what is said represents what Plato/Socrates thinks is true or whether it is said in order to play a specific role in the context of the dialogue. The answer, in general, is probably both. Of course then the key is to tease out the degree of each that seems most likely. Yet even if one does not accept such a principle of interpretation, the conclusion that the interlocutors draw from the cascade of imaging of books VI and VII can be extremely frustrating. Socrates seems constantly to be taking back—or at least qualifying—what he has previously said. For instance, in his summary of what dialectic accomplishes, we saw that he moves the boundaries, and thus the definition, of knowledge. Formerly it looked as though knowledge is what fell into the first, intelligible parts of the line. Now, anything that falls short of the ability to separate out *the good* from all other things (and to do this also in *logos*, in speech or argument) is not knowledge. Yet the whole movement, from the good itself in the middle of book VI to the end of book VII, is predicated on not really being able to lay hold of the good or to encounter it without intermediaries of any kind, but only to present it in speech– and scenario–*images*. At the point where we hope to cross over, to rise above the images, Socrates tells us that we cannot follow him—but also alludes again to the possibility that he cannot himself distinguish the image from the reality in an absolute sense, since he can tell us what the good is only insofar as it *appears* to him.[65]

At 534B, Socrates said that if someone knows how to separate out the good from all other things and appearances so that even we less accomplished thinkers can

[65] At 533A: "And, if I could, I would show you, no longer an image [icon] of and speaking about such things, but the very truth, as it seems [*phainetai*] to me."

distinguish it—even if that distinctness is not completely lucid and we thus have to call it opinion rather than knowledge, at least in some respects—we can scarcely deny that this person "knows the good itself." If the person lays hold of an image of it and *rests content* with it, that image is probably a simulacrum, an appearance that approximates the original but importantly deviates from it, and that deviates ever more from the original insofar as the person keeps dreaming the same dream, preserving the same, unvarying image, repeating the same account, the same formulas unchanged and without further testing. If the image holds up on the first try, then again on a second, a third, and a fourth, and if it is never allowed to become a fixed idea and a dead letter, it is more an icon of the good than a simulacrum. Part of its iconicity is that the very process of formation makes us better people—makes us good to a higher degree, and thus makes us be images or ideas of the good—because it engages us more fully with the amplitude of everything that is owed to the good.[66] Truth is no disengaged viewing!

This helps make clearer that we human beings can never be sure whether someone who gives us a good-sounding account has taken the final step to the ultimately knowable. We cannot even be sure, however carefully we have traced the emergence and limits of our own accounts, that we ourselves have arrived at an image for eternity. It is possible that *any* dialectical account we or Socrates or Plato can give of the good is to some degree a simulacrum that we individually dream up or an image–stereotype that the image makers of our community have instituted and constantly reinforce.[67] But we ourselves are images of the good. If we have learned our lesson, if we have through our pursuit of the good become more like it, disappointment in the limited accuracy of our previous examinations of a question will spur us to try again. If we are lucky we will today see more clearly the images that are nearest to us in the light of those that are more remote. Our "viewing" is an engagement in depth that does not necessarily make us a miniature of the cosmos but that preserves distances precisely insofar as we are engaged with the cosmos in the amplest ways that we can muster—ways that are nevertheless our own, because they fundamentally constitute us in the likeness of the good. That means that, insofar as we use the image of viewing, gazing, or regarding to understand this, we have to keep in mind that all the dimensions of depth, all the tiers on which images present themselves, are part or aspects of the proper "object" of intelligence. Our best way of being and knowing must be properly "placed." Perhaps that is the best that we of the human kind can attain. If it does not make us like unto gods, it is not nothing, and it would be a real accomplishment to achieve it. And, not the least important matter in Plato's account of ontological imaging, it would be an accomplishment that maintained proper proportions between us and the world.

Later culture has made the account of books VI and VII of the *Republic* a paradigm of Platonist thought, because it is said to show that all reality is derivative

---

[66] This combines themes from the *Republic*, the *Sophist*, the *Phaedo*, the *Meno*, and the *Symposium*.

[67] I am alluding to Cornelius Castoriadis's notion of the socially instituted imaginary; see Castoriadis 1994.

from the Forms and thus the reality of the material realm is inferior to that of the ideal. But if, for just a moment, we attend to the prevailing valence of the cascade and ascent of images of the good, we can argue another possibility that the Platonic conceptual topology offers. The good by its very nature as good shows itself in everything else; as such it is behind all the being and relations (including unrealized possibilities) that are established in the different realms of being. Appearance is a direct result of the kind of being that the good is, or has, or lends, or bestows—whatever the best term might be, pending future investigations. Being and appearance would then be a correlative mutuality, not a dichotomy. Or, rather, being with its unrealized as well as realized possibilities is what is correlative with the image-making power of the good on every level of the divided line. The correlative human powers of apprehending being–and–possibility on each level would themselves be images of the imaging power of the good. To drive this line of interpretation to its logical extreme: we human beings are what we are precisely insofar as we are image makers, in every human way of imaging, both active and receptive. To put it simply: the good is itself, and we are more perfect images of the good insofar as we make images of the imaging power of the good. The good itself is self-transcending in images, and we, when we follow it most closely, transcend ourselves likewise.

Let us take a more modest tack. On the divided line, appearance takes place not just in the segments of the visible but everywhere along the line. If in the allegory of the cave we start with the denizens of a dimly lit world who take physically produced shadows as the only (and thus the ultimate) realities, we discover, by following the path Plato's Socrates indicates as leading out of the cave, that the shadows on the wall derive their stability and knowability from something/somewhere else by way of imaging, and that those things in turn derive their stability and knowability from a yet higher or deeper realm, etc., until we finally "reach" the good itself. This is the transcendence in/of everyday life. Like *logos* in Heraclitus, it belongs to everyone, though few recognize it. The ontological imaging that derives from the good itself is not a derogation or loss of being: it is an expression of the dynamic structure of the entirety of being. This dynamic structure as a whole, in the dynamism of the sequence of images from sun to cave, is itself an image of the good. To try to see and understand the good, it is not permissible simply to transcend the visible or escape the cave. The visible realms and the levels of existence in the cave are themselves expressions of the good, and without them one would have a truncated—that is, *false*—conception of the good. Producing the full amplitude and range of appearances is as much the task of the good as is producing the full range of beings. In brief, appearance itself is a way of being.

Imaging is everywhere, it is ontological and not just epistemological or gnoseological. It is the sun that is the source of both being and knowing; each segment of the divided line corresponds by proportion to each other part of the line; any lower level in the cave allegory is an image of the next higher one. Consequently, each thing can be interpreted in light of the others, the lower in terms of the higher, the higher in terms of the lower. Each thing images others, "above" and "below." The ontological character of imaging—that beings as beings precisely image other beings, both "higher" and "lower" kinds[68]—supports the human cognitive ability to move in different directions of knowing and seeing.

Of course one can easily argue that all neo-Platonists recognized this, that it is precisely a development of this theme that led to neo-Platonist theories of emanations and hypostases. But of course the neo-Platonists understood the process as a descent toward nonbeing, and it was the destiny of the human being to reverse this as much as possible by reascending toward the unnameable unity. They certainly did not simply misunderstand Plato. They accepted the conceptual topology of different levels of being he had instituted. Each of these levels defines a place of being, and the whole constituted by all of them together in their dependence on the good establishes the total space of being and possibility. This is, in a fairly literal sense of the word, Plato's topology of the good and being. Yet the neo-Platonist interpretation of this topology, in particular their sense of the proper orientation of the human being within this space, is different from what is found in Plato's dialogues. The neo-Platonists were too strongly inclined to ascribe positive value to ascent and negative value to descent, to understand the cosmic structure as implying hierarchy even more than processual unity, and that the best place for human beings to reside would be the highest possible. This does not accord with the preferences of the good itself, or even of Plato's Socrates. If the theme of the human being's need to ascend appears frequently in the dialogues, so too does that of descent, once one has reached the heights. The human task is to inhabit many levels and to see a way from each to the others, but ultimately to return to where we always start from.

The *political* interpretation of the analogy of the cave makes one wonder why philosophers who had achieved the vision of the Forms/Ideas would return to the cave to govern. It is sometimes argued that Plato's Socrates believes they must be forced to return. Whether it is even plausible that those who have grasped the cosmos as a whole could be forced to do anything contrary to what they judge to be good is doubtful. (Could Socrates be forced?) They will "return" only if that is what their comprehensive understanding tells them is right. They could be persuaded, not forced, by an argument mindful of the totality of obligations. If one goes a step further and recalls that in the first instance the *Republic* aims to understand justice in the human soul, and introduces the city as an image of the soul, the degree to which the allegory of the cave limps or even fails as a representation of soul becomes clearer. In real cities, and apparently also in the city of the cave, those who rule have an interest in obfuscating the real situation of the city insofar as the truth would threaten their rule, and although their fate is tied to those who toil unaware of the true relationships of power in the deepest depths of the cave, the toilers will never be allowed to share power or to determine the socially accepted images and names. The image makers—or rather stereotype– and simulacrum–makers—do not in the

[68] "Lower" and "higher" have to be used with discretion and even irony, since they are too easily interpreted nihilistically—e.g., "sense perception and imagining are as nothing with respect to intellection"—and since they apply literally only to the cave rather than to the image of the sun or of the divided line.

last analysis worry about the harmony and happiness of the city as a whole; what they grant to the enchained is merely instrumental and done for expediency's sake. If they had any interest in the outside of the cave at all, it would be for the sake of tools and knowledge that they could promptly use to assure their status.

When the allegory of the cave is taken more directly as an image of the human soul, however, the significance of the different levels is altered. The problem at the lowest level of the cave is that the unfortunates chained there do *not* exercise the power corresponding to the fourth part of the line, *eikāsia* or image–perception. That is, they take the images simply as realities, and not as simultaneously the images of some other reality. They watch the television set, and what they take to be true, and how they are supposed to take it, is rendered in high definition on the flat screen. Their own lives they see according to how they are portrayed in the comedies, dramas, and reality shows flashing in front of them. In this respect the line rather than the cave offers a better portrait of human psychological powers. It also more clearly points to the fact that, for the individual human being, it is best to have the human cognitive/experiential powers both work on their respective levels and interact with one another from level to level.

In learning how to activate our intellectual powers—and this certainly must be learned in a way that is quite different and more effortful than the exercise of vision and image–perception—we have to look *through*, and that implies also *beyond*, the topological level that corresponds to our everyday living with others in a social and physical world where we trust and use things, and where often enough we take as true the ways in which they are commonly represented. Our aim should not be to transcend ordinary living by learning to live the life of pure intellect, nor to immerse ourselves in the "real" world of ordinary, tangible life to the exclusion of recognizing and thinking about patterns and forms. Our souls' powers are tiered and interdependent. We need above all to live commensurate with the specific character and "density of diversity" that these powers lend to our experience.

One can find further evidence for this goal of being able to live in and among the different levels of being in other dialogues, in particular in the *Symposium*. As part of an after-dinner, speech-making game at the house of the playwright Agathon, Socrates gives as his contribution an account of conversations he had with the priestess Diotima, which included a magnificent speech about how we can ascend to beauty itself. At the end of Diotima's description of that ascent, it turns out that the person who accomplishes this does not stay forever gazing upon pure beauty. "Or haven't you remembered," she said, "that in that life alone, when he looks at beauty in the only way that beauty can be seen—only then will it become possible for him to give birth not to images of virtue (because he's in touch with no images), but to true virtue (because he is in touch with the true beauty)" (212A). The body's eye can see beautiful things, but a higher power is necessary to isolate and thus clearly see the beauty that appears in and through all beautiful things: and only once one has accomplished this can one live—here and now—in a way that gives rise to acts that are beautifully virtuous, and not just an imitation of what one saw in a drama, or in the historical example of a great man or woman, or in some political or psychological theory of virtues.

But doesn't this passage militate against the positivity of images and imaging? At the moment that Socrates' retelling turns to the ascent to beauty itself he gives a warning: he says that Diotima spoke at that moment "like a true sophist." Is this the real truth that even sophistry is capable of, or an indication that the account is distorted? Certainly the conclusion she gives is defective as it stands. Her express words deny the image character of what we will produce at the moment of turning back to the world, but she should say instead that by beholding beauty itself we will be transformed so that in our own lives we will give birth to actions through which that beauty will be visible to others. That is, we and our actions will, at the appropriate levels, become ontological images of true beauty. What counts at this level is less the images we make than the images we become. Our virtues will truly be virtues precisely because they will image forth the beauty of what is highest—where "highest" is relative to how far we have been able to go.[69]

In book IX of the *Republic* we find that the most terrible thing for the tyrant, who seems to possess all power, is that he cannot cultivate wisdom and virtue but must devote his whole life to plotting how to retain power once he has gotten it. That is, he is as much a slave of the cave as anyone else. There is also the example we mentioned previously, the myth of Er, the account of the afterlife in book X. A man raised decently in an orderly regime, choosing his next life not out of any understanding of the good but only according to the habits cultivated in his city, selects the life of a tyrant. This virtually guarantees that, for him, the ordinarily unending cycle of reincarnation comes to an end, because as a tyrant he will be hurled into the deepest recesses of Hades, there to be eternally punished. These cautionary examples make clear that the best destiny of the human being is to exercise a differential awareness of all the places in which the good images itself, in the real and the possible, and to acknowledge and inhabit all those places appropriately, each in proper, proportionate relation to the others.

Yet more than these, the portrayal of the philosophical curriculum in book VII clarifies how multiple kinds and levels of complex imaging lead one to a deeper understanding of the image ontology of the good itself. The progression, we recall, starts with arithmetic, and passing through plane geometry, solid geometry, astronomy, and harmony, finally arrives at dialectics, which rises to a grasp of things without hypothesis. We rise, apparently, to the point where the power of intellection directly apprehends what is—although we have seen many reasons to doubt that this vision can ever be totally unmediated by images and beautiful appearance.

This rise to direct intellection represents in the first instance a traditional conception of what human rationality aims for, the gradual removal of the taint of the material realm in favor of the immaterial universal—with the corollary that images and imagining will simply be left behind. I have already cast doubt on whether this can truly be the ultimate goal, both because the dialogues always picture a descent following the ascent and because the good itself can be adequately known only in relation to everything it accomplishes. "Everything that it accomplishes" means, of

[69] See note 62, above, on the tacit relationship between the good and beauty.

course, all of experience, all that is rational *and* perceivable, all that appears, plus the entirety of the unrealized possible. And even the ascent from the cave represents the highest level as a world with different perspectives, not as a realm of pure ideals and abstractions.

There is something fundamentally defective in the whole notion that the goal is philosophical abstraction, because it is contrary to the conceptual topology Plato presents. If we look carefully to the *progression* of the philosophical curriculum, we can see that it is not properly describable as ever more abstract, at least not without a very nuanced understanding of abstraction. Part of the problem here is that abstraction, a Latinate rendering of the Greek *aphairesis*, is a concept used by Aristotle rather than by Plato. As we shall see in the next chapter, even the meaning in Aristotle is rather different from what it became in later philosophical tradition.

The progression of the curriculum corresponds to an ever more complex and thoroughgoingly accurate imaging of things in relation to one another, to things as part of a cosmos. Interpreted according to the logic of the divided line, the first five disciplines of the philosophical curriculum are mathematical. They require one to look at physical things, and representations of physical things, in light of their mathematical being (that is, things of the third and fourth parts of the line are seen as exemplifications of second-level things and relationships). One begins simply with arithmetic, that is, what emerges from the enumerability of things. One then looks upon things in two-dimensionality, and then in three dimensions. At that stage, one is in a position to portray the geometrical outline and schemas of enumerable and delimitable things positioned in the totality of geometric space. With astronomy one then puts all of these things into motion, and thus studies regularities that are not revealed in a static view. In moving on to harmony, Socrates points out that these previous levels have developed the *visibility* of things, and by turning to the *hearability* of things through the study of harmony he suggests that the major division of the line into the *intelligible* and the *visible* was somewhat misleading. He further suggests that there might be other qualities that could be approached similarly.[70] Does he mean the other sensible qualities, like touch and taste and aroma? Since he places harmony on the same level as astronomy, it certainly might be possible to derive further sciences at that same level that would show how the enumerable things in motion not only make sound but also exhibit all other basic appearances. This means that what the philosophical curriculum achieves is an ever more complex, more rationally understood, more fully elaborated representation of the cosmos in all its multitiered appearances, both sensible and intelligible. It is as though one ascends not to an ever more vaporous or abstractly ideal vision, but to an ever more comprehensive, ever more articulated appearance that produces in mind a perfect image—or rather icon—of the *cosmos*, viewed not just in its intelligible aspects but its sensible ones as well, with all aspects presented together. Thus ultimate

[70] At 530C–D, just as he is about to present harmony as the second discipline (besides astronomy) that studies motion, he says that "motion presents itself not in one form but several, as I suppose. Perhaps whoever is wise will be able to tell them all, but those that are evident even to us are two."

intelligibility is correlated with the ultimate manifoldness of appearance, grasped in its generative connection with the good itself.

This is an appropriate place to recall that the proportion between the mathematical section of the line and the section of the real is an *equal* proportion. In a sense, the mathematical does not exceed the real at all. Perhaps it is as close to an iconic image of it as is (humanly) possible. The question to ask, then, is whether Plato or Plato's Socrates was really suggesting, when he arrived at dialectic, that suddenly the mind would turn away from all representation whatsoever, from all *concrete* intelligibility of the totality of things, and see only an abstract ideality. The evidence weighs to the contrary. Dialectic, if it succeeds, will allow us to throw away the crutch of *taking for granted* what is and appears. Recall that the good is beyond being. This need not imply pure abstractness, but rather that we rise to the level of grasping the good in its nearly limitless productivity. We thus would no longer need the real as a crutch because, through dialectic, we would see and understand the real as it is, as the manifestation of the part of the good that has been realized, and not one of a vast number of mere theoretical possibilities. That is, to grasp the good through dialectic is to see in some distinct way everything it *does* and *can* give rise to, while maintaining a careful distinction between the *does* and the *can*. And that would be to develop to the humanly ultimate degree the power of seeing the good imaged in everything that is and can be, and to be able in light of the good to see what is and what is possible according to the various aspects that the good nurtures and illuminates. From the perspective of the multiple sensitive and cognitive powers of the human being the philosopher could see from any level to any other. He, or she, could move toward the ideas and away from the ideas, all in accordance with the limitations of the particular access to things that, as a human being, he or she is given. And he or she could see through the many ideas (that is, through the basic looks of all kinds of things) the idea or look of the good, and thus have possession of something that is next to the good itself, so that the ideas themselves would be a consequence of the good and an imaging medium through which to move toward it and away from it. To paraphrase the concluding thoughts of Diotima's speech in the *Symposium*, that would be no mean life for a human being to live. We would be living not in accordance with mere images; we would *be* the very kind of imaging that truly is, the imaging of the good, in the most unified way possible for human beings.

4.12 Conclusion

We have reached a crucial point in our historical narrative of imagination, but it is not a *turning* point. It is a point of constitution: the constituting and instituting (to speak with Castoriadis) moment of the image and the imagination and of their network of problems. Simultaneously, and precisely as the moment of constitution of the imagination and its problems, it is the moment of constitution and institution of the problem of rationality. I say "problem" in a sense at least partially observant of etymology. A problem is something thrown down before us. From this moment in the history of Western thought, imagination and rationality in their topological relationship confront in tandem anyone who wishes to understand the mind and soul of the human being. Yet, apart from any other shortcomings in understanding according to etymology, the sense of being before us, of confronting us, suggests that the problem is separate or apart from us. If we are confronted with the question of rationality and imagination, of what they are about, of their relationship to one another, of what they can in the end accomplish (if anything at all), it is not as an *object* that we can take it up. It confronts us in an unusual sense, in that it accompanies us everywhere we are and look. We can "ignore" it in the sense of blocking it from mind, of refusing to consider it thematically, of dogmatically declaring it resolved, or blithely ignoring it—but that no more eliminates it than breathing without conscious awareness of air eliminates the atmosphere.

At this point in the inquiry, we face a major obstacle: the prejudice of the learned with respect to rationality—even the learnedness that calls itself postmodern. When rationality is discussed by the learned—philosophers, scientists, scholars—it is usually with a sense of pride and achievement. For centuries, even millennia, we have assumed that rationality is superior to all other human powers, that its nature is obvious to the learned, and that once reached it is easy and always assured. We (and not only we) look back to the ancient Greek philosophers and see the birth in the West of this ideal. If we do not still share their specific vision of rational life, we nevertheless continue to dream of escaping the cave that binds us to the all-too-human, of reaching a level and kind of knowledge that raises us up at least to our meta-caves. Perhaps at the meta-level reason equates with the canons of logic, and the irrational is whatever fails to appear with the preferred type of clarity. Or, if we exult in the dream that there is nothing to guide us but the socially accepted and constructed, we may hope that another dream will displace the world's nightmare.

At its Western founding, however, philosophy and its rationality are problematic rather than thetic, presented indirectly rather than thematically.[71] That is, rationality is less set and settled (as reality and as ideal) than it is adumbrated as an encompassing problem, and thus a problem that has to be negotiated by living in the fields of experience in accordance with a conceptual topology that we gradually work out by differentiation and distinction. If the history of ancient Greek thought shows a tendency to identify the knowable with the totality of being—whether in Heraclitus' *logos*, or Parmenides' positing of the unity of being and thinking/knowing, or the Stoics' later postulation that the cosmos is governed by a rational world–soul—there is also a tendency to exhibit a restricted interpretation of both knowing and being. This restrictive tendency has been stronger in the followers of the outstanding thinkers than in those thinkers themselves. As followers they have had less a *first-person* experience of that which their master thought and witnessed than a *second-hand* experience through his authoritative concepts and propositions. In following words they often turned the master's distinctions into differences and differences into dichotomies according to a dogmatic logic. The masters named and described their experiences and established conceptual topologies; the disciples

[71] I am not suggesting that any other conception of rationality (e.g., Eastern) is an alternative, nor even that one can easily or simply categorize thinking by region.

absolutized the topologies and lost sight of the field whose contours the topology was meant to track. In particular, with the absolutization of reason, "image" and "imagination" came to represent something with the taint of the material, the accidental, and the unintelligible, whereas the names "idea," "concept," and "rational understanding" conveyed the universality that transcends all particularity and all attachment to realms of constant change and transient feeling. The durable, even the eternal of which we can have no experience, became the proper object of the philosophical pursuit.[72]

From this rationalistic perspective Plato's divided line placed the ideas at the origin of things and made intellect the essential human power. The allegory of the cave was thought to show that true freedom and life lie beyond the boundaries and ties of ordinary human existence, and that the goal of the best human being is to escape from the bonds that hold him (rarely her) to the busy-ness of society and the inertia of nature. As much as possible this kind of philosopher wants to live in tropes that only intellect can know and thus resists all attempts to drag him back into the cave.

Things look quite different from the narrative we have traced. Despite the tradition of translating *eikāsia* as "imagination," it is clear that Plato does not intend a human power of "inwardizing" things deep in the privacy of the mind or soul, nor is he "psychologizing" the appearances of things according to the categories of memory, reproductive imagination, and productive imagination. Rather, what he intends is evident in something as elemental as the human ability to recognize that light casts shadows as it illuminates objects. We recognize the shadows, reflections, or images as cast by the illuminated objects, as the *icons of the objects*, and that is already *eikāsia*. Moreover, although there is a tendency in Plato's texts, and an even stronger tendency in the traditional interpretations of Plato, to see a radical separation between the realms of the intelligible and of the visible, the details greatly complicate this. Socrates introduces the image of the sun and its light in explication of the good itself, the image of the divided line as an explication of the image of the sun, the story and image of the cave as an explication of the line, and the philosophical curriculum as an attempt to reconstitute, reconceive, and articulate as a whole what happens in the cascade of these images. By establishing connections between the different segments of the line and by insisting that the parts of the line represent a strictly observed proportionality, Socrates strongly reinforces the sense that the being of the line's different levels image one another and that the human powers of perception allow us to recognize these relationships between levels.[73] Call these the

[72]One of the most bizarre testimonies of this was the decision of Plotinus to postulate in human psychology a second, rational imagination that duplicates the contents of ordinary imagination without any trace of materiality. The ascending philosopher thus leaves behind the taint of sense and matter and retains only their intelligible forms. See Sect. 6.1, below.

[73]In fact the whole presentation also represents itself: that is, Socrates uses metaphors, analogies, and images—different levels of representation—to represent the good *as representing itself* at different levels. Since the explanation–representation shares in the character of the very thing it is explaining–representing, it is technically a *symbol*, or, in the sense understood by Orthodox Christianity, an icon.

Republic's doctrines of, respectively, the ontology of the intrinsic imaging of the good (and being) and the corresponding epistemological psychology of perceiving each realm in relationship to others. Common interpretations of Plato may claim that images are the least real things in the cosmos, but the extended imaging power deriving from the good shows that the cosmos *holds itself together* by imaging each kind of thing in other realms, both higher and lower. The good by its very nature is a productively imaging power. The human power of apprehending one level of being against the background of others is the fundamental way that the mind works: by eikastic, imaginative perception in and between places of existence and possibility. Unfortunately, human beings have a tendency to flatten out this experience into a reduced dimension and to focus on images and things as though they were ontologically isolated from one another. And thereby they forge chains that keep them in the cave.

Thus ends the first chapter of the overlooked tradition of occulted imagination: on the one hand the constitution of both rationality and imagination with the apparent superordination of the former to the latter; on the other hand, the coconstitution of imaging and intelligence, with the latter's dependence on the former for the human ability to rise—though without reaching total transcendence—to a thinking of *everything* that is/images. Whether and to what degree this thinking is simply equivalent to "understanding" itself has to be investigated. At any rate, such thinking and its concomitant understanding would be products and images of the good. Thinking in this sense always follows the tracks of imaging, and imaging provides intelligence with places to think through. There *is* a Platonic theory of the imaginative powers of the human soul, but it is subordinate to the ontological, agathological (pertaining to the good), and kalological (pertaining to beauty) placement of imaging.

This is the topological heritage within which the students of Plato's Academy learned to think productively. Among their number was one who took the heritage seriously enough to explore and amplify it, to whom we owe the basic inwardization and psychologization of imagination with which we have lived and thought ever since: a young man from Stagira, son of the physician to King Philip II of Macedon, the one who went by the name Aristotle.

References[74]

Arendt, Hannah. 1978. *Willing*. New York: Harcourt Brace Jovanovich.

Castoriadis, Cornelius. 1994. Radical imagination and the social instituting imaginary. In *Rethinking imagination: Culture and creativity*, ed. Gillian Robinson and John Rundell, 136–154. New York: Routledge.

Diels, Hermann, and Walther Kranz. 1974 [1903]. *Die Fragmente der Vorsokratiker*. 3 vols. Reprint of 6th ed. Berlin: Weidmann.

[74] If there is a second, square-bracketed date, it indicates the year the work first appeared in its original language.

Fink, Karl J. 1991. *Goethe's history of science*. Cambridge: Cambridge University Press.

Heidegger, Martin. 1927. *Sein und Zeit*. Halle a.d. Saale: Max Niemeyer. English editions: Heidegger, Martin. 1962. *Being and time*. Trans. John Macquarrie and Edward Robinson. New York: Harper & Row; and Heidegger, Martin. 1996. *Being and time*. Trans. Joan Stambaugh. Albany: State University of New York Press.

Kirk, G. S., and J. E. Raven. 1957. *The Presocratic philosophers: A critical history with a selection of texts*. Cambridge: Cambridge University Press.

Plato. 1968. *The Republic*. Trans. Allan Bloom. New York: Basic Books.

Plato. 1997. *Complete works*. Ed. John M. Cooper. Indianapolis: Hackett.

Pomeroy, Sarah B. 1971. Optics and the line in Plato's "Republic". *The Classical Quarterly*, n.s., 21: 389–392.

Rosen, Stanley. 1983. *Plato's "Sophist": The drama of original and image*. New Haven: Yale University Press.

Severino, Emanuele. 1982. Ritornare à Parmenide. In *Essenza del nichilismo*, 2nd ed. Milan: Adelphi.

Wood, Robert E. 1987. Image, structure and content: A remark on a passage in Plato's "Republic". *The Review of Metaphysics* 40: 495–514.

Chapter 5
Aristotle's *phantasia*: From Animal Sensation to Understanding Forms of Fields

As a student of the Academy, Aristotle had the opportunity to see and practice Platonic image–perceiving and –forming at first hand, and undoubtedly also the opportunity to think and talk about what makes it possible. From his writings we know, however, that the focus of his interest was different. Plato had concentrated on the ontology of imaging and the relationships between images on different levels of being. Correlatively, he named and placed in hierarchy different powers of soul that allowed human beings to recognize these various imaging relationships and to produce new ones. He understood, in particular, human speaking, the human production of *logos*–accounts, as itself a kind of imaging. Aristotle deemphasized the ontological considerations; he wanted above all to consider, name, distinguish, and define more particularly the fact of psychological images and the phenomena of imagination as an animal and human power. He strove to give a better-articulated account of imagination's operation and physical-organic character and thus to place it more exactly with respect to the other acts and powers of *psuchē*, of mind or soul. He aimed at, and thereby to a considerable degree achieved, a theory of imagination in the modern sense of the term. He placed imagination at the heart of the human powers of sensation and cognition, and what he said about it was foundational for what others said for nearly 2,000 years. Yet what he expressly wrote about it was sparse, and the underlying unity of the various passages where he discussed it was by no means obvious. The nature and meaning of imagination was elusive, and tantalizing. Those who came later could not help interpreting and overinterpreting Aristotle's theory; they inevitably expanded on what he said, adapted it in creative and even dubious ways, and often distorted it. As a result, there is hardly a more controversial topic in the entire body of his work.

Although Plato established conceptual and analogical approaches to the soul that are foundational for later work, Aristotle was the first to present the psychology of imagination systematically as part of natural science. Most of the major features attributed to imagination over the centuries have their at least distant origin in his philosophy. Already in antiquity his psychological writings took on authority among physicians and philosophers. The conquests of Alexander the Great, whom Aristotle

D.L. Sepper, *Understanding Imagination: The Reason of Images*, Studies in History and Philosophy of Science 33, DOI 10.1007/978-94-007-6507-8_5,

tutored, and the consequent spread in the Hellenistic world of his works (and of Greek culture more generally) helped ensure his durable philosophical and scientific influence, which extended into the Jewish, Islamic, and Christian middle ages and beyond. Because it was a late medieval version of his psychology to which early modern philosophers reacted as they tried to reconceive mind, Aristotle continued to shape modern theories even after his philosophy had been "rejected."

5.1 Aristotle's Physiologically Based Psychology of Imagination

Aristotle's psychology presents a theory of the intellectual apprehension of forms acquired by sense perception. Its aim was rational, but it gave due attention to the empirical, the physical, and the physiological.[1] This is reflected in two slogans used by medieval thinkers that are found, nearly verbatim, in his writings: whatever is in intellect was originally in sense; and there is no thinking without images (or *phantasms*, to use the more specifically Aristotelian term). It is especially the latter slogan that needs to be taken very seriously by any interpreter of Aristotle's psychology. Since Aristotle was the inventor of most of the founding topics and tropes of psychology, any serious investigation of its bases needs to grasp (rather than cavalierly dismiss) how his psychological theory actually works.

We must understand from the outset that applying the term "psychology" to Aristotle's writings involves a certain anachronism, even if the word looks genuinely Greek.[2] Analyzed according to the combined word roots, it means the *logos* of the *psuchē*, the reasoned word–accounting given of soul. Today we would say more concisely: theory of soul. But quite apart from the fact that "theory" suggests a specialized type of account that is narrower than *logos*, our understanding of "soul" outside of religious and spiritual contexts tends to be ironic. Some might argue that almost all our traditional psychological concepts are soul concepts, and thus more or less tainted by associations with outdated philosophies, untenable theories formerly believed to be scientific, and with religious belief.

[1]Labeling Aristotle as "empiricist" or "rationalist" usually says more about what those using these descriptors want to make of him than about Aristotle himself. There is no contradiction between Aristotle's "empiricism" and his "rationalism," if we leave behind modern acceptations of these terms. The experience we acquire from sensation is for him the source of everything we remember, imagine, and know, although the concepts we acquire from experience have a rationality and logic that we can recognize and elaborate. Our cognition begins with what we experience, and imagination (*phantasia*) is the essential mediator between sense and reason.

[2]The word is a learned sixteenth-century invention not attested in classical Greek or used in classical or medieval Latin. The Latin form, *psychologia*, was (perhaps) introduced in Germany ca. 1579 (see the etymology and etymological note under "psychology" in the *Oxford English Dictionary*) and became commonplace in the seventeenth century.

Aristotle is not, however, a thinker who can be dismissed casually, at least not when he is thinking about basic issues; indeed, in most respects he offers a model of clear thinking in such circumstances. *Psuchē* was an everyday Greek word shaped by various religious and popular traditions, but it was also used in more technical ways by his philosophical predecessors and contemporaries, Plato not least among them.[3] Aristotle always felt an obligation to assess what others before him had said about a subject. Although accurate truth was difficult, approximate truth was available to any reasonable person who seriously turned his attention to a question. When a topic was of everyday concern he consulted what common people said. Without predecessors, we inquirers would always have to begin from zero. We would lack the insights and concepts predecessors had devised and be condemned to repeat the very mistakes from which their example can save us.[4]

For Aristotle, imagination—for which he consistently used the term *phantasia*—has to be understood as only one soul power among many. His conception was not anthropological like our own, not human-centered; it was biological, that is, centered on the phenomenon of life. As such it was part of physics, the theory or science of nature (*phusis*). If someone today knows anything about Aristotle's theory of soul, it is likely to be that animals and even plants, not just human beings, have soul. If this strikes us as faintly ludicrous—whether we are scientists or common folk—perhaps the reason is that we have lost the fine art of making careful distinctions while keeping in view different levels of reality and possibility. Claiming that plants have soul means, in the first instance, thinking that they are alive. There is *nothing* naïve about Aristotle, though there often is in the judgments of those who are casually or deliberately ignorant of him.

A general rehabilitation of Aristotle is not our task. Yet just as much as with Plato in the preceding chapter, to understand how Aristotle shaped the inquiry into imagination we must take ample account of the context of his thinking. Thinking and the realities and possibilities thinking pursues are always networked. This is not a flaw in them or us or the situation but the very essence of how human beings conceive the world and bring it to speech: how we *bespeak* the world, to coin a phrase.[5] One of the hardest things is to be aware of what we are doing when we are engaged in conceiving and bespeaking things. Plato is among the greatest thinkers precisely

[3]Peters 1967, 166–176, distinguishes 36 senses in which the word was used in Greek popular, religious, and philosophical thought. It is by far the longest entry in the book.

[4]See, for example, how he analogizes the example of the progressive development of lyric poetry to the history of investigations into truth in *Metaphysics* II.1 (993b12–19). I cite Aristotle's writings by the chapter or book–and–chapter numbers (Roman numerals for books, Arabic numerals for chapters) and/or the Bekker page–column–line numbers (thus 993b12-19 is p. 993, column b, lines 12–19).

[5]Under the transitive uses of the verb "bespeak," the *OED* notes two. "To speak to, address" is marked as "chiefly poetic" usage; the other, "to speak of, tell of, be the outward expression of; to indicate, give evidence of" is last attested in the 1860s. My use tries to meld these two senses: we address, and thereby express, (things in) the world, and thus we become spokesmen for the world.

insofar as through his dialogues he constantly calls this phenomenon to our attention and compels us to think about it. And Aristotle is precisely the thinker who first gave the West a substantive account of the processes by which the world comes to awareness and becomes progressively more articulate: a process in which images and imagination are the fundamental, the central, the inescapable mediators—indeed the very medium—of experience and, in human beings, of understanding. If in some respects, even decisive ones, he was wrong, there is nevertheless a great deal to be learned from this story both about the history of imagination and about imagination itself.

"There is no thinking without phantasms." That is the crucial tenet of Aristotle's theory of mind (to use a modern term), a tenet that subsequently dominated the Western (and not only Western) understanding of human psychology and that structured inquiry well into the early modern period. Phantasms, the objects or products of *phantasia*, occur only in animals, by virtue of their sensitive or perceptive powers.[6] The tenet thus asserts that, although in human beings there might be something psychologically unique (viz., the power to think and understand), they are incapable of exercising this unique ability without the imaginative power they share with other animals. There is an essential connection between rationality and images, between intellection and imagination, between being human and being animal.

It is chiefly in the three books of *Peri psuchēs*, *On the Soul*, that Aristotle provides his theory of this essential connection.[7] *On the Soul* is the first treatise in Western thought that attempts to approach psychology comprehensively.[8] Aristotle understood psychological phenomena to be closely related to physiology. In this sense his psychology is an important forerunner of the kinds of theory that in the late twentieth and early twenty-first century are at the cutting edge of scientific research. The Aristotelian theory of mind/soul prospered in the ancient and medieval world in no small part because it could be (and was) adapted to the purposes of medical theory and practice. This is certainly a major reason that, as Islam spread through the southern and eastern portions of Alexander's former empire, Aristotle's

---

[6]I am using these two terms synonymously. The Greeks did not have an independent term corresponding to our "perception." In this chapter I will, for the most part, use "sensation" to render *aisthēsis*, since for Aristotle (in contrast to Plato) it encompasses everything from inchoate awareness of sense qualities apart from their objects to acute, intelligence-infused perception.

[7]The undertaking of *On the Soul* is extended in several other, shorter works that collectively go by the name *Parva naturalia*. We shall occasionally refer to some of these, in particular those on sensation and on memory.

[8]As late as the early nineteenth century Hegel commented that *On the Soul* and the *Parva naturalia* exceeded all more recent psychological works in scope and success and provided a model for his own comprehensive undertaking in *Philosophy of Spirit*, part 3 of the *Encyclopedia of Philosophical Sciences* (see Hegel 1840–1845, esp. sect. 387). In a 1923 lecture course, Heidegger agreed about the scope of the work but denied that it is psychology: Aristotle's *Peri psuchēs* "is not psychology in the modern sense but treats of the being of the human being (or of the living being *simpliciter*) in the world." Aristotle would thus be the founder not so much of psychology as of the phenomenological analytic of *Dasein*! See Heidegger 1994, 6 and 293.

writings continued to draw the attention of philosophers, natural philosophers, and physicians. It is no mere accident that the outstanding medieval commentators of Aristotle were medical doctors.[9] Aristotle's "lost" texts—lost, that is, to the West—began crossing the Pyrenees from Islamic Spain into Christian France in the later twelfth century, in a trickle at first, then in a flood. The first were translations from Arabic; the Greek texts followed, and by around 1270 the entirety of Aristotle's writings still extant was again available to Europe, both in the original Greek and in Latin translation.[10]

Even as late as the middle of the seventeenth century it was Aristotle's account of embodied soul that was taught in universities and that was presented in the medical schools of Europe as a philosophical basis for understanding human and animal physiology. Those who dissented from particulars, or even from the substance, of his teaching still typically approached issues using a set of questions and a vocabulary that had been shaped by him. This was especially true of physics, physiology, and psychology. Until quite recently, then, and perhaps still today, the basics of human thought, sensation, memory, imagination, feeling, desiring, and willing were conceived according to the conceptual topography that Aristotle had drawn. And so there was, and is, no understanding the psychology of imagination without Aristotle.

Even that large claim is a little too modest. Wherever imagination is of concern, it is not just imagination that is at issue, but rather the whole life of the mind, indeed the whole of human psychology. Plato used ontological and psychological imaging to unify the aspects of the cosmos with the apprehensive/cognitive functions of soul. It was Aristotle's genius, in turn, to give these different soul functions their canonical names and definitions and to identify their local habitations, that is, their basis in the organs, powers, and activities of the living human animal (and of other animals as well). The study of imagination, if not the totality of its history, is thus doubly founded on the question of the human psyche as a whole, understood as fundamentally embodied.[11] Aristotle's psychology, like our own today, was in essence a physio-psychology or psychophysiology, a psychology grounded in the body's specifically constituted materiality. Although it is always possible in Aristotle's psychology to give at least a first-approximation answer to what imagination (or any other psychological function) is in its own terms, the second approximation to the question always requires attention to how it is placed with respect to the other basic psychological powers and the particular forms of their embodiment. That sets our task in the following pages.

[9] To mention only three: Avicenna (ca. 980–1037; this is the latinization of ibn Sīnā) in Persia, Averroës (ibn Rushd, 1126–1198) in Spain, and Moses ben Maimon or Maimonides (1138–1204) in Morocco and Egypt. All were polymaths and masters of philosophy and medicine, and all wrote (in some cases extensive) commentaries on Aristotle's writings.

[10] See Knowles 1962, 185, and Dod 1982.

[11] The first founding was Plato's situating the soul and its powers with respect to the good and the levels of being, the second Aristotle's conceiving the soul as the unity of the whole physical organism.

5.2 Placing Soul in Aristotelian Context

Plato's discussions of the phenomena of imagination were almost always undertaken in the course of pursuing other questions, so it is perhaps not surprising that he did not standardize his vocabulary and concepts. Nevertheless, throughout his writings the image and the image-making powers play their role in a larger economy of the human soul and of the appearances of being and the good. Aristotle formalized, stabilized, naturalized, and inwardized the understanding of the human soul expressed in Plato's dialogues. Nevertheless, we have every reason to believe that Aristotle's basic criticism of past thinkers was directed at Plato as well: that none had given an adequate or consistent account of the specific kinds and limits of soul activities that are included under the general headings of thought and sensation, *noēsis* and *aisthēsis* (see, for example, the beginning of *On the Soul* III.3).

As in his other works, Aristotle aimed to give a more developed, particularized, accurate, and terminologically consistent account of what his predecessors had undertaken unsystematically. In *On the Soul* he took much further than Plato the claim that the soul is a certain kind of unity. Aristotle took that conception of unity as fundamental, but argued simultaneously that it had to be understood as a unity of parts and functions. What the word "part" means with respect to soul was itself a question to be resolved, as far as Aristotle was concerned, but one that could be properly addressed only if one first had a more detailed and well-defined understanding of the different activities and powers that were attributed to soul (411b6–31).[12]

Psuchē was an established term in popular Greek long before Aristotle adopted it. It was understood as what animated the living being. It was not unambiguously immaterial, and insofar as it was used for the "shade" in the afterlife (e.g., which Homer portrayed in the *Iliad* as fleeing the body at death and in the *Odyssey* as residing in the vaporous afterlife of Hades) it expressed more an extremely attenuated corporeal state than immateriality. It was not a specifically religious term, and both the pre-Socratic and post-Socratic thinkers had no qualms about adapting it to their more technical philosophical and scientific uses. One of the goals that Aristotle was pursuing in *On the Soul* was precisely to bring consistency and rigor to popular and technical usages of his contemporaries and predecessors.

Aristotle was able to use the term "soul" in a far less self-conscious and constrained way than we can, not least because his use preceded more than 2,000 years of development and, even more significant, because he deliberately and quite consciously provided the first technical or scientific definition after taking into account what the best attempts that preceded him had said. The definition comes at the beginning of book II of *On the Soul*, right after his examination of earlier theories and the questions they addressed in book I. The definition is straightforward for those familiar with Aristotle's philosophy but otherwise needs more than a little

[12] See n. 4, above, on citations from Aristotle.

explanation: soul is the first actuality (or the first level of being–at–work or being–in–operation)[13] of an organized body that is potential with life.

For now we can remark only a couple of crucial points. First, Aristotle defines soul as an activity or being at work *of a body*; that goes against the grain of the modern popular soul conception, strongly influenced more by religion than by philosophy and science.[14] One provisional conclusion to draw is that, for Aristotle if not for all Aristotelians, without a body there can be no soul. Second, the definition applies to the principle of activity of any living thing whatsoever, plant, animal, or human animal. For Aristotle and Aristotelians, everything living has soul—by definition, as it were! Yet the definition is not specifically tailored to fit any presumed peculiarity of the soul *of the human being* (rationality, immortality, or the like).[15]

One might suggest a first approximation to Aristotle's grasp of soul this way. If you can see and understand the difference between a plant that is dormant (in the heat of summer or the cold of winter) and one that is dead, you have grasped the difference between a body with an active principle of its specific ways of being and something that is inanimate. If you walked into a mortuary where two bodies lay covered to the neck by sheets, and suddenly one of them opened its eyes, got up, and asked what you wanted (the attendant on duty taking a nap), you would easily grasp the major difference between the two: one still has living bodily activity, the other has none at all. In first approximation, that is a grasp of soul very close to what Aristotle's definition intends.

First approximation gives way to second approximation, however, and with Aristotle (as with most philosophers of the first rank) that requires a more explicit understanding of the principles of his philosophy. That is not to say that the words taken in first approximation mean something different from their eventual meaning. Rather, there is a dimension of depth to them; they are part of networks of conception that provide them with stability and coherence by linking them to other phenomena according to basic principles of natural explanation. Plants and animals, for example, are instances of *ousia*, of substance; they are analyzable according to four causes or fundamental conditions (*aitia*, plural of *aition*) of their being—and in particular

[13]The term in question is *entelecheia*. It has long been considered a (near-)synonym for *energeia*, actuality or being–at–work, when (as in living things) the principle of actuality is internal to the thing that is in act.

[14]Even in modern academic and intellectual circles it is often assumed that soul is an immaterial ghost enclosed in a bodily container—what is sometimes, but also rather inaccurately, thought to be Plato's and Descartes's dualistic conception. People strongly influenced by the medieval Scholastic tradition would consider themselves beyond this unsophisticated image. But as someone who has taught generations of bright young Catholic undergraduate and graduate students, I can affirm that, when they have a clear notion of soul at all, it is more likely to be the ghost in the body (if not the machine) than the first actuality of an organized body.

[15]Aristotle did not think that the human soul was immortal, although he did think that the active intellective aspect of soul was unchanging and immaterial. In the middle ages of Islam, Judaism, and Christianity, this ambiguity gave rise to the most various, highly conflicting interpretations of Aristotle's soul doctrine.

according to their composition out of two of those fundamental conditions, form and matter. Like inanimate things, plants, animals, and human beings are formed matter or embodied form; and in fact Aristotle conceives soul as the term that indicates the form of *living* things.

Aristotle's basic conception of what constitutes a thing is, in the first instance, relatively commonsensical. In the *Categories*, one of the works of the *Organon* that was never lost to the West and thus was continuously influential there, he presents a relatively simple account of the being of substance and of the attributes of substance. "Substance" is the name for the truly fundamental kind of being. It has a primary sense and a secondary sense. Primary substance is, to use Aristotle's language, something that cannot be asserted of anything else or be present as an attribute or part in anything else. More colloquially if also less accurately, a primary substance is an identifiable thing that is individuated and relatively independent. This includes, in the first place, all the ordinary things of the natural world: grains of sand, rocks, plants, animals, stars, and so forth.[16] Such things cannot be predicated or asserted of other things (thus they cannot serve as P in statements of the type "S is P"); and they cannot be attributed to other things, at least not in the way that "six-feet-tall" can be attributed to a man or "red" to a wall surface.[17]

Secondary substance arises because not only do the fundamental things come as individuated, they come as individuals *within a species*. Socrates, Glaucon, and Adeimantus are all primary substances who fall within the secondary substance *human being* (we might say, more technically, *homo sapiens sapiens*). Obviously a secondary substance, unlike primary substances, *can* be predicated of other things. "Glaucon is a human being" expresses the kind of thing that the primary substance Glaucon is. Since different species can be gathered into a common genus and the genus name can in turn be predicated of the species and of the individuals in that species, one can say in a more remote sense that the genera, too, are secondary

[16] A caution: in at least one important sense, Aristotle did not clearly and unambiguously consider things made by human art to be substances in the proper sense. See *Metaphysics* III.4 (999b17–20), VIII.3 (1043b18–23), XI.2 (1060b16–28), and XII.3 (1070a13–20), and the painstaking analysis of them in Katayama 1999, 25–40. A major reason for Aristotle's doubts about artifacts as substance is tied to his conception of the being of things: they are, in general, formed matter, a composition of matter and form, in which the form is self-preserving. Natural things have a persistent intrinsic form, whereas artificial things have their form imposed on matter by human plans that do not persist by nature. On the other hand, when Aristotle gives examples of how the four causes contribute to the being of things, he almost always refers to artifacts like bronze statues!

[17] Again there are complications, more apparent than real. Just because a word can appear after "is" does not necessarily make it a predicate: "The man speaking to Plato is Socrates" can as easily be turned around, so it is an identification rather than a predication; and "Angered by the poet's song was Mycenae's prince" reflects an English word order allowable by poetic license, but "prince" remains the subject. That Socrates, Plato, and Glaucon can constitute a three-person committee of which Socrates is part is not an exception because the committee is not a substance: it is rather a relation among the substances Socrates, Plato, and Glaucon. An arm severed from a human body is not a substance because it no longer has the natural functions it had as part of a substance. Thus not all *things* are substances.

substances, although they need further determination or specification to identify the lowest (natural) species to which individuals belong. Glaucon is a human being; human beings are hominids; hominids are primates; primates are mammals; mammals are vertebrates; etc. Thus Glaucon is a hominid, a primate, a mammal, etc.; human beings are hominids, primates, mammals, etc.; hominids are primates, mammals, vertebrates, etc. Basic principles of logic derive from this inclusive (and exclusive) relationship between primary substances, species, genera, and their corresponding names (if A is B and all Bs are Cs and all Cs are Ds, we can conclude that A is C, A is D, all Bs are Ds, and the like).

Attributes—or, as the Latin rendering has it, accidents, things that fall to (*ad + cadere = accadere/accidere*) substance—are figuratively "things"; that is, they are the qualities, quantities, relations, and other characteristics that happen *in* or *to* substances. Thus they are things only in a loose sense. Red does not, for Aristotle, exist apart from primary substances that are called red. Destroy the substance that is red and the red is gone (*that* red, not the red of anything else or the species red). "Six feet tall" does not exist apart from things that have that quantitative determination. Committees as relations among several people–substances do not exist without the people who constitute them. Aristotle identified seven categories of these kinds of attribute of substance in the *Metaphysics* and nine in the *Categories*.[18] Perhaps that means there is no definitive enumeration.

For Aristotle, beings are, first of all, primary substances, which means relatively independent individuals.[19] Primary substances fall into kinds, so along with the primary substances there are secondary-substance terms that indicate the kinds. Moreover, individual substances have various attributes, some of which are essential to being of a kind (plants have to be alive and have to have sources of nourishment in order to be plants), others of which are not essential but typical (human beings have two eyes typically, but fewer or more eyes would not necessarily make them nonhuman), yet others that cannot be called necessary or typical but are one of several possibilities (like hazel eyes), and some even that might be unique (perhaps a special violet or a pattern of speckling in the irises has never occurred before and will never occur again). Thus attributes exist, but not independently of substance. They must always be attributes of some primary substance. The very fact that we can conceive of attributes as belonging (or not) to substances means that we can talk

[18] *Categories* 4 identifies the categories as substance plus quantity, quality, relation, somewhere, at some time, being in a position, possessing/having, acting, and being acted upon. *Metaphysics* V.7 gives the same listing with the omission of "being in a position" (perhaps reducible to "somewhere") and possessing/having (perhaps reducible to the other categories as collectively the ways that substance can have or possess attributes of any kind).

[19] My formula uses "relatively independent" because kinds of dependency can always be identified. A tree is a relatively independent individual that cannot be predicated of other things and is not a categorical part of anything else: but that does not mean it is not related to the ground, by its roots, from which it gets its water and nourishment. Thus Aristotle's formula that defines substances in terms of "not being predicable" and "not being present" (as an attribute or part) is more complicated but also more precise.

about the attributes per se (the various tonalities of red, for example), with the understanding that any real attribute has to be part of or inhere in a substance, a thing, a *res* (the Latin for "thing," from which we get the word "real"). We can thus also talk of possible attributes, that is, attributes conceived as possibly present in real or possible substances (a genetic engineer might want to try to produce violet eyes in an actual animal of a type that has never before displayed such eye color). The possibilities and placement of these attributes and their interrelationships will prove, shortly, to be of central importance for understanding Aristotle's notion of *phantasia.*

In order to understand Aristotle's definition of soul there is one additional point that needs to be made about his conception of the basic being of things: there is potential being (*dunamis*) and there is actual being (*energeia*). Talk of primary substances that fall into the kinds indicated by secondary substances and that have various necessary and possible attributes leaves out of account a chief feature of most things: they change. Change is characteristic of anything material. The only kind of thing that would not be subject to change would be something that had no matter, that was not a composite of matter and form but was instead pure form. Although Aristotle does ultimately think that there is substance that is pure form, he also frequently points out that the kind of substance we constantly encounter in our experience is formed matter, and in the *Metaphysics* he claims at several points that, if all substances were natural or physical substances—that is, composites of matter and form—then the highest knowledge possible for human beings would be physical knowledge, knowledge about nature.[20]

At the end of book VII and the beginning of book VIII of the *Metaphysics* Aristotle argues that, among physical things, what substance is is best answered by grasping that the form in substance is the cause of unity over time. Form is not just the particular organization we see at the moment of identifying a thing as a substance but also the developing form of the thing that guides its changes. Most natural substances emerge and develop; what keeps them the same through this development is having, in form, the principle of developmental change. Seeds mature into seedlings, seedlings into saplings, saplings into trees, which produce new seeds and eventually new trees. Human infants are human beings, but precisely as human they contain within themselves the capacity and the program or information (as we would call it) for further development.

The being of changeable things is thus an actual being, an actively being something here and now; but that actual being is also marked by regular developmental possibilities, by potentialities or potencies or potential being. The infant is a human being who is potentially a speaking human being; the 5-year-old is a speaking human being who is potentially a writing human being; the 10-year-old is a writing

[20]For example, *Metaphysics* VI.1, 1026a27–33. Aristotle of course "proves," in both *Physics* VIII and *Metaphysics* XII, that the physical universe would not be possible as it is if all there were were material beings. The proofs depend crucially on understanding the relationship between potential being and actual being in the process of change.

human being who is potentially the author of poems and novels; and so forth. This leads Aristotle, in particular in *On the Soul*, to differentiate three basic levels of potentiality and actuality. Human beings, being capable of speaking, can be called grammatical beings. An infant is grammatical only in the sense that it possesses the capacity or potentiality for grammar; since the infant never acts grammatically (in speaking or writing) it is not actually grammatical. But once a child learns to speak, it is clearly grammatical in an active sense (in speaking and in listening to others). But that active sense can be divided. In dreamless sleep, a mature human being does not exercise grammar, but an instant after being awakened she can speak and listen with full native mastery of grammar. When she is asleep, or when she is awake but not using grammar even in thought, she is in a state of what Aristotle designates *first actuality* with respect to grammar; when actively speaking or listening, or when thinking in words, she is in a state of *second actuality*. Thus second actuality is the state of full use, first actuality the capacity for use on demand, and potentiality the capacity for use if over time there occurs typical, appropriate development.

One thing this means is that, in order to describe adequately a being that changes, the being has to be understood here and now as having an actuality that by its very nature embraces or contains possibilities of change, that is, potentialities. The infant is, here and now, actually a human being (in second act), but that implies that the infant actually has potentials. Some potentials are for long-term development (like eventually being grammatical or being mathematical or being a runner), whereas others can become active on demand (the ability to cry or eat or smile); the former are merely potentialities, the latter are first actualities. Although this "rolling over" of potentiality into actuality, and of one kind into the other kind of actuality can seem ambiguous (with respect to second actuality, a first actuality is a potentiality!), the ambiguity is inevitably connected to the kinds of change that natural things undergo. To put it summarily: a material thing that is a primary substance is not just a composite of form and matter but also a complexly articulated composite of potentialities and first actualities that culminate in second actualities.

Finally we are in a position to give an explication of Aristotle's definition of soul as the first actuality of an organized natural body potential with life. Recall, first of all, that this is a definition that applies to all living things, not just human beings, so that the definition cannot count on any features connected specifically with human being or even with just animal being. The definition, to be applicable, requires a natural (as opposed to an artificial) body. Moreover, the natural body must have organs, which are parts of the body dedicated to specific kinds of activities that support or are part of the whole being (like lungs for breathing, digestive tract for absorbing nourishment, etc.). This excludes natural bodies that do not have organs (a puddle of water, a rock, a mountain, a pile of dust).

The relevant body has to be potential with life. Although the proper organization (that is, being divided in an orderly way into interconnected organs) of the body would seem to constitute a major part of this prerequisite, one cannot (despite Frankenstein stories) simply stitch together organs from corpses of various kinds and hope that the assemblage will come to life. Perhaps we cannot specify exactly what makes a body potential with life; we nevertheless know many things that would keep an otherwise properly organized body from taking on life. Today we are in a position to say that future genetic engineers will actually have to develop a theory of what makes an organized body potential with life in order to reproduce life forms nonnaturally.

But this line of interpreting "body potential with life" threatens to devolve into so many science-fiction scenarios. It implicitly treats the potentially living thing as though it were assembled from inert parts. There is a way of interpreting that is better aligned with Aristotle's conception. When we discuss the potentiality and actuality of grammar in human infants and adults we are talking about one of many capacities that the living human being possesses. Life and soul are not just about specific capacities, however; they are about a basic capacity of the living thing to appear or present as whole, as an interconnected whole of specialized organs, as the complexly articulated composite of potentiality and first actuality and second actuality.

Actualities, both first and second, have to be grounded in a being that changes, which means one that is characterized in its material actuality by potential being. To be and stay alive is already to have the potentiality of actual being developed to the point of a first, basic level of diversified actuality. Whatever more the living being can do beyond living, it has to live first, before it can exercise all its particular actualities and convert, over time, potentialities into actualities. Thus the first actuality we are inquiring about with respect to the organized natural body is not that of particular capacities but of the basic enacting of the organism in a way that allows us to say: it is a complex substance that is alive. The corpse on a morgue table, the plant that is desiccated because it has not been watered in 6 months, appear at first glance to be the right kinds of bodies and organized basically in the right way. But they are beyond the potential for life, whereas the hibernating animal and the estivating plant exhibit a basic level of actualization that allows us to say: alive. This kind of first actualization of such bodies is, exactly, what soul is *in the first instance*—at least according to Aristotle's definition.

To the objection that the presuppositions we have been elucidating here make Aristotle's soul concept too impossibly complicated to serve as a description of reality, much less as a definition, we need to muster an aggressive counterquestion: exactly how much complication should an adequate description or definition be allowed? Aristotle understands very well that often enough we can appeal to some more simply ascertainable characteristic a thing has as sufficient evidence that it falls within the scope of a definition. In II.2 of *On the Soul* he discusses some of the characteristics that, when any is present, assure that a thing is alive and thus has soul. A thing has soul if it has intellect, or the power of sensation, or the power of self-motion, or the power of nutrition, or the potentiality of deterioration and growth. As his conceptual and empirical analysis advances, others are added: appetite and desire, for example, and also all particular forms of the characteristics already mentioned, like vision, hearing, and tactile feeling as types of sensation. Given that Aristotle says that all living things have nutritive powers, it would have been possible for him to give the *criterial* definition, "Anything that has nutritive power has soul," where the presence of the power is evidenced by the organism's taking in appropriate kinds of food–matter from its environment. Aristotle intends his definition to be *comprehensive*, however, by which I mean that it aims both to be inclusive of all living individuals and to indicate, at least potentially, all their distinguishing characteristics, not just one or two criteria.[21] The issue that Aristotle was facing reflects Plato's question to Meno about what virtue is. We are searching not for a listing of different things we use to identify beings as alive, but for an articulable (i.e., definable by a *logos*–account) feature presupposed by all criteria that qualify something as living. This feature is being–organically–active. The point of the definition is to be general enough to apply to everything of the kind and specific enough to allow for all the determinations that it can receive. The measure of whether a definition is acceptable or not cannot, therefore, be an absolute minimum of complexity that is nevertheless sufficient to allow us to easily ascertain all the things to which the term applies. It must be neither more nor less complex than is required to capture the essential characteristics (plural!) of the thing.

So even before we have ventured upon specific inquiry into imagination in Aristotle we can say a few things that must be true of it. It will be one of those things that, if we find it in a being, will assure that the being is the kind of physical substance that is living and has the kind of form that we call soul (and furthermore the kind of soul that we call animal); it will be some kind of organic activity or closely related to organic activity; and it will be one among multiple powers exhibited by the organic activities of that living thing—and thus will be specifically interrelated with many of those other powers and appropriately "located" with respect to them and the corresponding activities of organs.

5.3 Aristotle's Imagination Conventionalized

The historical durability of Aristotelian psychology in many ways complicates making a straightforward presentation of his theory of imagination. Whatever the story of how they were produced and came down to us, Aristotle's texts left off in a state that invited not just interpretation but also development. In effect, it was less a fixed doctrine than a conceptual topology requiring elaboration. The fact that it endured recognizably in different traditions, adaptations, and modifications for nearly two millennia testifies to the persuasiveness and usefulness of its basic concepts, configurations, and phenomenal bases. Yet the main traditions of elaboration gradually overgrew the original, to the point that disentangling them is necessary but difficult.

Over the next few pages I will describe a simplified version of Aristotelian psychology. To give it a name, I will call it "conventionalized Aristotelian psychology."[22] Even though in many details it is problematic, there are several advantages to

[21] Later in this chapter we shall reconceive this distinction as the difference between *precision* (which absolutizes aspects or parts of things) and *abstraction* (which understands aspects as further determinable in particular respects—indeed, in respect to appropriate *fields* of possibility).

[22] It is based chiefly on contemporary renderings of Thomas Aquinas's interpretation of Aristotle, and therefore might also be called a conventionalized neoscholastic or neothomistic psychology.

presenting it here. It has an initial plausibility and intelligibility that a more textually or philologically accurate presentation may not. If it is not quite Aristotle, it is for the most part genuinely Aristotelian topography—Aristotle with many rough edges removed for convenient handling. Especially in versions that present the theory in terms of seemingly independent faculties or mind modules, it has a clean, analytic economy that gives every appearance of a smoothly achieved comprehensiveness about the human mind or soul. It gives a sense of the scope of Aristotle's concerns and opens up important themes discussed in his texts. Because it uses a familiar psychological vocabulary—a familiar variety of folk psychology, so to speak—it has an immediacy of appeal that does not require expertise in Aristotle's philosophy. And—not its least virtue—it provides a template that will make more easily discernible where and how it exceeds, diverges from, and even betrays what Aristotle conceived.

According to conventionalized Aristotelian psychology, human beings have three basic levels of power or faculty. They share, at least very generally, nutrition, growth, and reproduction with plants and animals. The sensitive faculties, the pleasure and pain associated with them, and the ability to move purposively they share with animals. In particular, human beings have five external senses and at least three internal senses (common sense—or, better, common sensation—memory, and imagination; a division of these into even more kinds occurred frequently in medieval thought).[23] Every animal, no matter how primitive, has at least the external sense of touch and (probably) also taste. More complicated animals have other external senses as well: smell, hearing, and vision. Of course, except for touch, no animal has to have any particular one of these or any particular combination in order to count as animal. Moreover, human beings have a basic power not shared with any other animal: reason or intellect, which allows them to know not just individuals but also universals and therefore to know in the strict sense.

Setting aside the powers of nutrition, growth, and reproduction, we can ask the question, how do the external senses work? First, each is independent of the others in its functioning, and each has a "proper object" that it shares with no other. Color is the proper object, the *proper sensible* of vision; that means that touch, taste, smell, and hearing cannot detect it.[24] Similarly, tone or sound is the proper sensible of hearing, savor or flavor of taste, aroma of smell. For touch we do not have any single agreed-upon name, with the possible exception of "feel"; but since that word is often used in a figurative or extended way for sensitivity in general, including feelings and emotions, let us call the object of the sense of touch "tactility" or "tactile quality."

To elaborate further: Each sense has its proper object, a *sensible form*. As the medieval Scholastic thinkers put it, the sense in act is the sensible in act. Aristotle understands the being of things in general as a kind of activity; substances like rocks,

[23] For a sampling, see Harvey 1975 and Steneck 1970.

[24] Obviously synesthesia, the phenomenon of (for example) a person's experiencing sound in viewing colors, complicates this claim.

trees, and animals are composites of form and matter, and the substantial form is a principle of activity that governs the genesis, development, and decay of the material thing. Besides the substantial form there are sensible forms in the thing, and these forms, which are themselves activities in the thing, are communicated to the sense organs of the animal when there is an appropriate medium between the thing and the organ (for example, in vision, transparent air illuminated by the sun or some other light source). So the red color of an apple is an activity of the surface of the apple, and when the eye perceives this color the form–activity of the apple's surface is communicated to the eye, where the same activity recommences, without the matter of the apple. This activity in the eye is the actual perception of color.

When the perception persists without the presence of the object we call it a phantasm or image, and the power of preserving and producing these phantasms constitutes what we call the internal senses. These internal senses, just like the external ones, have physical locations, and their functions correspond to activities in those locations—with the difference that they are internal to the body (in the brain, contrary to what Aristotle said—he thought that the brain merely cooled the blood, whereas the phantasms originating in the senses traveled to the region of the heart, where feeling/passion/emotion were united to the appearances of the senses).[25] The first internal sense is actually not imagination but common sensation, which is where the deliverances of the external senses are united into a common experience (that is, we do not experience five worlds through five senses but a single world with five sensory aspects).[26]

External sensation is capable of distinguishing all varieties of proper sensibles: vision perceives every distinct color hue, hearing every difference of tone, etc. But the fact that some given thing, say a granular powder, is white, sweet, and makes a crunching sound when compressed requires that the sensibles proper to different senses be coordinated and distinguished in a unified sensory experience, which is the level and function of common sensation. Moreover, there are in common sensation additional forms of sensation—the common sensibles—that assume full conspicuousness there, although they may be nonthematically present in the

[25]See Aristotle, *On the Parts of Animals* II.7, 652b4–7 and 17–27; and II.10, 656a14–656b7. Already in Greek antiquity many physicians argued that the brain was the seat of sensation and thinking. The internal-senses theorists located common sensation, imagination, and memory in the *ventricles* of the brain, hollows enclosed by the two hemispheres of the brain. The medium of communication between external sense organs and the ventricles was *animal spirits*, which were traced back to the heart; the theory of these spirits was developed in antiquity chiefly by the Stoics. For an account of the internal senses and their localization from antiquity to the middle ages, see Steneck 1970, Harvey 1975, Carruthers 2008, and Karnes 2011.

[26]"Common sensation" avoids the ambiguity of "common sense," which is used also for the supposedly universally available *knowledge* that all unimpaired human beings should possess (sometimes called "horse sense"—what even an animal is smart enough to know). There is nevertheless a connection between the two: without common sensation we would still experience only colors, sounds, smells, flavors, and tactile feelings, but not an articulated world of familiar things.

external senses.[27] The common sensibles Aristotle lists in *On the Soul* are motion, rest, number, shape, and magnitude; in *On Memory and Recollection* he adds time. They are called *common* because "each of them is not proper to any one sense but is common to all" (418a19–20).

Consider the example of motion. Vision, hearing, and touch can all detect motion (e.g., swirling colors, a passing siren, etc.). It is only in common sensation, however, that these changes at the level of the different external senses become unified and coordinated as, say, the motion of a ball across the floor.[28] The coming together of all the sensible properties into persisting things tracked through their positions and motions against a background is what common sensation provides. It brings together all the individual proper sensibles of real-world things and places them in a more complex sensation–experience. The multiple sensibles delivered by the five external senses thus come to be articulated in a common place and time, the field of common sensation. This common field provides a greater stability than the blooming, buzzing confusion of colors, smells, and sounds can. It is where complete or fully articulated phantasms or images of worldly things are constituted (for example, the phantasm of the sweet, white, grainy, and crunchy stuff that, applying a concept to what appears in common sensation, we call sugar). All animals have some form of external sensation; external sensation and common sensation (in animals that possess it) allow for quite complex, object-oriented behaviors from animals.

The complexity is raised even higher by memory and imagination. Memory is the internal sense power that allows animals to retain and produce on demand phantasms *of* and *as* what they previously sensed; it can be subdivided in various ways, for example into thing–memory and event–memory. The ability to retain and produce phantasms, though not as part of a remembered experience, is called imagination; it, too, can be subdivided, for instance into the power that *retains* appearance–forms and another that *re–evokes* and *recombines* them.[29] Aristotle also discusses but does not unambiguously name another internal sense power, later specified as the

[27] Objects moving in the field of vision, for example, produce moving colors in the eye, but the eye detects colors primarily and motion only in a secondary sense, whereas motion as the *object's* motion is experienced at the level of common sensation. Perhaps Aristotle was obscurely suggesting some kind of feedback mechanism between common sensation and the external senses, especially since the third kind of sensible, the incidental or concomitant sensible ("seeing," for example, that a white-clad figure over there is Diares' son—Aristotle mentions it before common sensation), adds some particular knowledge (our acquaintance with clothing and with Diares and his son) to what we perceive by sense (white).

[28] This is actually too conceptually dominated a way of bespeaking what happens. We might gain a better approximation by trying to imagine how a dog or a squirrel or a frog would experience the event (no *identification* of ball, floor, or rolling as such, yet with the ability to focus on a single object or action) or, alternatively, how we might experience it in a groggy or psychologically impaired state. We might see, for example, a kaleidoscopic array of color without even the beginning of conceptualization or any definite experience of unities—the latter are what common sensation, in contrast to proper sensation, supplies.

[29] See Steneck 1970, 13–16.

estimative power (and, in humans, the cogitative power, about which more, below): it is the animal's ability to experience what is presented to it as desirable or noxious. The classic medieval example is a lamb's response to the presence of a wolf: a very young lamb, without previous experience of a wolf, will somehow recognize it as a danger and behave appropriately (standing silent and stock-still, for example). This means that goodness and badness (what is good and bad for sheep, in this case) are naturally or instinctively associated with certain appearances or phantasms. This is the closest that nonrational animals come to predication, that is, to asserting of a thing S that it is P. Implicitly, by means of images, the lamb "knows"—not cognitively but sensitively—that the wolf is dangerous. Of course the wolf has its own estimative power, which determines other wolves to be its companions and sheep to be its food—without, of course, the wolf's having the explicit concepts *companion* and *food*.

The behavior of animals is governed by external and internal sensation, and the number and the capacity of their various outer and inner sense faculties is an index of the complexity of that behavior. The more external senses an animal has, the more discriminately it can behave with respect to objects in the world and its feelings and emotions of pleasure and pain, desire and aversion.[30] The more numerous the internal senses and the greater their diversity and capacity, the more nuanced, prolonged, and planned the animal behavior can be. An animal with internal senses has the ability to "reckon" with phantasms. The more phantasms it has stored and can reproduce—or possibly even recombine—the more complex are the phantasms that the internal senses can produce. The greater an animal's ability to form new phantasms, the more sophisticated and purposive its behavior can be. However simple or complex the phantasm, it can awaken the animal's desire or aversion without any corresponding object being actually present to the external senses.

An animal without memory, vision, hearing, and smell, if it is presented with food, can detect it only by touch, therefore only when its body is in contact with the food. An animal with smell or vision can pursue it from a distance proportioned to the effective range of its senses. An animal with short-term memory can move away from food and not lose track of the fact that it is there, and one with long-term memory can summon up phantasms of past situations and develop strategies accordingly. An animal with recombinative imagination can even come up with novel ways of acting.

Human beings sometimes behave solely at this basic level, for example when they are inattentive, drunk, overwhelmed by passions, or sick. They have in addition intellect or reason; yet, as Aristotle says, whatever is in intellect was first in sense, and there is no thinking without phantasms/images. This does not mean, however, that thinking is simply the having of phantasms, individually or in series, though when human beings dream that is often what their mental activity is like. The human

[30]This means, of course, that the sensation and its qualities constitute an articulated topological field (as explained in Chap. 2, above) that can be "navigated" by the relevant discriminating powers and their own topology.

being has, through intellect, *the power to abstract from phantasms what is called an intelligible species.*[31] This intelligible species can also be called "concept," and its formation "conception"; understood as the foundation of our ability to speak, it is called (especially by followers of Thomas Aquinas) "internal word." Aristotle illustrates the process by analogizing it to illumination. Intellection has an active part and a passive or potential part; call them agent intellect and potential intellect, respectively. When the human being is aware of a phantasm, it is aware of it not just in the sensory way of animals but also in an intellective way. Agent intellect, which is always active and always the same, is like a mental or intellectual light: it illuminates the phantasm and reveals the intelligible species that are implicit in it[32]; this illumination of the phantasm impresses the intelligible species into the receptive or potential part of intellect, just as an illuminated physical object impresses a sensible form or species in the receptive eye. An adult human being who has in awareness the phantasm of a squirrel associates and reassociates this image with other images the way a dog would, but he or she also recognizes that it is a squirrel or a pest, by virtue of agent intellect's illumination of the phantasm and the resultant appearance in potential intellect of the corresponding concept.

Thus we arrive at the point of the slogan that whatever is in intellect is first in sense. The complex of related phantasms already bears in itself the principle of intelligibility of the object that is the source of the phantasms, but this principle can be evoked only by intellect. What the intellect evokes, the squirrel being of the squirrel, was already *implicit* in the common sensation and the complex phantasm of the squirrel. Animals, by means of their internal senses, see resemblances and differences and act accordingly; but human beings have this special power allowing them to accurately conceptualize and formulate truths about things through a natural abstractive process, called induction (in Aristotle's Greek, *epagōgē*). The concepts we form through abstraction from phantasms are the mental form of what is already in things materially, by nature; the senses, external and internal combined—thus including imagination—gather the information necessary for the successful abstraction of the forms, natures, or essences of things.

Before leaving this account of conventionalized Aristotelian psychology there are three things to add. Because human beings have intellect, even their sensation of things is different, to some degree, from the sensation that other animals have. That some animals have sharper or duller senses than others is obvious to Aristotle: a bird of prey can spot a mouse a quarter mile away, a mole can do little more than see light and dark, and some other animals (like worms) have no eyes at all. But for

[31] In the next section we will question whether this kind of *abstraction* is really to be found in Aristotle.

[32] Eventually I will insist on the plural of "species" here: a species is not brutely *given* but is seen against the background of a field. But that is already far too complex for the conventionalized psychology, which is more likely to assert that intellect illuminates and accepts the *essential* species of the thing, its whatness (e.g., the squirrelness of the squirrel).

human beings, rational considerations, desires, and intellectually directed purposes are almost constantly conditioning the use of our organs and what we see. In some basic sense four people looking over a landscape see the same thing, but their experience, and in a basic sense also their seeing, can be quite different—if, for example, one is a land developer, another a painter, the third a geographer, the fourth a hunter. One might allow that in the first instant of seeing, say when the four of them reach the top of a hill, what they see will be quite similar, but after that moment their attention and what they see will be inflected very differently, according to their past experience and the kinds of knowledge and practices they have acquired.

The second thing to add explains this difference to some degree. Medieval thinkers conceived of animal estimative power (e.g., the lamb's recognition of a wolf as dangerous) as considerably expanded in human beings and sometimes gave it a different name, *vis cogitativa*, the cogitative power. What the cogitative is is perhaps most clearly expressed in Thomas Aquinas's alternative name for it: *ratio particularis*, particular reason. It is an internal sensation located within the brain, yet it is also where reason "touches" internal sensation.[33] It is that moment at which the phantasm is not simply an image networked with other images but where it receives a name indicating something related to its intelligibility: furry animal, buff-colored thing, etc., culminating in "squirrel." All the other external and internal senses *can* function without this effect of reason: even if our reason is impaired or clouded the external senses will be brought together into common sensation, images there will connect to images stored in memory, and a kind of image association–and–recombination can lead to novel behaviors, just as happens with nonrational animals in responding to their environment. But once the phantasm is named it evokes a course that is no longer subject to just the laws of association of the external and internal sensations.

The third thing to add needs to be put in the form of a series of questions, since the preceding narrative did not get so far: What happens with intellect or reason—what does it *do*—once it has abstracted a concept or intelligible species from phantasms? Is it still true that there is no thinking without phantasms? Once reason is in possession of pure concepts, is it not then capable of *thinking in pure concepts*, thinking *without* phantasms or images? Don't even strict-construction Aristotelians have to allow for this kind of thinking—the kind traditionally associated with Platonists and other rationalists? Perhaps one needn't become a hyperrationalist spinning out the total being of the cosmos and everything in it from pure concepts: but must not human beings be capable of achieving some kind of image-free thought, at least for a few moments? Isn't this one of the crowning achievements of being human?

---

[33] The ironic quotation marks are necessary because reason was considered to be immaterial and nonorganic, thus a literal touching was not possible. On the cogitative power in Thomas, see Peghaire 1943.

5.4 *Phantasia* Beyond the Conventions

Many quite different answers have been given to these questions, all well within the tradition of Aristotelian psychologizing; and, more generally, on almost every major point made in my presentation of the conventionalized theory there have been divergent and often conflicting variants and developments. For example, Thomas Aquinas was inclined to take the "no thinking without phantasms" slogan very seriously. For him, human beings cannot think, in any proper sense of the word, without the presence of a phantasm, even in the afterlife. This is not to say that thinking is simply the *having* of an image, but rather a conviction that human access to universals requires always and everywhere the forms implicit in images. Nominalists regarded apprehended images as the irreducible elements of thought; universals were nothing more than signs or labels that human beings apply at will or by convention to images. Followers of Averroës and most physicians were inclined to take with the utmost seriousness Aristotle's definition of soul as the actuality of a body. They regarded human thinking as an organic process, and therefore questioned Aristotle's argument that because intellect is universal in power it cannot be intrinsically organic. Those among them who accepted that reason was nevertheless a universal illumination had to locate it elsewhere than in the human body, for example by interpreting it as a light emanating from God or angels. Those who most strongly asserted the immateriality of the human soul, on the other hand, were sometimes inclined to assert that phantasms prepare the way for a purely intellectual thinking. If we focus instead on the number of internal senses, by choosing different thinkers in medieval Islamic, Jewish, and Christian Aristotelianism we can count as few as three, as many as seven. This should reemphasize for us that the conventionalized theory I have presented was not necessarily held by any particular thinker, and yet it can serve as the description of a kind of "average" conceptual topology of Aristotelian psychology.[34]

None of the traditions of Aristotelian psychology is simply a presentation of Aristotle and nothing but Aristotle. Later philosophers who drew from his writings were not attempting to reconstruct the past but trying to understand the human mind with the help of a living body of thought, much as a modern physicist might try to extend the work of Newton or Einstein not simply by recapitulation but by extending and updating it. Philosophers who borrowed from Aristotle's psychophysiology corrected what could no longer be sustained (e.g., his conviction that the heart rather than the brain was the central organ where the input of the senses was unified), seized on unexploited or insufficiently developed opportunities (e.g., by trying to determine more precisely the location of the internal senses), elaborated apparent implications that had no exact correlate in Aristotle's writings (e.g., the personal

[34] For an account of these and other matters concerning the history of Aristotelian psychology in Western and Middle Eastern philosophy, see Harvey 1975, Marenbon 1987, Kessler 1988, Park 1988, Park and Kessler 1988, and Karnes 2011. For the best analytic account of Aristotle's imagination within the context of his theory of mind, see Wedin 1988.

immortality of the human soul), and filled in empirical and conceptual gaps (e.g., with a detailed account of the sensible species impressed in sensation, and with a more complete articulation of agent and potential intellect and their relationship to one another).

In summary form, here are major divergencies between the conventionalized Aristotelian psychology and the conception one finds in Aristotle's writings that are of special interest for imagination theory. (1) The conventionalized theory virtually ignores the fact that Aristotle defines imagination as a *motion*. This not only reminds us that he regarded *On the Soul* to be a work about physics or the theory of nature, it also makes it doubtful that imagination can be fully and properly conceived as a psychological faculty or module.[35] (2) The conventionalized theory, like almost every theory of imagination, takes the image as a fixed and determinate unit and the canonical form of imagining as bringing and holding that fixed image before the mind, as one might view a finished picture in a museum gallery. Passages in Aristotle himself suggested this idea. But his prior discussions of sensation and its various forms (to which he expressly refers in defining imagination) and his definition of imagining as motion requires that the phantasm and *phantasia* be thought differently—in the first instance as being *incipient* and as having a certain *mobility*. (3) As a faculty theory, the conventionalized Aristotelian psychology presents intellect as superior to and separate from imagination. Aristotle's own discussions, by contrast, present the work of intellect chiefly as grasping and forming images, require that images be qualified in appropriate sensory fields, and even suggest that the intelligible species might themselves be images. (4) Aristotle's notion of intellect working with images is predicated on a field theory of imagination supported by a careful examination of the psychophysiological genesis of phantasms from the activity of sensation. Taken all together, these four points of divergence from convention can deepen our understanding of (a) how Aristotle thought through some of the implications of Plato's conception of images and imagining, (b) how his theories laid the foundation for much later developments that, from the perspective of the conventionalized Aristotelian psychology, might seem to have little to do with his psychology of imagining, and (c) why his psychology possesses the capacity to enrich psychological and psychophysiological theorizing even today.

5.5 The Perplexities of Imagination in *On the Soul* III: An Overview

Aristotle's fundamental discussion of imagination occurs in *On the Soul*, in the third book, and almost the entirety of his express treatment of the imaginative power appears in the third chapter of that book. He devotes more space to discussing what

[35]"Faculty" in the first instance means nothing other than "power," but over the centuries it came to imply that the power had independent, even modular status.

imagination is *not* than to what it is. He brings up imagination in the context of the question how our sensitive and intellective powers can make mistakes. Unlike several other perceptive and cognitive powers, which in their simplest uses tend to be truthful, imaginings can be either true or false—though more often false than true. More than half the chapter works to distinguish imagination from those other powers. In the process he describes many of the features that today we would distinctively ascribe to imagination. But little in the chapter prepares us for the conclusion that imagination is a *motion*. Scarcely anyone would hit upon that as a first- or even a second-degree approximation to the kind of thing imagination is.

Here is the definition, embedded in a single sentence of considerable syntactic and conceptual complexity:

> But since it is possible when one thing is moved for another thing to be moved by it, while imagination seems to be some sort of motion and not to occur without sensation, but in beings that sense and about things of which there is sensation, and since it is possible for a motion to come about as a result of the being–at–work of sensation, and necessary for it to be similar to the sensation, then this motion would be neither possible without sensation nor present in beings that do not sense, and the one having it would both do and have done to it many things resulting from this motion, which could be either true or false. (428b11–18)[36]

Once more with Aristotle, we find ourselves faced with an uphill climb before we can begin to make real sense of a definition.

The first thing to notice is that the statement is a long conditional, basically an if–then proposition: if (since) A, B, C, D, and E are the case, then V, W, X, Y, and Z are as well. If we focus merely on the "while imagination seems" clause, we notice that Aristotle here establishes a notion that will be a standard for more than 2,000 years, even to our day: imagination originates with and depends on sensation and is about things sensed. In the most widely familiar versions of this standard account, the images we acquire through the senses are stored intact in imagination and/or memory. But Aristotle says nothing about image preservation or storage here.[37] Moreover, the dependence on sensation is the *second* thing that Aristotle

[36]The translations from *On the Soul* and *On Memory and Recollection* are drawn, with slight emendation (for example, "sensation" for "perception"), from Joe Sachs's translations, in Aristotle 2001.

[37]Memory (along with time as a common sensible) is treated in the two chapters of *On Memory and Recollection*. Chapter 1 of that work briefly summarizes what Aristotle takes to be established about imagination in *On the Soul*. In particular, he reasserts that there is no thinking without phantasms; he conceives the beholding of an image as like sense-perceiving a picture that refers to what it resembles; after emphasizing that thinking requires something extended and temporal, and thus must be dependent on acquaintance with the "primary power of sensation," he remarks that "memory, even memory of intelligible things, is not without an image, and the image is an attribute of the common sensation power, so that memory would belong incidentally to the intellect, but in its own right it belongs to the sensitive power" (450a12–14); and he expressly attributes memory to the same power of soul that "is the very one to which imagination also belongs, and the things remembered in their own right are those of which there is imagination, while as many things as are not apart from imagination are remembered incidentally" (450a23–26). I shall return to this passage later in this chapter.

asserts of imagination. The first is that imagination is a motion. Motion is, in Aristotle's philosophy, a question addressed in the first instance by physics, the scientific investigation and knowledge of nature. Even once we remind ourselves that *On the Soul* is regarded as one of Aristotle's physical writings because living beings are an important class of natural things, ones that are self-moving, the average student of imagination, even one who is quite aware of Aristotle's role in the history of the concept, would not be likely to think of it as a *motion*.

There is, of course, as almost always, a first approximation that we are already capable of grasping provisionally. If imagination is a human (and animal) power, for Aristotle that means that it is, or is connected with, the activity of an organ. Activity in an organ suggests that there is some kind of motion involved. Aristotle would clearly agree,[38] as would modern brain sciences, for which any reasonable conception of imagining would have to invoke neurological activity and therefore some mechanical or electrical or chemical, etc., changes (at a synapse, in a field of electrical force, etc.).

Almost immediately following the definition of imagination, III.3 concludes with words that seem to bring the inquiry to an end: "So about imagination, what it is and the cause through which it is, let this much be said" (429a8). However, in III.7, after three of the most difficult chapters in all of Aristotle's writings—on intellect in its activity, its powers, and its simplest acts—he reintroduces images, in the middle of developing the notion that there is an analogy between the processes of sensation and of intellection:

> And for the soul that thinks things through,[39] phantasms are present in the way sensed things are. And when it asserts or denies that something is good or bad it flees or pursues. For this reason the soul never thinks without phantasms….Now the thinking power thinks the forms in the phantasms, and since what is to be pursued or fled from is marked out for it in those imaginings, even apart from sensations, it is moved when it applies itself to imagined things….And when the soul declares, as it would in the case of sensing, something pleasant or painful, here in this case too [of imagining a plan of action] one flees or pursues it, and so in all matters of action. But also apart from action, the true and the false are present in the same classes of things as the good and bad. But they differ in that the former are present simply, the latter in relation to someone. (431a13–16, 431b2–5, 431b8–13)

Then, in the summary given by III.8, which draws together the discussion of sensation (II.5 through III.2), imagination (III.3), and intellect (III.4–7), Aristotle concludes with this remarkable statement:

> Thus the soul is like a hand. For the hand is a tool of tools, while the intellect is a form of forms and sense is a form of what is sensed. The soul, then, acts like a hand, for the hand is an instrument that employs instruments, and in the same way the mind is a form that

[38] Aristotle understands all being as an activity, and matter as the underlying principle of what can change. In material things, change is always either motion of matter with respect to place, called *local motion*, or accompanied by it. See *On the Motion of Animals*, ch. 5, and *Physics*, VIII.7–8. We shall further discuss the kinds of change, besides local motion, below.

[39] That is, the soul as *dianoia*, discursive thinking, which joins term to term, as opposed to *noûs*, which (for example) apprehends the simple meaning of a single term.

> employs forms, and sense is a form that employs the forms of sensible things. But since—as it seems—there can be no item of experience apart from the extended magnitudes which are the separate sensible things, the intelligible things are present in the sensible forms, not only the so-called abstractions but all the active conditions and passive attributes of the sensible things. And on account of this, one who sensed nothing would not be able to learn or be acquainted with anything either, and, whenever one were to contemplate, it would be necessary at the same time to behold some phantasm. For the phantasms are just like the things sensed, except without material. And imagination is different from affirmation and denial, since what is true or false is an intertwining of intelligible things. So how do the uncombined intelligible things differ from being phantasms? But in fact these are not phantasms either, but are not present without phantasms. (432a3–13)

There is one more discussion of imagination in book III. III.9 raises the question of how living things move with respect to place, or, to put it a bit more narrowly, what makes possible the purposive bodily movements of animals. In III.10 he answers that desire and intellect cause such motion, "if one includes imagination as an activity of intellect, since many people follow their imaginings contrary to what they know, and in the other animals there is no intellectual or reasoning activity, except imagination. Therefore both of these are such as to cause motion with respect to place, intellect and desire, but this is intellect that reasons for the sake of something and is concerned with action, which differs from the contemplative intellect by its end" (433a10–15). Finally, III.11, starting out with the issue of whether animals that have only the sense of touch have desire and imagination, arrives at the conclusion that there is in all animals a sensory imagination (*aisthētikē phantasia*), whereas in beings that reason there is a deliberative one (*bouleutikē*, also referred to as *logistikē*, calculative)—"for whether one will act this way or that way is already a job for reasoning, and has to be measured by one criterion, since one is looking for the greater good, and thus is able to make one thing out of a number of images" (434a7–10).

So how does one start with an imagination that can view (fixed) phantasms, that is responsible for error, that is dependent upon sensation as a motion, and then end with an imagination that is the highest power animals have for purposefully directing their activity, that can in fact be called a form of intellection, that can undertake deliberation about good and bad, that is indispensable for reasoning and intellection, and that amounts to the very element in which human beings conceive of and think about what they do? And why, even if in the last analysis Aristotle asserts that the first or simplest intelligible things are not images but do require images, does he suggest the possibility that the simplest and most purely intelligible forms *are* images?

In "The Discovery of the Imagination," Cornelius Castoriadis argued that *On the Soul* presents two distinct theories of imagination.[40] He points out that, for the most part, the sequence of topics in *On the Soul* is quite logical. After book I reviews previous conceptions of soul and raises a long series of questions, books II and III

---

[40] See Castoriadis 1997. The French version dates from 1978 and is based on an earlier one in Greek.

advance Aristotle's own conception. Book II begins with his generic definition of soul (as the first actuality of an organized body potential with life) and a discussion of the criteria for calling something alive, whether plant, animal, or human being. It is followed by a presentation of powers shared by all living things (nutrition and reproduction) and then of the sense powers possessed by animals (a general accounting of the types of the sense objects—proper, common, and concomitant[41] sensibles—followed by a discussion of the five external senses, in the order vision, hearing, taste, smell, and touch, and concluding with reflections on the general character of sensation). Book III, after attempting to show that there can be no more than five external senses (because of the kinds of physical matter required),[42] discusses common sensation (in which the information channels of the five individual senses are united into the experience of a single sensed world), imagination, and intellection—one chapter each dedicated to common sensation and imagination, then five chapters to intellection. After three chapters on purposive movements and actions in animals (including human beings), the last two chapters summarize and draw further implications for understanding the life and powers of ensouled bodies.

The account of imagination in III.3 fits in neatly with this smooth progression of the treatise. It presents the imagination in a form familiar to us from more than 2,000 years of tradition, a power that follows and is dependent upon sensation and is transitional to other, higher powers. But then, in Castoriadis's words, the "ordering of the third book of the Treatise is brutally shattered on two occasions: first, by the sudden reappearance of the question of *phantasia* right in the middle of the examination of the dianoetic [i.e., discursive intellectual] potentiality…; then, by an insistent return of *phantasia* throughout the examination of the potential for movement" (Castoriadis 1997, 222).

Aristotle's III.3 version decisively shaped the conventional accounts of imagination in philosophy and psychology: as the power or faculty that is *intermediate* between sensation and intellection. It is portrayed as wholly dependent on sensation, and it serves as the necessary bridge between the two really fundamental

[41] This third kind is more typically known as "accidental" sensibles—in modern terms they might be called "things sensed by association." Castoriadis translates Greek *sumbebēkos* as French *comitant*, which Curtis, his translator, renders in English as "comitant." I will use "concomitant." In any case, it is long past time that the misleading "accident/accidental" be abandoned. The concomitant sensible corresponds, roughly, to the pseudo-Platonic definition of imagination: that one has a sensation (e.g., of a white-clad figure) plus an opinion (e.g., the white-clad figure is your friend). When we almost immediately identify the person coming toward us in white as our best friend, we say that we have sense-perceived the friend in the accidental or concomitant way. See Sect. 5.7, below.

[42] Aristotle's conclusion in III.1 is that, if earth, water, air, and fire are the basic material elements, then the five enumerated senses exhaust the relevantly possible combinations in the material constitution of the senses and their media of transmission. The conclusion depends on particulars of his (and ancient Greek) matter theory that by modern scientific criteria are quite wrong. Nevertheless, the argument provides an instructive example of how one should try to establish not only that one's descriptions and theories are plausible but also that they are complete and consistent with respect to other, more basic theories.

human powers. At least some of Aristotle's predecessors had managed to distinguish sensation from intellect; Aristotle's distinction of common from proper sensation and his insertion of imagination between sensation and reason served the purposes of those who came after him. Castoriadis calls this "second imagination." The second imagination helps explain the reproductive, combinatory, animal functions of imagination. Calling it "second imagination" indicates that it is second with respect to something more original, the first or primary or "radical imagination," which emerges in III.7 and III.8 but is no more than partially and implicitly developed. There, in the course of a discussion of intellect and the problems it raises, Aristotle comes to uneasy grips with the nature of an imagination that threatens to unsettle human psychology as a whole, in particular because it is irreducible to either sensation or intellection and therefore challenges their dual primacy in human cognitive being.

In Castoriadis's understanding of the history of Western theories of imagination, Aristotle's case is exemplary: those who have thought most innovatively about it rarely exploit their insights adequately, and they often end by backing away from the most radical consequences of their theories. The appropriation of these theories by their followers and other intellectuals almost always reverts to a homogenized, domesticated version of imagination, to a derivative power somewhere in the middle between sense perception and rationality and productive more of error than of truth.

One may have reservations about Castoriadis's dramatic claim that primary imagination "shatters" the order of the treatise, but there is little doubt that Aristotle's path from imagination in III.3 to imagination later on is more than a little puzzling. There is also little doubt that the conventionalized Aristotelian psychology, described in Sect. 5.3, above, derived from attempts by later thinkers and teachers to smooth out some of the conceptual difficulties in understanding Aristotle's psychology as a whole, to accommodate it to later (especially medical and religious) concerns, and in particular to provide a more precise, sequential, and circumscribed account of imagination around the canonical assertions of III.3. Castoriadis rightly asks whether their interpretations did not overlook basic motivating questions, phenomena, and concerns that Aristotle had in view. Castoriadis is right, I think, about the inadequacies of the derived traditions. In what follows I will show, however, that if we attend to these basic questions, phenomena, and concerns we can discover, *pace* Castoriadis's claims, the Aristotelian unity of primary–radical and secondary imagination.

5.6 The Imagination of *On the Soul* III.3: What It Is and What It Isn't

It is important to acknowledge that III.3 *is* a transitional chapter. Book II had dealt with the general definition of soul, with the criteria by which we say something is alive, and with features of sensation in general and the individual senses in particular. III.1 argued from the general properties of matter that there could not be more than five proper senses; III.2 then gave an extensive portrayal of how the individual senses are united, and common sensibles emerge, in common sensation. The elucidation of common sensation shows that there is at least one soul power that is "higher" than the individual senses but is still sensation. Indeed, the account of common sensation implies that the proper sensations are subordinated to it: they do not fully achieve what they are capable of until they are united there. Aristotle thus achieves a careful distinction of sensation kinds and begins to demonstrate a hierarchy in the powers of apprehending the world.[43] He has also begun making good the criticism that his predecessors had failed to sufficiently distinguish from one another the human sensitive and intellective powers. III.3 is meant to establish that imagination is not simply sensation, but also that it does not rise to the level of the intellective powers proper. That means that the sequence (1) proper sensation, (2) common sensation, (3) imagination is still not articulated enough to account for the cognitive powers of souls.

III.3 begins by reminding readers that earlier thinkers had attributed two different functions to soul, (a) motion with respect to place and (b) thinking, discriminating, and sensing. Deferring the question of whether and how the soul produces motion to the end of the book, he says that most earlier investigators considered thinking (*noein*) and judging well (*phronein*) to be similar to sensing (*aisthanesthai*). What they overlooked, he says, is what can be revealed by giving an account of errors due to sensing and thinking. Since many of those investigators argued that sensation works by "like perceiving like," error presents a dilemma: either appearances (*phainomena*) are always true, since they result from like perceiving like, or they somehow, in error, encounter the unlike. Aristotle had already argued that the senses are never wrong in the narrowest sense, with respect to proper sensibles—vision is not wrong about seeing the colors it does, hearing about tones, etc.[44] From this unwelcome dilemma it is evident to Aristotle that the simple identification of thinking and judging as nothing more than a kind of sensing is wrong, since thinking and judging *can* be wrong, whereas proper sensation is always true, even if common and concomitant/accidental sensation can be in error. Therefore the question about what other forms of true sensation and cognition exist, besides proper sensation, requires further exploration.

[43] This is a hierarchy predicated on different levels or planes, from and through which one can look to other planes and to world objects (whether real or possible). In this sense Aristotle builds his psychology on Plato's precedent.

[44] A weak way of interpreting this claim is that proper sensation is one of our sensitive and cognitive powers *least* subject to error. Since Aristotle distinguishes common from proper sensation, he clearly does not mean that, when we taste a powdery substance and it tastes sweet, we know that the substance is sugar or even that the substance itself is sweet—such judgments require more than proper sensation, since they presuppose the unification of the proper sensibles into a common experience and include the identification of kinds or species. The claim seems virtually unobjectionable if it is taken to mean that, in such cases, we are not deceived in thinking that the taste is sweet—sweetness *is* the appearance to taste as such—though virtually anything we say about it in a logical judgment might be in error (e.g., we might confuse meanings of words in our utterance). Even if this turns out to be a very limited claim about only the most basic sensations, it nevertheless justifies distinguishing proper sensation from other kinds of sensation and other forms of mentality that are more prone to error.

Aristotle immediately turns to drawing distinctions between the different thinking and discriminating acts of soul, and it is precisely at the outset of this task that he introduces[45] *phantasia*. Sensing and judging rightly (*phronein*) are not the same, he says, because though all animals sense, not all of them judge. Since thinking may be right or wrong, but sensation in the case of proper sensibles is always true and possessed also by nonthinking animals, neither can *sensation* and *thinking things through* (*dianoia*) be the same.

> Imagination is different both from sensation and from thinking things through, and does not come about without sensation, and without imagination there is no conceiving that something is the case [*hupolēpsis*].[46] And it is clear that imagination is not the same activity as conceiving that something is the case, for the former experience is available to us whenever we want it (for it is possible to make something appear before the eyes in the way people do who make phantasms to fit things into a memory-assisting scheme)[47] but to form an opinion is not up to us, since it has to be either true or false. Also, when we have the opinion that something is terrifying or frightening we immediately feel the corresponding feeling, and similarly if we think it is something that inspires confidence; but with the imagination we are in the same condition as if we were beholding terrifying or

[45] *Phantasia* had already made cameo appearances, in particular in the chapter on hearing, in the course of discussing the difference between voice (meaningful sounds) and mere sound (II.8, 420b29–32): "for not every sound of an animal is a voice, as was said—for it is also possible to make a noise with the tongue or in the way people do when they cough—but it is necessary for the part that causes the striking [of the air against the windpipe] to have soul in it and some sort of imagination [*phantasiai*] with it, since the voice is some sort of sound that is capable of carrying a meaning."

[46] *Hupolēpsis* is usually translated by "belief," sometimes "judgment"; but the former is too generic and the latter too definite, and it does not capture an *incipient* aspect implied by the Greek. The long phrase that Joe Sachs's translation employs, "conceiving that something is the case," comes closer to the phenomenon, especially if we keep in mind the actional, ongoing sense of the gerund "conceiving." Caujolle-Zaslawsky argues that we must interpret *hupolēpsis* against the background of the verb from which it is derived, *hupolambanein*, which suggests a sudden coming–upon or taking–hold–of from below; see Caujolle-Zaslawsky 1996, esp. 352–356. In the passage from III.3 quoted above, *hupolēpsis* is used as the genus of *epistēmē*, *doxa*, and *phronēsis*. Perhaps what Aristotle was remarking with the word was the phenomenon, whatever its specific modality, of the first moment of being struck by, or recognizing, the ways things are in an appearance, the moment when things appear thus and so—sometimes wrongly, as when we mistake what we see (e.g., take a red traffic light for an orange one).

[47] The first attestation of these imaginative memory-assisting techniques derives from a story told about the poet Simonides of Ceos (ca. 556–468 B.C.): right after the collapse of the roof at a party he had been attending, he was able to apply the memory technique he practiced (of associating things with locations) to recall who had been sitting at each place around the table. This technique of remembering according to places (*loci*, plural of Latin *locus*) is the central practice of the art of local memory, which was a staple of ancient rhetoric. For example, rhetors memorized speeches not by repeating words over and over but by associating the themes, elements, and even words of the speech with visual images, then arraying those constructed images in a familiar and thus easily remembered space that could be mentally traversed in a convenient sequence (say, attached to architectural features along a path through a public building they knew well).

> confidence-inspiring things in a painting. There are different ways of conceiving that something is the case—knowledge, opinion, understanding, and their opposites—but let the account of their differences be given elsewhere.[48]
>
> As for thinking [*noein*], since it is different from sensing, and since it seems that one sort of thinking is imagination and another sort of thinking is conceiving that something is the case, one ought to speak about the latter sort after having thoroughly distinguished what pertains to imagination. Now if imagination is that by which we speak of some phantasm as becoming present to us,[49] rather than anything we might call imagination in a metaphorical way, is it some one among those potencies or active states by which we discriminate something and are either right or wrong? Of this sort are sensation, opinion, knowledge, and intellect [*aisthēsis, doxa, epistēmē, noûs*]. (427b15–428a5)

Aristotle takes all the terms he uses here as more or less familiar to his audience. His object is to tease out the character of *phantasia* by contrast. The characteristics of imagination Aristotle identifies and differentiates in this passage are the following. (1) Imagination is distinct from sensation and from thinking of one sort, *dianoia*, but it seems to be thinking of some sort. (2) Imagination is dependent on sensation. (3) *Hupolēpsis*—conceiving or recognizing or being struck by the fact that something is thus and so, and an act that is the genus of all specific forms of being right or wrong—cannot occur without imagination (this is therefore a first approximation to the contention that there is no thinking without phantasms). (4) Imagination is nevertheless not identical to *hupolēpsis* and its specific forms, because there is a difference between an appearance that occurs to or for us (for instance, an image that pops unbidden into mind awake or asleep, or an image contrived for remembering a speech according to the art of local memory) and recognizing or conceiving the image–appearances as presenting something in a definite state, thus truly or falsely. This distinction means that some imagining is not false. (5) Imagining is something that we can undertake at will, whereas opining and *hupolēpsis* are not. (6) In contrast to situations where unpleasant real things present themselves to us or where we make a judgment in thought about pleasing things, in imagining we ordinarily are not overcome by feeling, passion, or emotion but rather witness the appearance with relative detachment. (7) Imagination is the coming into being for us of the phantasm or image.

These contentions are by and large consistent with the long tradition of conceiving imagination as "acquiring and holding a picturelike image in mind without the presence of the corresponding object." Although points 1 through 6 do not state what imagination is in a simple formula, they all capture salient features of it

---

[48] It is possible that *elsewhere* indicates the treatment of the intellectual virtues in book VI of *Nicomachean Ethics*. At any rate, Aristotle is putting aside the detailed account of the different species of conceiving something to be the case (*hupolēpsis*). *On the Soul* instead discusses only the *basic* powers of soul, which of course may also be combined into more complex or derivative functions.

[49] The verb here, *gignesthai*, in its most basic sense means "come into being." Thus Aristotle here marks the incipience of the image as the distinctive topic to be addressed.

(that is not to say that all later thinkers would agree with all of them).[50] Aristotle does give a kind of definition (which we have marked as point 7) at the end of the quoted passage, although it is clearly provisional and in need of refinement: "imagination is that by which we speak of some image as becoming present to us," with the exclusion of merely metaphorical uses of "imagination." What these metaphorical uses might be is, for the moment, unclear, and a proper definition of imagination has to wait until the end of III.3, after a discussion of how imagination differs from sensation, opinion, knowledge, and intellection. Yet the provisional definition strongly indicates that Aristotle understands imagination as starting with the *inception* of image–appearances, with the incipience of appearance.

Aristotle makes several points in differentiating imagination from sensation. Sensation exists in us either as a potency (the power of seeing) or an activity (actual seeing), but we can form images, as in dreaming (involuntarily) or memory art (at will), when neither the potency nor the activity of seeing is present. Some animals, like worms, may have sensation without imagination. Sensations (of the proper sensibles, at least, though Aristotle does not specify this here) "are always true, but most imaginings turn out to be false" (428a12–13). In a situation where we are trying to see or sense something accurately, we do not say that we are imagining it but that we do not yet see plainly what is true or false—that is, sense images moving or static have been incepted or received, but *hupolēpsis* has not yet supervened. And (repeating a point made earlier) visual images can occur to those whose eyes are shut.[51] Aristotle makes even shorter work of differentiating imagination from knowledge and intellection. Knowledge, precisely as being known, is always true, but imagination can be false, so they have to be different. He throws in intellection (*noein*) too as distinct from imagination, although at this point in the presentation he has not yet made clear what it is. But especially if one takes it in the narrow sense

[50] For example, in some versions of empiricism thinking would in essence *be* the series of images we entertain, making the distinction between thinking and imagination moot.

[51] I take this to mean *not* that, with our eyes closed, we can see reddened light or afterimage colors—both of which would be an actual perceiving of colors—but that we can close our eyes and (for example) imagine the face of a distant friend or a favorite but remote landscape. In *On Dreams*, ch. 2, Aristotle addresses the continuation of affections of the sense organs even after the perception has ended (for example, the procession of afterimages when we turn from light to darkness); he explains them as analogous to the continuing motion of a projectile after it has left the hand of the thrower or the continuing presence of heat in something after the heat source has stopped heating. An oddity is that—although chapter 2 in *On Dreams* runs to just over four pages in the Loeb edition, and although the end of chapter 1 (a) reminds us of *On the Soul*'s definition of *phantasia*, (b) calls dreams a kind of phantasm, and (c) concludes that "it is clear that dreaming belongs to the sensitive power, but qua imaginative [*phantastikon*]"—the terms *phantasia* (in the meaning "imagining act" rather than "power of imagination") and *phantasma*/*phantasmata* occur exactly once each in chapter 2, in the last half of the chapter's final paragraph. There the situation of a feverish person who sees patterned marks on a wall is contrasted with that of a person who *sees* the sun as a foot across (this is referred to as a *phantasma*) but by another power *knows* a truth that contradicts this imagining (*phantasian*). This and most of the other subject matter discussed in this chapter of *On Dreams*—which does not in fact treat of dreaming!—are thus sensations, not *phantasia* proper.

of a *noēsis* (thought) of a single meaning–term, say recognizing a single universal concept (e.g., "dog," "blue," "three-legged," rather than a dianoetic predication like "S is P," which could be either true or false), Aristotle's claim appears to be correct.

Aristotle then turns to the question whether imagination is opinion, a possibility not yet ruled out since both imagination and opinion can be either true or false. He quickly dismisses the possibility that it is simply identical to opinion, by reflecting that we always have trust or belief (*pistis*) in our opinions. Belief requires some kind of reasoning ability. Animals other than human beings, Aristotle points out, have imagination, but because they lack reason they cannot have belief. He reflects further that belief accompanies opinion, persuasion accompanies belief, and *logos* or speech accompanies persuasion; but except for human beings there is no evidence of speech in animals, whereas many of them quite clearly have the power of imagining.

Animal imagination thus also immediately disconfirms a definition of imagination that was attributed to Plato, that it is *opinion* associated with or added to *sensation*, but Aristotle elaborates the differences further by developing the consequences of this faulty Platonist notion.[52] The first criticism he makes arises from his distinction of different kinds of sensation, the proper, the common, and the concomitant. If imagination were opinion about sensation, imagining in its most primitive sense would have to be opinion about proper sensations like seeing white; indeed, the most primitive of all such opinions would be the opinion that the white one was seeing was white. This creates immediate complications if one tries to add on the other sensibles and any respective opinions (or associations) about them to reach the level of a complex matter of opinion.[53] What is yet more decisive for Aristotle's refutation of the Platonist definition is the problem that arises when something appears to sensation in a way that we already think is false. That is, there can be false appearances (false compared to the thing itself) with an attending opinion that is true. He uses the example of the sun, which appears to be small, but we opine it at the same time to be far larger than the earth (if we have made the proper kinds of astronomical and mathematical inquiries, we can even say that we *know* it to be larger). The difficulty arises like this: At one level we have a sensory appearance that seems to us to be a such and so, so we must opine it to be so, and according to the Platonist definition, that is an imagining. But, at the same time, we have an opinion about that sensory appearance that contradicts the first opinion; that, too, would seem to be an imagining (an opinion, albeit a second-order one, about the initial appearance).

[52]This conception of imagination is strongly suggested but never quite stated outright in Plato's dialogues. See the formulations in the *Timaeus*, 52A, and the *Sophist*, 264A–B. Its inadequacy as a definition of Plato's true understanding of imagination should be clear from our previous chapter; see esp. Sect. 4.4.

[53]Perhaps, in contemporary terms, we could say that the pseudo-Platonic definition, viewed in light of the different kinds of sensibles and the different levels and kinds of possible opinion about those sensibles, creates problems of recursion: how is an opinion about a concomitant sensation related to the proper and common sensations and the opinions about them, which are included in the concomitant sensation as parts?

Thus imagining is either contradictory, or somehow the opined appearance changes, in a way that Platonists do not explain, between the first and the second opinion. The Platonist definition of imagining as sense plus opinion once again adds unnecessary layers of complexity to the psychological phenomena.

Aristotle's point seems to be that these problems do not arise at all if we reject the definition that makes imagination a sensory appearance along with an opinion. Opinion always has to be either true or false because it is about something, and sensation is always of a sensed object, but an imaginative appearance, without regard for the object that appears, is just an appearance and does not need to be attended by any judgment or intrinsic truth as an object–appearance. That is the advantage of his notion of imagining over the pseudo-Platonic one. If human beings have the power of apprehending appearance–forms without the objects of appearance, then the definition of imagination as sensation plus opinion overlooks the point that there is a phenomenon, a "mere" appearance without an immediate object, that the definition fails to notice. That is, the power that Aristotle is concerned with falls outside the scope of the (supposedly) Platonic definition. But even for Plato, there ought to be a type of appearance–having that precedes opining and that is not sensation of an object.

If someone wants to define imagination as sensation–appearance plus opinion, that is his semantic right, but then, apart from being confused, it is something different from the imagination that Aristotle is isolating and defining. Aristotle's distinction here also separates him from Plato's theory of *eikāsia* (found in the eikastic art of the *Sophist* and in the *eikāsia* of the divided line). It is proper to *eikāsia* to see an image and to see it also as the image *of* something else; *eikāsia* begins with an appearance on the image–plane that seamlessly draws the mind to a thing on another plane. Aristotle insists, by contrast, that *phantasia* is just having the appearance.[54] If someone insists that it must still be the appearance of something, Aristotle can counter that the image need not be of any *other* kind of thing at all; an imagined red is a color, not a substance, even if the imagined red was once a perceived red. In the divided line Plato thus assumed, wrongly, that *eikāsia* always contains an implicit *hupolēpsis* that sees a this as a that (this shadow as the shadow of a tree). Only in the allegory of the cave did he entertain, though just implicitly, the possibility of appearances without reference to what produces them. This appearance without opinion must always be possible for Aristotle's *phantasia*.

It is at this point in III.3 that Aristotle makes the remarkable claim with which we began Sect. 5.5, above, that imagination is a kind of motion that starts in sensation. In fact he had already described sensation itself as a variety of motion in II.5 (418b34): sensation "consists in being moved and acted upon." We are within striking distance of understanding why he makes this shift to physics (theory of nature) in the strict sense. First, however, we must catch up on some implications of treating sensation and imagination in terms of motion.

---

[54] Aristotle here appears also to implicitly criticize Jean-Paul Sartre's conception of imagination; see Sect. 2.1, above.

5.7 Imagination, Sensation, Motion

Aristotle's III.3 discussion of imagination makes clear that it is dependent upon and cannot be understood apart from sensation. It constantly takes sensation and its character as a point of reference and of origin. The immediate context of III.3's portrayal of imagination is set up by a long discussion in III.2 of *common sensation*, and that follows immediately upon a nine-chapter-long presentation of sensation and its various forms that begins a third of the way through book II.

Sensation as Aristotle understands it requires the activation of a sense organ, and the sensation is, in the proper sense, precisely the activity of the organ. Before being exposed to an appropriate object (say a visible thing) with an appropriate medium between the object and the organ (a transparent medium activated by light, in the case of vision), the organ is in a state of readiness to act—the organ and the faculty are *dunamei*, in potency or potentiality.

What is perceived in the activity of an individual sense organ and only that organ is called a proper sensible; that is, the sensed thing that, in kind, is proper to that specific organ and to no other. Color is the proper sensible of visual activity because it can be sensed by the eye but by no other organ; similarly for tone with respect to hearing, flavor with respect to taste, aroma with respect to smell, and tactility with respect to touch. Shortly we will need to say more about how each proper sensible is differentiated—into many possible colors, smells, etc.—a differentiation that gives rise to a *field* of relevant phenomena, within which *phantasia* subsequently has a certain freedom of play.

In the middle of book II Aristotle contrasts the proper sensibles with two other kinds: the common and the concomitant. He specifies the common sensibles here as motion, rest, number, shape, and magnitude. The common sensibles are "not proper to any one sense" but are "common to all" (418a19–20). Motion, for instance, can be sensed by touch (at least in the case of things near to the sensing animal), by vision (color patches moving through our visual field), and presumably also by the other senses.[55] The concomitant sensibles have a greater complexity. Often something like this happens: we glance down a side street and notice some contrast of light or color, we look a little more intently and see a figure in shadow, and then recognize the figure as a person and finally as the son or daughter of a friend. In this

[55]"Presumably," but not certainly. One can of course detect motion, or learn to detect it, through changes in sound. We can, by moving our heads or bodies, often gain information about the source of a smell. But it is not clear at all how motion detection might function with respect to taste. The issue would not seem to be addressed by the fact that different flavors are sensed on different parts of the tongue. Insofar as motion requires time, however, the "blossoming" of a flavor or the alternation of flavors one after another might qualify as motion that is perceived by taste; and something similar could be said of aroma. But it would almost certainly require conceptual acrobatics to show that *shape* is perceived by taste (the tongue can, of course, feel shapes, but not by means of its flavor receptors) and smell. Perhaps what Aristotle really means, or should have said, is that a common sensible is detectable by more than one sense, but not necessarily all.

case we have moved from a proper sensible (light or color) to a common sensible (figure or shape) to a concomitant sensible (our friend's offspring); and often we say that we "saw" the friend's child from the first moment of noticing. That is, the color–and–shape appearance belongs to the substance that is the child of the friend, so concomitantly we sense the friend's child. What Aristotle says is this: "A thing is said to be concomitantly sense, for example, if the white thing is the son of Diares, for this latter is sensed concomitantly, because it is concomitant to the white that is sensed, for which reason nothing is acted upon by the concomitantly sensed thing as such" (418a21–24). The sense organ is affected or acted upon by white; Diares' son does not, as such, affect the organ. But something that we know, the underlying identity of this white thing that affects our vision, becomes concomitantly attached to the sensation.

Sensation in general "follows from being moved and acted upon," Aristotle says (416b34–35). Something similar could be said of anything physical for Aristotle: for a physical thing to undergo change, something has to move and affect it. More specifically, what the sense organ does, when it is active, is to be affected by the sensible form that is originally in the object and that is physically communicated from the object to the organ through a medium. To be more accurately Aristotelian, the activity of the sense organ and the activity of the sensible object become the same, although their essence is not the same (425b27). What this means (to use the case of visible objects) is that there is an activity at the surface of or within the object; this activity *is* the color, and it is somehow conveyed to the eye, which takes on that very same color activity, which in the eye is the *sensed color*. There is of course a difference: the activity in the object is an attribute of (a part of) the object and activity of its matter, whereas in the eye the activity is no longer part of the object's matter but has become something like an attribute of the material functioning of the eye. The activity–form in the sense (organ) is the same as the activity–form in the sensible thing, but without the thing's matter.

To make sense of this we need to do some work to see what is at issue. From the perspective of the modern sciences Aristotle's theory of vision seems antiquated. There is no color in the object. All there is is a disposition to interact with light rays or photons, that is, to absorb some and to reflect others. The reflected rays travel from the object to the eye and are focused upon the retina, where they have a disposition to interact with the nerve endings there according to the wavelength or energy level of the ray and the chemical character of the nerve receptors. The energy of the light is converted into electrochemical action that is transmitted from nerve cell to nerve cell and ultimately to the brain, where there is further electrochemical activity involved in signal processing by the visual areas of the brain. All there is in the entire nerve process is electrochemistry, not color. When the electrochemistry has done its work the experimental subject will report seeing colors. We thus say that color is in consciousness, not in any of the natural processes that occur. Alternatively, if we are pure materialists, we probably want to say that color perception is nothing other than precisely this entire process. Ironically, this is what Aristotle seems to say: the soul is the natural activity of the body's organized matter; in particular, sensation is the activity of the sense organ. We would want to amend that today to say that it is the activity of the visual *system*. But that may be only a relatively small difference physically, and even less metaphysically. Aristotle's physics is antiquated, but his physical commitments are not far removed from ours today.

As we saw in our previous chapter, Plato offered an ontology of imaging: images conform to physical things, physical things conform to mathematical things, mathematicals conform to ideas, and all are governed by the good itself. This conformation governs not just being but appearing. In Aristotle there is no single grand power from which all imaging proceeds; nevertheless, nature in general acts by forming matter. Existing forms are in individual beings (substance) and these forms shape and speciate the matter so that the substances act and appear in accordance with the possibilities defined by those forms. Form as activity literally and specifically (as *sensible* form) governs what is perceived by sensation. Activities are communicated in the process of sensation. They carry forward beyond sensation as *phantasia*, which resembles sensation and is an extension of the activity and motion that sensation actually is in its act of sensing.

As is typical with Aristotle, initial oddities of expression arise because something quite familiar needs to be put into its larger context in a formulation that applies to everything of its kind. What is sensation, sense perception? By his very understanding of soul as an actuality of an organized body, Aristotle is predisposed to think of specific functions as activations of the organs, which stand in readiness to act (thus are in the state of first actuality) after the growth and maturation of the animal has brought the organs to a developed state. The sense organs develop in the gestation of the newborn; at birth they are in readiness to be activated in the appropriate circumstances. Vision, the proper working of the eyes, becomes real rather than just possible, second-degree actualized rather than just first-degree actualized, in the presence of light; hearing, the working of the ears, becomes actualized in the second degree in the presence of sounding (even, as we know, in the womb).

Aristotle's physical understanding of what happens is different from ours yet not unintelligible or absurd. In the case of hearing, his theory is a first approximation to our own. In that sense they share a topology. There is an activity in the sounding thing in the world that is stimulated when it is struck, and the thing repeatedly moves—we call it vibration. This motion of the thing is conveyed by a medium (which is typically air but can be other media, like water) to the ear, where it becomes enclosed and concentrated in the ear canal and there gives rise to the hearing of sound.

Aristotle does not have a nerve theory, of course. This presents us with two specific problems in assimilating his theories to our own. First, he believed (contrary to many of his later followers and adapters) that not the brain but the heart was the place where the effects of all the different organ activities were ultimately united in common sensation. The brain, according to Aristotle, cooled the blood; to use an automotive metaphor, it was the body's radiator. Yet if Aristotle did not have a nerve theory proper, he thought that the motions set off in the body by the activities of the external organs were carried deeper into the body and taken up into further processes. This is of crucial importance for understanding the nature of *phantasia*, the motion originating from sensation. Nevertheless, he believed that the proper activity of the sense organs was, in at least some respect, complete in itself: color perception takes place at the level of the eye, hearing at the level of the ear, etc.[56] We certainly believe that some sort of event is complete at the level of the eye—for example, the effect of a photon on a cone or rod in the retina—but seeing or hearing cannot take place unless the nerves translate these events into the electrochemistry of nerve cell transmission, and these electrochemical signals move onward to the brain, where they are processed by the further electrochemical activity of nerves in different brain locations (vision, for example, is processed in the occipital portion of the brain).[57]

Yet it is not as though contemporary science has settled all the issues addressed by Aristotle. In fact, one of the impressions I want to leave is that, in an important way, Aristotle was the first to open up for scientific investigation these fields of interrelated phenomena, and he provided concepts to name and articulate the fields theoretically in a way that still makes sense and sometimes even informs contemporary research. Even when he was fundamentally wrong or incomplete in his theories and explanations (like the brain as cooling device), he provided the initial maps and surveys that allowed later investigators to more quickly familiarize themselves with the territories. And sometimes he raised questions with a kind of comprehensive intention that was set aside by later advances and specialization, a comprehensive intention that often has to be recovered later through a laborious historical process of rediscovery.[58]

I emphasize these points because a misunderstanding of their significance often blocks our appreciation of Aristotle—and something similar happens with almost all other past philosophers and scientists, as we come to take theories developed much later as obviously true and thus cannot understand how intelligent predecessors could have been so blind as to see things otherwise. Aristotle in particular

[56] This is one of the points of the first-paragraph discussion in *On the Soul* II.2.

[57] As usual, the more one looks into these matters the more complicated they become. One point is that Aristotle did not think that if you detached an organ from the rest of the body it could still perform its functions. This provides an opening through which a modern Aristotle could easily talk of a visual system rather than just an organ. Another point is that, in human beings at least, one does not have proper sensation apart from common sensation, except in rare or pathological moments. Perhaps something like a pulsing of light and colors in delirium or the brief swirling of light upon waking from a deep sleep—in neither case are we seeing things in a well-defined place and sequence—is the closest thing we have to proper sensation without common sensation.

[58] The point is sometimes made, for example, that for Aristotle and his contemporaries the sensible qualities of things (wetness, solidity, heat, and the like) were strongly connected to the theory of physical elements, and that this continued well into the early stages of the modern scientific revolution. The modern sciences, arguing that these sorts of qualities were secondary sense qualities and thus reside in the mind rather than in things, no longer felt an obligation to explain them at the level of physics or chemistry. With a more detailed and integrated scientific understanding, however, there has come a renewal of efforts to more strongly correlate physics, chemistry, and physiology with psychological events.

always strove in his explanations of things for both comprehensiveness and detail, at least as much as could be mustered given the state of inquiry. He pursued no area of knowledge just for its own sake; he saw the need to fit together everything that was known, and to investigate those things and fields about which too little was known.

Aristotle's theory of vision requires four basic conditions: the potentially visible object (not too small to be seen, nor completely transparent); the potentially seeing organ (the eye); the transparent medium between the object and the organ (air, or water, or something similarly transparent); and light. We can easily grant these things. If you have only three of the four elements you will not have actual seeing; all you will have is potential seeing, and potential visibility of the object. In particular, if you have a visible object, a healthy eye, and a transparent medium between them, there will be no actual seeing until there is illumination. What light does, for Aristotle, is activate the transparent medium. Once the medium is activated, it conveys the color activity in the object to the eye, and the eye takes on that same activity, without the actual material of the object. As Aristotle says, the activity of the visible thing and the activity of the organ are the same, although the thing and the organ activity have different essences—the former is a surface activity of matter formed into a substance and is restricted to being the color that it is, whereas the latter is the activity of the organ of a sentient being, an organ that can be set into as many different kinds of seeing activity as there are different kinds of color.

This is the basic schema that Aristotle sets up for all five senses. There is in the physical thing an activity–property; there is an organ in readiness to take on that activity and the same property formally, that is, without the matter; and there is a medium between the thing and the organ that itself needs to be activated so that it can convey the activity–property from thing to organ. The organ's activity will result in one of the many possible appearances it is capable of realizing. The eye will not be in a simply generic color activity but in a specific activity producing, for example, teal rather than aquamarine or azure. A few moments later, however, as the animal's direction of gaze turns, the activity in the eye will change to cinnamon brown, carmine, gray, or whatever other color is before it, whereas the color of the object turned from remains the same. In the final chapter of book II of *On the Soul* he sums this up in an enduring (but perhaps too mechanical, static, and oversimplifying) image, one that had already appeared in Plato's *Theaetetus*: sensation is like the impress of the form of a signet ring in wax. The same form finds its way into the wax, without the ring's bronze or silver or gold, just as the sensible form in the physical thing finds its way into the sense organ. What this image leaves out entirely is that having the form is an activity.[59]

[59] Beyond the form (the essential form) that accounted for the specific being of matter formed into a substance, later Aristotelians added various sensible forms to explain what is transferred from thing to sense organ. But that created an otherwise unexplained hierarchy of forms in the substance. I thus prefer to speak in terms of the less reifying and more authentically Aristotelian term, "activity."

5.8 What the Physics of Motion Implies

On the Soul is physics, an inquiry into nature. For Aristotle, physics is the study of change, which is also to say the study of matter, since matter is by its very nature changeable, and whatever changes must have matter (*Physics* 192a31–32). Nature (*phusis*) in the primary sense of the term is the principle that accounts for the motion and the stasis of things (192b21–23). In the *Physics* "motion" (*kinēsis*) is synonymous with "change" (*metabolē*; see *Physics* 201a10–11). In I.7 of the *Physics* (189b30–191a23) Aristotle argues that any generation or becoming of something requires a *pair of opposites* and an *underlying subject*. Motion involves a natural thing that undergoes a change from one real or virtual place to a relatively opposed one. What we consider to be the most basic kind of motion, local motion (*phora*)—that is, motion with respect to place—is just one type of motion, according to Aristotle (226a23–227b1, 243a6–11). The others are alteration (*alloiōsis*, change with respect to quality, as in the color change of ripening fruit) and increase/decrease (change with respect to quantity or size, as in growth). There is also another kind of "change" known as coming–to–be (or coming–to–be and passing–away, or generation and destruction); it is not, as are the aforementioned kinds of change, something undergone by an underlying thing, but rather the very origin of the underlying thing. Thus it is necessary to put "change" in ironic quotation marks, because, as Aristotle says, coming–to–be and passing–away are not actually changes.

Aristotle explains this contention at length in V.1 and V.2 of the *Physics*. It is a subtle but important point that does not jibe well with the standard meanings of the corresponding English terms. For Aristotle, change/motion in the technical sense requires that there be a single persistent thing undergoing the change, whereas in generation and destruction that very thing comes into or passes out of being, so there is not a single thing that persists throughout the process. Thus in the *unqualified sense*, as he is wont to say, generation is not change. But in a qualified or limited sense it is possible to say that the motions in or of a single underlying subject matter can be conceived as a kind of generation/destruction. For example, a newt or the leaf of a tree that goes from green to brown undergoes a "destruction" of the green appearance as the brown color is generated. In this way, one can extend to all types of change and motion the scheme involving three elements that book I of the *Physics* describes for generation/destruction: two contraries (like nonbeing and being, death and life, white and black) and an underlying subject (of which or in which the generation/destruction takes place).

As it turns out, in VIII.7 and VIII.9, Aristotle argues that local motion (*phora*) is the *primary* kind of motion and involved in the other two kinds (i.e., in alteration and increase/decrease). In *On the Motion of Animals*, chapter 5, he asserts outright that every change or motion *necessarily* involves local motion, motion with respect to place. If, then, we say that imagination is a *kinēsis*, it can be further specified as one of the three kinds of change: alteration, increase/decrease, or local motion—with the understanding that some kind of local motion will be associated with it even if it is one of the first two. This perhaps allays somewhat our surprise that imagination and sensation are defined as a motion: the motion could be the qualitative alteration or the increase or decrease of some underlying thing, and if it is not simply identical with the local motions of matter it will involve such other motions.

One last roadblock to understanding the motion of imagination is Aristotle's conception that motion must have both an origin and a destination. It is relatively easy to explain but hard to justify in light of the modern scientific concept of motion, which in effect reduces all motion to directed local motion. Aristotle considered motion to be the key question to address in order to resolve the mysteries of change. The universe is well ordered as a whole and in its parts. The best-ordered things, like the motions of the stars, seemed to Aristotle to be the least subject to dissolution, change, or irregularity; the stars keep the same relation to one another, night after night, though they appear to circle the North Star once each day. Things on the earth are more diverse and more subject to change and disorder, yet even they do not change just randomly. Aristotle was of course not the only one to notice these facts—it was part of the intellectual heritage going back to Thales. Aristotle argued, from the principles of the matter theory that he inherited (but also further developed), that the kind of local motion that occurs depends on the kind of matter in a thing. The bodies in the heavens are made from the most perfect kind of matter, aetherial matter, which is not subject to alterations, increases, or decreases but only to local motion; and because it is so perfect, aetherial matter moves in circles. Uniform circular motion, of all the local motions, seemed to him (and also to Plato) least in need of explanation, because at every moment of the motion the moving thing maintains the same speed and the same distance with respect to the circle's center. That is, the thing is always at the same distance from the center and always moving at the same rate along the circumference of the circle. The earth and its surroundings, on the other hand, consist of diverse mixtures of four other elements, earth, water, air, and fire, the sublunary elements; these in turn consist of different proportions of two pairs of contrary sensible qualities, hot/cold and moist/dry. A change in the qualities, taken far enough, changes the element. Earth, which is basically cold and dry, by having its heat greatly increased becomes hot, which drives out any moisture mixed in it and eventually turns it into hot and dry fire. This kind of change could thus account to some degree for phenomena like burning and, more generally, explain why in contrast to aether the mundane elements are subject to dramatic changes and corresponding appearances.

The four sublunary elements, unstable and imperfect, also exhibit a less perfect form of local motion than aether: their natural tendency is to move in a straight line. Why is linear motion less perfect? Because, without further specification, it is headed to any and every point along the infinitely extendable line of motion, and thus one can say that it is headed nowhere in particular. Uniform circular motion is fully determined once you know its center, its radius, and its speed. A straight-line motion does not change direction, but it has no determinate end except in the accidental sense that it might eventually stop somewhere. For linear motion to be well defined, Aristotle thought that it needed both a specified beginning and a specified end.

The kind of linear motion we encounter in everyday life does tend to have a relationship to an origin and a goal—for instance, a baseball pitcher's throws have the pitcher's mound (and his hand) as the origin, home plate and the catcher's mitt as its goal. By nature earthy matter has a natural tendency to go from wherever it is resting toward the center of the earth. Earth sinks in water, so the natural place of water must be above earth. Air bubbles up from water; its natural place is above water. And flame ascends in air, so its natural place is above air, though once it reaches the aetherial stuff in the heavens it rises no further. The straight-line motion of these elements taken in pure form is thus accounted for. When any one of them is found out of its natural place, it strives to move to that place along a straight line. In the impure, highly mixed forms of ordinary objects, the natural motion will still be linear, toward or away from the earth's center, although the speed and natural place will vary according to which elements, the heavy or the light, preponderate. Aristotle never offered a theory in detail of how the blended elements produce the qualities we see in everyday macroscopic objects. One unfortunate consequence was that the basic element theory of Aristotle set him up for easy refutation once the notion of inertial motion emerged—that all kinds of matter move in the same way, and that motion tends to continue along a straight line unless some force stops it. Galileo's famous, even if perhaps apocryphal, experiment of dropping different objects from the clock tower of the Pisa cathedral (the Leaning Tower) showed in principle that objects of different kinds fell at the same rate, regardless of their composition.

But the theory that motion must have a starting point and an ending point makes considerable sense when we talk about artifacts (things made by art) and complex natural things like living organisms and their actions. In hunting, a cat begins from a motionless crouch, then springs on its prey. In a cell, the attachment of a molecule to a cell–membrane receptor sets into motion a chemical process that leads to ions moving into the intercellular space and (for example) passing into a neighboring cell, then reversing and starting all over again. For Aristotle, motions that are self-sustaining are complete and cyclical; those that are not are a transition from an initial situation to a final one. "Initial" and "final" are relative notions here, relative to the underlying subject and its situation.

Aristotle's motion–and–change theory is the context that begins to make real sense of the definition of imagination. Although we have been concentrating on motion with respect to place, the structure of motion for Aristotle always requires an underlying subject or substrate (*hupokeimenon*) and a pair of contraries (*enantia*, pl. of *enantion*). The pair of contraries provides the extremes between which motion is possible. Motion in general presupposes a starting point or origin and a point of arrival or destination. The definition of imagination tells us what the origin is (the activity of the sense organ induced by the perceptible object) but not the destination. Nor does it expressly give us the subject or substrate in which the process inheres.

Here is one way to explain things more clearly. There is an organ, say the eye, that is set into its specifically characteristic activity by a physical transfer process to it from visible objects in the world (through the medium of air illuminated by a lamp or the sun). What is transferred to the eye is the materially based color of the visible, real-world thing. The coloration of the matter is not an inert state but an activity. The eye that apprehends the color is also material, though of a different kind than that in the object: it is a material capable of receiving *any* color activity, not just that of the object of the moment. Between the object and the eye (presuming that there is light present and that the intervening stuff is transparent) there is the ongoing process of transference to the eye of the thing's activating principle of color. This activating principle is not color when it is in the intervening medium, but at the beginning and the ending points only.

The eye is not an end point in an absolute sense, however. Presumably the animal that sees has to respond to at least some of what the eye actualizes. A fly may respond to a moving shadow by flying away from it, and a frog to a horizontally moving dust mote by shooting out its tongue. Since the eye is material, the seeing of something must involve a local motion, that is, some motion of matter; and thus the activity in the eye can give rise to some further local motion. In particular, the local motion in the organ of sense can set into motion matter in the animal body that is adjacent to the organ. Although Aristotle knew nothing of the nervous system, it would actually facilitate Aristotle's explanation, since by virtue of its operation we can specify a pair of contraries (rest and activation, or inhibition and stimulation) with a substratum or underlying subject (the material organization of the cells along the nerve fiber). Since in animals all the sensitive and motive powers are organic, that is, exercised by organs, and all motion that is not circular must have an origin and a destination, if the motion of the external sense organs is not final in an absolute sense, then it can achieve relative finality only by being communicated to and reactivated in other organs or places.

Touch would be absolutely final, for example, if the touch were felt and there were nothing further than that. If motor activity is affected by touch, however, then there must be some motion initiating in touch that is finalized, at least relatively to the earlier motion, in another bodily place. Higher animals have common sensation, which means that, at the very least, the activities of the particular senses have to be communicated to where common sensation takes place.[60] If there is memory, the motions that were set off in common sensation work their way to whatever organs or places are involved in memory. Perception by the external senses guides an animal's behavior, so in some manner the workings of the organs of sensation have to communicate with the parts of the body responsible for physical response. Thus the being–activities of physical things in the world outside the animal are communicated in physical ways to the receptive external organs of the body, and the activities of the organs set loose other motions that move deeper into the body, to other organs and places and their respective activities. The ultimate result is the whole organism's behaviors and actions in the world.

---

[60]Nowhere in the discussions of III.1 and III.2 of *On the Soul* does Aristotle say that there is an *organ* of common sensation. See especially III.1, 425a14: "it is impossible that there should be a special sense organ [*aisthētērion*] to perceive common sensibles." To have a special sense organ would be in essence to have another organ of *proper* sensation. But there is some kind of unification of the proper senses in common sensation, and Aristotle's discussions in other works suggest that this unification or interaction occurs in the heart or in its vicinity. See *On the Parts of Animals*, II.1 (647a24–32), II.10 (656a28–30), and III.4 (666a11–13), and *On Sleep and Sleeplessness*, I.2 (455b34–456a5).

One can argue that two distinctions, of activity and motion on the one hand and of motion and appearances associated with the motion on the other, have an at least rough correspondence to a much more modern understanding of animal (and human) neurobiology. It was not uncommon in the seventeenth century for scientifically oriented thinkers to argue that sensation was a vibratory or a push–pull mechanism—examples are Hobbes and Descartes. This meant in particular that they conceived an essential continuity of motion from the sensed object, say a vase, through the retina and the optic nerve to some location in the brain, with the occurrence along the way of certain transductions of the motion from one mechanical form to another (for example, a light particle would collide with a nerve receptor in the eye and thus transduce particle motion into pressure or vibration along the nerve fiber). Our own contemporary science understands this process differently, as electromagnetic in the physics of the external world and electrochemical in the body. All along the visual pathway there are cell activities, understood largely in terms of molecule movements and chemical interactions; then there is a local motion that transmits the result of this first cell's activity to the next one, where a new activity followed by new local motion takes place in turn, etc. At a higher level of organization we understand what happens analogously: there is processing in an organ (in terms of chemistry and local motion), the result is passed along to another organ, where a new activity of processing takes place. In a generic sense—that is, leaving behind the specific details of the process—Aristotle's conceptual topology of organ activity–motion followed by a local motion transference followed by activity–motion in another organ is a more accurate description of what we understand happens than is the mechanical understanding introduced in the seventeenth century. We often conceive the seventeenth-century scientific revolution as rightly "rejecting" and "overcoming" Aristotelianism. But the argument can be made that, by today's scientific standards, it is necessary to reject that rejection.[61]

5.9 From Motions of Sensation to Structures of Imagining

Aristotle creatively adapted and altered the conceptual topology that Plato had bequeathed and thereby radicalized Plato's theory of the imaging process and of the specifically human power of appropriating images as such and forming new ones.

[61] A discussion of the rights and wrongs of these and similar judgments require independent investigation. One might begin by recognizing that "global" repudiations of theories often throw out babies with bathwater by rejecting a more basic conceptual topology they share. The aspiration of many seventeenth-century thinkers to a total reduction of the understanding of nature to simplistic mechanics and kinematics was never an adequate approach to all the phenomena of physical and chemical matter, much less to living matter. As Immanuel Kant, no enemy of mechanical science, wrote in §75 of the Third Critique, there was never going to be a Newton of the blade of grass—that is, living organisms would never be exhaustively explained by particle–and–force mechanics.

For Plato, the process was in essence ontological. A psychic structure, commensurate with the tiers of being, allowed human beings to apprehend things on one level in relation to things on other levels (including causal relations). For Aristotle the ontology of imaging and its causality had to be translated into a physical process, with crucial nodes or places (typically body organs) where there could be a renewed formation and corresponding appearance. Like Plato he resorted to *forms*, but now understood as *activities* in individual substances that are communicable through appropriate media. The images proper, however, were not fully constituted until they had activated the nodal places and become psychological or soul activity. That is, the activity of the soul had to occur as the activity (or, in common sensation, the interrelated activity) of organs whose functional cooperation *is* part of the whole life activity of the body. Sensation occurs first in the external sense organs. If an animal is capable of having such appearances without the presence of corresponding real-world objects currently affecting the sense organs, then it has imagination in the proper sense of the term. Imagination is the ongoing motion–activity–appearance that originated in sensation.[62]

On the Soul enumerates five proper senses, each of which has its proper sensibles. These proper sensibles are basic qualities of sensed things, and they occur in their one proper sense and no other. Thus Aristotle takes even further than Plato the mandate to give a scientific account of every kind of sensation.[63] Aristotle does not think that the five proper senses are adequate for sensing all sensible qualities, however. Although they are sufficient for perceiving red, loud, sour, rough, and the like, they do not have a direct perception of qualities that are common to two or more of them, like place and time.

It is worth recalling that Aristotle's account applies not just to human beings but to other animals as well. Suppose we investigate an animal that possesses touch (all animals must possess touch, he says) and the ability to discriminate light from dark, perhaps with a preference to move toward the dark (as cockroaches do when you turn on the lights). Some type of directionality and thus a certain aspect of a positional or place sense would seem to be implied, but without further evidence it is hard to assert outright that the animal experiences three-dimensional space as human beings do, even without the directionality of up, down, left, and right. If, for

---

[62] According to a summary Aristotle gives in the first chapter of *On Memory and Recollection*, both imagination and memory belong to the power of common sensation. The reactivation of the appearance thus does not need to actively involve the external sense organs, and the source of the motion responsible for the *re*activation would thus come from elsewhere in the body. (The later inclination to locate the potential for reactivation in other places or organs was thus a development of the topology Aristotle laid out.) This difference would seem to be sufficient to account for why the images of imagining are not typically perceived as though they were *actually* being sensed (if they were, they would be hallucinations).

[63] The mandate comes just after the account of astronomy in book VII of the *Republic*, where Socrates points out that astronomy is a science of what is seen and that every other sense might similarly have its own science, though he goes on to discuss just one more: hearing's science, harmony.

example, the animal sees light and dark but not *objects* as such, it would be hard to attribute to it a sense of near and far. If we discovered that the animal can detect motion in its field of view, that would imply a further sophistication of its experience.[64] It might not yet imply the experience of individuated things, however, since there is quite a jump from possessing light–dark and motion detection to having the ability to grasp the unity of a mass of matter as a thing. But there does appear to be a real advance in sensibility insofar as the animal notices a light or dark or colored patch as moving. Something like it is needed for a frog to be able to flick out its tongue at a small object moving across its field of vision, an object that is likely to be a fly but that will elicit the same behavior if it is a small plastic pellet.

Whether or not Aristotle intended such a specifically gradualist acquisition of sensitive powers (part of a "great chain of sensation"), his conceptual topology implies that graded steps like this will be found in the animal kingdom. Touch, which all animals have, implies only a minimal sense of space and place, but space as we know it requires more, and more sensitive, sense faculties, vision above all. Yet vision per se does not necessitate the perception of all the features of vision with which we are familiar (as the housefly's compound eye and the colorblindness of many animal species prove). Aristotle's *common* sensibles begin from the proper sensibles but involve their complication and their appearance in a field of overall perception whose "dimensions" not only contain the proper sensibles in copresence and interrelation but also the higher-level features of unity, shape, position, motion, and time.

Aristotle suggests rather than strongly urges this gradualist progression from a field of proper sensation to the comprehensive "place" of common sensibility, of ever more numerous and ever more acutely and accurately located and perceivable qualities.[65] His approach emphasizes instead the correlation of information from the five different proper senses and the emergence of new properties in the place of common sensibility. Vision perceives white because white is a color, taste perceives sweetness because it is a flavor, and touch perceives graininess because it is a tactile feeling, but the how and where of perceiving a white, sweet, grainy substance that we call sugar requires much more. Vision cannot judge of flavor and tactility, taste cannot judge color and feel, etc., so it appears that there must be another organic level of experience where this kind of sensory information is united and cross-sensory discriminations can take place.

The unification of different channels of sense information in temporally and spatially positioned things produces different objects of sensation, both simultaneously and successively, in a highly structured and articulated field of appearances. Sugar, salt, saccharine, tiny dried flakes of white paint, etc., can then be simultaneously and successively discriminated in the connected moments of world experience.

---

[64] Cockroaches can detect motion from slight air currents that disturb fine hairs on their legs.

[65] "Place" is in quotation marks because it implies far more than three spatial dimensions. Aristotle's differentiation between *space* and *place* is in any case not just a peculiarity of the Greek language.

If common sensation is to serve the perception and distinction of the variously visible, audible, smellable, etc., characteristics of things, it must first place them together in the ways of time and space.[66] For human beings and at least a few other animals, common sensation provides a unified field of proper and common discriminations.

Both the five proper sensations and common sensation operate in the presence of the object. In human beings, five channels of particular types of sense information gathered from the object are brought together in the common sense. With a view to what later Aristotelians developed from this basic phenomenon, one might describe this as a conception of faculties or (with an eye to some twentieth-century theories) information processing modules, even if it is relatively clear that Aristotle did not think of it in this way. It is clear enough, however, that Aristotle's conception, in contrast to Plato's, made sense appearances completely at home in the soul or, as we prefer, the mind.[67]

But a narrow conception of the external and internal senses does not sufficiently articulate how the senses constitute a *place* for images. Without emphasizing the placement structure of sensation that underlies Aristotle's theory of the senses, his understanding of sense perception and the phantasms that arise from it will be confused with modern empiricism, with its idea– or impression–units acquired by experience filling an otherwise unstructured mind. For Aristotle, appearances are not absolute, but rather relative to *fields* of appearance, and each field is capable of or potentiated with every possible sensible quality that it can realize. Each sense is thus more than just a container filled with discrete and otherwise unrelated items called images or image–appearances. Vision's object is more than the sum of all the colors one has experienced. The potentiation of vision—especially vision as taken up into common sensation—is directed toward *all* possible visible qualities. This implies two things: that the field of the visual has an essential structure or organization prior to any particular experience, and that the activation of the field is intrinsically labile and changeable according to these structures. This lability means that both sensation and imagination are fundamentally about the inception and incipience of appearance–activity.

Aristotle understands the sensibles as by nature being *organized by contrariety*—a conception made fully explicit in a work that develops themes of *On the Soul* with

[66] Is this Aristotle anticipating Kant? Yes, in a very weak sense, and no, in a much stronger one. If theories are propositions, then the answer is no. But if propositions attempt to articulate conceptual topologies, one can easily say that understanding Aristotle's common sensation in detail requires asking the question of the emergence and status of time and space (which would in no way diminish Kant's innovations or his difference from Aristotle).

[67] I hesitate to say that he makes them "internal" to the soul, if only because of the modern tendency to dichotomize the internal as subjective and the external as objective. Yet there is no doubt that Aristotle understood these processes as both physical and psychological. According to his physics and physiology, the physical processes that communicate themselves to the external sense organs become ever more physiologically internalized; at the same time they serve to inwardize the form–activities of things perceived in the external world. The objective-subjective dichotomy is an option for *differently* situating the basic topology of the Aristotelian theory of sensation.

respect to sensation, *On Sense and Sensible Things*. It is in fact a specific application of the general theory of contrariety that is fundamental to his physics and metaphysics. Change requires contraries in a substrate medium. Contraries, as Aristotle explains in the *Metaphysics*, are extremes within a genus or kind; between these there are usually many intermediate states, sometimes a limitless number. In vision, black–and–white is the basic contraries–pair that structures visual experience (though there are other contrarieties not expressly mentioned by Aristotle).[68] Moreover, as he says in III.12 of *On the Soul* and explains in more detail in chapter 2 of *On Sense*, sensation works by determining ratios between the extreme contraries. The organ has a motion of adjustment or accommodation to the object's sensible properties (which, we recall, are activities transferred through the medium between the object and the organ), a motion that is the physical expression of this ratio. He explains further that this process is the reason that sense organs can be damaged by excesses of stimulation. Excess stimulation temporarily, and sometimes permanently, impairs the organ's ability to respond with the necessary proportion of motion.[69]

In modern theories black and white are called achromatic, but in Aristotle's theory the hues that we consider to be chromatic are arrayed as different ratios of black and white; they are the extremes between which all other colors are arrayed. Whether or not contemporary scientists consider Aristotle to be wrong about the nature of color, they do not in principle disagree about the existence of various (virtual) aspect spaces with respect to sense qualities. Color is scientifically determinable, even quantifiable. Hues are differentiable according to the oppositions of the red–green and blue–yellow systems, but also according to retinal cones sensitive to shorter- or longer-wave light (another contrariety); they are variously matte or glossy according to characteristics of the surface; they have measurably different degrees of saturation and of lightness (sometimes called *brightness* or *value*) that seem to correspond to the light–dark contrariety of special concern to Aristotle.[70] Contrarieties as extremes or

---

[68] Aristotle points out in *On the Soul* and *On Sense and Sensible Things* the existence of several different sets of contrarieties for sound, taste, and tactility. See especially his remarks in *On the Soul* II.11, 422b17–33. We now, for example, would routinely distinguish in colors the contrarieties of light–dark, blue–yellow, red–green, and matte–glossy. The *Metaphysics* account of contrariety is in V.10.

[69] See *On the Soul* II.12, 424a25–33.

[70] Whether Aristotle meant that we can simply mix black and white pigments to obtain chromatic hues is doubtful; given the overall "actualizing" tenor of his physics, it is more likely that he meant something more dynamic. In the last half of the twentieth century, semipopular accounts of color vision began emphasizing that, although the retinal cones are divided into three kinds by differences in maximum spectral sensitivity (short, medium, and long according to whether they are maximally sensitive to short-waved blue, middle-waved green, or long-waved red), color determination occurs at a higher level of neural processing according to the networking of the cones into the contrary pairs of the red–green and blue–yellow systems plus the contribution of the light–dark system of the rods. Although it is not part of my brief here to discuss, much less to settle, how far Aristotle's conceptions are compatible with these matters, it seems to me important to insist on the plausibility of his topological orientation in the context of recent science. For an approachable introduction to contemporary color science, see Hardin 1988.

endpoints thus create a virtual space of sense qualities reflecting activities in the organ, and since the space between the extremes is structured according to the less and the more of the contraries, there will be a well-defined ordering of all the possible sense-quality appearances between the extremes. The senses work in particular by alteration (*alloiōsis*, one of the fundamental kinds of change or motion in Aristotle's physics, motion with respect to quality; see *Physics*, 226a26–27).[71]

We are accustomed, for another example, to take the keys of an 88-key piano as presenting the effective range of musical tones. Between the lowest key (an A) and the highest (a C) there are 86 intermediate pitches marked, but of course pitch is capable of nearly limitless variation and nuance.[72] There are similar virtual spaces of sense qualities according to each such contrariety. Each contrary pair creates, in the first instance, a linear space that has the contraries at the extremes with intermediate positions (arranged discretely, though sometimes conceived as indefinitely distinguishable) arrayed between them in a sequence ordered by the less and the more of whatever property differentiates them. This is a "tensive" orientation: the intermediates are well-delimited positions of a motion held in tension between one extreme and the other. Every additional contrary pair adds a dimension, a new sense of betweenness, and a new tension that can incept all possibilities comprehended between the contraries. The placement, the virtual space, of the sensed or imagined appearance rapidly becomes more complex, and the interrelationships of the different contrarieties of the single sense produce a multidimensional vector space, a space of directed tendencies, where the possibilities of pursuing a path from one quality to another follow trajectories just as complex as the differential topology of the virtual space.[73] Even if we call the space of placements set up by contrarieties *virtual*, we must remember that in perception it is an *actually phenomenal* space.

For the active field of sensation to be altered, there must be a motion that is a qualitative change (*alloiōsis*) between a pair of extremes accompanied by some local motion (*phora*): that is, a motion corresponding to the realization of that

[71] More specifically, "a change within the same kind but with respect to the more and the less is an alteration; for a change from a contrary or to a contrary is a motion, whether unqualified or qualified" (226b2–4). Compare *On Generation and Corruption* I.7, 323b33–324a3: "it is a law of nature that body is affected by body, flavor by flavor, color by color, and so in general what belongs to any kind by a member of the same kind—the reason being that 'contraries' are in every case within a single identical kind, and it is 'contraries' which reciprocally act and suffer action." And *Metaphysics* V.14, 1020b9–12, where *poion*, quality, is defined as "all modifications in moving primary beings (such as heat and cold, whiteness and blackness, heaviness and lightness, and the like), in terms of which bodies change, when they are said to be altered."

[72] We must beware of categorically asserting that there are infinite pitches, since human hearing has not only upper and lower frequency limits but also, between these limits, a finite ability to discriminate between variations in vibrations per second. Nor should we think of this as an unfortunate limitation, for precisely such limitations make hearing and music not vibration detectors but the phenomena they are.

[73] Thus these relations are not simply additive-linear (as, for instance, most criticisms of Aristotle's theory of color as a "mixture" of white and black assume).

quality in the substrate or field.[74] If several such pairs are involved and variable in the same process, the variations create a virtual space of a higher order. A familiar example, from color science since the late eighteenth century, is three-dimensional solids (often spheres) devised for the sake of conceiving each color as a function of three variables, for example hue, saturation (degree of admixture of gray), and brightness (degree of admixture of white), with each of these three varying between two extremes. And, just as the perception of motion is prerequisite for perceiving a moving thing, an active awareness of the mobility of these qualities of hue is necessary for understanding their field character, a field character that is expressed but not always actively perceived (e.g., when one stares at a stable scene).

What is also germane to the complex virtual spaces constituted by multiple contrarieties in a single sense is that features identifiable in the virtual space can be a guide to the actual division of the sense power, both physically and phenomenally. For example, most scenes we look at display many different shades of color at the same time, and our power of sensation takes in this variety simultaneously but differentially (that is, not in a confused way). In simultaneous contrast, a broad expanse of one color in a part of our visual field conditions and alters colors we see elsewhere in the field. When we experience the phenomenon of successive contrast—color afterimages, for example by fixing our gaze on a colored square and then shifting our view to a white or light-gray field—the hue we see will be the diametric opposite or complementary of the original. In cases like these, oppositions and relations we actually see correspond to the matter and functions of the eye or visual system.

Although the gathering of proper sensations into common sensation may appear at first to introduce a kind of sensory redundancy (color is seen at the level of vision and again in common sensation), common sensation actually yields a greater richness and complexity than a simple linear sum of the parts could produce, and precisely by bringing the proper sensibles into correlations. Objects as objects are no more than implicit in hearing or even in vision alone; objectness as such is not constituted in its peculiar phenomenal richness and complexity short of common sensation. But even that richness and complexity would not be very dense or textured without powers other than those of proper and common sensation. An animal without memory and imagination might have the ability to perceive some kind of object, but as soon as the object disappeared from the sensory field it would vanish from that animal's world and thus would be incapable of entering into and sustaining significant behaviors according to the relations of qualities in its sensory field. A frog, in order to catch a fly with its tongue, has to have some sense of motion and

[74] Recall that it is a fundamental tenet in Aristotle's physics and metaphysics that for there to be change of any kind there must be contraries and a substrate (or *subject*, to use another common rendering of *hupokeimenon*); see Sect. 5.8, above. In the conceptual topology of Aristotle's psychology, then, one is justified in searching for the substrate, field, or plane in which appearances change, both in the real-world object and in the physiological organ. This makes him, with Plato, the founder of the notion that images are located in a substrate/field (for Plato, a level of being). In addition, for Aristotle the existence of different sets of contraries in what appears to a sensitive or cognitive power increases the number of field dimensions—thus of subfields—in which one can perceive and imagine. See the paragraphs that follow for explanation.

perhaps of spatial depth, but that sensing pales in comparison to the vision of a bird of prey and its integration into the bird's flying and hunting.

This makes clearer why we can say that sensation is itself a motion, and prepares the way for understanding quite precisely why *phantasia*, too, is a motion. Sensation is motion (*kinēsis*); in particular it is an alteration (*alloiōsis*), because it involves changes of quality. When we see first black and then white, the black quality must be "destroyed" and displaced by virtue of a motion—in this case, a motion from one extreme to the other. Because all the forms of *kinēsis* are also attended in some way by the type known as local motion (*phora*), there must be some kind of local motion in the alteration. This makes all the more sense when we recall that sensation, as a "part" or power of the soul—the soul that is the first actuality of the body that has organs—is an organ activity. Since bodily organs have matter, and all kinds of bodily change involve the local motion of matter, this local material motion can, potentially, be channeled elsewhere in the body. It has the potential of being carried further inward (inward from the external sense organs) to other organs and places of interaction. There is no reason in principle that the further inwardized motion cannot reactivate something of the appearance that showed itself in the original sensation, as long as that motion is channeled to an organ or place where there can be appropriate activation–appearance. Memory, imagination, and even common sensation all depend on the forms of appearance transiting from one organic location to another, where they not only appear again but also interact in ways and with features unprecedented in their first organic place. There is no reason that the reactivated motion and appearance, placed in a new context more deeply embedded in the organic activity of the body, cannot to a certain degree be productive, even in a sense creative. This is evidenced already by common sensation, in which the proper sensibles are not merely recapitulated but recombined in a manner that gives rise to the appearance of the common sensibles. The so-called concomitant sensibles, by which we see not only whiteness or a spatially unified body but the son of our friend Diares, are further evidence of the emergence of qualities and more complex virtual spaces of the psyche through the inwardizing of the active forms of appearance.

About these motions and processes Aristotle gave precious little detail—probably because, given the state of the ancient Greek knowledge of physics, anatomy, and physiology, there was not much more that he could say. Such detail is precisely the sort of thing that modern psychologists and neurobiologists achieve. Yet if he did not and could not provide detail, his understanding of the conceptual topology of human psychophysiology foreshadowed at least the possibility of a deeper, more sophisticated grasp and coordination of both the physical and physiological sciences and the phenomenology of soul appearances.

5.10 What Aristotle's Definition of Imagination Means

Let us return, at last, to the definition of imagination in III.3.

After establishing that imagination is not any of a number of other psychological powers, Aristotle suddenly turns to a discussion in terms of his physics of motion.

Here, beginning with the long, definitional sentence we quoted above, in Sec. 5.4, is the crucial passage:

> But since it is possible when one thing is moved for another thing to be moved by it, while imagination seems to be some sort of motion and not to occur without sensation, but in beings that sense and about things of which there is sensation, and since it is possible for a motion to come about as a result of the being–at–work of sensation, and necessary for it to be similar to the sensation, then this motion would be neither possible without sensation nor present in beings that do not sense, and the one having it would both do and have done to it many things resulting from this motion, which could be either true or false. This last point follows because, while sensation of its proper objects is true or has the least possible falsehood, there is in the second place the sensation that those things that are concomitant to the ones sensed are in fact concomitant to them, and here already it is possible to be completely mistaken, not mistaken that something is white, but that the white thing is this or that other thing. And in the third place there is sensation of the common attributes that accompany the things concomitantly sensed, to which the things properly sensed belong (I mean, for instance, motion or size), about which most of all it is possible to be deceived as a result of sensation. And the motion that comes about from the activity of sensation, stemming from these three ways of sensing, will be different in each case, the first sort being truthful while the sensation is present, while the others could be false whether it is present or absent, and especially when the thing sensed is far away. If, then, it is nothing other than imagination that has the attributes mentioned (and this is what was being claimed), imagination would be a motion coming about as a result of the being–at–work of sensation, and corresponding to it. And since sight is the primary sense, imagination has even taken its name from light, because without light it is impossible to see.[75] And because imaginings remain within and are similar to sensations, many animals act in accord with them, some, the beasts, because of not having intelligence, but others, humans, because their intelligence is sometimes clouded by passion, disease, or sleep. So about imagination, let this much be said about what it is and the cause through which it comes about. (428b10–429a9)

Aristotle presents this extraordinarily condensed and tortuous passage on the expressly stated condition that the (preceding) discussions of sensation, motion, and cause are correct. After our prolonged treatment of the physical background, however, it is far less puzzling than it was initially. Sensation is an organic activity of qualitative alteration accompanied by the local motion of matter. Aristotle never totally reduces the activity of the organ to motions of matter[76]; he also regards all motions of matter to be incomplete or imperfect in themselves. Material change leads to further material change, both in the thing and beyond it; one material activity almost always leads to another. Because material change does not cease until an end is reached, the activity of sensation as such can be an end in itself only relatively speaking insofar as it is the activity of just one or a few organs within the entirety of the human or animal body. The activity in sensation sets off changes that have to arrive somewhere else in order to reach a proper end; and if this end, when achieved,

[75] Aristotle is saying here that the term *phantasia* is derived from the Greek word for light, *phaos*. This etymology is considered basically correct.

[76] Both because there is sense appearance and because there are three causes in addition to matter, form, goal/end, and efficiency. The "efficient" cause, more precisely the fundamental condition that is the source of the motion or change bringing the thing into being, points back to whatever that source is, and not just to the material of the thing brought into being.

involves further material activity, it is again only relatively speaking an end, a transit station rather than a terminus, an end only for the time being.

One reason that Aristotle does not say more about the motion mentioned in the definition of imagination is that among animals there are many paths along which motion might proceed in order to reach further goals. These are as many as the ways in which sensation can be activated, refined, developed, analyzed, synthesized: in a modern word, *processed*. Bees, moles, and owls all see, but the characteristics of their seeing are radically different. Moreover, the more complex the activity of sensation is, the more specified must be the corresponding motions. For example, in a passage of *On the Soul* that is easily overlooked because it occurs in the chapter on smells, Aristotle remarks that in human beings it is the sense of *touch* that is the most refined (II.9, 421a17–27) and contrasts it with vision. What vision in itself communicates is light, dark, and all possible intermediate colors; properly speaking, one does not see objects in vision (they appear as such at a higher level of processing, in the common sensation) but only differently colored or illuminated surfaces (the underlying subjects or substrata) in the visual field. Touch appears to discriminate far more characteristics (that is, sets of contraries) in things than vision does (II.9, 422b24–34).[77] Thus one can speculate, in particular on the basis of Aristotle's other biological writings, that saying that an animal has vision, or hearing, or smell, or taste, or touch leaves only minimally determined what that animal's visual, sonic, aromatic, savoring, or tactile experience is like. Quite apart from the fact that animals may have comparatively sharper or duller senses, they can also experience different qualities of things that are closed off from animals that have the same sense power but fewer qualities and contrarieties available through it. This means that the imagining of the animals will be different as well, since the animals' imaginations are in the first instance precisely the motions following on the sensations that they actually have.

We have mitigated to a considerable degree the initial strangeness of Aristotle's conclusion that imagination is a *motion*. The definition is *unqualified*: true absolutely, that is, when we disregard all further specifying characteristics that attend the phenomenon in reality. The motion is the motion that follows sensation; thus imagination can exist only in sentient beings, that is, in animals. What happens beyond this basic assertion is left to be determined. For example, if there is no further "sensation" produced by the motion set off by sensation—that is, some reappearance of the appearance–forms that first occur in sensation—then there is no imagination proper. If we conceive of an animal with a nervous system so simple that it is not capable of evoking past images, then the only follow-on motion from sensation would likely be motor activity coordinated with the sensation (for example, flight in the face of pain, pursuit in the presence of food). It is easy, after reading Aristotle's differentiation of imagination from activities like scientific knowing and opinion, to forget that what he says in III.3 is not just about imagination as a uniquely human

[77] As I noted earlier (n. 68), Aristotle quickly concludes that there are in fact many more contraries also in the other senses than we typically count.

capability. *On the Soul* is a book not about human beings per se but about living things and their powers.

But then we face another conceptual obstacle. In the long first sentence of the quotation given at the outset of this section, Aristotle notes that because sensation produces the motion, the motion necessarily resembles sensation. It is not entirely clear whether a twenty-first-century person must assent to the idea that a motion caused by an X will resemble X—we are accustomed to thinking that effects do not need to resemble causes. Moreover, if we take motion in its barest sense, it would appear that Aristotle is saying that (for example) the motion set off by seeing red is red all along whatever path it takes. To this it is probably necessary to respond (in a typically Aristotelian fashion) that, whatever motion is set off by seeing red is red only potentially, until it reaches a destination (whatever it may be) where it can become an actualized red.

Consider the following example. Someone is eating strawberries. The motion set off by the activity of red in the organ of vision must be passed onward to appear alongside other sensations: sensations produced with the motions that have been set off by the activities of the distinctive sweet taste, the firm, grainy feel of the strawberry's surface produced in the fingertips and the luscious coolness of juice on the lips, the blossoming aroma sensed by the nose, and the associated sounds produced by eating. When these motions reach the place of common sensation—for Aristotle, near the heart—these will be unified with one another and more precisely situated with respect to the common sensibles like unity, place, and time. All these proper, concomitant, and common sensibles are associated with the real-world activity of enjoying strawberries—or, to put it more strongly, all of these united *are* the experience of eating strawberries. The activities of each sense produce the *phantasia*–motion that is potentially part of a similar experience until the motion reaches a place where it can be actualized, to whatever degree the physiological circumstances of the relevant paths and organs make possible. That would mean that, beyond what Aristotle explicitly states, his theory would seem to imply that *phantasia*—in this case the motion set off by the original proper sensation that can, in the right circumstances, produce a new organic alteration that will give rise to the appearance–form once again—is required for there even to be *common* sensation in the first place.[78] Reactivating the appearance–forms of sensation without the presence of the object, in imagining proper or in remembering, would in turn require yet another communication of the material and qualitative motion to the appropriate organic place. According to chapter 1 of *On Memory and Recollection*, the destination of the communication would once again be the place of common sensation. In II.8 (420b29–32) of *On the*

[78] Perhaps, more accurately, we might have to say that something like *phantasia* must commence even before common and concomitant sensations are constituted. Yet we might hesitate about calling it *phantasia* in the strict sense, because the object being sensed will still be present while the motions originating in the activity of the external senses (and thus corresponding to the definition of imagination) are on their way to being taken up into common sensation. Perhaps it would be serving a function something like Kant's transcendental imagination. See n. 60, above, on whether common sensation has an organ.

Soul, Aristotle distinguishes the mere sounds animals produce (for example, human beings smacking their lips as they enjoy strawberries) from voice by noting that voice is a sound along with some imagining (*meta phantasias tinos*) produced by a being with soul. Although he gives no further indication of how this happens, it suggests that some *phantasia*–motion goes out from the place of common sensation to affect the organs that the human animal uses to make significant sounds. Even if there were no place or organ of common sensation—say in an animal that had just touch and no other external sense power—a tactile appearance–sensation would not be an absolute finality, because it would typically, as pleasurable or painful, lead to some further organic development (perhaps ingestion).

In the twenty-first century we accept neurological processing as fact. The information derived from input at one level of processing is carried on to the next level of processing with that information more or less intact, that is, with the same information content as it had when it left the previous processing node or level. If the activity in the place where the input is originally processed results in phenomenon A, and that activity also produces an encoded transmitting motion that arrives at a new place properly arranged, it can easily yield a similar process with similar result (phenomenon A, or at least A′). This may in fact amount to a modern adoption of Aristotle's contention about *phantasia*. If Aristotle means something similar when he says that the motion of imagination *resembles* its cause, we can more easily pass from the claim that the motion of imagination *potentially* resembles the sensation experience with which it began to the further claim that imagining will *actually* be qualitatively similar to sensation. The physical motion is not an appearance when it is in local motion from the eye or the ear or the nose to wherever it arrives and is reprocessed—at least not until it gets there. There is no reason to assume that Aristotle is thinking here in a way obviously incompatible with modern neurology.

Aristotle notes at the very end of the III.3 discussion that it is not surprising that, whatever the animal that possesses imagination, it is capable of being affected and of acting in many different ways. To understand this requires looking more deeply and questioningly into the rest of *On the Soul*, especially the end of book III, where he talks about how having and making images is related to animal and human actions and desires, and even beyond *On the Soul* (in particular to related works on sensation, memory, and the nature, parts, and motions of animals). What the claim means is that there are manifold ways of follow-on development implicit in the definition. If imagination is what follows from sensation, there will be other *functions* and *powers* that follow, depend on, and (perhaps partially) reactivate the appearance–forms of imagination. To speak in a more contemporary way, the processing of the input of information in sensation does not stop with imagination; imagination is an intermediate stage, and what it does is prepare the way for fulfillments elsewhere in the organic economy and activity of the animal. That "elsewhere" in animals can be a kind of thinking (that is, the imaginative reckoning they are capable of, like enough to thinking that Aristotle expressly counts it as a variety of thinking in the opening of III.10, 433a10–11) and also purposive bodily movement. This is not only the actual order of presentation that Aristotle follows, it is also a necessary order. The purposive activities of animals cannot be explained without first establishing what they are sensorially, imaginatively, and memoratively able to respond to and how. Even an animal that did not have imagination in the ordinary phenomenal sense would have to move its body in response to what it senses. This is one of the chief reasons to speak of *proto-* or *pseudo*-cognition, a kind of experience that is at least analogically like cognition and that leads to at least minimally purposive bodily activities.

To modern sensibilities—by "modern" here I mean educated sensibilities over the last four centuries—it can seem simply confused to talk about a motion that is associated with an appearance, a confusion that modern philosophy and science devastatingly criticized, in the first instance in Descartes's dualistic argument that extended matter (and its motions) had nothing in common with thinking. Yet any research program that claims to explain consciousness and appearance, whether now or eventually, as a product of or an association with physiological processes still holds to a strong conception of an articulated association between physical events and psychological events. This is true even for those who identify thought *as* certain brain processes: there has to be *sufficient* similarity–with–differentiation between thought and brain events so that they appear as *distinct* terms A and B in a proposition of the form "All As (thoughts) are Bs (brain processes)" or "Thought *a* corresponds to physical brain event *A*."

Thus in the first instance Aristotle need not be dismissed as passé, unmodern, or unscientific insofar as he talks first of a motion originating from the activity of sensation and then says it is distinct from the three kinds of sensation he identifies, proper, concomitant, and common. The appearance associated with that motion is the same in form wherever it appears (because it is an appearance–form). The motion that commences beyond the activity of the sense organ—*phantasia* in its strict definition—is only potentially an appearance when it is in passage from one organic center to the next. One might say then that what Aristotle defines as imagination is a motion that is potentially an appearance. Any aspects of that motion that are irrelevant to appearance would not correspond to what sensation, imagination, memory, or thought actually experience. Aristotle is, at any rate, no more clear or confused than modern neurologists, who would not want to claim that optic nerve impulses conveyed to visual processing areas of the brain are already *actual seeing* before they arrive in the brain. A trauma that severs the nerve just short of the brain assures that there will be no seeing at all. Aristotle is simply expressing relevant factors regarding the appearances of sense in an articulate, differentiated, and initially plausible way.

Aristotle began III.3 by chastising earlier thinkers, who, although they had tried to identify what made thinking and sensation true, had not developed a parallel theory of what makes them false. Some scholars argue that III.3 is in fact a theory of error.[79] Imagination has almost always been associated in Western thought with error and illusion (including fiction and fantasy). But for Aristotle this association is hardly a necessary one. Imagination, he says, can be either true or false, even if it

[79] For example, Caston 1996.

is for the most part factually false. As the invocation (at the beginning of III.3) of the image-making techniques of memory art suggests, in certain circumstances images are neither true nor false. If, to help remember the themes of a speech and the major points you will make about them, you form an image of a personified justice holding a balance scale, armed with a shield, without a blindfold, and looking into the distance over the heads of a group of supplicants who are offering symbols of money, power, and flattery, the image is neither true nor false. It just helps you remember.

One could of course say that this image is false because justice is *not* that way; but one could as easily say it is *true* insofar as the image is a symbolic representation of a speech. It all depends on the background upon which the image is projected. Even when images appear to be simple they can have complexity. A simple red bears complex relationships to other hues. A red strawberry can be viewed according to its simple redness, the degree of redness as a sign of ripeness, a visual signal to animals foraging for food, etc. Which of these is relevant in animal perception is not a question that can be answered absolutely. Notice that in the definition-giving paragraph Aristotle differentiates between the truth/falsity of proper sensibles (never false), concomitant sensibles (sometimes false), and common sensibles (often false). This complexity of and in images once again offers the possibility of diverse kinds of further "processing" of the information, depending on the animal, the specific receptivity of its sense organs, and the other organs and powers that become involved. A cat can respond to some differences in color or brightness, for example, and thus must have corresponding proper sensibles; it can notice the distance at which these appear, so it must have common sensibles; and if it distinguishes something as prey or as a danger it has a kind of concomitant sensible. Later Aristotelian interpretations especially identified the concomitant sensibles as pertaining to the estimative/cogitative power, by virtue of which reason and phantasms "touched." I see an expanse of black in my field of vision; I see that it is a black cloak; it is a cloak concealing Diares' nephew, who is up to his usual nefarious business. Seeing black has to do with a proper sensible; seeing a cloak and Diares' nephew are associated or concomitant sensibles. In human beings, who are able through reason to look upon proper and common sensibles and categorize them further in innumerable ways and according to diverse purposes, images can be seen, or projected, against a limitless number of backgrounds, and the forms of intelligibility that these complexes of phantasms against backgrounds offer is equally limitless.

Aristotle says that both proper and concomitant sensibles are surer than common sensibles. Clearly Aristotle needed to analyze this further, since a limited expanse of color, insofar as it is a figural determination, is a proper sensible further determined as part of common sensibility, and it is impossible to identify blackness as belonging to a cloak unless it is first seen spatially. In terms of being, common sensibles have to precede concomitant sensibles. In terms of degree of truth, however, which is the specific question posed by Aristotle in the passage, there is a good deal to be said for his conclusion. An example dear to his heart is that of the sun, which appears to be (say) the size of a coin but is actually huge. At the level of sensation, we are much less likely to be wrong in a first-approximation identification of what a thing is than in determining its size or position.

5.11 Is Imagination the Same as Intellect?

The traditional theory of the intermediacy and role of *phantasia* is deeply grounded in III.3 of *On the Soul*: far more grounded in that chapter than in, say, Plato's *eikāsia*, the image-perceiving power of the divided line in the *Republic*.[80] Almost in passing Aristotle mentions some attributes of imagination; the last is that animals do many things in accordance with images, including the human animal, in particular when the human rational power is weakened or out of commission, as in dreams, hallucinations, passions, or illness. Although III.3 places imagination between sensation and intellect, it also places it, as Plato does, in a certain *contrast* to reason. That is why it is at least unexpected, and perhaps downright amazing, that in III.7, and without any intervening development of the theory of *phantasia*, Aristotle asserts that there cannot be any thinking at all without images. And the last sentences of the concluding discussion of intellect (in III.8) seriously entertain, if only for a moment, the possibility that the primary or fundamental things known[81] might themselves be phantasms. The human power that, it seems, ought to rise above images, *intellection*, thus is *radically dependent* on imagination and even in its ultimate form might, in certain circumstances, seem to be an image. That is, Aristotle at least wonders whether intellection *is* imagination, the simple *having* of an image. Intellection is itself, like imagination, an appearance–activity, and if it is not simply the appearance of an image it is, nevertheless, an intellective appearance that requires another appearance, an image. Perhaps intellection is an appearance *in* or *through* images (rather than *of* images). Yet even this qualification could imply a subordination of intellect to images.

It is immediately after III.8's inconclusive summary of *On the Soul*'s discussion of intellect that Aristotle investigates purposive animal activity. Why has it had to wait so long? Might it not have been more illuminating for him to move from the statement, at the end of III.3, that animals do many things in accordance with imagination to the discussion of purposive animal activity in general, rather than turn to intellection, an activity that only the human animal exhibits?

It is certainly conceivable, as a different possibility, that Aristotle might have interpolated parts of the two long chapters that constitute *On Memory and Recollection* in order to extend our grasp of the imaginative psychology of animals. Yet, although this might have made *On the Soul* more comprehensive, it would have improved only slightly the explanation of intellect and of purposive animal and human behavior that rounds out book III and the work as a whole. The explanation

[80] This does not mean that Plato's theory was without effect on tradition. It typically goes without saying, for Aristotle as much as for anyone else, that images are recognized as x-like appearances, so that in phenomenological terms one must say that conceiving an image i bears within its conception an intention toward the x of which it is an image. For Plato, this is an essential characteristic of images of all kinds.

[81] These things, the first or primary *noēmata*, include the most fundamentally intelligible of all intelligible concepts.

of animal behavior really needs to appeal only to the fact that some animals have the ability to retain phantasms they have experienced and in some cases to produce new ones. The details of the memorative process are not necessary for this; and, in any case, Aristotle's theory of memory and recollection is not refined enough to make any great difference in the explanatory outcome.[82]

Perhaps an even more important reason for the existing order, however, is that Aristotle had first to clarify the nature and action of imagination by differentiating it from intellection. Recall that a chief complaint he makes against his predecessors is that they had insufficiently distinguished the sensitive and cognitive powers of soul, and often enough they had confused them. The differentiation of imagination from various cognitive powers hinged critically on the criterion of truth: knowing is always true, opinion is always taken to be true, but imagination can be either true or false. Thinking, however, just like imagining, does not always have to be true—it is not even always noncontradictory! Thus if thinking is itself an appearance or a reappearance and always requires the presence of an image, one needs to exclude the possibility that there is no difference between thinking and imagining before going on to describing animal activity. If the proper object of thinking as such *is* a phantasm, then it would be hard to resist the conclusion that thinking is "just" a form of imagining. Let us look more carefully, then, at the difference between intellection and imagining.

Aristotle presents III.8 as a summary of and conclusion to everything discussed since the introduction of sensation in book II. Here it is, in its entirety:

> And now, bringing together what has been said about the soul under one main point, let us say again that the soul is in a certain way all beings, for beings are either sensible or intelligible, while knowledge in a certain way is the things it knows, and sensation is the things it senses; but one needs to inquire in what way this is so. Now knowledge and sensation are divided up into the things they are concerned with, and there is in potency knowledge or sensation to be divided into the things that are in potency, and knowledge or sensation at-work-staying-itself that is divided into the things that are at-work-staying-themselves; so what the sensing and knowing capacities of the soul are in potency are the same things that are either known or sensed. This has to be either those things themselves or their forms; it is certainly not themselves, since a stone is not present in the soul, but its form is. Thus the soul is like a hand, for the hand is a tool of tools, while the intellect is a form of forms and sensation is a form of sensible things.
>
> But since—as it seems—there can be no item of experience apart from the extended magnitudes which are the separate sensible things, the intelligible things are present in the sensible forms, not only the things said to exist by abstraction but all the active conditions and passive attributes of the sensible things.[83] And on account of this, one who sensed nothing would not be able to learn or be acquainted with anything either, and, whenever one were

[82] At the very beginning of *On Sense and Sensible Things*, which follows immediately after *On the Soul* both traditionally and according to its opening sentence, Aristotle makes clear that he thinks that "memory, passion, desire, appetite, and pleasure and pain belong to the more particular study of living things" and belong to both soul and body (436a9–11). They would all have to be incorporated into a detailed theory of purposive activity, but most of them are not necessary to account for the principal sources (*archai*) of such activity that he treats in *On the Soul*.

[83] The distinction Aristotle is making here will be discussed in the next section.

> to contemplate, it would be necessary at the same time to behold some image. For the things imagined are just like the things sensed, except without material. And imagination is different from affirmation and denial, since what is true or false is an intertwining of intelligible things. So how do the uncombined intelligible things differ from being images? But in fact these are not images either, but are not present without images. (431b20–432a14)

What is curious is that although, in the final analysis, it is clear to Aristotle that the simple intelligibles are not phantasms, he does not explain the difference. It must not be simple enough to do in a sentence or two, since here he says nothing more; but neither does he explain it elsewhere. What are we to make of this flirtation with the possibility that *noēmata* are images? Why does Aristotle end his discussion of intelligible things by puzzling over, and failing to resolve by argument, how the primary concepts, the fundamental intelligibles, relate to images?

5.12 Parsing the Phenomenon of Thinking

Aristotle's presentation of thinking is notoriously complex and murky, so any firm answer to these questions will be controversial. There are nevertheless key features that are clear enough. (1) Unlike Plato, or at least conventional views of Plato, Aristotle claims that the principle of intelligibility is actually in the things we sense. (2) Intellect or thinking is like sensation in that there is an intelligible thing (corresponding, by analogy, to the sensible things in sensation) that activates the intelligence. Intellect, just like sensation, is based on change from a passive or potential state to an active one. (3) The result of the activation of intellect and of sensation is a form: a form that actually and originally exists in real-world material form, but that in the intellection and the sensation exists as a form—a form in the substrate of thinking and sensation, respectively. (4) What the intelligibles do is think something in, about, or with respect to the *appearances* of things, their phantasms; they require images but are not themselves images.

In very simplified versions of Plato we tend to say that ideas form an autonomous realm separate from things of the ordinary world. That this is an oversimplification is evident from the account we gave of the divided line in Chap. 4. In particular, human knowledge works by recognizing that each level of being, coordinated with a corresponding human cognitive or sensitive power, is proportionally related to others; and this is fundamentally connected with the projective appearance of the good itself. There is, nevertheless, a natural tendency to think of these levels as relatively isolatable from one another; and whether or not the philosopher ever fully achieves the contemplation of pure forms or remains for very long at such a level, the images, analogies, and metaphors Plato uses suggest some at least relative kind of separateness for the ideas/forms. Aristotle, by express contrast to Plato, understands forms as present in the things of the world. Substances, the most basic beings, are composites of form and matter; they are matter showing a form. If one is looking for the cause of the substance and its intelligibility, one has to look precisely to the form–in–matter.

What Aristotle understands further is that the form is, simultaneously, (a) what unifies the substance in all its aspects, in the present and over time, and (b) the principle of intelligibility that communicates itself to beings capable of knowledge. These are not merely Aristotle's assumptions or hypotheses; they are conclusions arrived at by extensive dialectical argumentation about how things appear and how they are. The first part of this understanding is achieved in the middle of the *Metaphysics*, at the end of book VII and the beginning of book VIII. The second part of the understanding is in effect established by *On the Soul*, although it is less thoroughly argued than the metaphysical claim—and thus it seems to modern sensibilities even more problematic than the metaphysical claim. A major source of the problem is that he does not even take the time to work out themes and questions that might better bridge the transition from sensation to intellection. In the Middle Ages, the great Islamic and Latin interpreters tried to bring precisely this aspect of Aristotle's psychological writings to a greater perfection, especially with their more fully developed theories of the internal or inward senses (and corresponding brain anatomy) that "prepare" the phantasm, derived from sensation, for the act of intellective understanding. That is, they tried to address questions that arose from the Aristotelian theory by appealing to experience and other texts and by thinking out the problems for themselves.

So, for example, one can read *On the Soul* in light of the account of *epagogē*, or induction, at the end of the *Posterior Analytics* (II.19, 99b17–100b17). Induction is understood by Aristotle as a natural process by which human beings can, over time, grasp the intelligibility of what they sense. The *Posterior Analytics*, taking for granted the development of syllogistic logic in the *Prior Analytics*, presents a theory of scientific or epistemic knowing. Syllogistic allows one to arrive at true conclusions from true premisses; but the further question is whether the premisses are properly relevant to (as *causes* of) the matters at hand and how we know that they are true.[84] If every truth has to be proved by the principles of syllogistic reasoning from other truths, there will be an infinite regress; we will never find a stable basis from which we can claim to know things. Induction corresponds to this need for a basis that is true and properly related to things. It is a natural capacity by which human beings (and animals) can recognize orderliness in their sensations, so that what is sensed becomes *experience*. It presupposes that sensations in some sense persist.

> Where sensation does persist, after the act of sensation is over the one sensing can still retain the sensation in the soul. [Notice that, in light of *On the Soul*'s account, this persistence and preservation must presuppose *phantasia*.] If this happens repeatedly, a distinction immediately arises between those animals which derive a coherent impression from the persistence and those which do not. Thus sensation gives rise to memory, as we said; and repeated memories of the same thing give rise to experience; because the memories, though numerically many, constitute a single experience. (99b35–100a3)

Experience, Aristotle says, is the universal that comes to rest as a unity in the soul.

[84] This is precisely where the inherence of the premisses in the same field or plane of concern is relevant, indeed necessary.

A little later in the chapter Aristotle restates the conclusion in terms of stages or "stops." As soon as some single thing comes to a stop in the soul (say, the many squirrel appearances one sees), this is the first stage of the universal. More such stops occur, and gradually relationships of universals to one another appear as well (for example, squirrel, mammal, and the inclusion of squirrel within mammal). It is the inclusion and exclusion of classes of things from one another, of course, that is the foundation of syllogistic logic: if all squirrels are rodents, and all rodents are mammals, then by being a squirrel a thing is a mammal as well. But this account of induction and experience is clearer only in its immediate context, where the leading question is knowledge rather than the potentials and activities of basic soul powers. It appeals to but does not explain the activities of the soul. *On the Soul* supplies the explanation of these but also introduces a complexity to the process that raises more difficult, and more elemental, questions.

Understanding a thing is not achieved simply by sensing it, yet without sensing there can be no understanding. So how does understanding occur? There are two elements to the answer. The first part is that there is an analogy between sensation and intellection insofar as each has to be activated by an appropriate object. The second part is that the activities of sensation and intellection are linked—alternatively, one might say that they achieve a tangency—insofar as sensation ends or results in something that is the beginning point for intellection. That point of tangency is the phantasm. If *phantasia* is the power of imagination and is in essence a motion (or the potentiality of activating the appearing quality along with the associated local motion), the phantasm is the appearance that the corresponding motion produces, and this production of appearance is the near or proximate goal of the motion.

Let us assume for the time being that we human beings are presented with a phantasm. What happens? "The thinking power grasps in thought [*or* noetically thinks] the forms that are present in things imagined," Aristotle says in III.7 (431b3) of *On the Soul*. In III.5 he analogized this activity to the physical illumination of light. Light turns potential colors (for example, the colors that an object in the dark would display if it were illuminated) to active colors by activating the intervening medium. More precisely, the active form in the object's surface that is communicated to the eye in the presence of light is what produces an activity in a previously passive eye, and that activity is seeing the color. Similarly, the active aspect of intellection illuminates the phantasm and produces intelligibility, an intelligible thing, in that receptive part of the intellect that becomes now one thought, now another.[85] In Greek these different aspects of intellect were called (following III.5) *noûs poētikos*, making or poetic intellect, and *noûs pathetikos*, suffering or receptive *noûs*; comparable

[85] An updated version of this analogy could use x-rays instead of light. Agent intellect would be the x-ray emitter, the potential intellect would be the x-ray film or detector, and the phantasm would be the physical body placed between the two. The x-rays of agent intellect could then be said to impress on the film of potential intellect the intelligible structure of the phantasm—its skeleton, as it were. But this version probably introduces too much isolated fixity to both the elements and the results of the process, and furthermore it suggests that, ultimately, the concepts persist independently of the phantasms. For Aristotle, if there is no phantasm, there is no thinking, where thinking is recognizing the phantasm's intelligibility.

terms from the Latin tradition are agent or productive intellect and potential or passive intellect.

We cannot tarry long trying to parse the nuances of this conception, much less its interpretations. Historically, there is probably no single theory more variously commented on in the entire extant writings of Aristotle—not least because it gave rise to stark conflicts over whether either or both, potential intellect and active intellect, properly belonged to the human soul, and whether the theory allowed for the personal immortality of the soul. Averroës, for one, interpreted both the active and the potential aspects as extrapsychic—perhaps belonging to and issuing forth from a transpersonal or divine mind—so that the very highest, most sophisticated soul power that the human being actually possessed was imagination. Avicenna thought that the potential intellect belonged to human souls but that the active did not. In the Latin-speaking Christian middle ages there were followers of both, although the solution preferred by orthodoxy was that both active and potential intellect belong to the individual human soul. For our purposes, however, what is even more important is that the various doctrines of intellection all centered on the conceptual topology surrounding the phantasm.

The process by which intellect grasps something intelligible from a phantasm was typically called *abstraction*, from Latin *abstractio*, a drawing or pulling away from something. This Latin term rendered the Greek *aphairesis*, which means a taking or carrying away. There is a problem with the Latin term's use, however. Aristotle typically used *aphairesis* only for the act of mind that allows us to grasp geometric entities and geometric space.[86] Geometric things are physical things completely dematerialized and left only with geometrical dimension: extensionless points, lines with length but no width, plane surfaces with two-dimensional extension but no third dimension, and the like. Medieval Latin-speaking philosophers, however, used *abstractio* to refer more generally to any act of deriving or drawing a concept or intelligibility away from phantasms, and in particular for the abstraction of the essence of a substance.

This terminological extension becomes a very large problem if it is assumed that the concept has a being independent of the phantasm, a large problem worsened when it is argued that, as a result of the process of active intellect's abstraction of "intelligible species" from phantasms, these species are impressed and then preserved in the potential intellect.[87] Every step in this direction moves closer to a realm where concepts rule without the copresence of phantasms, a realm of rationality that is pure and divorced from the taint of materiality.

---

[86]"In this way one thinks the mathematical things, which are not separate from matter, as though they were separate" (*On the Soul*, 431b16–17). Books XIII and XIV of the *Metaphysics* develop at length Aristotle's conception of the being of mathematicals along these lines. See Philippe 1948.

[87]This is, formulaically, how the doctrine is often presented in modern neoscholastic thought. The use of the alternative "productive intellect" can further contribute to the sense that the result of the process, conception or intelligibility, is reified in an independent product. The notion that the intelligible species impressed in the receptive intellect are simply retained there runs up against 450a12 of *On Memory and Recollection*, where Aristotle expressly states that there cannot be memory of intelligibles without phantasms.

There is another problem of interpreting as abstraction the active/poetic intellect's illumination of the phantasm. It is a problem encouraged by Aristotle's explanation of induction at the end of the *Posterior Analytics*, where the accumulation of experiences gives rise to universals that relate individuals to species and genera. In accordance with this notion, abstraction was understood paradigmatically as the abstraction of the intelligible essence, the essential form, from the object of experience. One looks at or imagines or remembers a squirrel; the phantasm is illuminated by active intellect to yield an impression of the concept squirrel in the potential intellect; and then the potential intellect preserves this concept/essence. Yet *On the Soul*, Aristotle's preeminent treatment of the powers of thinking, does not discuss or even present examples of intellect grasping the essential or substance-giving forms of things. Instead it gives examples like understanding Cleon as pale-skinned, experiencing *flesh* versus *being flesh*, conceiving the straightness of a thing versus the straightness of a line, and the like.[88] That is, the objects of experience are, in terms of their intelligibility, complexes of aspects that can be grasped according to different kinds of intellectual focus. Just as the biologist with a microscope can isolate different parts of his sample and focus on different planes within a slice of nearly transparent tissue he has sectioned,[89] the intellect can focus on different aspects of sensed, remembered, and imagined appearances. The physicist looks at a molecular pattern and asks what forces have caused it; the chemist looks at it and wonders whether it will be impervious to acids and bases; the mathematician looks at it and sees it as exemplifying a set of *n*-dimensional algebraic transforms.

One example that is especially revealing is the "snub nose," which appears several times in *On the Soul*, in particular in the discussions of intellect in III.4 and III.7. Likely Aristotle recognized that this example would evoke Socrates, who had such a nose. Its more immediate purpose is to illustrate how thinking can make distinctions and divisions with regard to images and thus recontextualize imagining. Aristotle starts by wondering whether recognizing something as flesh and understanding what it is to be flesh are exercised by the same or different powers. It might be two, he says; or, alternatively, it might be a single power in different relations to the thing in question. "Now since a magnitude is different from being a magnitude, and water is different from being water (and so too in many other cases, though not in all, since in some cases the two are the same), being flesh is distinguished either by a different power from the one that distinguishes flesh, or by the same one in a different relation. For flesh is not present without material, but like a snub nose, it is this in that" (429b10–14).

The comparison to the snub nose indicates his preference for the second explanation. In a snub nose, one can *distinguish* the concave shape from the fleshy matter

[88] The difference between experiencing a fleshy thing and the being flesh of the thing is certainly relevant to the question of essence; but rather than highlight the being flesh as essence, Aristotle emphasizes that the two experiences involve different ways of having intelligence of the phantasm's intelligibility. Neither is achieved by leaving the phantasm behind.

[89] Those who have been asked to sketch what they have observed through a microscope will easily recall the difficulty of locating the plane of focus that reveals the structures the instructor wants them to see.

(the substrate within which different qualities and magnitudes appear), but the snub nose itself is a *unity* of flesh and concavity. Still, as a "this in that," the snub nose can be considered in several ways: as the curvature (which happens to be in the flesh of a specific face you are considering); as the flesh (which happens to have concave curvature at its tip); or as this curvature in that flesh (the actual nose you see, whether abstracted from or actually on an individual's face). All these possibilities presuppose the appearance, in sensation, in memory, or in imagining, of "snub nose." The recognition of shape, as such, requires the kind of intellect that can focus on the *magnitude* in the sensory appearance. That is the beginning of abstraction in the properly Aristotelian sense: *aphairesis* ultimately grasps geometry apart from materiality. But that abstractive power of mind does not imply in any way that the geometrical object actually exists apart from materiality.

Our mental focus can lie elsewhere, on what it is to be flesh, or on the fleshiness of *this* appearance, or on some other aspect of its presentation. Through the proper sensibles we can focus on things like color and texture; through common sensibles on the shape, the unity, the motion or lack of motion, the temporal aspects. Through concomitant sensibles we can say that it is fleshy or is a nose or a dog's nose or a philosopher's nose, and it is due to our being intellectually endowed animals that we can think such things. Whether one calls this the work of a single power or several powers is perhaps less important than calling them all intelligibilities provoked through the phantasm. That Aristotle thinks it is a single power used in different respects rather than multiple powers is evident from *On Memory*, where, in summarizing the lessons of *On the Soul*, he ascribes precisely these kinds of distinction to different ways of thinking the same phantasm.

One can, as it were, think the same phantasm in multiple ways, then. To begin with, the snub nose is taken as flesh, it is taken as having a certain shape, and it is taken as being simply a nose. Following further along this same line of conceptual topology, one can begin to proliferate ways of thinking it. For example, one can take the shape in all its individuality (this is a unique snub nose never before encountered by humankind); one can take it as a particular type of snubness among many others; one can take it simply as having the very general shape called snub. One can concentrate on its flesh, or on the fact that its fleshy conformation is supported by a cartilaginous structure. One can think the nose as a sense organ, as a facial feature, as matter in which the skill of a plastic surgeon might be exhibited, and myriad other ways. One can think the nose–phantasm as an end (fascinated as one is by Socrates' strange nose) or as a means (for the sake of understanding how flesh covers cartilage or understanding how many nose types there can be). In every case there is the same object–phantasm, the same direction of attention, the same mind powers; but they are focused, related, and contextualized differently. The intelligibilities, intelligible species, concepts, or whatever one calls them are not necessarily, and not in the first instance, essences of substances.[90] Infants, as they first acquire concepts and language, are not abstracting the essences of substances, but rather features of form like the softness, roundness, redness, and shininess of a plastic teething ring; and in

---

[90]In any case, a nose is not a substance for Aristotle, but rather a part of the substance called human being.

the first instance these are less full-blown abstract concepts than a nascently abstract conceiving of appearances. Something similar can be said even of adults who are already comfortable with concepts as they discover things they have not encountered before.

It is possible now to give a more definite view of the difference between the approach of *Posterior Analytics* and that of *On the Soul*. The aim of the former is a theory of detailed scientific knowing according to real causes. In modern terms, the *Posterior Analytics* invokes psychology for epistemological purposes. The little bit of psychology in II.19 serves only to specify how it is possible to arrive at premisses that are not themselves demonstrated by syllogistic reasoning from causes—truths that can serve as first truths, *archai* not deduced but rather proved (i.e., tested) through and by experience.[91] The account elucidates what is needed to complete the account of knowledge by explaining how ultimate premisses might be known without being deduced. In *On the Soul*, how one knows scientifically or epistemically is only a derivative concern. It focuses instead on the basic classification of the powers of living things, including the powers of the human animal. Thus the first question is not how we form an inductive truth but rather how the soul powers interoperate, and in particular the role *phantasia* plays in the intelligibility that is a distinct aspect of the unified activity of the (human) soul. Intelligibility is, of course, a crucial preliminary to scientific knowing but not to be confused with it.

There is another distinction between the two works that further clarifies their difference. The intellectual universe of the *Posterior Analytics* is discourse about the knowable essences and the essential attributes of substances. Individual substances fall into a "lowest" species, the lowest species falls into a genus, that genus falls into a more comprehensive genus, etc. Substances have certain attributes that are always (or nearly always) true of them. Aristotle is a human being, and a human being is a mammal, and a mammal is an animal. As a human being Aristotle must have a body and be mortal; and, unless he has been seriously impaired, he will have rationality, memory, imagination, common sensation, vision, hearing, etc. These are the kinds of basic truth about beings that make really scientific knowing possible. In *On the Soul*, however, the focus of II.6 to III.3 and III.9 to III.13 is a very detailed account of how animals have awareness of things by way of their senses and how imagination and the persistence of images make that awareness and the behaviors predicated on it richer and more complex. Scientific knowing, it turns out, is not necessary for animal behavior. Knowledge *is* part of the awareness of the human being; but before knowledge comes intelligibility of phantasms.

Intelligibility does not lead straightaway to knowledge. There is an intermediate stage that is almost wholly overlooked by the traditional theory of abstraction. This stage comes into sharper focus by again considering what taking note of snubness, that is, the concavity in a nose, requires. If we think about this in primarily conceptual terms, that is, in abstraction from real snub noses or phantasms of snub noses,

[91] This seems to be related also to the account of indemonstrable truths about particulars in *Nicomachean Ethics*, VI.8.

there is little that seems surprising. We can analyze in terms of conceptual features and list true propositions. Noses are made of flesh, flesh is body, body is physical reality. Like other macroscopic physical things noses have qualities of appearance: color, shape, position, and the like. Snub noses in particular have a concave curvature.

If we take seriously Aristotle's dictum about the necessary relationship between thought and phantasms, however, we must beware of believing that one can understand things like snub noses in a purely conceptual way or in words without reference to things and their images. A formula uttered without the direct sensory or indirect imaginative or memorative presence of an appearance that the formula is about is empty, and cannot be considered understanding in anything but a purely potential sense. For example, we have said that a snub nose has a concave curvature, but where? How does that curvature conform to other curves in the flesh of the nose? If the concavity of curve is along the top ridge of the nose, running from the bridge at the eyes down to the tip, then the nose is not snub but a Cyrano-de-Bergerac nose! To be a snub nose, the concavity has to be at the tip. None of this makes any sense unless one is thinking of an at least potentially visualizable schema of a nose,[92] and a nose placed at least approximately with respect to a few other features and conformations of a typical human face. Without something thus approximately and incipiently imaged, the snub nose's features are nothing more than a list: concepts without an object.

What, then, is required for active understanding? What is involved in recognizing a snub nose as snub, as nose, as flesh, etc.? I have pointed out several times that Aristotle understands change as requiring contraries (*enantia*) and a substrate (*hupokeimenon*). He also frequently says that *knowledge* always extends to contraries in a substrate.[93] In one sense, perhaps the one that interpretations of Aristotle most emphasize, this means that knowing a substance requires placing it in a genus (which is analogous to matter or substratum, he says) according to differentiating factors (which within the genus are contraries). The person who best knows what a squirrel is also knows what it is not, in particular by being aware of other things that fall into the same genus but are specifically different. According to modern biological taxonomy, squirrels belong to the family Sciuridae (meaning "shade–tail"), which includes tree squirrels, ground squirrels, chipmunks, marmots, and flying squirrels. These kinds correspond to subfamilies, like the Sciurinae, which embraces both tree squirrels and flying squirrels. To reach common American squirrel species, for example the eastern gray squirrel, one has to descend through the tribe Sciurini, the genus Sciurus, and the subgenus Sciurus to reach Sciurus carolinensis. Real knowledge of the eastern gray squirrel—I assume that a zoological taxonomist

[92] If this sounds Kantian, that does not mean it is a false imposition on Aristotle. Good history of philosophy is not just a matter of finding the same words and parsing them, but even more fundamentally the thinking through of a problem in the contexture of its topology. And definitions serve no purpose unless one recognizes what things and world situations they can be true of.

[93] For example, *Metaphysics* IV.2, 1004a10–32. The importance of *Metaphysics* IV.2 for understanding the project of the *Metaphysics* can hardly be overemphasized.

can genuinely claim to have such knowledge—thus implies also familiarity with the structure of speciation and the differentiating factors that account for it. For the average person most of these words are mere placeholders of possible meaning; for a sciurologist they are rich with articulated content.

Something similar is true for intelligibility in general, or even mere familiarity. Even slight knowledge or the first dawning of intelligibility is *situated*. The background of multiple experiences of domestic and other animals enables a child in the early stages of language acquisition to recognize and distinguish cats from dogs and squirrels from both. A variety of hair color and tail bushiness, a repertory of typical motions and variants, general body configuration, etc., produce a situated sense of the range of possibilities called squirrel. Often a little experience can go a long way in developing this situatedness, this sense of the field of possibilities of the thing, this sense of its substrate, its *hupokeimenon*. Some of the possibilities can be varied continuously; others are more discrete. For every term of a conceptual analysis, situated experience presents the genus–substrate as a field in which the indicated features are placed, with ranges of variation between extremes of features that are typical of the species in the genus. If the overall body shape is too elongated, if the rear legs are too short and insufficiently muscled, if the hair color is too brown or white, if it is more arboreal or terrestrial, etc., etc., etc., the animal in question will not be an eastern gray squirrel, or not a gray squirrel, or not a squirrel at all. This substrate–*field* differentiated by multiple contrary–pairs allows one also to anticipate speculatively other individuals, varieties, or species that have not yet been encountered in actual experience. And together they constitute a complex imaginal field.[94]

An imaginal field is grounded in the familiarity of a field of real experience and therefore bears its specific conformation, resistances, textures, and topography. Just as with a real landscape that a geographer is preparing to map for the first time, the imaginal field is initially undescribed but describable. The geographer already has in his possession a well-developed conceptual topology, shared with all his fellow professionals. The topography of this field is the conceptual topology in action. Even when new situations or unprecedented features in a familiar situation are encountered, the attempts at topography will be guided first by existing topologies. One marks features and highlights landmarks, then relates other aspects of the field to them, for example in a real geographical field by using continuous curves (like isobars on a map of atmospheric pressure) that connect all points having the same elevation above sea level. Through the graph, the diagram, the figure, the image, one has in mind the thing it represents and the specific resemblant[95] features that are

[94]This term is not to be assimilated to "imaginary field"; the latter term suggests that we are in the realm of the fantastic, unreal, or impossible, whereas "imaginal field" presupposes simply a field in which images take place.

[95]Some readers will balk at this term and insist that there is no resemblance between a curved line and a landscape directly presented to the eye. That there might well be a sense-perceivable feature highlighted within an image in the example could be illustrated by recalling how such topographical maps are made: by mapping a landscape's plan from above, then superimposing on it the contours that have been measured with surveyor's equipment and can be experienced by walking a path that neither rises nor falls.

highlighted in this particular genre of representation. Unprecedented features can gradually be incorporated into established practices of representation, even if this incorporation does not manage (or manage fully) to acknowledge what is new or unique. And if someone notices a new field—for instance the continuous field of hue that Isaac Newton noticed in the prismatic spectra he produced while trying to figure out how to manufacture superior lenses—the first attempts at a new mode of description are a kind of topography that bears within itself seeds of a new conceptual topology. The description of something new, or of a new aspect of the old, can begin to conceive—more properly *incept*—a new substrate. If multiple attempts allow some consistency and coherence of topography to emerge, one may be participating in the development of a new substrate–field with its peculiar conceptual shapes, its characteristic topology.[96] Conceptual topologies are therefore both abstractions from the concrete and (when they are used to mark an actual situation) concretions of the abstract. The imagination as Aristotle conceives it is a double abstraction, insofar as it separates an appearance–form from an active sensation that, in its turn, drew the appearance–form from a real-world object. Insofar as imagination can re-incept appearance–forms and vary and intertwine them, it is a kind of reconcretion. This, it seems to me, is what Walter Benjamin had in mind when he described imagination as requiring in the first instance de–formation, a release of the forms of appearance from the immediate and exact constraints of their original situation into a field of variant possibilities and development.[97]

---

[96] The first person ever to conceive of a map was in process of originating a conceptual topology by way of conceptual topography. Marking features of a place—by signs, words, or symbols in any way whatsoever—is already conceptual topography, the writing up of place by writing it down (even if one has noted these things only mentally, that is, in a schema of imagining). Whatever in the topography is repeatable in similar circumstances becomes conceptual topology, which is a network of concepts articulated by contrarieties against a background of a typical substrate–field.

[97] See Sect. 3.3, above. The abstraction in question here is neither Aristotelian *aphairesis* nor medieval *abstraction*, but it is still easily understandable. If the reader prefers, he can reread the preceding paragraphs with "model" substituted for "field." But that would obscure to some degree the very distinction I want to draw between *imaginal* fields and fictional *imaginary* fields. The latter is, or starts as, a model; the former is the conceptual topology that presently deepens our experience of the familiarly real. Not all models or imaginary fields have a specific density sufficient to place us familiarly in a field of experience, which is to say that they will not lose the aura of weightlessness and unreality characteristic of arbitrary fantasy. The field of experience, which is a substrate for articulations according to contraries, must appear to be a plausible "cross section" of the world, the contextural context against which intellect places a previously noticed phantasm. Perhaps a working definition of imaginal field, then, is that it is what commences with how things strike us and that we judge to be so (the first part of the predicate clause corresponds to Aristotle's *hupolēpsis*) and that is brought to fruition according to how intensively it permits us, in amplified and developed form, to inhabit and experience the world as world. In either case, whether we are talking about the imaginary or the imaginal, it is a biplanar experience, in the sense developed in Sect. 3.8, above. See also the next note.

5.13 Thinking Imagination

It is doubtful whether fields, be they real, imaginal, or imaginary,[98] can be explained simply by the concept *association* and its three major Humean forms, resemblance, contiguity, and cause. The Humean conception of experience tends to maximally atomize it. Ideas/images are weakened, remembered forms of sense perceptions or impressions. The original sense impressions can be associated and reassociated in virtually limitless ways.[99] The category of resemblance would seem to assure that, for example, green will be preferentially associated with other greens, then with colors, then with surfaces and kinds of objects, etc.—but resemblance is in the eye of the beholder, so there is no sufficiently strong reason to assume that every human being will naturally develop a strong sense of colors as closely related, much less a scientific taxonomy ratifying the taxonomies produced by everyone else.

It is precisely insofar as Aristotle's theory of mind is psycho*physiological* that it makes the Humean account implausible.[100] The working of the visual system shows black, white, and other colors if it is functioning at all. Colors and their interrelations constitute a natural psychophysiological subfield of visual experience. So if one's thoughts are already in that field, it is less surprising that one can "move" from one color possibility to others in typical ways—for example, by interpolating a new shade of blue one has *not* encountered between two one *has*, or by sequencing all the blues one can conceive. To do this, you have to have a familiarity with the typical configurations and features of the quality or object you are interested in. The thing must already be situated with respect to basic features that make it

[98] Later, especially in Sects. 8.1 and 8.2, I shall present imaginary fields as fictionalized imaginal fields. But that, in turn, requires some conception of how the fictional is contrasted with the real. It will not be sufficient simply to invoke the putatively real as a hedge against (for example) the epistemological threats of fiction. Wherever we confront a fiction with reality—whether it is *Crime and Punishment* set against an actual case of murder, a scientific hypothesis being tested in a laboratory, or a lie discovered by the testimony of witnesses—we have already begun schematizing the real situation according to what we consider an immediately relevant conceptual topography. Without yet having laid a basis for it (although this section begins that task), here I will only assert that the real is what can be sectioned into innumerable, variably dense fields, and that what is real is marked as a kind of maximum implicitly measured by the scope, textures, aspects, and specific densities of such fields.

[99] Hume makes an exception for mathematics, of course, because, he says, it is based purely on relations of ideas. But what I have been suggesting here and earlier is, in effect, that Hume's presentation needs to be absorbed into a theory of a vastly greater number of substrates/fields of appearance than pure mathematics requires. I should also remark here that Hume's explication of imagination in the *Treatise* is far more subtle and profound than the later, simplified version of the *Enquiry*. That the latter is more frequently read and taught by philosophers has deeply shaped the tradition of imagination.

[100] British empiricism after Hobbes approached epistemology without reference to physiology; see Locke's disclaimer in the *Essay concerning Human Understanding*, bk. 1, ch. 1, sect. 2 (Locke 1690). But Hobbes and the so-called rationalists always considered both, and that is a mark of superiority.

distinguishable (and thus intelligible at least in part), even if it is not fully or even adequately *understood* as a result. Familiarity with the landmark features and the variational possibilities of a phenomenon rather than expressly conceptualized understanding is what imagination requires in order to function. The features and variations allow for imaginative sequencing and order according to the more and the less, and sometimes to the possibility of determinate measurement, where we find a correlation that lets us specify a unit measure. The basic features of colors in their typical character, grounded as they are in human (or animal) psychophysiology, support the flexibility of productive imagining. For every phenomenon type among the sensibles, whether proper, common, or concomitant, there are many other such features similarly grounded.[101]

Through the more particularized experience of worldly things, which depends on the contingencies of what we encounter and do, we become familiar with concomitant sensibles like flesh, tail, hair, bushiness, and, more generally, species appearances that permit us to perceive an individual *squirrel* as having all these features. Once this unity of features is recognized, it is subject to voluntarily directed imaginative reproduction, variation, and new production. Aristotle claims that it is by induction, *epagōgē*, that we are able to recognize things once we have experienced them often enough. One is free to say that it all happens by association, of course. But in doing this one must not become ideologically blinded to the fact that the things and features in experience are already associated with one another (1) in natural fields grounded in the psychophysiology of proper and common sensation and (2) in acquired fields largely based on and correlated with proper and common sensation—fields of concomitant sensibles, to use Aristotle's term—and with which we have become familiar by being situated in social and physical worlds that have been articulated and bespoken before us.

These fields and subfields of experience allow one to survey ranges of possibilities and to variably locate the current object of attention (whether a thing in the world or a phantasm) with respect to those fields. Whether a snub nose shows itself as being an individual nose, being specifically a snub nose, being generically a nose, being flesh, having a shape, being an instance of an essence, or being in need of a plastic surgeon is a question of circumstances; none of these can be privileged as the default response of the thinking mind, although any and all are possible. But what Aristotle portrays throughout book III of *On the Soul* is our ability, in the context of a situated thing or phantasm, to begin to think about it and to place it differentially with respect to the many fields of possibility. This means that we are dealing with *thinking imagination*, the kind of imagining that is guided by intelligibilities that can be seen by reason in the phantasms and the phantasmal substrate–fields.

---

[101] Neurophysiological studies, for example, have demonstrated the importance of boundary detection in the constitution of the configuration of space and objects in the visual field. Boundaries in the field of vision, which are common sensibles, are constituted by light–dark contrasts and contrasts of colors, which are matters of proper sensibles. There is no reason to think that locating contrarieties in substrates *à la* Aristotle is incompatible with contemporary research.

To see a nose as snub is to recognize already that a nose, although part of the human face, can be considered with a certain degree of absoluteness ("absolute" literally means to be "loosened from") or, to use a technical philosophical term, prescinded from[102] (literally "cut away from"), its ordinary attachments—in this case, from being attached to a face. On first glance this might seem to be a falsifying move: noses don't exist apart from the bodies they are attached to, so by thinking of a nose in such a way you are thinking falsely. Yet the ability of intellect to take an object or its form of appearance and to separate it decisively from its ordinary settings is crucial to the human ability to think persistently about something over time and to understand it in detail. A certain irrealism is necessary to the discovery of real possibility. The truth or falsity of what is learned by prescinding one thing from another to which it is invariably joined depends on what you do with that precisioned learning. For example, precision (or prescission, to make the sense clearer) allows cosmetic surgeons to create catalogs of possible nose shapes—imagine 200 types of nose configurations drawn from different perspectives but unattached to a face—or a series of nose statuettes presenting sample noses. It would be a bad cosmetic surgeon who did not at some point think about specific nose shapes, in various sizes, in relation to the kinds of faces it would flatter, although as a technician his major concern during surgery would be how to endow a given nose (largely prescinded from the patient and the rest of his or her body during surgery) with a predetermined shape from the catalog. Any prospective patient would be rightly

[102] *Prescind* (*praescindere*) gave rise to the terms *precision* and *precise*, which etymologically suggest being cut away from something, as if by a very sharp blade. *Praescissio* (there are various spellings), precision, is often explained as a variety of abstraction, but that invites misunderstanding. Thomas Aquinas discusses the difference in his short metaphysical treatise *De ente et essentia* (*On Being and Essence*), ch. 2, par. 6–7 (Aquinas 1968, 38–40). An abstraction is a concept understood as necessarily subject to further determinations in order to be real, whereas a precision is a concept treated as though its referent exists really and absolutely, apart from further determination. When we treat body and soul as though they were in themselves two complete things that can be compounded together, we are talking about both *with precision*. We *abstract* if—observing Aristotle's understanding of soul as the first activity of a body and all materials of a body as already possessing some form before they are taken up by the body—we recognize that when we talk of body it always has further essential determinations and forms. For example, it will be animate or inanimate, carbohydrate, fat, or protein, red, blue, or green. When we talk of soul, it does not exist apart from the body but only as the activity of body. Similarly, when we treat the nose as though it were something complete in itself, we *prescind*; when we recognize it as part of a face and a human being's whole body (even if at the moment we do not highlight that fact) we *abstract*. Precision might be described, then, as abstraction followed by absolutization or reification (but then it is no longer properly abstract). Abstraction always keeps in view (possible) background relationships while focusing concern on a foreground plane. Precision does not necessarily lead to falsehood, except when the mental difference, the difference in the way a thing is taken by intellect, is ontologically absolutized. Aristotle's *aphairesis*, which isolates the spatial aspect of material existence, is a form of precision rather than of abstraction that is nevertheless controlled by the fuller knowledge that geometrical forms never actually exist outside the material world. Reductionisms, by contrast, are all precisions.

concerned about the competency of a doctor who in consultations about esthetics did not notice anything but the nose. Thus treating the nose precisely, as isolated, and treating it abstractly, as necessarily involving further determinations and situationing, are not incompatible in the long run. In the short run they are quite different ways of taking the image, of placing or situating it.

Although Aristotle does not use the later vocabulary of abstraction and precision, he clearly understands that intellect, which grasps the forms presented by phantasms, can grasp them in different ways. Part of that difference of grasp is undoubtedly related to the total phantasm of a single thing taken as a whole. One can, for example, try to form and hold in mind a very clear visual image of a single "snapshot" pose of a squirrel, say one that lives in the tree in one's backyard. But *your* incipient squirrel image (in northern Europe, say) will not be the same as mine in Texas, even if we limit ourselves to typical attributes like size, coloration, bushiness of tail, pose, etc., of specimens of a similar species and set aside the inevitable differences of each person's experience. The images will be even less similar if we take into account that your squirrels will be sitting beneath a London plane tree, whereas mine is sitting in the branches of a mesquite tree (which looks more like a bush); yours is overlooking a vegetable garden whereas mine is above an expanse of cement; you are an expert in rodents, I am a hunter; etc. And that is not even to raise the question of further "associations" that are natural or likely: seeing the squirrel as moving rather than stationary or thinking of it in relation to other squirrels or animals that inhabit the trees and yards nearby.

Notice how this quickly turns us to the question whether we are thinking/intellecting the squirrel or whether we are imagining it. But that is an important point, both thematically and historically. In the narrow sense Aristotle defines a basic phenomenon of imagination: that beyond the original sensing of something in the presence of the thing, there is an organic motion that continues onward, deeper into the body, where it can give rise again to similar (though not necessarily identical) appearances. Human beings are the most complex animals, so where that motion and reappearance end in human beings is far more complex and multiple than in other animals. That means that in an extended but legitimate sense—extended, that is, beyond the definition of imagination as the motion immediately originated by sensation—it is right to call these more remote, complex, multiple, and recursive processes and instances of reevoking appearances *imagination/phantasia* and the appearances that occur through these processes *images/phantasms*. No matter how narrow the definition of imagination as a kind of motion may seem at first glance, it is directed toward ever more complex forms (and formats) of appearance. It would be an exaggeration to say that more than 2,000 years of the proliferation and complication of imagination was implicit in Aristotle's definition, yet it is no exaggeration to say that, against its background in his physics and metaphysics, it potentiated detailed elaboration into a more complex and sophisticated soul power. In a large but still specific sense, one can say that the Romantic conception of imagination is a radical expansion, even a hypertrophy, grounded in the conceptual topology that Aristotle laid down—even if we need at the same moment to recall that the transition from imagining to intellection by way of activities in variable substrate–fields was foreshadowed by Plato, and that the Romantics were inspired by Kant, Fichte, and Schelling, not Aristotle.

Thinking imagination can more than terminologically bridge the gap that Castoriadis described between conventional, second imagination (of *On the Soul*, III.3) and the radical, first imagination that appears in the later discussion of intellect.[103] Imagination begins as the motion of appearance–alteration that originates in sensation. Sensation itself is a kind of motion, and since it involves a reactivation of the same activity that is in a real-world thing (though without the matter), it is the beginning of the abstraction of form—with "abstraction" understood here quite specifically and narrowly as "treatment apart from a certain relevant kind of matter." In Aristotle there is matter and there is matter. The red of an apple is removed, in the sensation of red, from the matter of the apple, but the sensation is in its turn an activity of a different matter, that of the eye and the visual system. *Phantasia* in its most basic sense is therefore the immediately next stage of abstraction from matter: the formative activity of the sense organ becomes an articulated motion that moves on to be realized in other parts of the body's matter.

We are so accustomed to conceiving of imagination as giving concreteness to thought that we easily overlook the fact that it begins—in Aristotle and (I would claim) in fact—in and as an abstraction. It is a form of abstraction that, in Aristotle, can reconstitute the original appearance *elsewhere*, in an appropriate organic location, which he calls the first or common place of sensation. The activity in common sensation then originates a richer *phantasia* that carries elsewhere the appearance-producing motions not just of the proper sensibles but also of the common sensibles (unity, position, motion, temporality, and the like); it also bears information regarding the concomitant sensation that marks the complex phantasms of common sensation with significations like "Diares' son" (when the white, noisy, extended, moving thing you see is that person). Imagining is a reevocation of the original appearance (a reactivation of the initial sensation without the direct action of the external senses). But it is by no means evident that those reevoked images have to reproduce exactly what was experienced before. Aristotle's complex analysis of common sensation's integration of the proper-sense phantasms and the emergence in this integration of the common sensibles and concomitant nameable features of the world suggests quite the opposite: that we can reevoke as much or as little as we like, at least when imagining is being guided by thinking. We may evoke a figure as rich as the original phantasm; or we can schematize it as a mere shape, exactly or approximately, or produce it with color but no sounds, or sounding and moving but no colors, etc., etc., etc. One phantasm thus leads to other phantasms as surely as, in Peirce, sign leads to other signs. Because all of these features are positioned in their respective virtual (and, in some cases, real) substrate–fields, the produced phantasms can further diverge in all such virtual respects from any so-called original. There is thus a freedom implicit in Aristotelian imagining that has no correlate in

[103] See Sect. 5.5, above.

modern empiricism: a freedom not of randomness but rather one exercised within the organ-based and extended fields pertaining to the image as controlled by one's purposes.

Anything new that one experiences, a new object, can in analogous ways immediately be subjected (in the form of its phantasm) to a similar kind of variational reevocation and recontextualization. To use a mathematical model that I think quite exactly describes this process: the phantasm is constituted as a function of multiple variables, and we can thus subject it to corresponding varieties of *partial differentiation*. Through this differentiation of a new object we explore the variations this partial differentiation effects and become aware of the region of phantasmal spaces occupied by the object. As we become more familiar with this region of virtual space, we can then exercise a kind of analogue of *partial integration* that allows us to array the object and its variants in an enriched, more complex phantasmal space. Finally, such imaginative work deepens our sense of the real object from which the phantasm arose, and sometimes we can then imaginatively project our partial differentiations and integrations back into the "real" world—for the sake of more carefully conceptualizing the object, more elaborately conceiving its environment, or even making and placing something new in the world, technically, artistically, or practically, according to a plan that our thinking imagination has conceived.

It is because human beings can decontextualize and recontextualize appearances in radical ways, with many more possible features and differences of aspect, that they are far more complex in their functions and behaviors than other animals. We can accomplish decontextualization and recontextualization because we can see many different forms of the thing and its relationships in phantasmal fields, and we can form and reform phantasms and their surroundings in ways corresponding to the forms we have seen and grasped. This thinking or intellectualized imagination appears to be distinct from the kind of imagination that animals exercise. The latter is purposed by desire and memory, which could be called two degrees of freedom, but not with the virtually limitless degrees of freedom that the form-grasping-and-contextualizing power of intellect makes possible. The justification of using "intellectualized imagination" is reinforced by reflecting that to call it "imaginationalized intellect" would be in essence an unrevealing redundancy for Aristotle—since there is no thinking/intellecting activity without phantasms.

Thus I think it becomes clearer why Aristotle did not feel an immediate need to discuss the varieties of imagining at greater length in book III of *On the Soul*. He does, however, give at least one very pregnant example, in III.7, that typifies how we should think about imagination's more complex behavioral and cognitive workings. Right after remarking that "the thinking part thinks the forms in the images," he says that what is to be pursued and avoided is determined for the thinking part by sensations (when the object is present) and moved in a similar way when there are phantasms before it rather than things sensed.

> For instance, sensing that a signal light [light representing a proper sensible] is a fire, and observing by what is common to the senses that it is moving [a common sensible], one recognizes that it is an enemy [a concomitant sensible]; but sometimes, by means of the imaginings and thoughts in the soul, just as if one were seeing, one reasons out and plans

what is going to happen in response to what is present. And when the soul declares, as it would in the case of sensing, something pleasant or painful, here in this case too one flees or pursues it, and so in all matter of action. (431b5–11)

One can quarrel about how clearly and enduringly any given individual can imaginatively picture a battlefield at night, but Aristotle clearly believes that our thinking always takes place about or with respect to senselike forms actually appearing. One might easily conclude that the efficacy of thinking is in many respects strongly correlated with how carefully and accurately one's thinking is directed to possible appearances. A sloppily conceived image of the expected battle formation of the enemy is not going to elicit a good counterstrategy. Nor is one that is too detailed, trying to conceive the exact position, height, and weight of every soldier.

If someone thinks that imagination is thus confined merely to the practical actions of human beings, Aristotle immediately points out that his reflections are not confined to actions. Questions of the true and the false, even necessary truth and falsity, still fall under the same genus as actions, that is, the genus of the good and the bad. Questions of the theoretically true or false are per se unqualified, although they are qualifiable and applicable in manifold ways. A given real-world situation like that of the lights advancing across the plain at night is *actually* qualified, in the particular ways manifested by sense perception in the immediately present situation. Intellectual imagination can shift qualifications, change or alter them, and it can think about such situations in generic or precise ways. But this does not mean that there is thinking without phantasms against an appropriate background. It is immediately after this observation that Aristotle mentions the objects that we think by abstraction (*aphairesis*), the mathematical objects that can be isolated from physical things, and also the case of the snub nose. That is, by the sequence and character of the examples he is affirming in dramatic fashion the exquisite flexibility of intellect's placing the phantasm in a context and then taking hold of the forms that it perceives in the situation.[104]

Aristotle concludes III.7 of *On the Soul* with the remarkable statement that "in all cases the intellect, in its being–at–work [actuality], *is* the things it thinks. Whether it is possible for something to think any separated thing without itself being separate from extended magnitudes, or not, must be considered later" (431b16–19). The first sentence could be interpreted in a variety of ways, but in the immediate context of III.7 and with III.8 being devoted to summary, it must have a fairly definite meaning. It is not that when we think a squirrel or a squirrel phantasm

[104]In chapter 1 of *On Memory and Recollection* Aristotle says that we can look at a picture either as a picture or as an image of an original, and that when we move from the experience of the picture as picture to the recognition of the person pictured there we move from an instance of (simple) thought to memory. He also expressly points out the many contexts in which we can draw or view a diagram in geometry: a drawing of a triangle is of a definite size, but one can use it even if the triangle being conceived has no definite size; "and in the same way, one who is thinking, even one who is thinking of something that is not a quantity, sets a quantity before one's eyes, though one does not think it *as* a quantity, but if the nature of it is among things that have a quantity, but an indefinite one, one sets out a definite quantity, but thinks it just as a quantity" (450a3–7).

our intellect becomes the squirrel in an unqualified sense; rather, the intellect becomes the squirrel precisely in the way that the intellect takes up the form it is thinking. Although there is a strong temptation to think this means that the intellect must take on the *essential* form of the thing, that is by no means self-evident, especially in light of how book III's discussion of intellection focuses on the *various* intelligible *aspects* that phantasms present. My suggestion is that what Aristotle means in this concluding passage of III.7 is, to use an example, that if one is trying to remember whether one has seen this particular squirrel with its peculiar coloring in the yard before, the intellect becomes, with a questioning inflection, the possibly–remembered–squirrel–with–these–colorings–in–this–location; if one recalls that one has seen this particular species of squirrel before, anywhere, the intellect becomes the squirrel–as–representative–of–species–actively–recalled–as–previously–perceived. The snub nose seen as flesh, as snub, and as nose indicates three different, though closely related, active states of being of intellect. This means that Aristotle's conception of intellection is far more nuanced and complex than most rationalisms would conceive. It is no more or less nuanced and complex, however, than the nature of being and the corresponding phantasmal thinking require.

Although Aristotle does not discuss *phantasia* in his ethical writings, he does not really have to, since his ethics is implicitly a paradigmatic deployment of situated thinking with regard to phantasms. The *Nicomachean Ethics* is structured by his theory of virtue or excellence and its many specific kinds. These virtues or excellences are defined in general as habits (*hexeis*, plural of *hexis*), which are first-actualized "havings" or dispositions. A virtue is a habit of choosing well, a choosing that is based on finding a mean between two extremes of possible practice, a practice that follows the example of those who possess the virtue (those who have excellence and wisdom with respect to practical matters). The extremes are a pair of contraries. For example, the virtue of bravery is a habit of choosing between cowardice and rashness. Bravery thus takes place in a field of possibilities arrayed between the extremes cowardice and rashness, a field of educated feeling and articulated action that has been shaped by the upbringing and education of the young in the city.[105]

There are many different fields in which these virtues are placed. They are fundamentally structured by human psychophysiology, though that is apparent only if one notices that the extremes between which the virtues make their choices are contrarieties of pleasure/pain and desire/aversion. Aristotle explains this in the chapters immediately preceding the general definition of virtue (which occurs in *Nicomachean Ethics* II.6). Human beings are born with natural inclinations, for example experiencing pain or displeasure in, and consequently aversion to, fright-inducing dangers, or pleasure in, and consequent attraction to, overrich foods. Some individuals have a stronger natural pleasure or displeasure, others weaker, and

[105] And similarly for other virtues. This means that examples of virtues one has actually seen and the possibilities one can imaginatively entertain are not identical from city to city, only similar. In some cities brave actions might typically be accompanied by verbal display, in others by taciturnity.

the upbringing of each child must take account of these differences. The city, in particular a parent in the city, educates children to learn over time how to "persuade" passions, emotions, and desires according to reasoned consideration of what middle degree of affect and attraction is appropriate in given situations. There is in such practical matters no theoretically correct answer to how exactly one should act. There are too many determinations that are particular to the situation and the moment. Thus one needs to be able to discriminate the relevant particulars and adjust one's actions according to the more and the less with respect to one's individual sensibility for pain/pleasure and aversion/desire. One's education habituates actions and thus, to a certain extent, makes them automatic, although that does not make them irrational (no more than one's ability to move mouth and facial muscles in order to speak can be considered irrational). This habituation inculcates the right kinds of desire in a typical variety of orientations to specific situations, but that frees one to attend as needed to the relevant particulars that differentiate every present case from any rule one might try to prescribe in advance. One can think, sometimes very quickly, in terms of past examples, whether they were witnessed in living memory (by oneself) or are historical or fictional. These are scenario–images that provide dynamic models, models that can be differentiated and varied according to the relevant features of the situation one finds oneself in. The models are located against the background of a complex of fields that have personal and communal dimension. Sometimes certain aspects that are ordinarily left in the background may have to be brought to the foreground when there is time and need for deliberation. The individual virtues and the major aspects of their fields of deployment are cross sections of one's character that have been formed by the ways of life of the city.[106]

It is therefore not surprising that thinking phantasmally, intellectualized imagination, is *constantly* demanded in practical life. That is not to imply that in the theoretical realm phantasms are left behind. What is different is that the theoretical often deals with what does not change and thus can be expressed about all possible situations without qualification. The arithmetic mean of 2 and 4 is 3, now and forever, without qualification; a contradiction is a contradiction, no matter that one contradiction takes place in Persia and the other in the Peloponnese.[107] In practical life, the mean (and the means) is always approximated to and qualified by the conditions in which an end is being pursued. Aristotle remarks that most people will eat less than a wrestler in training but more than (say) a jockey trying to make his weight for a race. The specific amount and kinds of food you need must take into

[106]There are fairly obvious affinities here to the later Wittgenstein's language–games as forms of life.

[107]It is important to recognize, however, that applying (that is, qualifying) such truths requires imagination, and their truth can be seen only in light of phantasms (since there is no thinking at all without them). The principle of noncontradiction is an especially illuminating example, since it can be violated only when there is no difference in time or respect. The appearance of a contradiction therefore is, in the first instance, an invitation to situate the formula in an appropriate variety of substrates with the aim of locating possible differences of time or respect (aspect) that would make the contradiction only apparent.

account many other (and, *theoretically*, perhaps limitless) considerations. Typically there will be no exact, scientific answer to the questions they raise; moreover, one teaspoonful of food more or less usually makes no ethical/practical difference. The major difference between the practical and the theoretical context is not, however, that one is approached imaginatively and the other not. Every kind of thinking takes place "in view" of phantasms in a situation, with respect to the real, or the historical, or the fictional, or the contextually conceived thing taken in some specific respect. Understanding something theoretically, no matter how abstract, requires situating it in typical circumstances in which it might be actualized, with all the attendant variability of features in the relevant contextual field, conceived now more, now less *abstractly*—which is simultaneously now less, now more *concretely*. Practical affairs usually have so many intertwined considerations of contrariety (and all their intermediate possibilities) that the most effective way of thinking proceeds by looking for examples and appropriately deploying them—that is, differentiating and varying them to resemble one's own situation. By contrast, thinking situations theoretically is simplified by the abstraction or precision of determinations. In determining the arithmetic mean between 2 and 5 there are no considerations of the matter of the things to be considered. However, it does make a great deal of difference if one is talking about bushels of wheat, or automobiles, or children, or liters of water to be added to liters of alcohol.[108]

At the end of *On the Soul* III.7, Aristotle says that he will take up later whether a thing separated from matter can be thought by a being that is not separated from matter—that is, by a human being. One commentator writes that "no treatise on this subject is known to exist; so either it was written and lost or it was not written."[109] A separated thing would be something that has no materiality *at all*. In the *Physics* and the *Metaphysics* Aristotle arrives at the conclusion that there is a kind of substance without matter, form that is in activity having nothing potential about it—no matter at all—and therefore is always fully and exactly what it is. Whether such a thing could be thought by our intellect, unless our intellect itself were separate from all extended magnitudes, that is, physical, sensible, imaginable things, is the question. If it is strictly true that, for human beings, there is no thinking without phantasms, then it is difficult to see how Aristotle could allow that we can think pure form purely. Actually considering this possibility would have shed further light on whether his dictum about thinking and phantasms is true in a fully unqualified sense, in *any* and *every* respect.

---

[108] A unit of water added to another unit of water gives two units of water, and similarly for alcohol; but one unit of water added to one of ethanol yields less than two units of liquid.

[109] Hippocrates G. Apostle, in Aristotle 1981, 179, n. 30. Joe Sachs, in his translation of *On the Soul*, renders Aristotle's "later" as "in the next chapter." Only the last sentences of III.8 possibly just raise the question, however, and although there Aristotle asserts that the simplest intelligibles are not themselves phantasms he fails to present an argument. In the *Nicomachean Ethics* Aristotle presents a vision of the happiest human life as one of the pure activity of thinking the fully and always actual highest being (i.e., true divinity). It is clear, however, that even if this is possible, for human beings such activity cannot last for long.

5.14 Conclusion

I have been arguing a maximalizing interpretation of imagination in Aristotle, less for the sake of recognizing what is clearly and fully expressed and developed in his works than what is partially developed or suggested and susceptible to further development. I have also aimed to understand Aristotle as having taken up and innovated upon Plato's heritage with respect to the *location* of images and of imagination. If we try to sum up what has appeared from the foregoing, we might say that Plato and Aristotle established the topic of placement as intrinsic to imagination.

For Plato reality is mimetic. There are different levels of being, and in general lower levels of being form themselves through taking on the forms of (and thus imaging) higher levels. By means of sensitive and cognitive capacities the human being is able to recognize these relations, that is, to take one level as representing or imaging another. Crucially, the mind or soul is thereby capable of being on two levels at the same time. The human being ascends to a recognition of higher levels from lower ones; from those higher levels he or she can then more knowingly produce images of the higher levels on the lower levels. The experiencing soul can be schematically understood as a hierarchy of corresponding powers of recognition. If image–perception proper is the recognition of a shadow, reflection, or other kind of physically produced image, simultaneously it involves seeing the image's relationship to its original on another level, and all sensitive and intellectual acts that in an analogous way put a lower level in participative–and–resemblant communication with a higher one can be understood as a kind of imagining. The soul can cognitively move both up and down. Moreover, the human being who has learned to bring forms from a higher level into a lower one has discovered a variety of productive imagining that ontologically mimics the good itself.

If we expand our interpretation of images and imagination beyond the *Republic*, for instance to the *Theaetetus* (as well as back to the *Sophist*), we will discover that Plato has a conceptual grasp on issues of the inward psychological operation of these imaging powers; but, as is typical of Plato and Plato's protagonists, these issues tend to be formulated in analogies from nature and art. For example, beginning at *Theaetetus* 191C, especially 193B–195A, truth is conceived in terms of "lining up" present and past soul impressions as we might physically line them up if they were impressions in wax.[110] But it was Aristotle who fully psychologized the imagination. He did this on the basis of a psychophysiological theory of animal soul. The physics of activity of natural things in the world is conveyed to the sense organs where the same activity is formally induced in the matter of the organ; this gives rise to sensible appearance. This sensible appearance–activity initiated in the organ of sensation detaches itself from the sense organ as it is conducted deeper into the physiology–and–psychology of the animal. What happens from that point depends

---

[110]This analysis in the *Theaetetus* presupposes that knowledge proceeds by comparing image–impressions in different planes of memory.

on the organic constitution of the animal and the physiological paths forward or inward. In any case, this continuation of motion and its ability to resummon the active appearances of sensation without their matter is imagination proper and the *sine qua non* of subsequent psychological appearances *of all kinds*.

Even at the level of proper sensation, however, it is not just a question of acquiring discrete sense data but of the activation of the fields of sensation that are defined by the fundamental contrarieties discoverable in the respective sense powers or abstracted from them. In contrast to animals, whose highest possible sensitive power is the protocognitive sensory imagination (the ability to have, recognize, and produce images within the virtual spaces of sense appearances, but without intellect), the human being has a seemingly limitless power by virtue of intellect to recognize similarities and differences in topologies not immediately presented by sensation. Every act of the intellect must have a significant relationship to images situated in substrates, since there is no thinking without phantasms. To the question, "Where are the images in human beings?" Aristotle has a psychophysiological answer: within and with respect to the operations of the sensitive and cognitively relevant organic places of the human body. Intellect, of course, is for Aristotle not proper to any organ, since having a particular instantiation in matter would limit its scope. The scope of intellect is universal.[111]

But an argument can be made that, in light of there being no thinking without phantasms, human intellect's true function is precisely to raise images out of their original space or field and to abstract them into new contexts or spaces. These new contexts or spaces still have a certain phantasmal character, even if they may be ever more abstracted from their original home in the world of real things. If the foundational event of imagination is the separation of the motion and appearance in the sense power from the material circumstances of sensation, that is only a first abstraction, from the most concrete kind of matter. It does not imply that the intellect ever totally abstracts from *phantasmal matter*, at least not with human beings. Intellect always abstracts from some but not all particularities of matter by highlighting, naming, and recontextualizing the appearances of formed matter in new substrates. Thus, imagination in its more profoundly developed human forms *is what intellect does with respect to images against the quasimaterial background of substrate–fields as articulated by contrarieties*. One image always leads, by way of rationality, to others. And since the mind can in deliberation form a new image out of many others that leads to action or production, imagination can, as in Plato, be the instrument whereby images work themselves back into the world through human arts and practices.

Conclusions like these are, in the first instance, an attempt to locate a plausible common ground, the underlying conceptual topology, for the interpretations of Plato and Aristotle given here. That does not mean it is a minimalist interpretation, however, nor is it a least common denominator achieved by diluting Plato's and

[111] After frequently raising the question whether intellect has an organ earlier in *On the Soul*, he finally gives this answer (and makes the argument about matter and universality) in III.4 and III.5.

Aristotle's "theories of imagination" so that some agreement can be found. Instead, conceptual topologies offer a different way of approaching the history of philosophy. Most historiography falls into narrating either a series of philosophical opinions (Kant's remark about "the scandal of philosophy" in essence falls into this class), or topical comparisons and contrasts, or conflicts of positions or interpretations, or a dialectical-logical progression, or Heideggerian *Seinsgeschichte*. As an element of historiography each of these has a role to play. But the first task in understanding philosophers of the first water is to immerse oneself in the element of their thought, which means immersing oneself in what their thought engaged and how it engaged it. That is to say: one must immerse oneself into the combination of phenomena, field, and articulation where their thinking took and had its place: in the conceptual topology that they shared (and can continue to share) with other thinkers who were trying to find their way in the place of thinking. This is far more ambitious than an attempt to ascertain the "object" of someone's thought, and if the search for the object of a philosopher's or a poet's or a scientist's thought is to be successful, we must first locate that object in its place and grasp the ways the object is at home there.

It is one thing, however, to show a common topology for Plato and Aristotle, master and student; it is quite another to show that the topology, hidden from conventional view, nevertheless continued to be effective, even at a distance of 2,000 years. That is the larger task to which we now turn.

References

Aquinas, Thomas. 1968. *On being and essence*, 2nd ed. Trans. Armand Maurer. Toronto: Pontifical Institute of Mediaeval Studies.

Aristotle. 1981. *Aristotle's "On the Soul"*. Trans. Hippocrates G. Apostle. Grinnell: Peripatetic Press.

Aristotle. 2001. *"On the Soul" and "On Memory and Recollection"*. Trans. Joe Sachs. Santa Fe: Green Lion Press.

Carruthers, Mary. 2008. *The book of memory: A study of memory in medieval culture*, 2nd ed. Cambridge: Cambridge University Press.

Caston, Victor. 1996. Why Aristotle needs imagination. *Phronesis* 41: 20–55.

Castoriadis, Cornelius. 1997. The discovery of imagination. In *World in fragments: Writings on politics, society, psychoanalysis, and imagination*. Trans. David Ames Curtis, 213–245. Stanford: Stanford University Press.

Caujolle-Zaslawsky, Françoise. 1996. L'Emploi d'υποληψις dans le De Anima, III, 3. In *Corps et âme: Sur le "De Anima" d'Aristote*, ed. Gilbert Romeyer Dherbey, 349–365. Paris: Vrin.

Dod, Bernard G. 1982. Aristoteles latinus. In *The Cambridge history of later medieval philosophy*, ed. Norman Kretzmann, 45–79. Cambridge: Cambridge University Press.

Hardin, C. L. 1988. *Color for philosophers: Unweaving the rainbow*, 2nd ed. Indianapolis: Hackett.

Harvey, E. Ruth. 1975. *The inward wits*. London: Warburg Institute.

Hegel, Georg Wilhelm Friedrich. 1840–1845. *Enzyklopädie der philosophischen Wissenschaften im Grundrisse*. 3 vols. Georg Wilhelm Friedrich Hegels Werke, ed. Friedrich Förster et al., vols. 6–8. Berlin: Duncker & Humblot.

Heidegger, Martin. 1994. *Einführung in die phänomenologische Forschung*. Gesamtausgabe, Division 2: Lectures 1919–1944, vol. 17. Frankfurt a.M.: Klostermann. English edition: Heidegger, Martin. 2005. *Introduction to phenomenological research*. Trans. Daniel O. Dahlstrom. Bloomington/Indianapolis: Indiana University Press.

Karnes, Michelle. 2011. *Imagination, meditation, and cognition in the middle ages*. Chicago: University of Chicago Press.

Katayama, Errol G. 1999. *Aristotle on artifacts: A metaphysical puzzle*. Albany: State University of New York Press.

Kessler, Eckhard. 1988. The intellective soul. In *The Cambridge history of Renaissance philosophy*, ed. Charles B. Schmitt and Quentin Skinner, 485–534. Cambridge: Cambridge University Press.

Knowles, David. 1962. *The evolution of medieval thought*. New York: Vintage.

Locke, John. 1690. *An essay concerning human understanding*. London

Marenbon, John. 1987. *Later medieval philosophy (1150–1350): An introduction*. London: Routledge & Kegan Paul.

Park, Katharine. 1988. The organic soul. In *The Cambridge history of Renaissance philosophy*, ed. Charles B. Schmitt and Quentin Skinner, 464–484. Cambridge: Cambridge University Press.

Park, Katharine, and Eckhard Kessler. 1988. The concept of psychology. In *The Cambridge history of Renaissance philosophy*, ed. Charles B. Schmitt and Quentin Skinner, 455–463. Cambridge: Cambridge University Press.

Peghaire, Julien. 1943. A forgotten sense, the cogitative according to St. Thomas Aquinas. *Modern Schoolman* 20: 121–140, 210–229.

Peters, F. E. 1967. *Greek philosophical terms: A historical lexicon*. New York: New York University Press.

Philippe, M. D. 1948. Ἀφαίρεσις, πρόσθεσις, χωρίζειν dans la philosophie d'Aristote. *Revue thomiste* 48: 461–479.

Steneck, Nicholas Hans. 1970. The problem of the internal senses in the fourteenth century. Ph.D. dissertation, University of Wisconsin.

Wedin, Michael V. 1988. *Mind and imagination in Aristotle*. New Haven: Yale University Press.

Chapter 6
The Dynamically Imaginative Cognition of Descartes

The goal of Chaps. 4 and 5 was to understand the topology of imagination and its functioning in a Greek master and his master student, who together established the founding topological structures of Western psychological theory. The present chapter and the next one intend to show imagination and related powers, in theory and at work, in two modern refounders, René Descartes (1596–1650) and Immanuel Kant (1724–1804). As with Plato and Aristotle, I will be interested in showing important connections between them. But unlike the founders, these moderns are heir to complex traditions of conceiving imagination and the human psyche, traditions that had been cultivated for two millennia. Moreover, Descartes and Kant were not master and student but figures separated by more than a century of newly vigorous thinking about imagination.

The theories of reason and imagination that came down to Descartes were largely conventionalized. They were distant descendants of the ancient texts, but the lines of descent had been crossed by too many other influences to bear a simple resemblance to the originals. If Descartes read the originating texts of Plato and Aristotle, he left no direct indication of it. That he was familiar with some version or versions of the conventionalized Aristotelian tradition is, by contrast, clear.

If what he came to understand of imagination involved little actual reading of Plato and Aristotle, it nevertheless had a great deal to do with coming to see unexploited possibilities offered by the conceptual topology they had produced. To expand on this figuratively: the two Greek masters had labored over a series of interrelated maps of a territory showing the populations, the political divisions, the flora and fauna, in general the "lay of the land" of the human soul and its relations. What their followers derived from it were variations, amplifications, and simplifications that often lacked the metaphysical, physical, and psychological amplitude and density of the originals. Simplified schemas displaced the originals and were accepted as adequate representations of the phenomena. Even when the relationship to Plato or Aristotle was more direct, the rich conceptual topology they had elaborated usually lost out to other concerns of interpreters. To recover a richer sense of the topology

D.L. Sepper, *Understanding Imagination: The Reason of Images*, Studies in History and Philosophy of Science 33, DOI 10.1007/978-94-007-6507-8_6,

it took someone like Descartes, who found the intellectual maps he inherited so unrepresentative that he had to explore things for himself. Many of the things he discovered, as original as they in fact were, were consequences of his rediscovery of elements of the old conceptual topology.

From an early age Descartes did not merely theorize about imagination, he practiced it. After a youthful flirtation with the hope of Renaissance humanism that poetic imagination might reach even the highest spiritual things, he turned to the philosophic-cognitive use of imagination in mathematics and physics. If there is any truth to the claim that Descartes "mathematized thought," it has to be understood more precisely as an imaginization of thought.[1] This imaginative mathematics and physics provided the conceptual and methodological structure of his first publication (in 1637), three scientific essays (on the optics of refraction, the physics of atmospheric phenomena, and analytic geometry) and the more famous writing that served as preface, *The Discourse of the Method for Rightly Conducting One's Reason and Seeking the Truth in the Sciences*. That was not the end of his imaginative practice and theorizing, however. The meditational approach of the 1641 *Meditations on First Philosophy* was modeled on the religious and philosophical method of using memory, imagination, and intensive cogitation to arrive at fundamental truths, and the work itself probes the limits of imagination. And in his last publication, the *Passions of the Soul* (1649), he conceived imagination as an act closer to will than to intellection and provided imagination with an important new noncognitive task: to help manage the feelings, emotions, and passions, with the goal not of suppressing them but rather of learning to use them in order to "taste as much as possible the sweetness of this life" (AT XI.488).[2] In this perspective, there may be more than mere symbolism in the fact that the last work he composed before his death in February 1650 was a *masque*, a combination of dance, music, and drama presented as an entertainment for the court of Queen Christina of Sweden.[3]

That this is not the Descartes presented by traditional historiography of philosophy should not trouble us. If Cornelius Castoriadis's judgment about the history of imagination in the West is correct, we should expect gaps and occultations in historians' accounts. Castoriadis argued that the followers of the most innovative thinkers homogenize and conventionalize fundamental insights, and that even the innovators rarely draw all the consequences they could. One should therefore not expect to find the profoundest insights in standard accounts or the philosophy of schools.

---

[1]This is an ugly neologism, to be sure, but no more so than "mathematization." The thesis that modern rationalist-mathematical science has "disenchanted" the world, but that the disenchantment can somehow be reversed by "going back" to imagination, is hollow. There is no golden age to go back to, and the proposed remedy is just another symptom of the failure to understand reason, imagination, and their relation.

[2]References to Descartes's works are given by volume and page number in the 11-volume Adam and Tannery edition, Descartes 1964–1976, indicated by the abbreviation "AT." All translations are my own.

[3]The work is lost.

The conventional "truth" that Descartes mathematized thought impedes our access to a deeper truth. The mathematics he invented, analytic geometry, the unification of geometry and algebra, was the most rigorous and active use of the imagination ever conceived. Moreover, several other thinkers and mathematicians who were in close contact with his thought (Leibniz, Malebranche, and Pascal, to name the three most important) agreed that mathematics and natural science were preeminently imaginative undertakings. But by the middle of the eighteenth century, less than a century after Descartes's death, the notion that mathematics is imaginative work was largely displaced by the opinion, the cultural commonplace, that mathematics was the product of rational intellect.

I say *opinion* advisedly. No theory of mathematical intellection was advanced, no effort to come to terms with and to overthrow the understanding of great mathematician–philosophers like Descartes, Pascal, and Leibniz was made. The opinion simply came to prevail without evidence or argument, and it has been with us ever since. There is an historical irony involved: one of the reasons that the agreement of Descartes, Pascal, and Leibniz could be so easily contradicted is that their plainest statements about it were in writings available only posthumously. Pascal, to take one instance, was strongly influenced by Descartes's conception; some of his philosophical and religious writings contain profound reflections on the consequences. Pascal in particular emphasized the moral temptations imagination was subject to and condemned the pride it exhibited, even as he acknowledged its positive use in the sciences. The problem for us, however, is that most of the crucial relevant passages were not available to his contemporaries, and in fact many were not published until nearly two centuries after his death. Leibniz was clearer about the role of imagination in his private letters than in his philosophical and scientific writings; not until much later did the former come to light. With Descartes, the fundamental role of imagination in mathematics and science was most richly documented in his earliest notes and compositions. These writings were for the most part unknown to his contemporaries (Pascal, Malebranche, and Leibniz, who had access to some of them, were exceptions) and were not published until generations, even centuries, after his death. When they appeared they seemed to be little more than curiosities.

The mathematical and scientific role of imagination was clearly and publicly enough stated by Malebranche in *The Search After Truth* (1674–1678), but that came in the middle of a very long book arguing that, outside of mathematics and natural sciences, imagination was deceptive. When all is said and done, Malebranche in his own right had little mathematical or scientific authority, and his predominantly negative evaluation of imagination, in a book that was widely read and greatly influential, had far more effect.[4] That gives a decided edge to the notion that we are dealing with an impaired tradition that, had it been cultivated, might have had glorious successes. Instead it suffered occlusions and occultations; it lapsed and sank from view.

[4]See McCracken 1983.

6.1 Imagination After Aristotle and Before Descartes

Although this is not a historical survey of Western theories of imagination, it will be useful to have in mind a schematic overview of the more than 1,900 years between Aristotle's death and Descartes's birth.

The influences of Plato and Aristotle lived on and became intertwined, especially in neo-Platonism, which was the most successful philosophical school of late antiquity. It was, in particular, respected by educated Christians and adapted by them to give philosophical support to Christian theological tenets. Neo-Platonists typically tried to show that Plato and Aristotle were compatible with one another. They regarded Aristotle's *Organon* (which contains not just logic and rhetoric but also the basic elements of his theories of being and knowing) as an introduction to philosophy proper. His other writings they treated with respect, especially those, like his writings on nature, that had no equivalent in Plato's corpus. The psychological theory of *On the Soul* and the shorter essays of the *Parva naturalia* were readily adapted to Platonic purposes, not least because the conception of the hierarchy of soul powers was similar. Moreover, insofar as Aristotle's theory of intellection was predicated on the noetic power's capacity to recognize intelligible forms in images, it offered a naturalistic account of the ascent from the sense-apprehended material realm to the intelligible realm of mathematics and ideal forms.

In more strictly materialist schools the image-bearer theory that we first saw in Empedocles continued to prosper (for example, in the notion of the *eidōlon* in Epicurus, which has come down to us chiefly through *De rerum natura* of the Roman poet Lucretius). The Stoic school, which in an important sense counts as materialist (rational soul was regarded as the finest and most mobile kind of matter, capable of directing the motions of grosser matter), played a crucial part in reconceiving the critical moment when the images conveyed from the physical world appear at the threshold of mind. There, in the vestibule of the brain's organ of rationality they called the *hēgemonikon*, the appearances had not yet been internalized and so could be put on trial with respect to whether they were true or illusory; only the former would be allowed to enter the domain of reason. Since Aristotle had presented common sensation, imagination, and memory as physical functions preparing phantasms for intellectual purposes, the Stoic and Aristotelian theories could thus coexist in a very simplified topology of the human soul.

The tradition most immediately important for understanding early modern developments nevertheless remains Aristotelian. Aristotle's conceptual topology of imagination and soul exercised its influence well into the seventeenth century. Basic conventionalizations of its conceptual topology were accepted even by those who were not particularly well disposed to his overall philosophy.

After correcting Aristotle's preference for heart over brain as the location where all the special or proper senses are united in common sensation, the physician–theorists of Western antiquity easily adapted his philosophy to their concerns. Medical theorists pursued in more detail the proper localization of soul powers in the brain and other organs of the body; a few even contradicted Aristotle's exemption of intellect from localization by assigning it a brain place. Philosophers who were not physicians resisted for the most part the physical localization of reason, but otherwise the ratification by medical doctors of Aristotle's organic psychology reinforced the authority of his theory and encouraged philosophers to continue developing psychology within its framework. Many of the greatest commentators of Aristotle in the middle ages, Muslim, Jew, and Christian, were themselves doctors or engaged in the study of nature—for example, Avicenna (ca. 980–1037), Averroës (1126–1198), Maimonides (ca. 1138–1204), Albert the Great (ca. 1200–1280), and Roger Bacon (1214/1220–1292). Even though most of the works of both Aristotle and Plato were lost to the early middle ages, at least in the West, their reputation among the learned remained strong. If for centuries the natural philosophy of the soul was no more than lightly cultivated (at least before the great Islamic philosopher–physicians), it was typically done in a blend of Aristotelian "faculty theory" and Platonic metaphysics.

One briefly observed early case will serve as an example. Plotinus (ca. 204/5–270), the greatest of the neo-Platonists, placed an Aristotelian division of human psychological powers within a Platonistic metaphysics of emanation. Emanation, a spilling over of the being of the ineffable Oneness that was the source of everything, produced eight progressively lower levels. Matter (where evil is located) was the lowest of all, soul immediately above it, and intellect above that. This metaphysics radicalized and further elaborated the tendencies of the good's communication of itself into all levels of being that we saw in Chap. 4's explication of the *Republic*. As with Aristotle, soul was understood as a form distinguishing living things from nonliving matter. Its aim was to rise above its material, organic conditions and, through the purifying ascent to intellect, to move closer to the One.

One of the curiosities of Plotinus' psychology is a duplication of certain lower soul powers. There is, for example, memory of what happens to us bodily but also an independent intellectual memory. Similarly for the imagination. One imagination simply reproduces what is acquired by the senses, the other is oriented toward the spiritual; the spiritual reflects and preserves the purely spiritual aspect of the lower kind of phantasms. This allows Plotinus to retain the Aristotelian principle that there is no thinking/intellection without phantasms without having to retain any direct connection to materiality. The phantasms of the intellectualized imagination and memory are nonmaterial, and the soul, in its ascent to intellect, is purged of all traces of matter (it retains the pure forms of what was learned in the lowest levels of experience). Thus Plotinus could assert the unqualified primacy of intellect over sense and bodily imagination and memory—indeed, a total separation from them. This would not be the last time that philosophers would turn their backs on matter and try to make a home among the pure forms of reason.[5]

With the progressive recovery in the West of Aristotle's works—beginning roughly in the middle of the twelfth century and essentially complete by the

---

[5]On Plotinus's double *phantasia*, see Brann 1991, 48–50, and Warren 1966. The central Plotinian discussions of conceptual imagination are at *Enneads* I.4.10 and IV.3.30, and of sensible imagination at IV.3.23.

middle of the thirteenth—the kind of internal-senses doctrine that I explained in conventionalized form in Sect. 5.2 became a staple of Western theories of the soul and its powers. Albert the Great, who wrote commentaries on many of Aristotle's nature treatises and himself engaged in first-hand study of natural phenomena, proposed a five-part theory of the internal senses (following an interpretation of Aristotle by Avicenna), whereas his student Thomas Aquinas (1225–1274) proposed four (favoring an interpretation by Averroës). There was no single theory or number of powers that could be called canonical for the Western middle ages, however. Aristotle himself had left the question open, and there was no little argument over how different places in or around the brain ought to be correlated with the internal sense powers—a question that Aristotle had also largely left open. Already in antiquity anatomists had identified not organs but hollows in the brain—called *ventricles*—as the places of common sensation, imagination, and memory. There was no unanimity about the exact enumeration of the ventricles; often they were conceived in a way that suited the theorist's preferred number of internal sense powers. Positionally the ventricles could be easily enough divided into anterior, medial, and posterior, but more refined theories of internal sense functions could subdivide them as needed. For example, if common sensation was typically attributed to the anterior ventricle, the one closest to the eyes, nose, and tongue, one might still argue for a differentiation of its functions depending on whether the activity took place in the front or rear of that ventricle; similarly for the ventricles of imagination and memory. Furthermore, internal senses theories also eventually incorporated an important element from Stoic philosophy, a very fine and active form of matter called spirits filling the ventricles (and, as knowledge of the nervous system emerged, filling nerve fibers as well). It was thought that spirits could take on and transmit evanescent forms of appearance; thus they were used to further elaborate the physical theory of images/phantasms.

As I mentioned in the conventionalized version of scholastic theory in Sect. 5.3, the internal senses engaged in a kind of "phantasm processing" intended to explain purposive animal behaviors apart from reason; in the human being they exercised similar functions but also prepared the phantasm for abstraction by intellect. The increasingly complex anatomical and physiological processes hypothesized by internal sense theorists were doubtless encouraged by improved knowledge about the brain and nerves not available to Aristotle,[6] but it responded more directly to questions that the general topology of Aristotle's psychology raised. Aristotle had gone so far as to claim that the organic processes of both remembering and imagining took place in the primary place of common sensation, but he did not even begin to formulate the physical complications of how to conceive them anatomically or physiologically (recall that he denied that common sensation was the activity of a specific organ). When he said that we grasp the forms in phantasms he did not ask

[6]One should neither overemphasize nor minimize advances in medical knowledge in antiquity. If Aristotle was one of the first to recognize the need to incorporate techniques of dissection into the study of living things, he and his school did not take great strides in this direction, but in the following centuries Alexandrian researchers brought a greater sophistication to this work.

whether this occurred the same or differently in immediate sensing, imagining, and remembering. When at the end of the *Posterior Analytics* he described how similar sense experiences could reach a "stop" in the inductive recognition of some kind of thing, he did not ask questions about the anatomy, physiology, and phenomenology of that stop.

Nevertheless, Aristotle had blazed a trail that led to the questions he did not ask, because in principle if not in detail he had conceived the powers of common sensation, imagination, and memory as part of a complex psychic and physiological system that organically coordinated and integrated phantasms. The topology he had established not only suggested but impelled later inquiries and conceptions. Seen in this light, for example, II.19 of *Posterior Analytics* suggests that, before you arrive at a concept by induction, you store up sense experiences and sensory forms in memory, which are reawakened by related future sensations; and one day, finally, you see the same kind of thing again and, suddenly, the present and the past experience are pulled together into a sharply focused grasp of what the perceived thing is. It becomes conceptualized, and a name can be assigned to it. Even if medieval doctors and philosophers could not spell out in detail how ventricles and spirits interacted in a way differentiated by sensation, imagination, and memory, in principle this psychophysiological approach allowed them to stipulate physical processing and sequencing of phantasms corresponding to psychological differences. And then, at the end, intellect could take these vigorously processed phantasms and derive—abstract—from them concepts, intelligible species (as *Posterior Analytics* II.19 had argued only phenomenologically).

Without reminding ourselves of this background, Descartes's (and other early modern) attempts to coordinate nerve and pineal gland motions, spirit flows, and the like look arbitrary and idiosyncratic. They are nothing of the kind; they are extensions and radicalizations of the psychophysiological conceptual topology that had prevailed for nearly 2,000 years. All sixteenth- and seventeenth-century thinkers, of whatever traditions or intellectual commitments, were aware of, and usually schooled in, this topology, and new discoveries in anatomy and physiology were incorporated into it. Descartes in particular began his philosophical inquiries by trying to adapt a simplified version of the standard psychophysiology to his own methodological, physical, and mathematical inquiries. He would give radically new meaning to the "preparation of phantasms" by his invention of new, highly intensive uses of imagination for cognitive purposes.

6.2 Descartes's Starting Point

If you begin with the extant earliest writings of Descartes, none of which were published in his lifetime, you discover that imagination was fundamental to his mathematics, his science, and his conception of method. It was fundamental precisely insofar as Descartes worked through and revolutionized the conceptual topology of imagination he had inherited. He was interested in the question not just of how we have images, but even more of what we do with and to them. The conceptions he formed of *idea* and *thinking*[7] were radicalizations of image and imagining, respectively, though in thinking through the problem of imagining he also came to recognize the limitations of images and imagination. Unfortunately very few interpreters have grasped the implication of these things or even noticed them.

Considering the relatively small number of writings that are preserved from the earliest period, which begins shortly before 1620, when Descartes was still in his early twenties, it is surprising how often imagination and images/phantasms are mentioned and discussed. In fact they show themselves to be central to his conception of the acquisition and the dignity of knowledge. The earliest evidence of this work is the most precisely dated. In the autumn of 1618 Descartes was in Amsterdam. One day he and another man both stood reading a poster advertising a contest to solve a problem in mathematics. He remarked to the man that he could solve the problem and many more like it. They struck up a conversation that quickly led to a years-long friendship. We know the story because the other man was Isaac Beeckman, a scholar–scientist who kept a voluminous journal of his ideas and experiences. He was an advocate of explaining nature by using corpuscular matter theory and mathematics, what he called physico-mathematics. He found in the Frenchman someone who was carrying on a similar kind of research. The 22-year-old Descartes was as good as his word in demonstrating to Beeckman's satisfaction the sophistication of his problem-solving abilities.

As a New Year's 1619 gift Descartes presented Beeckman with a work about music theory, the *Compendium musicae* (*Compendium of Music*). At the outset Descartes laid down a set of postulates that present the sense of hearing as governed by simple proportions. This was an old standby of aesthetic theory, in virtually all the major philosophical traditions.[8] Medieval scholastic thinkers in particular had followed Aristotle in asserting that the sense organs themselves were *proportional means* or *middles* between the extremes of their objects and that sensation itself was a *determination of proportion*.[9] Descartes was scarcely unconventional in including among his postulates the thesis that the quality experienced by the sense organ was a kind of proportion, and furthermore that the pleasure or displeasure one feels was correlated with the proportionality or disproportionality of the object to the organ (for example, a very bright light will be unpleasant because it is disproportionate to the organ's capabilities) and of the various qualities to one another (in hearing, tones that have simpler proportional relationships are pleasing, like the octave and the major fifth, whereas dissonances are not). Nor was he breaking new ground in

[7] Even though historical research has made the story far more interesting and complicated, it is still basically true to say that Descartes gave impetus to a fundamental historical shift in the learned use of the terms *idea* (French *idée*) and *cogitatio* (*pensée*). I shall say more about this below.

[8] See, for example, Eco 2002.

[9] Once again, we see in Aristotle and Aristotelians the understanding that sensation is organized between extremes in a relevant qualitative field. In *On the Soul* the organ of sense is called a mean at 423b27 and 424b1, and sensation is called a mean at 424a25 and 426a27. See the brief discussion in the note to the passage marked "423b27 ff." in Aristotle 1993, 112.

claiming that, even if harmonies were pleasing and dissonances displeasing in themselves, a perceiver would be wearied by hearing only harmonies. The listener is pleased best, he asserted in agreement with long tradition, by a variety of tones, harmonies, and disharmonies, so that the goal of music is not to provide harmony at every moment but rather a pleasing impression overall.

The *Compendium* is richly provided with geometric and other figures to express and summarize the proportions that hold between sounds. This, too, is not particularly original, although perhaps the degree to which such figures populate a relatively short work is unusual.[10] It shows that the young Descartes was already quite conversant with proportional geometry and arithmetic, and that he had mastered the art of compressing information into geometric figures and images. But the most striking feature in the work occurs in an early passage that explicitly attributes to *imagination* the function of perceiving a musical composition *as a whole* by joining part to part to part in a kind of calculation of the song's proportions.

In the very first section after he lays down his postulates, Descartes explains how we recognize the time, or rhythm, of music. While we are hearing the present beat we recall what we have heard before in proportional relation to it, and we progressively hear our way through the piece, extending all the proportions that we have heard right up to the present moment. This is not merely sensing what is immediately present plus remembering what is past but rather actively and continuously synthesizing a whole, here and now, out of the present and the past and moving conjecturally into the future:

> For then, when we hear the first two members, we conceive them as one; when [we hear] the third member, we now conjoin that with the first ones, so that the proportion is triple[11]; thereafter, when we hear the fourth, we join that with the third so that we conceive [them] as one; thereupon we again conjoin the two first with the latter two so that we conceive these four simultaneously as one. And thus our imagination [*imaginatio*] proceeds all the way to the end, where finally it conceives the entire song as one thing fused out of many equal members. (AT X.94)

Without this synthetic power of imagination we of course would be able to hear what was sounding now, we might recall hearing sounds in the past, and perhaps we might be able to expect a new note in another fraction of a second. But in order to have a sensibility for a song as a unified whole, we have to perceive a progressively expanding unity through the experience of the parts. That is the work of imagination. Although Descartes does not bring up imagination expressly when discussing relationships of pitch, there is no reason to believe that imagination does not perform a parallel though more complex function in unifying successive and simultaneous harmonies and dissonances.

---

[10] For example, in a recent modern edition of the *Compendium* (a French translation alongside the original Latin), excluding tables of terms there are 21 figures in 34 pages of Latin text; see Descartes 1987.

[11] It is this shift from duple to triple rhythm that suggests the synthetic process of imagining is projectively conjectural and self-correcting. Whether the meter is double or triple is fully determined only after one has heard more than the first three beats.

It is hard to overemphasize what an extraordinary idea this is. One measure would be the boast that Immanuel Kant made more than a century and a half later. In his critical philosophy he argued that imagination had more to do than merely follow upon sense perception, it actually helped constitute perception. One of the so-called transcendental functions of imagination was to "unify the manifold of sensibility" in a way that connects moment to moment and articulates the potential chaos of impressions into a well-ordered experience to which we can apply basic conceptual categories. Most historians of philosophy and psychology would more or less agree with Kant that this notion is his innovation. In one of the footnotes to the work in which he announced it, the first edition (1781) of the *Critique of Pure Reason*, he remarked: "Probably no psychologist till now has thought that the imagination is a necessary ingredient of perception itself." One can certainly argue that a couple of sentences in passing in a piece of writing not meant for the general public by a young man who had not yet made his mark on the world hardly counts as much of an exception, especially since the idea is confined to listening to music, indeed just to rhythm. Nevertheless, it is still an extraordinary thought. It indicates that, at the beginning of his philosophical career, Descartes recognized a possibility of imagination that had been no more than implicit in previous developments of its conceptual topology.[12]

Even more important, there is every reason to think from other roughly contemporary writings of his that Descartes recognized analogous functions for imagination in other tasks. This holds even more strongly for his strictly mathematical thinking and his efforts to apply mathematics to physical problems. Sensation gives us the data of the present, memory gives us data of the past, but the ability to see relations in and between the data requires the synthesizing power of imagination, which sets the present situation against the background of the past and tries to generate new appearances necessary for grasping what is at issue, and ultimately for solving problems and answering questions of almost any type. This kind of thinking is not just temporal, synthetic, and projective, it is also fundamentally biplanar, to recall a term that we introduced in Sect. 3.8, above, and have used of both Plato and Aristotle.

Biplanarity would be present even if it were just a question of synthesizing a presently heard note with past notes: one is setting the present note against the past as background. A trained musician will be able to do this in a more nuanced and ample way than a novice will. The novice might be able to notice basic rhythm or key changes when they occur, but the experienced musician will also hear them (imaginatively) as part of the entire sequence of key changes that have already occurred and will, furthermore, anticipate other changes to come. As he listens, the musician can frequently shift the momentary focus of attention, to highlight now the rhythms, now the keys, now the relations between the tonal and rhythmic style of this piece and another by the same composer, and then between the style of this composer and some other. Thus perception is never without memory, and never

[12]But one can also certainly argue that the seed of the idea could have been spurred by combining the biplanar eikastic imagination of Plato and the Aristotelian-scholastic conception of the preparation of complex phantasms by the internal senses.

without the constant ability to recontextualize, to change backgrounds, foregrounds, and middle grounds, and thus never without this contextualizing work of the imagination. For the young Descartes, then, there would be no perception or memory without imagination, and there is no imagination without the establishment of various planes and the mental ability to shift attention between them.

Descartes's acquaintance with theories and uses of imagination should not be entirely surprising. Starting at about the age of 11 he attended the Jesuit Collège Henri IV at La Flèche.[13] As at all Jesuit colleges of the time, philosophical instruction was central, and the basis of philosophical instruction was Aristotle. Moreover, it is said that the Jesuit Fathers rather indulged young René. His health was delicate, so they allowed him to stay in bed till late in the morning. Because of his talents he was allowed access to books that were not part of ordinary instruction. What he actually read and when he read it is conjectural. But the Jesuits were receptive to new developments in the sciences—natural philosophy, as it was still called. Especially at their university in Coimbra (in today's Portugal) they were hard at work producing extensive commentaries on Aristotle's writings that simultaneously summarized current speculations and discoveries, in natural philosophy and all the other parts of philosophy. These would doubtless have been available in the library at La Flèche.

The pages of the Coimbran commentaries on Aristotle are laid out like commentaries on Sacred Scripture: the original Greek text on the left-hand page, a Latin translation on the right-hand page, with commentary forming a U-shape in the wide margins surrounding these texts, on both sides and below. The volumes were in essence compendia of primary texts plus brief discussions of major interpretations and criticisms, from ancient to modern. Anyone using one of them—for example, the commentary devoted to *On the Soul*, and in particular the passages about *phantasia*—would have gleaned not just a clarification and elaboration of the principal text but also introductions to alternative conceptions through the ages and up to the time of the commentary's publication.

Historical studies have shown that the Jesuits in the sixteenth and seventeenth centuries emphasized the importance of human psychology to philosophizing; this affected their manner of interpreting Aristotle's work, especially his logic and other writings included in the *Organon*.[14] One plausible reason for their psychologizing interpretations of Aristotle was the mandatory Jesuit practice of spiritual exercises, a practice that was guided by the work under that name written by their founder, Ignatius Loyola. The spiritual exercises of the Jesuits were founded on the long-established ancient practice and medieval theory of meditation and contemplation. The medieval high point of theoretical development came relatively early: in the twelfth century school of the Abbey of St. Victor in Paris. Hugh of St. Victor (ca. 1096–1141) laid the foundations, which were further elaborated by his student Richard of St. Victor (d. 1173). The theory they developed observed a progressive

[13] Geneviève Rodis-Lewis, who has done more than anyone to disentangle the confusing documentation, argues that Descartes matriculated in 1607 and left in 1615. See Rodis-Lewis 1998, 8–10. La Flèche is on the Loire river, halfway between Angers and Le Mans, about 220 km southwest of Paris.

[14] See Risse 1970, 2: 14–47.

discipline of the internal senses. Although the corresponding practices were often used to meditate on passages in Sacred Scripture and truths of faith, and sometimes to strive toward mystical illumination, the theory was not intrinsically religious. Rather, it was about the methodical use of human faculties for the recognition of truths that reveal themselves only to a properly prepared soul and mind. As we have seen, both a tradition following Plato (the Victorines were Augustinians, and thus their basic philosophical orientation was Platonist) and one following Aristotle (whose theory of the necessity of phantasms for thinking had become part of the common philosophical and medical understanding of inward senses) would acknowledge the importance of using images in order to arrive at deep truths. Accordingly, the Victorine theories had emphasized the need to work persistently with and through the forms of sense in order to arrive at intuition.[15] If in the last analysis the goal was to reach a profound, intellectually apprehensible truth, this could be accomplished only by mulling over again and again what one sensed, remembered, and imagined.

If the *ultimate* goal of this process had something mystical about it—a feature that came especially to the fore in the sixteenth century meditational practices of John of the Cross and Teresa of Avila, for whom the goal was to "silence" the busy activities of all sensitive and cognitive faculties so that one might apprehend God's glory and be overwhelmed by His light and love[16]—the process itself required extraordinarily intense and active imagining. The goal of Ignatius's spiritual exercises in fact went beyond cognitive results. The Jesuits were a "militant" order dedicated to activity in the world, with the aim of converting it. Their practice of the exercises intended to form the human will. One strove for a deep knowledge and love of God in order to bring about in oneself the resolution always to do what is pleasing to Him and thus to be directed solely by His will (in particular by obeying one's Jesuit superiors).

There are many ways in which this practice might have influenced the young René, each somewhat speculative when considered by itself but overwhelmingly likely when taken in sum. Students at La Flèche, even though they were not destined for holy orders, practiced certain abbreviated forms of the Ignatian spiritual exercises. Even in the shortest forms of spiritual exercises (as opposed to the full 6 weeks presented in Ignatius's *Spiritual Exercises*) René would have been taught to make intense, proliferative use of imagination. For example, in meditating on the passion and death of Christ, one is directed to imagine being pricked by a pin, and then to imagine what it would be like not just to have a single pinprick applied but

[15] Sometimes contuition was the term used, the etymology of which suggests gaining a fundamental insight by bringing many things together in a single view. In any case, the –tuition terms both implied an intensive intellectual recognition of the unity of what was presented to the mind in a series of more or less complex phantasms. On the Victorines, see Sepper 2000.

[16] Thus John's "dark night of the soul" was not, as it is sometimes presented, a crisis of existential anguish, but a state preparatory to the vision of God. Metaphorically speaking, it is achieved by shutting off all the "lights" or appearances in the soul, including those of reason. Just as the light of the sun obscures the stars, the busy activity of sensation, memory, imagination, and reason hinder the apprehension of the light of God. All other appearances in the soul are to be "shut down," so that God can become "visible."

a crown of long thorns pushed deep into one's skull. One is supposed to feel and see the bloody red rivulets trickling down one's forehead, into one's eyes, the bloody salt taste on the tongue, etc., and the racking pain. But of course this was only the beginning of Christ's passion, only the beginning of trying to supply sensory content to events narrated in the gospel accounts and thus to give proper amplitude to the significance of the words. Imagining these things was not the end of the exercises, but it was the constant means.

There was another relevant practice the Jesuits employed that might have influenced René's conception of imagining. They were masters of the *emblem*, of producing and interpreting visual images that condense and symbolize information and doctrine. In a sense the emblem has always been an aspect of religious, mythological, and historical art, but it was cultivated with special intensity in the Baroque period, the art of which was flourishing in Descartes's lifetime.[17] One place where the emblem played an especially pervasive role was in the frontispieces of books, which can often be seen as figurative representations of what the book is trying to convey. At the very least it would have provided a stimulus for exploiting the instructional value of figuration, even of a more mathematical kind (for example, the figures he used in the *Compendium*).

In the seventeenth century alone, tens of thousands of Jesuit-guided students around Europe and the world performed some version of the exercises without their developing a philosophy like Descartes's. Why it might have taken so differently and uniquely in his mind and soul is unknowable in any final sense. But his earliest extant notebooks do provide possible evidence. The most immediately germane is a note in which he says that at school he was in the habit of picking up a book, reading its title, and trying to bring to mind what the content must be; the note remarks further that he was successful in the majority of cases. Even if we allow for self-flattery and the fact that book titles (and frontispieces) in the sixteenth and seventeenth centuries were far more elaborate than our own, the claim suggests that at an early age he had become accustomed to exercise formidable anticipative imaginative powers in all kinds of ways.

6.3 The Imagination of Notebook C

The *Compendium of Music* is only one piece of evidence about imagination in the early Descartes. The richest source is notes dating from 1619 to 1621, kept in a now-lost notebook designated **C**.[18] The still extant notes were published in 1859

---

[17] For an overview, see Adams and Harper 1992.

[18] Notebook **C** was included in the effects sent back to France after his death in February 1650 in Sweden. That the notebook survived at all was close to a miracle, since the ship carrying the effects sank to the bottom of the Seine, and the chest containing the effects had to be salvaged. Leibniz transcribed notes from it during a visit to Paris in 1676. It is not clear what proportion of the notebook's contents was ultimately preserved, although scholars have been able to reconstruct how it must have been organized. See especially Gouhier 1958, 11–18.

under the title *Cogitationes privatae*. Many are scientific and technical, examples of what Beeckman called physico-mathematics.

For example, in order to solve problems about how far and fast a free-falling body travels, Descartes tries (incorrectly, as it turns out) to use a right-angled triangle to correlate speed, time, and distance traveled; small changes in each he correlates with adding incremental, proportionally related segments in order to progressively increase the size of the triangle.[19] In order to solve other problems, both physical and mathematical—and sometimes just to see what happens—he describes forming mental figures (or sketches them in the notebook) and then proceeds to add to or subtract from them, to vary them, to manipulate them.

For example, Descartes visualizes two-dimensional geometric figures as increasing or decreasing in size, as sliding through the plane, as rotating around a point, or as rotating around a line and thus producing a three-dimensional solid. He pictures and tracks processes of division or analysis that never end but still approach a limit. He looks for ways to construct from existing figures a unit of length in terms of which all other lines and figures he is using can be expressed as whole-number multiples. He devises sketches of simple machines consisting of sliding line segment sides and pivoting points, all the parts of which move in a well-regulated, interactive way when a force is applied to a single part. Many of these he conceives as imaginative versions of possible real-world instruments for calculating problems and constructing figures, the capabilities of which would far exceed those of comparable instruments available to ancient mathematicians.[20] He imagines pencils being attached to moving parts of these devices and considers the paths they would sketch out as the device is operated. He even begins to conceive of systematically employing marks, symbols, and other representative forms to stand for the information embodied in the figures he drew and imagined, and to use these representations to express proportions, equalities, and inequalities, and thus serve for finding new ways of manipulating figures to construct solutions to mathematical and physical problems. This work was the beginning of what he gradually transformed into analytic geometry. The term that he used to describe all this kind of activity was not intellection or reasoning but *imagining*.[21]

Other notes of **C** go well beyond mathematical and physico-mathematical concerns. A number of them reflect a conviction that those searching for truth are brought into touch with higher things by their resemblance to lower ones; the instrument of this ascent is, once again, imagination. One of them appeals to a Renaissance,

---

[19] The increasing length of the vertical side of the triangle stood for speed, the increasing length of the horizontal side for time, and the increasing area for distance traversed. Beeckman recorded Descartes's faulty solution and then offered a correct one of his own. For a detailed account, see Shea 1991, 35–60.

[20] This is not to say that Descartes was the modern pioneer of such devices. Many others, including Galileo, had preceded him in thinking about and even building some. But this is to say that the natural approach or "tack" of Descartes's mathematical, problem-solving imagination took this kind of course.

[21] See also the discussion in Sepper 1989, esp. 383–384.

quasi-Platonic sensibility for the spiritual symbolism of nature: "sensible things are fitted to conceiving Olympian things: wind signifies spirit; motion over time signifies life; light signifies knowledge; heat signifies love; instantaneous activity signifies creation. Every corporeal form acts through harmony" (AT X.218). Beyond reinforcing the notion that there is an expressly poetic and aesthetic character to the young Descartes's conception of imagination, it suggests that he entertained the possibility of there being a natural concordance and active metaphysical harmony between things accessible to the senses and things of the spirit.

One of the notes goes so far as to suggest that reason is unable to keep up with sensory imagination and intellectual imagination as they work on different levels of existence:

> As imagination uses figures to conceive bodies, so intellect uses certain sensible bodies to figure spiritual things, like wind, [and] light: whence, philosophizing in a higher way, we can by cognition raise the mind into the sublime.
>
> It can seem amazing that weighty meanings [are to be found] in the writings of poets more than in those of philosophers. The reason is that poets write through enthusiasm and the force of imagination: there are particles [*or* seeds] of science in us, as in flintstone, that through reason are drawn out by philosophers, [but] that through imagination are struck forth by poets and shine out more. (AT X.217)

This note is especially pregnant with consequence, not just because it tries to establish by analogies the very principle of analogy but even more because it presents imagination under a twofold significance that was central to Descartes's early conception. The very first clause of the first sentence treats imagination as the power of making corporeal things conceivable through figures and images; this is a very general way of describing the imaginative power that his mathematical and physical problem-solving notes were investigating. Next he analogizes to it the capacity of intellect—or, given its function, it might more accurately be called *intellectual imagination*—for using corporeal things as figures of spiritual or intelligible things. There is a lower imagination, and there is a higher imagination. The lower one figures or images physical bodies, it is physical imaging; a higher one uses bodies symbolically to express spiritual significance or presence. There is more than a little influence here of the kind of topology seen in Plato. The last sentences of the note are particularly surprising insofar as they argue that imagination can bring us more directly in touch with spiritual and intelligible things than can reason (*ratio*). Reason is plodding; imagination strikes sparks and shakes free the particles of science toward which reason plods. Whether, without the anticipative capabilities of imagination, reason would even know in which direction to plod is doubtful.

In this note there are two notions of imagination: image- and figure-making imagination in the strict sense, and then the projective, poetic-cognitive imagination that uses things and their representations for higher, spiritual purposes. If the expressly poetic concern for the most part drops out of Descartes's later philosophizing, the cognitively driven projections of imagination remain a constant. Even when we simply see a thing, we immediately desire to make more out of it; we mentally schematize or simplify the thing in a figure and use the figure to conceive the thing and its motions, actions, and possibilities. This might well be a further, visual development of the kind of synthesizing and conjecturally projective *imaginatio* that Descartes thought was at work in music listening. Perception is not passive, or at least it is not finished just by presenting us with something. If it does simply present us with something, still, in the very next moment, we can take hold of it in a new way. The first look can cause us to intensify our attention in a second look, or we can take the appearance, the image, the phantasm of what we saw a moment ago and proceed to portray it in a new medium—for example, we can mentally conceive it in a geometrical configuration. That geometrical configuration then becomes in its turn a new object of attention, though it still implicitly refers to the phantasm of the original thing. (We shall return to the implications at the end of this section and the beginning of the next.)

But there is more to the imagination of the thing than just conceiving it mathematically. Things are related to others, and they are signs and symbols of other things on other levels of being. Thus, with intellectual rather than corporeal imagination, one can symbolize spiritual and intellectual matters with corporeal images, and thus one can think about them by thinking in terms of their images—that is, by intellectual imagining. Here the Platonic heritage is strongly in evidence: for example, when Socrates in book VI of the *Republic* adumbrates the governance of a city by talking about who is the best person to pilot a ship, or in book X presents cosmic justice by describing a soul's journey through Elysium and Hades.

Imagination can therefore be conceived in both narrow and broad ways. Image making in the narrow sense is the ability to form, divide, and recompose images of corporeal things. Something like it is at work already as we exercise sense perception (we see something and as we look upon it we conceive it figuratively) and explicit in productive and reproductive imagination, in memory, and in mathematical and other kinds of figurative representation. Imagination in the broad sense is the generalized power of using one kind of appearance in order to reconceive and understand something else "figuratively," as we say. It is based on the human capability, emphasized by Plato, of seeing through things and images to other things overarching them.

Another note shows the degree to which Descartes had begun to conceive of the work of imagination not just as ad hoc but as capable of grounding a method of investigation. The note begins by saying he was reading a book on the art of local memory. We encountered local memory very briefly in Sect. 5.6 (n. 47). It is a technique for remembering complicated matters by producing striking images and symbols and mentally positioning them in a familiar, remembered place. Descartes then explains to himself how his own methods improve on the arbitrariness of the memory art.

> Reading through [the book]...I thought that everything I discovered could easily be grasped by imagination: It occurs by leading things back to causes; when all those are finally led back to a single one, it is evident that there is no need of memory for all sciences. For whoever understands causes, will easily form anew in the brain by the impression of the cause the altogether vanished phantasms. This is the true art of memory and it is plain contrary to the art of this sorry fellow: not that his lacks in effect, but that it requires the whole space that should be occupied by better things and consists in an incorrect order: which [right] order in this is that the images be formed from one another as interdependent. He omits this—I don't know whether advisedly—which is the key to the whole mystery. (AT X.230)

That is, if one understands causes, one can generate image from image from image using causal understanding, rather than spend one's ingenuity coming up with laboriously constructed, arbitrary images according to the so-called art of memory (or the related Jesuit-Baroque techniques of constructing symbolic emblems).

Descartes concludes his reflection on his new art of imagination (as opposed to the art of memory) by hypothesizing yet another technique of cognitive imagining. One could take several related images and generate a new one, either common to all or otherwise generated from previous ones taken all together. In this way, he says, each would have a determinate relation to every other: "not only would there be a relation to the closest, but also to the others: so that should the fifth relate to the first by way of a spear thrown on the ground, the middle one [would be related] by stairs from which they descend, the second one by a dart projected toward it, and the third by some similar analogy [*ratio*]" (AT X.230). Although the exact course of his thinking here is obscure, it looks as though he believes that concrete symbols can be used to symbolize and even to generate an elaborately detailed, proportional correspondence between real things. Quite clearly Descartes had hopes that the imagination could be deployed in a far more cognitively active and productive way than the imagination of memory art did. The kinds of figures he had used earlier, in the *Compendium of Music*, to embody information were perhaps too modest: he wanted to be able not merely to summarize information in figures but to generate new relations, new determinate proportions, and correspondingly new knowledge.

Plato had of course understood that we can image a thing in different ways and at different ontological levels, and in at least one passage about the practical/ethical use of imagination Aristotle had explicitly argued that human beings (and perhaps some other animals) have the ability to form a new image out of many existing ones. Both of them had understood there to be an interdependence between the different sensitive and intellective powers of the soul, and Aristotle had presented imagination as a kind of proportionalized movement that allowed for its involvement in other psychological activities. Yet they never attempted to show imagination in its detailed psychological functioning, and there is no strong reason to think that, if they had done so, they would have ascribed to it the continuous dynamic processing for which Descartes argues.

Far more than the great Greek thinkers did, Descartes conceives the human mind to be constantly, energetically active, productive, and inventive. It takes in information through sensation and then immediately reforms it in re–presentations. It makes new images from existing ones by applying rules and relations, and thereby it gains new insights. It analyzes aspects and parts of appearances and synthesizes new appearances. It can take different appearances and extract from them what is common, and this leads to its ability to generate series of new appearances from familiar ones. It is hardly surprising, then, that its functions are capable of being applied methodically and systematically to everything that appears, because imagination is founded on and supported by the nature of images and appearances and the proportions that govern and hold among them. The question then becomes what the limits of it are. If imagination poses and solves problems and can even address the spiritual and intellectual realms, is there anything it cannot do?

6.4 Imaginative Representation and Manipulation

In the first and second parts of the *Discourse* Descartes tells readers that he was not happy with the way people applied their minds to understanding things. Although he says that he was educated at one of the best schools in Europe, it is clear that he was not satisfied with what he learned and how he learned it. And so he began searching for an alternative way, which culminated in his method and the discourse he wrote about it.

This "and so" conclusion is drawn too hastily, however, and it short-circuits understanding what is really at issue. The method was a response to a *question*, to a *problem*—the term that the Latin-speaking Descartes used was a traditional one in medieval thought, *quaestio*. What was the problem to which the method responded? It was that the intellectual culture Descartes found himself in was predicated on the existence of knowledge, yet precious few people knew how to show that they actually possessed it, much less knew what possessing knowledge means. To put the problem more pointedly: people claimed to have science, but what they really had was the art of disputation.

As Descartes says in both the early *Regulae ad directionem ingenii* and the mid-career *Discourse*, someone who knows should be able to show others what he knows and what makes it knowledge. The only people who regularly, but not always, do this are mathematicians. Why? Not because mathematicians are better or smarter than anyone else, not because only mathematical things are knowable, but because mathematics deals with things that are easy to understand or can be *reduced*, which here means *brought back*, to easily understandable things. When a researcher arrives at what is easy to see, there is really nothing more she can do than "see the truth" of the thing. Anyone who can do this also has the key to leading others to see the truth. If someone lacks the power to see simple truths, however, then there is simply nothing to be done with him.

Throughout his career, in various ways, Descartes asserted that every person with the least bit of reason has the ability to see simple truths. But from very early in his career he claimed that not everyone is willing to do the work necessary for easy seeing. Heraclitus in Greek antiquity had declared that people prefer their private reason to publicly accessible *logos*; Descartes's claim is a little bit more generous. Yet ultimately he came to believe that it was far easier to teach peasants to see easy truths than the already well educated.

The legacy of antipsychologism in the past two centuries has made Descartes's claims even harder to credit insofar as it has thought that private intuition and introspection are suspect. A truth may be "clear and distinct" to you or to Descartes, the antipsychologist can say, but what if it's not clear to me? There follows the counterclaim that truth ought to be public and conform to an objective standard. But "objective standard" really means "objectively verifiable standard," and that raises the question of who does the verifying and how. Objectively verifiable standards are usually standards arrived at by training human beings to accept a common standard—but someone (presumably everyone who is trained) still has to *see* that the standard is met. The standard is intersubjective. There is no escape from the fact that someone—what we call a subject—is going to have to see that something is true, that something is this way and not that. And that repeats, with respect to antipsychologism in the twenty-first century, what Descartes had to say in the seventeenth with respect to most of his philosophic predecessors and contemporaries. The secret to mathematics was not that it was mathematics but that it had simple objects and simple standards to apply. If one could specify how one gets from ordinary experience to complex representations of it, find ways to analyze and simplify those complex representations, then judge them in light of the analysis, one could achieve something similar for nonmathematical experience as well. For human beings, the representations taken from experience in the first instance are images or phantasms of the experience. Once the thing is no longer in front of us we have no recourse other than dealing with the phantasms of it. So learning how to explicate the relationships of phantasms to one another and how to generate new phantasms from existing ones is likely to be useful in understanding all kinds of things in general. And something of this lesson from mathematics can be applied to virtually every thing of every kind, in particular when we can evaluate things in light of the more and the less, the larger and the smaller, the less intense and the more intense, etc. This is Descartes's rediscovery of Plato's conception of fields of experience, and of Aristotle's notion that the fields are articulated by the more and the less between extremes.

At any rate, at some point between late 1619 and the mid-1620s Descartes began formalizing his early insights into imagination into a full-blown method. The first evidence of this is a work he never published or even finished during his lifetime, the *Regulae ad directionem ingenii*.[22] The intensive methodological reflection that Descartes began in this work eventually led to the 1637 *Discourse* and the three scientific essays to which it served as a preface.

As with the *Meditations*, which appears to speak slightingly of imagination, there is a "tradition" of selective quotation from the *Regulae* that allows imagination to be dismissed from further consideration. A favorite line to quote in this spirit occurs in rule 3, where Descartes says that the most basic way of knowing, intuition (*intuitus*, which is being defined in the passage), is "not the fluctuating faith of the senses or the fallacious judgment of a badly composing imagination" (AT X.368). This is the first mention of imagination in the work, and the long phrase certainly makes it seem that imagination (as well as sensation) is not to be trusted. Quite apart from issues of the immediate context, however, a major task of the work as a whole

[22] The work was translated and first published in Dutch, in 1684; the publication date of the original Latin was 1701. The title was assigned the work by its late seventeenth-century editors. Descartes began work on it possibly as early as 1619, though most scholars place the start date in the middle or late 1620s. He apparently abandoned it around 1629. The conventional translation of the title into English is *Rules for the Direction of the Mind*, which unfortunately does not at all convey the specificity of *ingenium*, which is in fact one of the work's key terms. Scholars have suggested alternatives like "native wit" and "native intelligence"; perhaps even better is the etymological cognate "ingenuity."

is to show how imagination can compose things accurately and well. A more subtle point is that the phrase "fallacious judgment of a badly composing imagination" leaves open where the fault for the fallacy lies. It seems clear that the imagination does the bad composing, but the fallacious judgment could be either imagination's fault (the subjective genitive) or the fault of some other power that judges the badly composing imagination fallaciously (the objective genitive). If it is the fault of another power, then there are two points to be addressed: how and why imagination composes badly and how and why the judging power is mistaken. Descartes says later, in the middle of showing the proper uses of imagination, that "badly judging intellect" is responsible for error (see rule 14, AT X.443)—which, by the standards of those who selectively quote the earlier passage to dismiss the importance of imagination for Descartes, ought to lead them to dismiss intellect's importance as well.

Another line of attack is drawn from the fact that in the mature account of method, the *Discourse*, imagination is conspicuous by its absence. As Leslie J. Beck pointed out in his book-length analysis of that work, however, even if the role of imagination is not given detailed attention, Descartes nevertheless does expressly mention it. It is just as central to the *Discourse*, argues Beck, as it was to the *Regulae*, and in the same ways. Immediately after stating the four rules of method near the end of part 2 of the *Discourse*, Descartes says that he had discovered that the best technique for applying them involved representing the parts and relations of problems by lines,

> because I did not find anything simpler, nor anything that I could represent more distinctly to my imagination and senses; but, in order to retain them, or to understand several together, it was necessary for me to explicate them by certain symbols, as short as possible; and, by this means, I would borrow all the best from geometrical analysis and from algebra, and would correct all the defects of each by the other. (AT VI.20)

This is not a bad summary of the lesson of the *Regulae*. Thus, once one recognizes that for Descartes mathematical representation and problem solving is an imaginative activity, it is impossible to miss these clues. That they are still widely overlooked is undoubtedly an index of the degree to which our culture assumes without evidence that mathematics is basically a rational activity. Of course it is rational, because it produces *rationes*, the setting of things in determinate relations to one another. But, for the most part, such setting things in determinate relations has to be done in fields of images.

The *Regulae* presents an art of problem solving that is supposed to be especially well adapted to the human being's psychological capacities. The senses, memory, and especially imagination are to be deployed in aid of the intellect. Most errors people make, Descartes asserts in rule 14, are due to the intellect, especially when it makes judgments without reference to an imaginable object (i.e., there may be thinking without images, but it usually goes astray because it is not thinking about anything in particular). Descartes not only follows the late Renaissance tendency to reduce the number of internal senses to a minimum (common sensation, memory, and imagination), he in essence reduces all of them to functions of imagination. In particular, the *ingenium*, the "ingenuity" or "mind" or "native wit" that is to be directed by the rules,[23] is defined in rule 12 as the knowing force (*vis cognoscens*) "when it at one moment forms new ideas in *phantasia* [the organ and place of imagination], at another applies itself to those already made" (AT X.416). But ideas in *phantasia* are images; and thus *ingenium* is the power of conceiving, recalling, varying, and developing images—as one does in his mathematics, but not just in his mathematics.

The *Regulae* was to consist of three parts, with twelve rules in each. The first twelve deal with the method of solving problems in general, chiefly by discussing the human powers that are best fitted for understanding and solving problems (this part is essentially complete, though a few rules seem to have gaps). The second twelve were to deal with "perfect problems," those that are sufficiently well defined to provide everything needed for a solution (thirteen through eighteen exist in fairly complete form, nineteen through twenty-one have headings only, twenty-two through twenty-four are nonexistent). These rules were to show how to use figurative and symbolic representation of the givens of a problem and then to break them down and combine them in the course of problem solving. The figures were chiefly geometrical line segments and plane figures produced from line segments; the symbols were marks or names of points, segments, and figures. Part two as it was left hardly gets further than showing how to add, subtract, and multiply line lengths, and the proper use of algebra is hardly more than mentioned before the work breaks off. Part three, which was to consist of twelve rules regarding "imperfect problems," does not exist at all.

What was Descartes trying to do? In rule 14 he says outright that all the problems being solved by the representational and manipulative techniques he introduces will be using imagination, because quite literally one will be making and transforming figures and images and generating symbolic representations that will track and anticipate the transformations. It appears that Descartes is trying to give a systematic account of the problem-solving imagining he had been using ad hoc in his earlier notes and the *Compendium musicae*. The clearest sign of this is that he advises using only points, lines, and plane figures in the representation of problem elements and calculations; that is, one should avoid three-dimensional figures. This advice is a response to the difficulty mathematicians had had since antiquity in conceiving the multiplication of more than three numbers. Advances in mathematics since the sixteenth century, to which Descartes himself contributed in no small part, introduced the techniques that we use today: you multiply two of the numbers, you multiply the resulting number by the third, that result by the fourth, etc., as many times as you need to. This is easy because we take numbers as absolute: they are defined in terms of nothing but themselves, and using any operation to combine two of them gives just another number.

Before modern mathematics, however, numbers were regarded as attached to what they were the number of. In pure mathematics, number was considered to be

[23] Granting that the title of the work was provided by later editors, they did not choose badly; it is one of the most frequently occurring terms in the work. In terms of debates about method in the early modern period, Descartes was casting his lot with those who (especially in the rhetorical tradition) thought that presentation had to follow the dictates of invention rather than vice versa—and *ingenium* was the power that guided invention.

the measure of geometrical figures. The measure of a line segment was not the same kind of measure as the measure of a surface area. Whenever you multiplied two simple numbers you were understood to really be creating a plane figure. Multiplying *m* times *n* was taking a line *m* units long and another *n* units long and producing the rectangle that had the line of length *m* as one side and the line of length *n* as the adjacent side. The product of *m* and *n* was therefore an *area* measured in square units. In order to multiply three numbers you had to create a three-dimensional figure—a rectangular parallelpiped, to give it a name. And multiplying by a fourth number was technically impossible, because it would have required entering a fourth spatial dimension. Ancient mathematicians had developed "workarounds" for this, but they had no systematic justification.

It is no accident that Descartes's earliest mathematical thought concerned problems that required shifting back and forth, from one to two to three dimensions and beyond, for example by making a triangular area representing distance out of two line segments representing speed and time, and using the figures to solve problems of proportional relations between the parts. The theory of proportionality that was so much on Descartes's mind was the heart of the ancient methods of calculation. If you could not directly compare a square area to a linear length, you could set up a relation between two square areas that was proportional or equal to the relation between two line lengths (square area #1 is to square area #2 as length of line A is to length of line B); and *indirectly* (*alternando*, in the Latin terminology used for this kind of proportion) you could say that the proportional relation of the first square area to line length A was the same as the proportional relation of the second square area to line length B. To us, who are used to algebraic imagining, this is an overcomplicated way of saying that if you have the equation "area #1/area #2 = length A/length B," then you can multiply both sides of the equation by "area #2/length A" to get the new equation "area #1/length A = area #2/length B." Notice that what allows us to do this is the postulate that we can treat both area and length indifferently as absolute numbers. This violates the traditional sense that a number cannot be so cavalierly separated from what it is a number of. For the ancient and medieval mathematicians, only measures of the same kind could be compared directly.

The fact that Descartes proposed a not entirely uncomplicated alternative shows that, although he was not committed to the strict limits of the older methods, he was not quite ready to treat numbers as absolute. His alternative techniques allowed one to convert any line length to an area or any area to a line length, although they took line length as the authoritative or canonical form of measure because it was simplest. Thus if you needed to multiply four numbers you would represent each by a line length, combine two into a rectangle to get the multiplied area, apply the conversion technique to reduce this to an equivalent line length, multiply this new length by the third original length in a new rectangle, convert this rectangle to an equivalent line length, multiply this new line by the fourth original length in a third rectangle, then convert this rectangle to a line that (finally) represents the product of multiplying all four numbers! Addition and subtraction were comparatively easy, but division of two numbers and the taking of roots was complicated.

Even with our modern computational techniques there are remnants of this ancient problem that we still have to deal with. If you have twelve apples to distribute among six people you divide the twelve apples by the six people, get two, and proceed to hand two apples to each person. But what does it mean to divide apples by people? Giving it some thought, we see that we separate out the numbers from what they are numbers of and perform the division, then we recall that the answer is neither apples nor persons but "apples per person." If this seems like a mere technicality, it is no mere technicality for natural scientists when they multiply acceleration by elapsed time to find out how much faster something moves after that number of seconds has passed. Acceleration is (say) meters per second per second, or m/s^2, and the elapsed time is seconds, so the product is m/s, which is velocity (or speed, if we ignore direction—another complication that we don't always attend to!). But the genius of modern as opposed to ancient arithmetic and algebra is that we can separate the reckoning of the numbers from the reckoning of the units that those numbers express, only to combine them again once calculations are at an end. Descartes was precisely on the cusp of the change from the old conception to the new.

All of this is dealing with *ratios*, and thus if any kind of thought deserves to be called *rational* it is this. Yet it is simultaneously all about routines for comparative imagining. The reader probably has a sense of relief that in order to multiply and divide we don't have to worry about these complications any more—and because of electronic calculators we often no longer need to recall the algorithms for the operations on absolute numbers but can just punch the buttons. That is, we are very happy to be relieved of the need to think or imagine these things according to the older ways. What exactly the mental operations are that correspond to mathematical calculation more completely eludes us the more we use machines to do the reckoning.

The first thing to say about these mental operations is that, even today, there has to be some theory of what representation is *of* and what representation *implies*. That is, we need to deal with the ontology and epistemology of representation rather than focus on practical techniques of algorithmic imagining. The "we need" has to be taken a bit loosely, of course, since not that many people feel such a need. Descartes, historically alive to the conceptual topologies of his heritage, had a keen sense of all these concerns. Insofar as such questions are to us a dead letter, we are satisfied with imagining (rather weakly) that someone, somewhere actually understands and has justified what most of us do, so most of us can work in secure ignorance. But that means we have historically recapitulated the problem to which the method was a response. Analogous to medieval and early modern scholars, we use "rational" techniques that conceal our ignorance of or indifference to the specific character of truth.

6.5 The Dynamism of Imaginative Ingenuity

The first rule of the *Regulae* begins with the reflection that the light of reason shines on everything knowable and that there is a single method of knowing, in accordance with this light, whatever comes before the mind. The method has to acknowledge the nature and limitations of the human powers that make knowing possible and recognize what sorts of things are most knowable. Mathematics can provide a model for knowing because the things it studies are "so pure and simple that they make no assumptions that experience might render uncertain" (AT X.365). This does not mean that only mathematical things are knowable. Rather, what is simple is more knowable than what is complex, and when we are faced with complex objects we must face up to this difference and develop appropriate ways of relating what is complex to what is simple.

Descartes in rule 3 argues that there are only two acts of mind that are useful for knowing, intuition and deduction.[24] Intuition is the simple, clear, indubitable recognition of truth by "a clear and attentive mind." The examples he gives are of a person's intuiting that he exists, that he is thinking, that a triangle is bounded by just three lines and a sphere by a single surface. It is something that you can see all at once when you have properly prepared the view (about which preparation there will be much more to say in the later rules). When something simple is brought before the mind—whether it is an image or something more than an image—any person sensitive to the light of reason will see, will *intuit*, simple truths about it. Deduction is in essence a series of intuitions: we see first one truth, then a second, then a third, etc., and recognize that the last comes from the first, step by step. He compares this to knowing that the first link in a chain is joined to the last by inspecting, rapidly and one after another, each of the intermediate links. In rule 7 (AT X.388) he calls this movement of thought, from one thing to another in a continuous sweep, an act of the imagination. In a sense, the imaginative process that the *Compendium of Music* had analyzed as necessary for synthesizing the rhythm of song has been generalized to "seeing the truth" of any complex situation that we are able to articulate into a series of clear elements. In rule 16 he goes further: he suggests that at least in some cases what we originally come to know by deduction, in a fast but still step-by-step sweep of one intuition after another, could itself come to be apprehended by a *single* intuition.

How would one arrive at the point of being able to do all this mental activity as well as possible? Let us take advantage of the traditional name of the work, *Regulae*, rules, to point out that the work teaches how the imagination can be ruled and regulated (provided with a measuring stick) by the knowing power, by reason, by rationality; and that rationality works by setting one thing into a determinate relationship, into a *ratio*, with another. I believe that, rather than playing with words, this is an evocation not just of what is meant by Descartes but what is implied by the entire history of the conceptual topology of imagination and reason in Western thought.

Traditional logical demonstration, as well as mathematical proof modeled by Euclidean geometry, proceeds step by step. Descartes's complaint was not with the stepwise advance, because it was characteristic of his own method in the *Regulae* as

[24]*Intuitus* and *deductio* in Latin. Etymologically *intuitus* is a power of looking clearly into a thing—not far removed from "insight." The usual English rendering of it as "intuition" is less than perfect, especially if it is thought of as a rather mysterious power of anticipating things before there is evidence for them. I shall nevertheless use it here. *Deductio* etymologically suggests drawing or leading something down from something else, for which the English "deduction" is an adequate translation when it is a matter of drawing truth from other truths.

well. What he demanded was what they could or did not provide: the opportunity for insight, for seeing or intuiting the truth. Aristotelian-style logical demonstration was fine for reminding yourself of truths you already knew (all men are mortal, Athenians are men, therefore Athenians are mortal) but not for discovering anything new. His complaint about earlier and much contemporary mathematics was that it was filled with obscurities, both in concept and expression; even when what mathematicians presented was true, the clarity of presentation left a great deal to be desired. He countered that simple truths can be recognized by everyone possessing the least bit of rationality. He thought that people—especially educated people—neglected easy and commonplace truths in the pursuit of grand truths and mysteries, and thereby they impaired their ability to recognize more elemental and productive kinds of truth. The *Regulae* therefore counsels its readers to practice looking for truths in all circumstances of life, and especially advises looking for principles of organization and order no matter how humble or trivial they might seem. For example, they should examine the orderliness in the construction of a watch or a balance scale, the various weaves of fabric, and tactics useful for solving games and puzzles (see rule 9, AT X.401–403; rule 10, AT X.404; rule 7, AT X.391). The point is this: if you find a principle of order in which B or C follows A, the mind knows where to go when it comes upon A. Seeing that something is A and that it is followed immediately by B is one of the simplest acts that intuition performs, and the recognition of the organizational power of orderly series of things is the result of many intuitions put into series, which rule 3 calls deduction. The mind's motion is thus ruled and regulated. In Aristotelian terms, the intellect sees the orderly forms in the phantasms.

It is no accident that rule 4 emphasizes order and measure as the principles of method, and it is no accident that, when rule 6 spells out "the whole secret of the art" of solving problems in an orderly way, it focuses on putting things into orderly series. The leading principle explained there is the degree or proportion in which a thing contains or participates in a *nature*. The rule, however, dismisses thinking of the nature as an essence that constitutes a thing, because such an approach is largely useless for solving problems. Some examples (not in the *Regulae*) will help illustrate what this means. A farmer planning his spring planting will not be assisted by reasoning from the essence of farming, the essence of plants, or the essence of spring. A general will never get around to organizing the battalions of his army if he has to reflect on the essence of an army or the essence of man. The essences of such things may not be *totally* irrelevant—for example, tactics require that soldiers accept their orders, and to accept and understand orders they must be rational animals—but they are ordinarily only of tangential relevance to what is being sought in the *quaestio*. It is not so much that essences do not exist as that they don't usually help one find an answer to a given question. Essences can, for the most part, be taken for granted. A person solving a problem must, in the first instance, know enough to pare down the considerations to what is essential *for solving the problem at hand.*

Descartes first mentions specific examples of natures when he defines things that are called "absolute": "whatever contains in itself the pure and simple nature that is in question: as [for example] everything that is considered as independent, a cause, simple, universal, one, equal, similar, right, or others of this kind" (rule 6, AT X.381). He contrasts to the absolute the "respective" (*respectivum*, usually translated as "relative"), "whatever participates in the same nature or at least in something from it, in accordance with which it can be referred to the absolute and deduced from it through some series," as examples of which he gives what is "dependent, effect, composite, particular, many, unequal, dissimilar, oblique, etc." (AT X.382). This second listing consists of opposites or contraries of the first list. What does this all mean?

We must recall that, in his earliest notes about imagination, Descartes conceived of taking something from experience and re–portraying or re–presenting it. The thing has already been received into mind: now we step back and take it in. Even if we only try to preserve the thing as received, something new will be added to the experience. There will be something dramatically newer if we "change the take."[25] This not only creates the different levels or planes for imagining, it also means we are taking the thing in a certain respect, relative to a certain portrayable or imageable "nature" that we want to highlight. Rule 6 emphasizes not the representation or portrayal per se but the thing's relation to the nature of which the experience is a specimen, and this allows us to put many things and their representations into a discrete series (the beginning of a matrix) in light of their "containment of" or "participation in" the nature. Once one has a nature in mind, one can put things or objects in a series with respect to the degree of participation in the nature, up to the extreme of wholly containing it. In any given series the "absolute" is the first member of the series, the thing that most participates in the nature, at least in one's own experience; those that are more remote are called "respective."

In rule 6 Descartes does not give many examples of what he means,[26] but it is not hard to construct a few plausible ones. A person considering a career change may

---

[25]This is my phrase. It could be as simple as the difference between looking and then taking notice. You enter a room and look right where you see someone moving; you notice how large the room seems; you then remark that you are seeing your own reflection because the wall to the right is mirrored, and thus you adjust your assessment of the room's size and furnishings. Those things all happened without a marked change in attitude. But now you can easily "change the take," that is, take things in a different modality: for example, by considering the quality of the (image reflected by the) mirror. You have changed the level and context of the experience; it now plays out it in the phenomenon of mirroring, optics, fashion, etc.

[26]He does discuss some problems of extended proportion. If we want a third number that is to the second the same as the second is to the first, and the first two are 3 and 6, it is simple to determine that the third is 12. The equation is $3/6=6/x$; multiplying both sides by $2x$ gives us $x=12$. It is a little harder, when given the first and the third, to find the second; that is, given 3 and 12, finding the geometric mean x such that $3/x=x/12$ requires us to solve $x^2=36$. Even harder is being given the first and the fourth in a four-number series and then deducing what the second and the third must be (e.g., if the first and fourth numbers are 3 and 24, the values of x and y must solve the equation $3/x=y/24$—there is no unique solution). But if we are given the first and the fifth as 3 and 48, determining the third (which is 12) turns out to be no harder than determining the 6 that comes between 3 and 12 (the second example of this note): we break the problem into parts, first finding the number x such that $3/x=x/48$ ($x=12$), and then we find the numbers y and z such that $3/y=y/x$ and $x/z=z/48$ ($y=6$ and $z=24$). Although the examples involve only simple arithmetic, the point is that the element or "nature" 3 is contained in each number of the series to a different but definite degree; because the series follows a rule and is well ordered, the exact degree of "participation" in three is easily determinable.

be seeking autonomy; she will accordingly consider the options according to how much each participates in autonomy. If she is practical she will notice that the more autonomy a job has, the more initiative it entails, and perhaps also the more variability with respect to income. These are comparative relations that can be put into roughly parallel sequences, and to some degree they might be quantified and thus measured (this would, by the way, be an imperfect problem, in which the information is not complete enough to generate perfect solutions—the kind of problem part 3 of the *Regulae* was going to address).

If we are considering the purchase of a house and have visited ten, we may well actually put them order, from one to ten, according to their participation in various natures. First, there is the series of the asking prices; the nature being participated in might be called simply price, with a convenient unit of measure, the dollar ($328,000), or perhaps the unit will be 1,000 dollars ($328 K). But no house buyer will stop with that. For example, he will compare the houses according to their participation in the nature *area*—there is a convenient unit measure (square feet or meters) that allows us to put them in a very strict series. The real-estate agent will probably have informed him of the *average price per square foot* in each neighborhood visited (these can be put in a series); thus the buyer will have been provided with mathematical proportions that allow the production of a new series: what the houses should cost if they sold at the average square-foot price for their neighborhoods. This series (call it the expected price series) will probably be at least slightly different from the asking-price series. The buyer can create yet another series by subtracting the expected price from the asking price: this will give a comparison of how much above or below the expected price sellers are pegging their asking prices. This series might then be informally coordinated by the buyer or the real-estate agent to certain psychological traits of the sellers: pride in how well they have maintained the house, financial realism or irrealism, greed, etc.

Of course if we are good house buyers we are not yet finished. Whether a house is a real bargain depends on other factors beyond the discrepancy between asking price and average expected price—for example, we will create a series corresponding to refurbishing costs. Probably we will need to break down refurbishing into different factors, each of which will give rise to a new series—roof repair, HVAC replacement, painting—that will lead to ever more intricate comparisons. If we are worried about heating and air-conditioning costs we can compare the houses in volume; this series will not diverge much from the area series, except insofar as the houses have different ceiling heights. But volume is just one of the factors that enter into heating costs, so we might have to sketch out some other quick comparisons of window age and quality, insulation, blower capacity of the furnace units, etc.—and we might finally devise a formula that takes all these factors into account (contrariwise, we could just do an "eyeball" estimate). We will compare them according to the number of bedrooms or baths. We will compare according to less clear-cut features or natures, like *expandability* or *brightness* or *airiness* or *comfort*. We can easily imagine coming up with some kind of measure for expandability and brightness, but airiness is a little harder—it depends on feel, though it is probably also related to ceiling height, overall room proportions, and admitted light (which is related to brightness). Comfort is probably the least tangible of all, though one could probably specify certain factors that enter into it, like ease of movement through rooms and the appropriateness of the layout of the house to one's preferred ways of occupying a living space. In each of the series we create, the absolute is the member of the series that participates most in the nature, and all the other members of the same series will be called respective or relative.

We should not think that by using the example of house buying we are trivializing the procedure, as though only solemn acts of scientific, mathematical, and philosophical reason are eligible for consideration. Descartes's complaint is that people simply do not consider how many different types of order, nature, and measure are exhibited to them every day, in every way, and how the various natures and measures are interrelated. If they do not see it in ordinary things, they will not know how to find it in the complex. Descartes points out (in rule 12) that sometimes an artificial or fictitious "nature" can serve for ordering the materials of a problem. The key is less whether the nature is solidly real than whether, considering things "in respect to" the possibly fictitious (purely imaginative?) nature, it provides the ability to put the things into an articulated series.

Organizing things into series according to natures and dimensions is a kind of preparatory work. Each problem we face requires us to deal with such series. Moreover, every series we have created in the past serves as background information and knowledge for addressing future problems. They become part of our imaginative "tool kit." Of course geometry and arithmetic/algebra problems are usually far more accurately solvable than real-world problems because the natures involved are already quite simplified. Geometrical things all participate in the nature extension or spatiality. One way to order them is according to the traditional dimensions of spatiality: points have no dimensions, lines have one, planes have two, and cartesian or euclidean space three. If these are the natures one is considering, then (for example) every closed plane figure wholly contains the nature of being two-dimensional and closed. Another way they can be ordered and compared is according to their size in the appropriate dimension; that is, we put into series line segments by their measured length, two-dimensional figures by their measured area, solids by their volume.

Most series one forms are a means rather than an end, and often previous series do not have to be explicitly called back to mind and expressly presented in their full extent. Each serves as a limited but extendable topological field matrix with marked positions into which we can project a new problem or, more likely, some aspect of a new problem. So if one is posed a geometry problem that specifies three line segments, an angle, and an area, and then asked to construct from them a four-sided plane figure of the specified area with two of the line segments meeting at the specified angle and the third joined at one of its endpoints to the open endpoint of one of the other two segments, one may well not actually array the three line segments from shortest to longest, but instead one will try to remember orderly procedures one has used in the past to solve problems of the same general type. In this way one interrelates different series, and in effect creates a matrix of two, three, or more dimensions. One will also very likely apply some of the other techniques that

Descartes presents in the *Regulae*. For example, one will use letter designations for the line segments (a, b, c) that are known, the unknown side will not at first be drawn but its length will be indicated by a special letter marking an unknown (x), and one will use various formulas (equations) one knows from geometry, trigonometry, and algebra to express the area of a quadrilateral consisting of sides a, b, c, and x. We will then be able to manipulate the equations to give us insight into the nature of the problem and its solution. Perhaps the equations will give us a direct solution in a single grand algebraic manipulation; more likely we will use only as much algebra as we need at a given moment, constantly correlating it with some aspect of the problem at hand. And even if we do arrive at a single grand solution, we will still need at the end to actually construct the solution to prove that the equational solution corresponds to something real.

Problem solving according to series ordered by the degree of participation in natures does not, in the process of arriving at a solution, have to respect the intrinsic essence of a thing—not even when the thing is mathematical (say, a circle or a triangle). Think about this for a moment in terms of our everyday methods of problem solving. When asked to divide twelve oranges among six people we for a moment just forget about the oranges and the people and calculate the numbers (twelve divided by two). If we were asked to divide twelve oranges among five people and realized it was 2.4 per person, we could apply an axe to the oranges. But it would at that point be wiser to remember something about the way oranges are (their essence?): they come naturally in sections. To solve the problem in certain circumstances (say, everyone wants to eat the oranges right now) it would be advisable to give two whole ones to each of the five people and then to peel the two remaining ones and divide them into sections. We might hope that those remaining two naturally divided into fifteen or twenty sections so we could give three or four to each person, but even if they did not, everyone would probably be satisfied by an approximation. If we were making orange juice, on the other hand, all this subdivision would be irrelevant to the solution. In the process of problem solving, we set aside the particular essences of things for the time being and think only in terms of the natures and aspects relevant to the problem at hand, and we do our calculations according to the series and measures we have established. This goes back to the basic situation of the *Regulae*: we have finite powers for understanding things, there is a limit to how many things we can consider simultaneously, we must break down complicated problems into parts or aspects for easier solution. We do keep track of what comes from what and what kind each thing is, but we do not actively consider these at every moment. This is the working situation of intelligent imagining.

Descartes is careful to note that, even in the case of expressly mathematical problems, we do not necessarily work the solution in terms of the kind of object we have before us. If the givens of a geometry problem are six regular solids with particular dimensions and we need to order them with respect to volume, for the sake of problem solving we may decide to convert the volume of each into a rectangular representation or into straight line lengths or pure numbers; the deciding factor is convenience. As I have already noted, Descartes recognized further that we could put a numerical symbol (or some other symbol) to stand for the line segment or the rectangle, and that, if we did not know a given factor but knew or suspected that it existed, we could represent it by a "dummy" representative, like the letters x or y or z—the classic representatives of unknowns in analytic geometry. Indeed, one can argue that his essential innovation was to treat unknowns in the process of problem solving exactly the way knowns are treated, the point being that you keep manipulating the formulas and the figures until you can directly determine the actual value of the unknown. In terms of equation manipulation, if you start with $x^2-6x+9=0$, you want to be able to manipulate it so that it reduces to $x=3$. Thus mathematics problems are not different from other problems: one chooses the way one will represent the knowns and unknowns of the problem not according to the nature or essence of the givens (for example, that one is dealing with squares or heptagons) but according to whether the way is convenient for getting at the desired solution. For any particular problem, it is possible—even likely—that there will be different sets of representations and/or different approaches that will provide one with the (same) solution.

In rule 12 Descartes points out that if we are dealing with colors we might want to represent the colors white, blue, and red, respectively, using a series of figures (see Fig. 6.1). This is a case of suggesting a fictional principle of ordering. Descartes is not asserting that this ordering–patterning is physically correct. What stands in its favor is that color is produced by light from an object striking the eye; we can easily imagine or conceive that some two-dimensional pattern might be impressed on the retina (which Descartes calls "the first opaque membrane"), and so colors might well be distinguished from one another naturally by such pattern differences. This type of pattern representation further suggests the possibility that there might exist a "system" of such patterns that would allow us to "calculate" or predict the result of adding different colors to one another. He is not arguing that this representation is the right one or will lead to such a system, but rather pointing out that, among the almost limitless number of possible two-dimensional patterns, some set might work in the way he suggests. In the long run, we are likely to discover a set of imaginative representations that allow us, in the process of calculation, to set aside thinking and experiencing the actual colors long enough to do the calculation. This is, in very rough approximation, how modern systems of color representation work. Clearly Descartes is presenting the question of two-dimensional patterns for colors not as a solution to a "perfectly understood" problem but instead as an example of how we can imaginatively approach a problem when we have ideas that are only plausible. We look at the phenomena we are interested in as a field (colors in general), we note their discrimination from one another (as hues), we try to find some other orderly field we are acquainted with (two-dimensional patterns), we note that in such a field we can create more complex patterns by combining simpler ones (a pattern of parallel vertical lines can be overlaid with a pattern of parallel oblique lines to yield a cross-hatched pattern), and we see whether we can mark more explicitly and determinately the orderliness of one (the hues within the field of colors) by representing them in the field of the other (the line patterns). If the system we come up with works, then we have a solution to the problem. It may be final, or it may need further refinement; it may be real, or it may turn out to be artificial. But when we next

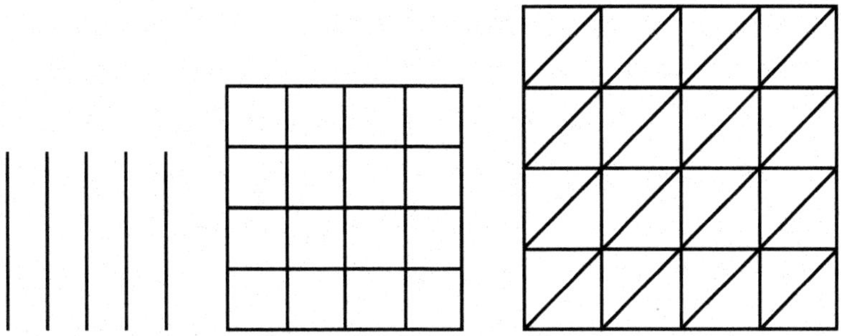

Fig. 6.1 Descartes's hypothetical representations of *white, blue,* and *red*

address the situation we are better off than when we started, because we have some principles of order, organization, and representation to call upon. Moreover, we may discover complexities in the colors that cannot be adequately represented by existing line patterns, and that can push us to advances in our understanding of lines patterns, eventually even apart from colors. And if some day it occurs to us that the patterns among hues in some respects resemble the patterns among tones in an eight-step musical scale, we might be tempted (as Newton was in his optics) to use well-understood mathematical representations of the latter to help order our understanding of the former. This is how Descartes's biplanar or dual-field imagining works.

Finally it is possible to understand more clearly how intuition and deduction function in the method of the *Regulae*. Intuition, I said earlier, is a simple act but not necessarily directed to simple things. Seeing green may be an incredibly complex physical, physiological, and psychological phenomenon, but simply to see that something is green or looks green is a matter of seeing, and to ascertain such seeing is a matter of making it as clear as possible. If we are trying to determine the color of an object through rain or fog, if it is behind a screen, if the illuminating light is dim or colored, we will need to do some work so that the phenomenon becomes clearer. The apparatus of a mechanical watch or a computer may be complicated, but someone experienced in their construction can clearly and distinctly perceive how it (as a whole) and each part (as distinct from others) are functioning when the rest of us only see "a lot of things" there. Clear (or perspicuous) and distinct seeing and understanding can be trained and even taught. An interior designer as well as a color scientist can teach us things about how we can make colors stand out more sharply; the watchmaker or computer designer can clarify and distinguish for us the parts of a computer and how they go together.

Although seeing in the first instance appears to be just registering what is there, as soon as there is a problem to be solved we have to mentally reconfigure and recontextualize the thing, the parts, and their situation. The very act of taking a look at something, re–presenting it, and setting it into distinct relationship with other things is a work of clarifying and distinguishing, of making things clear and distinct.

Each moment of clarified and distinguished seeing is a moment of intuition. Every time we take a step from one moment of intuition to another, we are engaged in deduction, or at least an attempt at deduction. This is the central work of imagining in Descartes's method.[27]

6.6 How the Knowing Power Recognizes Itself in Imagining

The two-imaginations note of notebook **C** had not described the exact relationship between figurative imagining and the intellective use of figures and images. By the time (presumably a few years later) that Descartes formulated the matter in the *Regulae* there is no doubt about it: as rule 12 explains, imagination is not separate from intellect or the knowing power, it is a special kind of highly active work of presenting, representing, organizing, and manipulating that the knowing power does in and through the medium of the organ of imagination in the brain. All by itself, this makes it evident that Descartes would have a harder time than Aristotle ascribing imagination in this sense to animals without also ascribing to them the intellective power that directs this kind of work. Descartes did, however, accept that what produces the physical impression on the eye sets off a chain of physical/physiological actions and reactions; he possibly even accepted that certain animals are able to perceive the hue in color, and that in combination with the activity of the organ of *phantasia* and memory locations in the brain this perception might produce a kind of sorting of experiences that would lead to an appropriate response.[28] None of that, however, could be deliberative or even conscious, at least in the sense that human beings can (for instance) perceive colors as colors. The animal would not be capable of consciously placing the experience at one level in relation to another (which in fact begins when one sees teal as a blue, and blue as a color; predication, stating that S is P, is a biplanar act). This is another way of saying that, for Descartes, animals are complicated stimulus–response devices, and thus any consciousness they might have is certainly not biplanar. Their sense organs and brains can acquire image–impressions, but they have no ingenial power of manipulating them. Only the human being has intellect, and intellect is precisely the power that can take impressions in the brain not merely as appearances but as images. Once intellect takes an image as an image, the world of re–presentation and re–imaging commences.

---

[27]The phrase "clear and distinct" is rare in the *Regulae*; "clear" appears more frequently in association with "perspicuous." Nevertheless, well before the phrase became a stock part of cartesian vocabulary, arriving at the clarified and the distinguished were essential goals of his method. As I shall argue below, it is a fundamental misconception to think that objects, ideas, or other things are clear and/or distinct per se; it is rather perceiving, conceiving, and portraying that are. This misconception has long encouraged excessively rationalist interpretations of Descartes.

[28]The single best account of the real complexities of Descartes's understanding of consciousness and how it might be compatible with (animal) body is Baker and Morris 1996.

ability to compare them. Intellect recognizes this, not by having one figure or two figures in mind but rather by having a figure in mind, noting its extension, then taking that figure as representative of all figures, as such noting that all figures must contain extension, recognizing that taking a figure as figure and taking it as extended reflect a difference in the taking of the presence of natures, then seeing that the different ways of taking the natures in the same figure is due not to the figure but to the knowing power. Only the knowing power, the intellect, can perform this differentiation. Thus one clearly and distinctly intuits the difference between imagining and intellection and recognizes that intellection is not per se a forming and holding of an image but rather also the stepping away from any image as image.

Perhaps I am too much belaboring the point about the quantity and quality of activity involved, both imagining and intellective, in thinking what is not imaginal. But I do this because conventional conceptions of cartesianism[31] underestimate the degree of activity involved in the (clarifying and distinguishing) work of thought. The cartesians—that is, followers of Descartes—and the post-cartesians—that is, later philosophers who, in various ways, responded to Descartes—came to speak not of clear and distinct *perceiving* but clear and distinct *ideas*. Ideas are not, however, intrinsically clear and distinct. Rather—and to take very seriously the definition of idea that Descartes gave in his reply to the second set of objections to the *Meditations*—the idea is the *form* of what appears to consciousness, but the total appearance is form plus "matter." Here that "plus" has to be understood as indicating the actual total appearance in consciousness, with consciousness understood as having some medium in which appearances vary from moment to moment; this medium or receptivity is like the "matter" of consciousness that can be, and is, constantly formed and reformed. The idea is thus not simply a static form but a formative agent in this medium of receptive consciousness.

The upshot is that one cannot simply "insert" into mind a "clear and distinct idea"—of extension, of thinking, of ego, of God—and without further ado think it precisely as such. It is certainly easier to think extension or self or God clearly and distinctly after one has done it before, but in every instance of such thinking it still requires preparatory work by the mind: the mind's clarifying and distinguishing activity. Thinking for Descartes is not simply "having an idea." What the mind thinks always occurs in a context. This is a generalization of his principle that "givens" are precisely what they are as givens *of a problem*. In one problem a line segment represents speed, in another it represents degree of pain, in a third it represents the unity of God. Thinking is seeing appearances in a context, against a background, taking different approaches to them, trying to vary them, trying to situate them against new backgrounds, etc. It is only in this way that they can become ever clearer, and it is only by being set in contrast to other things against a background that allows them to become ever more distinct. Most of this human work of thinking

[31] I use the miniscule or small letter to keep open the possibility of retaining "Cartesian" as an adjective meaning "genuinely characteristic of Descartes" rather than characteristic of his purported followers, the cartesians, who are of course cartesian but often not Cartesian.

is in the imaginative mode. But it is only by working imagination as hard as is humanly possible that one begins to genuinely glimpse the possibility of a thinking that is other than imagining.

When Thomas Hobbes was asked to comment on Descartes's *Meditations*,[32] this partisan of the notion that thinking is nothing but the having of sequences of images and assigning them names took the author's "idea" as a synonym for "image." Hobbes criticized in particular Descartes's use of "idea" for the thought of God. According to his philosophy, we have no image of God but only a name. Over and over Hobbes argues that we have no idea of God, only a name; over and over Descartes responds that we do have such an idea, although it is not an image, and Hobbes's use of the word "God" is a sure indication that, at least in some unclear way, Hobbes has the idea.[33]

Whoever engineered this set of objections and replies into a fictional disputation produced the appearance of an increasing irritation of the philosophers with one another's stubbornness. The debate, if it can be called that, would be more amusing than enlightening were it not for an aside that Descartes makes about why he chose the word "idea" in the first place (in the reply to Hobbes's fifth objection). If you want "idea" to be used only for "the images of material things depicted in our corporeal *phantasia*," then it is true that we do not have an idea of either angels or God. But especially in the passages Hobbes objects to, Descartes says he was careful to use the word idea "for everything immediately perceived by the mind, so that, when I will and I fear, because I simultaneously perceive myself to will and to fear, this same volition and fear are counted by me among ideas." Then comes a surprising remark: "And I used this word because it was already commonly used by philosophers to signify the forms of perception of the divine mind, even though we recognize [there is] no *phantasia* in God" (AT VI.181). That is, "idea" signifies, by analogy with God, the forms of perception of the human mind; but strictly speaking the analogy works only if God has a corporeal imagination, which he does not!

Descartes does not draw out the counterfactual comparison any further, but it shows as clearly as could be desired that even if there are ideas that are not properly images, ideas are the divine analogue of human images, the forms of God's imagination. Ideas are images raised to a higher power, and even if God does not have corporeal imagination it is not nonsensical to think of them as a higher kind of image. Since the divine ideas were also, under the influence of Augustine, understood as the exemplars according to which all God's creatures were made, they have to be understood not merely as passive shapes of creatures but as dynamically formative. If images for human beings are distinguishable from ideas that cannot be imaged, the ideas are nevertheless conceived as imagelike. Both the idea and

32 The result was the third set of objections to the *Meditations*, to which Descartes provided a set of replies.

33 He points out in his reply to Hobbes's fourth objection that if a Frenchman and a German can use different words to discuss a matter it is because the different words regard the same thing: "for if he admits that something is signified by the words, why does he not want our reasonings to be about what is signified, rather than about mere words?" (AT VII.179).

the image are re–presentations in a plane different from that of what they represent, and the being that they re–present is, in principle, fuller and more ample than the re–presentation actually shows. Yet even the re–presentations are actualizations in the appearance–potentialities of their proper planes of representation. They are less static views than generated and generative appearances. Neither thinking nor imagining is passive, because both operate within and between planes, even when they try to hold something constantly before the mind.

Forgetting the work of variation and recontextualization, of placing and replacing appearances to consciousness in different kinds of situations and explicitly recognizing these as occurring in and between different planes, is to entirely miss the nature and character of what Descartes conceives thought (and imagining as a type of thought) to be. To miss that is to miss the nature and character of *cogitationes* and *cogitare*. In the philosopher who is remembered as the proponent of the *cogito*, of the *cogito*–argument, that is to miss nearly everything.

For Descartes, the real actor in imagination is the knowing power, intellect. Even in the work of imagination we can recognize that imagining, having and working images, is not the same as intellection. That nonidentity is true even if there is never any thinking totally apart from images, because we can recognize that some of what we do mentally is not merely the having of images or the working of images or the having or taking or conceiving of images in some particular respect. What *does* the having, the working, the taking, the conceiving, is a power that exceeds images proper. The directed mobility of imagining comes from some other source than the images. If animals have images and a change or movement of images, it is nevertheless different from what happens in human beings, because animals cannot recognize the source of the mobility as other–than–image. The human power that works and holds images moves not only between images and image fields but also away from and out of all images. Even if it can never fully rise above them, it can clearly and distinctly recognize the limits of images and imagining.

6.7 The Limits of Imagination

Between abandoning the *Regulae* and publishing the *Discourse* with its scientific essays, Descartes's thinking about imagination began to acknowledge another kind of limit. The *Regulae* operated on the basis of a presupposition: that method reflects reality. More precisely: it presupposed that the orderly method of organizing appearances according to their participation in natures, along with the notion that the natures somehow combine or compound in complex experiences, will eventually be shown to correspond to the way things are, physically and metaphysically (Rule 12, AT X.418). At the end of rule 4 Descartes in fact remarked that the purpose of writing down the *Regulae* was to secure its precepts as preparation for more difficult philosophical tasks lying ahead. Presumably these included not only addressing actual problems presented by the world but also understanding the ontology of physics and the metaphysics of creator and creatures. The *Regulae* presented a comprehensive theory of the orderly processes of intelligent imagining that regulate human ingenuity, which can image and reimage things and rework and represent them in accordance with principles of order and measure, in particular by employing the simplest figural representations of order and measure. The ultimate warrant or guarantee for this process of efficient, problem-solving representation according to the degrees and measures of participation in natures had to come from a different kind of inquiry, an inquiry into the nature of natures.

Rule 12 began to present—but did not, and probably could not, complete—the ontology of natures. Natures there are divided into three basic kinds: material, common, and intellectual (AT X.419–420). The *material* ones "can be known only in bodies, as for example figure, extension, motion, etc." Of the *common* natures, which can participate in both material and intellectual things, Descartes mentions just a few examples: "existence, unity, duration, and the like." The *intellectual* ones get a longer account:

> Purely intellectual are those that, by a certain inborn light, and without the assistance of a corporeal image, are known by intellect: for it is certain that there are some such, and that no corporeal idea can be made that represents to us what knowledge is, what doubt, what ignorance, the same for what the action of will is that is usually called volition, and similar things; all of which, nevertheless, we truly know, and as easily as possible, for which it suffices that we be participants in reason. (AT X.419)

The intellectual natures are thus present and appear in the acts (or act–states) of intellect or soul. No image can per se enact doubting, although a human being can have doubt about anything that is an image or connected with one (for example, whether a particular painting of Christ's postresurrection appearance to his apostles manages to convey the doubt of Thomas and its resolution). Not even a text (consisting of word–images) can enact doubt like that expressed in the *Meditations* of Descartes; it is only when a real human mind enacts the words represented on the page and thinks their objects that doubt occurs.

Rule 14 had argued that the ordered techniques of imaginative representation could be used of any problem that was subject to *mathesis universalis*. This is not quite to say that they can be used only of corporeal things, things that share in corporeal natures, precisely because the common natures participate in *both* material *and* intellectual things. There is, however, a limit to the representability by points, lines, figures, symbols, and equations when one is representing intellectual natures. Recall the use of line patterns to stand for colors: although Descartes does not believe his particular representation truly represents the differences between hues, he points out that light striking the eye must produce some actual patterns. So even if the particular representation he chooses is false, the true one is of the same general kind, a pressure– or impact–pattern. If, however, we try to represent one act of will by a line segment and another act of will by a second segment, we quickly run up against obstacles. Insofar as there is a unity (a common nature) in each act of volition, we can legitimately say that each participates in unity, and there is nothing false about using something that has unity to stand for it. But if we think that the lengths of the lines express something further about their nature, for instance their duration or their intensity, we are at serious risk of confusion. What, after all, is the duration of a volition? One might well be able to say, "At 2:55:40 p.m. I decided to go to the Renaissance Fair, and that volition remained equally active in my consciousness for 10 seconds, and then began to fade erratically until there was no trace of it left in my mind by 2:56:30 p.m." But if consciousness of some kind is essential to making an act of volition, it is not clear that the volition lapses when one no longer has it actively in mind. If the Fair doesn't start until 8:00 p.m., I do not have to consciously renew the volition repeatedly over the next 5 hours. Descartes would probably argue that there is a duration involved in volition, but it is not intrinsically measured by clock time. Perhaps volition as intellectual or spiritual is more akin to character traits: it does not make much sense to ask how long they last, either.

On the other hand, the intensities of acts of will look like they could be sufficiently well represented by the relative lengths of line segments. It does not seem wrong to say something like this: "My will to go to the Fair is not as strong as my will to please my family." But we can't go on to say that anything more than "stronger" or "weaker" is represented by the lines. With line patterns representing hues we expect that something in reality corresponds to the particulars of the pattern, its size, its orientation, and the like, but they need to be determined by future work. We do not have any such expectation with a two-inch segment representing one act of will and a one-inch segment representing another about half as strong. We don't actually know much about measuring will intensity or what the proper unit would be. Although acts of will can come into conflict, it is not like forces or impulses in space; we lack a general theory of how and when volitions act and interact. We know that a very strong will (to eat healthily) can give way to what is a passing velleity (which induces us to gobble down half a pound of Belgian chocolate). In some circumstances volitions don't interact at all (visiting a Renaissance Fair as part of one's professional activities takes nothing away from one's family per se), in others they can conflict mildly, moderately, or enormously (the last when I selfishly insist on going in violation of a solemn promise). What this illustrates is that, although both corporeal and intellectual natures can be represented by *mathesis universalis*, elemental universal mathematics, with corporeal problems we expect there to be some more extensive reflection of reality in the representation, whereas the representation of spiritual things is superficial and "figurative." If, according to Descartes's early two-imaginations note, corporeal things can represent spiritual things, like wind standing for soul, and if such poetic tropes can be strikingly insightful, in the last analysis (and that is undoubtedly the right word) this kind of imagining cannot be taken very far, at least not within its original terms. At some point one must simply focus on the intellectual or spiritual phenomenon as such. A painting may give us insight into a moment of doubt, but to understand the nature of doubt we must look to actual doubting rather than to images that try to express it.

There was another, more important reason to set clear limits to imagining, however, and that was the question of infinity. Though not mentioned in the *Regulae*, the theme appears in mathematical writings dating from the same period (the 1620s). In those writings Descartes attempts to conceive and manipulate unending processes that nevertheless arrive at a determinate mathematical result. He says, for example, that imagination can conceive a limit to an unending reiteration of a procedure or the reapplication of a concept (e.g., drawing a limitless series of parallels to an original line, AT X.75, or performing endlessly more refined divisions of space or time, AT X.73 and 75). In another mathematical work of the period, "Excerpta mathematica," Descartes drops the use of *words* signifying imagination but *shows* that indefinitely expanding algebraic representations of series of sums and differences, displayed in increasingly complex tabular form employing continued fractions,[34] can be used to express the length of the side of any regular polygon inscribed in a circle with unit radius. By proceeding in this way, one can easily derive a series of ever more accurate approximations to the value of π, which is the proportion of a circle's circumference to its diameter. (By extending the number of terms indefinitely one can determine the length of the side of an inscribed regular n–gon for arbitrarily large choices of n ; see AT X.285–297.) This kind of physical and mathematical work must have convinced Descartes that pattern-deploying imagination could quite easily handle (a countable) infinity by employing an orderly method.

Two letters to his Paris friend and correspondent Marin Mersenne from 1629 and 1630 (that is, very near to the time when it is thought Descartes abandoned the *Regulae*) signal an important shift in Descartes's thinking about infinity. In a letter of 20 November 1629 he responds to Mersenne's inquiry about an author (referred to only as "Monsieur Hardy") who claimed to have devised a universal language. Descartes judges Hardy's proposals to be less original and less useful than they at first appear. In conclusion he adds something that, he says, he is sure Hardy has not thought of, because it requires the true philosophy and depends on ordering all the thoughts that human beings can have. Just as one can learn in a day "to give an infinity of names" to numbers, one might give names to all other human thoughts.

> And if someone had explicated well what are the simple ideas that are in the imagination of men, out of which everything they think is composed, and if that were received by all the world, I would dare to hope as consequence a universal language very easy to learn, to pronounce, and to write, and what is the principal thing, that would aid judgment, representing to it all things so distinctly that it would be almost impossible for it to be deceived; instead of which, to the contrary, the words that we have have almost only confused significations, to which the spirits of men being long since accustomed, is the cause that people understand almost nothing perfectly. (AT I.81)

Yet less than 5 months later, in a letter of 15 April 1630, Descartes complained of those who speak of God as though he were Zeus (a finite god) rather than attempt to understand his total infinity. This is the earliest evidence of the distinction he commonly drew in his mature work between "ordinary" infinities—like that of space, the counting numbers, or the divisibility of a line segment—and the infinity of God; speaking strictly, he would no longer call the former kind "infinite," but "indefinite" instead.

These nearly contemporaneous passages do not necessarily contradict one another. Yet his earlier confidence that we might perfectly understand an infinity of

[34] A continued fraction is generated by an unending recursive process that adds at each step a new term to the fraction. The well-ordered series of resulting formulas produces an imaginative pattern—and it is truly an *imaginative* pattern, because the series displays it, makes it appear.

thoughts according to their composition in imagination is displaced by the claim that he has learned something new about human beings and about God, and that he has found arguments even more persuasive for demonstrating metaphysical truth than mathematical proof (which he had been conducting with the aid of the regulated imagination, the subject matter of the *Regulae*).

The difference may well be explained, at least in part, by another metaphysical discovery Descartes announces in the letter of 15 April 1630, the *mathematical truths* (which Mersenne calls "eternal truths"; the name that scholars tend to use is "created eternal truths"). He had concluded that "the mathematical truths, which you call eternal, have been established by God and depend entirely on him, just as much as all the rest of creatures" (AT X.145). This announcement appears to be behind a new development that he mentions in the letter, that he had begun a new approach to physics. This approach would eventually culminate in *The World*, a work completed in early 1633 but not published in his lifetime.[35] In one sense at least *The World* is merely an extension of the project of the *Regulae*: it develops the seeds of knowledge about proportionalized relations that are native to, inborn in, our minds. The technique of imaging by using proportionalized series of the *Regulae* was the foundation of the elemental universal mathematics that Descartes had called there *mathesis universalis*. But the *Regulae* had bracketed (or omitted) several basic questions: whether *mathesis universalis* required a foundation—thus whether there was some more ultimate frame or horizon within which it functioned—and in particular whether what *mathesis* discovered about the *relationships* between conceived things and natures actually corresponded to the *reality* of the things, especially physical things.[36]

6.8 Imagining the Cosmos

In principle the imagination can imagine in any way it likes. It can populate its spaces with the gods of Olympus, satyrs, demigods, fairies, Alpha Centaurans, or brains in vats; it can proceed in any direction (literally or figuratively) that the imaginer likes; it can jump backward and forward in time and space with little or no logic guiding the development. (One need only look ahead to Descartes's description of dreaming in the *Meditations*.) However much Descartes knew of such imagining, it was not the cognitively useful, directed imagining of the *Regulae*. The *Regulae* implicitly recognized the continuous motions of imagining, but the method it proposed for regulating it occurred by discrete steps and by the patterns of order we

[35] Descartes declined to publish *The World* after hearing that Galileo had been condemned in Rome for advocating the heliocentric conception of the planetary system, which Descartes understood as a necessary consequence of the divisibility and motions of matter in his physics.

[36] See the discussion of "natures" in rules 6 and 12, AT X.381–382 and 418–424. For a fuller account, see Sepper 1996, 195–197.

already know. Continuous imagining might be approximated, or even in some cases achieved, by learning how to traverse the steps so rapidly that the imaginative motion became a continuous sweep. The goal of problem solving was not to set up a real-world scenario and then set it in motion (say, to set two trucks traveling in opposite directions at different speeds, one from city A, the other from city B, and then to see where their paths cross) but to analyze the givens of a problem, to find in them some pattern capable of being represented, to use the simplest kinds of geometrical figures or other images to represent the givens and patterns (thus as much as possible to keep the figuration to no more than two spatial dimensions), to label the representations for more convenient use, to put the labeling symbols into formulas expressing the patterns and proportions of relations equationally, and to manipulate and calculate the equations until an unknown is expressed totally in terms of what is given or derivable from the given. Once the form of representation was selected, no realism in the movements and manipulations of the representations or direct emulation of real-world activities and motions in the original problem and the things it was about was necessarily implied. But once Descartes started addressing metaphysical questions in Holland in 1629 he recognized that one could employ an alternative, more physically real form of imagining.

The *Regulae* never required a preliminary understanding of the world as a whole, only attention to whatever parts and aspects were immediately relevant to solving a problem. The overriding consideration was that all relations can be put into proportionalized form. After carefully posing the problem in a set of consistent terms in the plane or field of the real world, one would translate it into a field of figures by representing the measure and degree of nature–participation with lines and other images and then work out solutions by manipulating them. To assist in this one could also resort to a third field, that of calculation, by using symbols to stand for the figures and their measures and manipulating those symbols according to rules of algebra. One did precisely as much, or as little, representation and manipulation in and between as many fields as one needed to solve the problem.

The World, by contrast, asked from the outset about the entirety of physical reality. Unlike the *Regulae*, *The World* is predicated on grasping any particular problem as part of a *world* situation. The world is *indefinitely* extendable and *indefinitely* divisible. In order to solve a problem one must place it in the world situation and be able to track its circumstances and its evolution. The limits one imposes depend on how much of the world situation one needs to take into account. If there are physical impulses coming from afar that affect objects of interest, one has to take them into account and represent them. If all the motions and impulses of relevance are local or can at least be accurately represented in terms of their local effect only, one needs to picture only that immediate vicinity.

Although the element in which *The World* operates is still imaginative—the model of the world is constructed in "imaginary space," as he says in part 6—Descartes assigns an even more decided and directive role for intellect as providing the fundamental *parameters* for cognitive imagination. The chief issue is that it requires intellect to recognize that all motions, whatever paths they actually take, are at every moment based on straight-line tendencies or impulses (the actual motion is the result of myriad such impulses being applied to the same bits of matter); furthermore, the total motion in the universe is a constant. The latter point is basically what we would call a conservation law: the total amount of motion is conserved, so is neither greater nor less from moment to moment.[37] The former, straight-line requirement is a fundamental restriction on how to represent the components of an object's motion and the various forces affecting that motion. The two requirements together privilege understanding natural situations not according to step-by-step analysis but rather by imagined scenarios of continuously changing motions. It was a consequence of his conception of created mathematical truths: out of all the possible worlds that God could create, he created one with elemental truths that governed all events of the physical world and that could be known as such. As a result, in the spaces of our imagination we can set imagined representatives of the world's objects going and imagine them moving exactly the same way.

The approach of *The World* was not entirely new for Descartes. Recall that in notebook **C** he had imagined manipulating, rotating, sliding, etc., geometrical figures in all sorts of ways. The relevant imagining was always determined ad hoc, however, for the sake of solving some *individual* problem according to the particular principles of ordering one recognized in it. In *The World*, by contrast, he places figural imagining into a larger and far more dynamic imaginative context. After stating and discussing the intellect-discovered rules of motion, he recommences in part 6 by creating "in what philosophers call imaginary space" a world that duplicates the real one in three dimensions and the dimension of time. That is, what he had done with individual things, according to the two-imaginations note of **C**—take them and conceive them according to an image or figure—he now does with the totality of things: in imagination one produces a simulacrum or model, potentially of the entire universe. The universe is not conceived as actually infinite, yet it is indefinitely extendable in any direction one likes. It is infinitely divisible, both spatially and temporally, but because of the way the rules of motion work one does not need to break it all down actually to infinitesimals. To solve a problem about real or really imaginable things, one needs to set up in this imaginary space the situation that holds at some moment and then let it evolve according to the rules of straight-line impulses and straight-line tendencies to motion.[38] The perfect icon defined by the Eleatic Stranger in Plato's *Sophist*, the image that perfectly reproduces all the proportions of the original, is realized in Descartes's *World*.

---

[37] Strictly speaking, it was motion's speed rather than velocity (which is speed in a definite direction) that was conserved. This was a principal source of the inadequacy of Descartes's rules of motion: total speed was preserved in collisions, but direction was not. One consequence of the variability of the direction of motion was that it ultimately allowed the pineal gland to redirect spirit flows in the brain without violating the rules of motion.

[38] Straight-line tendencies and inclinations do not typically lead to straight-line motions, however. Given the indefinite divisibility of matter, at any given moment there are huge numbers of impulses affecting every point of the continuum. The resulting motions of the total interaction will typically be curved rather than straight.

This mathematical approach led almost seamlessly to the more sophisticated mathematics that Descartes began working on in 1632–1633 and that culminated in the analytic geometry of the *Geometry* (published in 1637 as the last of the three scientific essays accompanying the *Discourse*). Analytic geometry can in principle tell us the path that every point of this world's stuff, matter, or extension will follow when it is subjected to any number of inclinations, impulses, collisions, and the like. It solves real problems by raising them to a conceptual space—or, rather, an imagined space—though now (unlike in the *Regulae*) the solutions are not developed step-by-step but evolved continuously from given starting conditions by imagining matter in motion, in duplication of real motions of real matter in real space.

This is not to say that the new method completely abandons principles of the *Regulae*. In analyzing and setting up a problem for solution we still need to apply the part-by-part, step-by-step comparison of givens (for the most part done just two at a time). Nor is it the case that every solution will be a simulation of a real-world, three-dimensional, temporal scenario. As always, when it comes to problem solving Descartes remains an opportunist. Some of the classic problems of ancient mathematics that cannot be solved by the toolkit of Euclidean geometry, the straightedge and the compass, can be solved by real or conceptual "machines" that he had been thinking about since notebook **C**. Just as much as in the *Regulae*, one analyzes the elements of the problem, represents them in simple labeled forms, generates formulas/equations for the sake of algebraic manipulation that can be translated back into figures, and so forth. One will abstract these elements from the original way that they are given and incorporate their measures into the imaginative devices one has conceived. There is therefore a great deal of artifice in setting up the solution. But then, when problem-solving time arrives, one takes that device—which may be mathematically equivalent to a second-degree algebraic curve sliding long a fourth-degree curve and producing another curve by the moving point of intersection of a tangent to one of those curves and yet another line—and sets it into motion. With a real device that is a real-world motion: one that simulates a possible state of the world analyzed according to dynamically imagined mathematics.

The *Geometry* develops a mathematics that gives us the *possibility* of tracing the movements of things to any degree of analytical complexity we need. However, what human beings can *actually* imagine or do is restricted by the complexity of the problem, the finitude of human capacities, and the practical limits of time and resources available for working out details. Some motions are of a complexity that requires equations of algebraic order far beyond the human capacity to analyze them in actuality. Although *de jure* our techniques apply to knowledge of any complexity whatsoever, *de facto* what we know will have different degrees of certainty. God alone can track the infinitely fine divisibility of matter subject to limitless collisions and impulses throughout indefinitely extended space and duration. The mathematical truths Descartes announced to Mersenne in April 1630 do imply, however, that we can clearly and distinctly perceive that all motions can *in principle* be analyzed according to the mathematics of analytic geometry, even if there are motions—in fact an infinite number of motions—that no human being can ever come close to grasping according to all their particular causes.[39]

Looking back to the *Regulae*, one might say that its regulated, step-by-step motion of thought was a less decisive break with the philosophical past than it at first seems. Previous philosophies had for the most part modeled truth on the apprehension of unchanging, eternal things. Practical and technical matters were degraded forms of knowledge because they concerned themselves with what was changeable. Because the truly knowable things were stable, coming to know them was comparable to arriving at a state of rest, because once one knew the truth one's inquiry would cease.[40] The method of the *Regulae* is a kind of compromise between rest and motion. It produces a kind of knowing that "hiccups" from step to step, from stopping point to stopping point. One intuits a truth (stop); then one remembers to look to the problem one is trying to solve to search for the next place to look; and one keeps looking until something else is intuited (stop); and the cycle begins once more, until finally the solution to the overall problem is reached (stop).

However, Descartes's claim that *intuitus* could learn to sweep through the steps of a *deductio* to become continuous suggests that even early in his career the *goal* was to transform stepwise thinking into a continuous flow of imagining, and that he recognized that step-by-step motion, no matter how rapid, gives no more than a poor imitation of continuity.[41] The *Geometry* provided real continuity, not a poor imitation. It made continuous motion of points and lines fully and accurately trackable by virtue of the translation of regulated motion into algebraic formulas. An algebraic formula does not per se have a beginning or an ending point: any substitution value is as legitimate as any other, and the "point" of any substitution is its continuous relation to nearby values. The true account of a motion is one in which the mind traces the evolution of the formulable curve[42] that is the continuous path of a moving point. The *Geometry* thus fully implemented the mathematical truths that Descartes had announced in 1630 and that were implied by the physics of *The World*, in a manner that made their dynamism fundamental both to world and to mind. The things of the world move in a way that is accurately describable by the mathematics that the mathematical truths, created by God, establish and regulate, and the mind can always understand and imagine these motions in principle, even when it is not possible to imagine them in full factual detail. Knowing was not achieving rest from

[39] I hope that the reader recognizes that with this mathematics Descartes in effect anticipated what has only recently become possible with electronic computation (which is a much more powerful simulator than any he could build or even imagine). But even though computers have a much greater capacity to carry out the kinds of simulations *The World* and the *Geometry* foresaw, they are still finite means.

[40] Platonic, Aristotelian, and Stoic traditions all shared this view, and it continues to have an effect, even in unexpected venues. For example, Peirce's pragmaticism postulates that inquiry begins with an irritation and ends when an answer is found that brings the irritation to an end. But of course Peircean semeiosis is infinite, because ever new sources of irritation arise.

[41] This is one of the most decisive respects in which he rejected Euclid's *Geometry* as a model.

[42] That is, the geometric curve expressed by algebraic formula.

inquiry but being able to track in space (real and imagined) everything that a formula–solution implicitly contains.

What this means in its ultimate development, Descartes thought, was that all mathematics and all physics could be represented by rigorously imaginable figures and their motions, and these motions could be correlated with algebraic formulas. There was a fundamental intertranslatability of the geometrical and the algebraic. This mathematical realm was the *res extensa*, the "extended thing," conceivable as existing in imaginary spaces according to *The World* (AT XI.31–32) and described at the end of Meditation 5 as "the whole of that corporeal nature which is the subject matter of pure mathematics" (AT VII.71). Descartes thought (incorrectly, as we know)[43] that this approach would be truly comprehensive—that algebraic equations would be sufficient for tracking all actual motions and for solving all problems capable of solution. But because this mathematical and physical knowledge was grounded in the mathematical truths (because God had created them and sustained the world in accordance with them), we truly know that this approach to real, worldly things is correct, even when our human limitations keep us from analyzing all the detail. From the arguments in the *Discourse* and the *Meditations* we know that certainty about this correctness ultimately depends on our clearly and distinctly perceiving the difference—the real distinction—between extended things and thinking things on the one hand and between finite thinking things and the single, truly infinite thinking thing on the other. This is, of course, the reason that God becomes the focal point in the *Meditations* as the best known thing of all. He is best known, above all, because without him the rest of our knowledge is not truly knowable. The *Regulae*'s equal knowability of all truths had thus been radically and thoroughly displaced by the differential knowability of things represented in imaginative knowing and the ultimate but unimaginable knowability of God.

If we think this ultimate knowledge of mathematics and nature is solely rationalist[44] we are sadly mistaken, in Descartes's view. This kind of problem solving is a complex activity involving the several levels of sensation, imagination, memory, and intellect. It is true that the *intellect* notices what is or is not the case in problem solving, but the objects to which it attends are not pure ideas but the elements of problems represented by images and symbols. Intellect guides the whole procedure by shifting its attention from one thing to another and noticing the proportions that

[43]Descartes knew that there were mathematically possible curves that could not be represented algebraically; he called them imaginary (as opposed to mechanical, that is, those that could be expressed by complex machines representable by algebraic formulas). Apparently he did not believe that they could be real problems, that is, could describe real situations in the physical world governed by God's created mathematical truths. Unfortunately for the sake of his ambitions, many problems in nature cannot be expressed using algebraic equations, but at best only approximated.

[44]If it is not rationalist, it is still rational. Rationalism conceives of pure reason set free from the limits of this mortal coil. By contrast, recall the elemental meaning of *ratio* explained in Sect. 4.7, n. 46, above: it is the putting of one thing into determinate relation with another. Mathematical proportions are rational in this sense, as also are propositions; and so is the imagination that imaginatively realizes a determinate possibility with respect to an imaginative field that is articulated by features that can be differentiated as more or as less.

hold between them, but it takes no steps without the help of images and symbols. There is no intellectual problem solving without imaginative figuration.

As we know from the *Geometry*, one of the three essays that accompanied the *Discourse*, the algebraic approach to geometry—analytic geometry as we call it—is predicated on the following insights. First, the rigorously interconnected motions of points, curves, and lines generate new curves and lines. The various aspects of these motions can be expressed by algebraic symbols and equations and can be used to solve any problem based in mathematical proportions. The mathematical figures and equations, in their turn, can be used to represent the motions of actual bodies in space. Analytic geometry is therefore the fulfillment of the hope that Descartes expressed in notebook **C**, that unlike in the memory art he would find a way to generate images from other images, once the principle of their causation and proportionality was evident. As much as had the method of the *Regulae* or of his early notes about imagination's power, this ultimate expression of Cartesian method required great mental agility. The mind had to be able to move from plane to plane, from field to field, from space to space. Original problems had to be translated (literally "borne across," "carried across") into terms of analysis with shared or interrelated dimensions; bare lines standing for quantities of representable natures had to be carried over into a system of interrelated lines in the space of analytic geometry, marked positions of points and lines had to be translated symbolically into well-formed algebraic formulas, the calculations of algebra had to be translated back into the movements of the points and lines in geometric space: back and forth and back and forth until the imagined solution could be rendered back into the original terms of the problem and its corresponding real-world situation.

In comparison, the Socrates who makes and analyzes lines, squares, rectangles, and triangles in the *Meno* had noticed just the tip of the iceberg of imaginative mobility. Descartes had not only rediscovered the existence and the virtues of imaginative fields and the ability of mind to move between them, he had discovered in this multiplicity of fields of imagining a flexibility, a dynamism, a cognitive power that had never before been conceived. What nearly four centuries of development more has shown is that this method retains its power even when used by those who can give accounts neither of rationality nor of imagination.

6.9 Imagination in the *Meditations*

Why, then, is our first inclination still to think that Descartes places human beings above and beyond imagination?

When it comes to imagination in Descartes, readers will first of all remember two passages in Meditation 6. One argues that human beings cannot properly imagine a chiliagon, though they can easily understand it clearly and distinctly. The other claims that understanding but not imagination is part of our essence.

Taken by itself, the first shows a difference between imagining and understanding with respect to both clarity (in that the vagueness in one's imagining of a chiliagon is as evident as is the precision of understanding that a chiliagon has exactly one thousand sides) and distinctness (in that imagining and understanding the same thing, a chiliagon, are set against one another to show a sharp contrast).[45] If we ask ourselves how clearly and distinctly we imagine and understand (respectively) a triangle, it is likely that we will not notice a major difference. Indeed, one of the advantages of starting with the triangle is that we easily conceive the situation as follows: in both imagining and understanding the triangle, we begin by first imagining a triangle and immediately "seeing" that it has three sides. In the clear imagining of the triangle there is simultaneously a clear understanding of it as a three-sided figure.[46] Even if there might be a doubt or two here about whether we have expressed this quite rightly, it looks in first approximation as though understanding the triangle is either (1) nothing other than clearly and distinctly imagining it or (2) something that must immediately follow our clearly and distinctly imagining it. But if we repeat the experiment over and over, with a four-sided figure, a five-sided, etc., we will see that we were deceived. At some point we will find that it is harder and harder to distinctly imagine the figure we are trying to picture, whereas understanding that you are adding one to the number of sides is no harder when you move from nine hundred ninety-nine to one thousand than from three to four. Thus understanding an n–gon as an n–gon is different from imagining it. It does not require a clear and distinct image of the n–gon, and even when a clear and distinct image of a figure hovers before our mind there is still a difference between imagining as an act of the mind and understanding as an act of the mind.

The chiliagon passage is less definitive about imagination than it looks, however. First, in the *Meditations* Descartes tends to conceive imagination in a very narrow, physiological sense. In Meditation 3 he says that he uses "image" to mean the image or figure that is formed in the pineal gland, toward which the mind turns its attention (in imagining) so that it has an experiential idea of the imaged thing. The chiliagon example does not intend "image" and "imagination" so narrowly, since it is less about the figure formed in the pineal gland than about the experiential idea of the figure as it appears to us in ordinary geometric consciousness. The passage therefore does not justify the conclusion that we understand the chiliagon as such without any help from the imagination *whatsoever*. The argument is presented in comparative mode. It is not that there is nothing going on in the imagination when

[45]Recall, however, that the *Meditations* describes a prolonged and often repeated process of arriving at insights, so one must constantly beware of assuming that one can pop ideas into one's consciousness and immediately intuit everything about them. So part of the clarity and distinctness of the difference between imagining and cognizing the chiliagon is remembering that in contrast one can quite easily imagine a triangle or square, and that one can in comparison as easily conceive a one-thousand-sided figure as a three- or four-sided one. That does, however, also presuppose that one has learned arithmetic and geometry—so an infant cannot arrive at this insight of Meditation 6, although eventually he will be *able* to, even if he or she never actually does.

[46]The understanding we are talking about here in each case is simply that the thing in question is a triangle, or a rectangle, or a pentagon, or in general an n–gon. The difference between imagining and understanding the same figure also becomes clearer when we reflect that even a good geometer understands far better than he imagines that an n–gon has the sum of its internal angles equal to 180° times (n–2).

one thinks the chiliagon, but rather that the success or failure of the imagining is not essential to the understanding, no more so than if we were trying to imagine a myriagon (with its ten thousand sides). Imagining, we see, requires a harder effort of a different kind than does understanding (intellecting).

But doesn't the argument show that understanding is independent—redundantly one might say "completely independent"—of imagination? Even the one phrase that is most suggestive of an intellection so pure that it has no need of imagination in any sense whatsoever is not decisive. It occurs in the following sentence:

> If in fact the question were about a pentagon, I can indeed understand its figure, just like the figure of a chiliagon, *without resource of imagination*; but I can also imagine the same, viz. by applying the sharp edge of the mind to its five sides, and at the same time to the area contained by these; and here I manifestly notice there is need for a certain exertion of rational soul [*animi*] peculiar to me for imagining, which I do not use for understanding: this new exertion of rational soul shows clearly a difference between imagination and pure intellection. (AT VII.72–73, emphasis added)

The question is whether the phrase *absque ope imaginationis*[47] that I have emphasized can mean, purely and simply, that the imagination (or sensation) is totally irrelevant to understanding. Of course one can say "I know that a pentagon has five sides" as easily as "I know that a chiliagon has one thousand sides." Yet, already in the *Regulae*, Descartes had stated that it is properly the intellect that makes mistakes, not imagination, and that the intellect is especially prone to error when it does not provide itself with an appropriate image. Here in the *Meditations*, just two paragraphs earlier at the end of Meditation 5, the meditator congratulates himself on coming to realize that all knowledge depends on God, then says that "now indeed innumerable things can be plainly known and certain to me, both about God and other intellectual things, and about that entire corporeal nature that is the object of pure *Mathesis*." In conformity with his mature conception of the essential identity between the entirety of mathematical space and real space, in these concluding words of Meditation 5 he is dividing all human knowing into the purely intellectual and the purely corporeal. The latter realm, whether virtual (mathematical) or real (physical), is known by the imaginative methods of the *Geometry*. And of course Descartes had used the word *mathesis* 20 years earlier for the kind of orderly, imaginative, universal mathematics that he presented in the *Regulae* as the foundation of mathematics and scientific knowing. It is inconceivable that Descartes would almost immediately assert, with the chiliagon example at the beginning of Meditation 6, that the intellect in understanding something mathematical (a triangle, a chiliagon, a myriagon) could accomplish this in complete and total abstraction from *extension*, whether perceived or imagined. At the very least, the truth intellectually apprehendable about the thousand-sided figure must contain some reference to the fact that it is a geometrical *figure*, that it has *straight sides*, that it is *plane* and *closed*, etc., etc., etc. Insofar as it does not implicitly refer to any such things—that is, insofar as such things are not implicated, enfolded, in understanding

[47] The phrase is translatable in various ways: among others, without assistance/help/aid of imagination, without the power of imagination, without the resources/treasure of the imagination.

it—the formula "a chiliagon is a thousand-sided plane figure" is vacuous and meaningless. Given all these considerations, it would be better to read the phrase *absque ope imaginationis* as meaning "without any *special* aid of imagination."

It is of course possible that Descartes's formulation indicates some shift of position. A later passage from one of his letters to Princess Elisabeth supports the notion that he is talking about an understanding totally devoid of imagining, when he says that "the body, that is extension, figures, and movements, can also be known by the understanding alone, but much better by the understanding aided by the imagination" (28 June 1643, AT III.691). That this is unintelligible in light of a never otherwise repudiated conception of mathematics as *imaginative in essence* should make us at least pause before conceding. If there is an ultimate solution, it seems to me it requires conceiving this understanding without imagination as the immediate result of a remotion[48] (negating motion) of thought like that involved, in the *Regulae*, in the pure intellect's understanding that figure is not extension. That is, it is not that the imagination is put completely out of action but that the understanding is not intrinsically a matter of clearly perceiving an image or something that an image directly shows. To recognize that figure is not extension one must begin with something extended, see it in its figure and in its extension, and notice that being figure and being extension are not identical. This is intellect's remotional recognition that figure is not extension. Understanding in such a case is neither having a specific image in view nor staring at a formula in the absence of any reference. Neither having the image nor repeating a formula is *thinking* about figure or extension. Analogously, having a particular image in mind or even any finite series of images is not all by itself an understanding of body. Understanding is something that the intellect brings to the experience of the world so that we can see what does not directly or immediately show itself as such.[49]

The second passage in the *Meditations* that discourages ascribing importance to imagination is the claim that we would still be thinking things, *res cogitantes*, if we did not have imaginations (at AT VII.73). This, too, turns out to have less force than first appears. Though true, the claim needs to be seen in the context of the *Meditations*' method and set against what Descartes says later in Meditation 6 about the total nature of the human being. Most of the inquiry of the *Meditations* is conducted by the thinking being precisely insofar as he/she/it is a thinking thing, prescinding[50] from any other possible character of that being and (at the beginning of

---

[48] See Sect. 6.6, especially at n. 30, above.

[49] Anyone who presumes to settle the question of what pure intellect means in Descartes has to resolve this question—and resolving it probably also requires having something more than just a historical understanding of the matter. This is a moment in Descartes's thinking where his development of the conceptual topology of imagination and reason breaks down. It also prepares the way for a solution: the transcendental functions of imagination in Kant.

[50] Prescinding and prescission are explained in Sect. 5.13, esp. n. 102. Prescission treats as absolute a difference that is actually relational. In the paragraph above I am arguing that Descartes's method, insofar as it demands the maximal division of problems into parts and allows the inquirer to use artificial expedients when real ones fail, is based on prescission rather than abstraction, but a prescission which (in theory, at least) does not lose track of the relations in a falsifying manner.

Meditation 6) from the being of any other being besides God. Descartes's entitlement to this kind of prescission is tightly tied to his method, which says (1) to accept nothing that is not clearly and distinctly seen to be true, (2) to divide problems as thoroughly as necessary, (3) to construct complex knowledge from simple in an orderly way, and (4) to ensure (at the end) that nothing of relevance has been left out. In particular, wherever one clearly and distinctly perceives some difference, that means that a division is possible in thought, and if it is possible in thought then it is possible for the infinitely powerful God to make things actually different in such a way.

This prescissionary method is precisely why the images and ideas of things can be divided from the actual existence of the things, because our experiences of being wrong about sensation, dreaming, remembering, imagining, etc., provide the differences that doubt exaggerates. Our senses show us one thing when another is true, we can dream an entire world that, when we awake, we see does not exist at all, and so forth. Descartes goes through these things in an order that corresponds to setting aside our confidence in, and where possible "turning off," one after another, the external senses, the internal senses (in the forms of dreaming, memory, and imagining—including mathematical imagining), and then trying to do the same with intellect itself. The fundamental failure of each power, when it fails, is reflected in our inability to know that something real corresponds to what appears to us. Eventually the meditator expresses this difference in its most fundamental sense by making the terminological contrast between the formally real (to use the term Descartes settles on in Meditation 3 for "really existing," "existing in a thinglike way") and the objectively real (existing as present to mind in an idea–appearance). But with intellect, with my act of thinking, the arguments of doubt fail. The reason that the attempt against reason fails is that as long as I am conceiving, thinking, proposing the meaning of "I am, I exist" and trying as hard as I can to doubt it, I cannot be nonexistent, precisely as a conceiving, thinking, proposing, and doubting being. I may be nothing more than a thinking being, but the self-activated experience of thinking in the specific form of doubting leaves no doubt that, precisely as a doubting and therefore thinking being, I exist.[51]

But this does not at all settle the status, relevance, or use of imagination for human beings. Throughout the *Meditations*, and even more emphatically in his replies to objections, Descartes is careful to say that the conclusions drawn in the course of the work are carefully qualified and often only provisional, so they may not be cited as unqualifiedly true unless they are said to be so at the end. The imagination words are used quite narrowly in the *Meditations*. Imagination per se is the ability of our minds to focus on a figure produced in the pineal gland. But if we have no bodies (as our doubting has led us to think possible), then we have no pineal glands and no figures produced there to be focused upon. Thus it is possible that

[51] It is almost universally pointed out that the familiar form of the cogito argument, "I think, therefore I am," does not occur in the *Meditations*. That sentence, in particular its appearance of being a deduction from premisses, has nicely obscured Descartes's manner of argument and meaning for more than three and a half centuries.

I have the appearances that I call imagining without there being any real things corresponding to them. So I would be having thoughts that were like what people call images, but I would not be imagining. Without a pineal gland and images there I would still be the same being I am, a being that thinks. My essence as a thinking being is thus unaffected even if there is no such thing as imagining proper—precisely because I can imagine myself, picture myself to myself, as existing without imagination! Once again the meditation must engage imaginatively in remotional thinking in order to see a truth that directs our attention to what lies beyond our imagining, even if we never manage to think it with our imagining totally annihilated.

As I will point out momentarily, this conclusion about imagination is not the last word. Understanding this requires that we go beyond the question of our essence as thinking beings (which in fact is just an aspect of our being) to the question of our human nature (which is about our total being).[52] And, as things turn out, this means that in the last analysis, in order to understand imagination in Descartes we have to go beyond even considerations of the Aristotelian internal senses, beyond our merely cognitive powers, whether sensitive, imaginative, or rational, to an even ampler and more fully adequate understanding of what it is to be human.

In *Descartes's Imagination* I argued at length that the entire *Meditations* is meditational, and that means that it is as centered in imagining as its medieval and early modern forebears (meditations and spiritual exercises) always were. I even claimed that the cogito–proof and Meditation 3's proof of God's existence required both positive and negative-remotional uses of imagination (in the sense of "remotion" explained in Sect. 6.6, above). For our purposes here it is not necessary to go quite so far. Meditation 1 is, by any measure, the work of insistent, repeated, concrete imagining (of course, as always with Descartes, at the direction of intellect). We bring to mind different situations in which we have put trust in the senses but been deceived, we think about what madmen claim to experience, we think about what we take to be real in our dreams and see it can be every bit as mad as what madmen say, we wonder about ways of dividing up our experiences into components so that, even if how everything appears all together is not right, we might nevertheless find that the components are real and true, we see how confidently we assert and see the truth of mathematical claims but then recall that we make many mistakes, we come around again to assert that at least the simplest elements of mathematics must be true, then we realize that we do even make simple mistakes of counting and addition and then wonder fearfully whether our minds might not always "slip a cog" when

[52] Descartes is nothing if not faithful to the method that he learned in the *Regulae*. Considering ourselves precisely as thinking beings is, in the *Regulae*'s terms, viewing a thing (ourselves) with respect to a nature (thinking). There are many true things about ourselves that we can thereby discover with respect to what we are as participating in thinking. Thinking of ourselves as sensing or imagining is to consider a thing (ourselves) with respect to natures other than thinking in the strictest intellectual sense, and so the truths we discover thereby will be different—even though (or especially because) it is also true that, when we examine the interparticipation of natures, sensing and imagining turn out both to be species of thinking.

we try to add simple numbers or count the sides of a figure, and so forth. When the Meditation 1 doubt thereby reaches the threshold of purely intelligible things (e.g., God's being, in particular whether he could be a deceiver) the meditator pulls up and changes tactics, because the grounds for legitimate doubt become less clear than before. The meditator at that point has to devise an expedient, the evil genius dedicated to deceiving the meditator in every way possible. This is, of course, an imaginative device, and in particular it is motivated by a resolution of will. The meditator near the end of Meditation 1 has noticed that the earnestness and success of doubting fades as one becomes fatigued, and after resting one has lost the vivid sense of doubt one had good arguments for earlier, so that after a short while one is ready to accept again, without reason, what one has found reason to doubt (in particular the evidence of the senses). This is one of the first indications in Descartes's published writings that imagination is somehow connected with will more than with intellect, but not until the 1649 *Passions of the Soul* does Descartes explain in detail what this means.

The *Meditations* is more centrally concerned with the relationship between will and intellect than between either of them and imagination. Intellect, it turns out, is finite, or rather one might say indefinite: although there is no limit to how much we can know there is always indefinitely more that we do not know. Moreover, the fact that intellect can be easily misled shows that it is imperfect. Of will, however, the meditator says that it is perfect in its kind, that is, precisely as will (in Meditation 4, AT VII.56–57). Our knowledge about many things is and remains obscure, and since perceiving is the chief characteristic of the knowing power, obscure perceiving is a flaw. But any object that can come into our consciousness in any way or to any degree, obscure or clear/distinct, is equally well a potential object of our volition. We might greatly desire something we do not clearly see or understand because we do not grasp it clearly; if it were seen clearly we might see it as problematic. In such a sense will might almost be called genuinely infinite in us, and, insofar as at the end of Meditation 3 the meditator invokes the Judaeo-Christian theological doctrine of man's being made in the image and likeness of God, it is more with respect to will that man resembles God than with respect to intellect. This supremacy of will is a theme that is further developed in the 1644 *Principles* and the 1649 *Passions*: in particular when Descartes argues that thinking consists of both perceptions and volitions. Perception is a passion of the soul, whereas volition is an action; every perception is the passive side of the volition to know; and since things are more properly named in terms of their actions than their passions, the thinking thing is more properly understood as the willing thing than as the perceiving/understanding thing. Insofar as imagining is preeminently a volition—something clearly stated and argued in the *Passions* (AT XI.342–343)—that suggests that, contrary to what the *Meditations* says, imagining may be closer to the essence of the thinking thing than is intellectual perceiving.

The Meditation 4 reflections on intellect and will conclude that the disproportion between them is the main cause of our errors. We want more things to be true than we see to be true. This is important to Descartes's conception of both truth and error, since he distinguishes (clearly and distinctly, he doubtless would say) between what we perceive things to be and our affirming (by will) that things are the way we perceive them. Thus wrongly seeing things does not force us to commit errors; it is, rather, the fact that our will wants them that way (for example, in cases when our intellect does not perceive clearly and distinctly). But since the thinking thing that is the human being can train intellect and will to become differently balanced—that is what the meditator achieves by persisting in the full course of his meditation, his spiritual exercises—God bears none of the fault for our errors.

In the course of these considerations Descartes entertains the possibility that God could have made us to perceive things quite differently than we actually do. We might have had almost perfect clairvoyance about some or all of what we direct our attention to, for example, or an intelligence like that of some other kind of thinking being, or one that operated in a way we cannot even begin to conceive or imagine. Not far from the surface is the contention that God could have given us perception and will in many different particular ways, even, for example, with a kind of knowing that depended not at all on imagining or sensing (like the angels of Christian doctrine), or on a radically different kind of imagining and sensing. As perceiving beings we would still be essentially the same, only the objects and typical certainty of perception would be different. This is the substance of Descartes's statement that, precisely as thinking beings, God might have made us without imagination, and in that sense imagining and sensing are not part of our essence as thinking beings.

But that is not the end of the matter, when in Meditation 6 Descartes turns to the question of what our *total nature* is as human beings. It is at that level that God gave us sensation: not for the purpose of cognition, but for the purpose of staying alive—or, as the meditator puts it, maintaining the unity of thinking and extension, the unity of our soul and body. That unity and everything that provides for and maintains it is our nature. Intellectual perception is made for cognition, but sense perception is for self-preservation. Willing is superior to perception, so human beings are made not just for knowing but even more for acting properly in accordance with volition, whether those actions are thoughts or real-world actions.

In modern philosophy there is a tendency to treat volition as a question of ethics and politics rather than as part of fundamental anthropology, psychology, and epistemology, much less as part of metaphysics—in fact will is virtually irrelevant to modern epistemology.[53] Descartes's discussion, however, is more strongly related to medieval discussions that asked which of the two faculties, intellect or will, is nobler. The answer depended, to a great extent, on the thinker's conception of intellect and will in God. Of course the medievals understood these (and all other positive powers and qualities) as unified in God, who was conceived as radically one and simple. Like all other creatures, human beings were more diverse than God, so that one could at best think of the existence of intellect and will in the human being

[53] Not, however, to all modern epistemologists. I will mention only Hume, who, in the demanding and original form of the *Treatise* (Hume 1739–1740; see esp. bk. 2, sect. 2), understands all mental activity as having both a direct object and an indirect object (typically affects and passions, though one can say in general that the indirect object is the self—as long as one does not insist on too unitary a notion of self!). This dimension of mental activity is missing from the *Enquiry* (Hume 1748). It is the latter that has been more influential among professional philosophers.

analogously to their radically simple and unified existence in God. Some theologians asserted the primacy of will, especially insofar as love/charity was conceived as the culminating act of human existence; others argued for intellect, not least because the second person of the Trinity, the Son, was understood (following the opening sentence of the Gospel of John) to be the creative word and wisdom of God. But both sides in the debates recognized that every act of intellect implied an act of will and vice versa. In either case, the answer was not just a matter of psychology but also of anthropology, and not just of both of these but also of metaphysics, insofar as the existence of such powers in the human being implied an orientation toward their ultimate fulfillment in the metaphysical destiny or telos of human beings: loving God with the will entirely turned toward him, and turning toward God in proper apprehension and understanding of what precisely he is and thus being filled with the infinite intelligible species of the divine—something not accessible to human beings in their earthly state.

6.10 Willing, Images, and Passions

Descartes's regulation of imagination for cognitive use culminated in understanding geometrical space as identical to the essence of matter and opened the way for an apparently thoroughgoing reduction of physical reality to the mathematically imagined mechanics of motion. His last work, published less than a year before his death, *The Passions of the Soul,* takes a different, indeed quite surprising tack. As I have already noted, after dividing thinking into its active side, volition, and its passive side, perceiving, he defined imagination as an action or volition of the soul, rather than a passion.

The *Passions* gives Descartes's most detailed published account of human psychophysiology. It is not only important for imagination or psychology more generally but also deserves fully canonical status in presenting Descartes's philosophy and assessing his achievement.[54] Helpful in interpreting its significance is his correspondence with Princess Elisabeth of Bohemia (1618–1680; also known as Elisabeth of the Palatinate), since it was in the course of addressing her questions that Descartes worked out what was essentially a first draft of the *Passions*.

Elisabeth appreciated the metaphysical, scientific, and mathematical implications of the *Meditations* and the *Discourse* and accompanying scientific essays, but she had many questions about them as well as political, ethical, and anthropological concerns she did not find addressed. In her very first letter to Descartes she expressed dissatisfaction with his account of the relationship between mind and body. She could not understand how a physical account in terms of extension could be

[54] Unfortunately it has often been treated as a kind of afterthought or simply reduced to a specialized application of mechanist reductionism. A major exception is the magnificent study by Kambouchner 1995. Kambouchner situates the work in the history of the theory of passions and shows how it resolves Cartesian problems and brings Descartes's philosophizing to fulfillment.

reconciled with the acts of a soul entirely lacking extension. She asked for more precision, and in particular for a definition of the substance of the soul apart from its actions (thoughts) and a more rigorous account of causality between mind and body. In his response Descartes introduced the theory of "primitive notions."[55] By primitive notion he meant something that could be experienced and known only through itself, and not by trying to divide it up into simpler components out of which the whole would supposedly be grasped.

Two kinds of primitive notion are hardly unexpected, for they correspond to his division of things into thinking things and extended things. There are those that belong to the soul alone, then those that belong to the body alone. These and their difference, he explains to Elisabeth, are what his earlier works had distinguished. What is more surprising is the last kind he identifies, notions concerning soul and body *together*, for which "we have only that of their union, on which depends that of the force the soul has to move the body, and the body to act on the soul, in causing its sensations and its passions" (AT III.665). It would appear at first glance that the third kind of primitive notion must and should be analyzed into its two components, soul and body, that is, into the other two kinds, and then understood from their combination. But that is precisely what Descartes says is *not* possible. No matter how clearly and distinctly we understand thinking and body separately, there is nothing in that understanding that explains, articulates, predicts, or otherwise accounts for the how and the why of their being together. Perhaps this should not be surprising after all, however. Thinking and extension/corporeality share no trait in common. There is no reason to think that perceiving them clearly and distinctly, each apart from the other, could lead to understanding how they are united.

Just before introducing the primitive notions Descartes tells Elisabeth that she is assuming he has already tried to explain in his works how body and soul go together. Almost everything he has written hitherto tries to show the real distinction between body and soul, he says, whereas with respect to how they "act and suffer" together he has actually said almost nothing at all![56] Is the *Passions of the Soul* the work in which he finally treats them together?

[55] 21 May 1643, AT III.665–666. There are in effect four kinds of primitive notion introduced by this passage. First are "the most general [notions]—those of being, number, duration, etc.—which apply to everything we can conceive." Then come those "as regard the body in particular" (extension, which entails shape and motion), followed by those "as regard the soul on its own" (thought, including intellectual perception and inclinations of will); "lastly, as regard the soul and the body together, we have only that of their union…" (AT III.665; for the continuation see the next paragraph of the main text). To recapitulate: there are four kinds of primitive notion: (1) the notions of the first kind, which apply to everything we can conceive, which includes soul, body, and soul and body together; (2) the notions that apply (a) some to body alone, (b) others to soul alone; and (3) those of the last kind, which belong to the union of soul and body. What are usually referred to by scholars as the *three* kinds of primitive notions are 2a, 2b, and 3.

[56] "There being in the human soul two things on which depends all the knowledge that we can have of its nature, one of which is that it thinks, the other that, being united to the body, it can act and suffer with it; I have said almost nothing about this latter, and have exerted myself only to make the first well understood, because my principal design was to prove the distinction there is between the soul and the body; for which only the first could serve, and the other would have been harmful to it" (AT III.664–665).

Reading Descartes's remark in the introduction to the *Passions* that he composed it *en physicien*, "as physicist" (given seventeenth-century usage, one might even expand that to "physicist, physiologist, and natural philosopher"), it seems more likely that he was viewing the passions solely in their corporeal aspect. On the other hand, at the outset he delves into an abbreviated account of prerequisites from his metaphysics and psychology, and throughout the work he parallels the account of what happens physiologically in the various passions and emotions with psychological descriptions and occasional pragmatic and ethical considerations. As Kambouchner suggests, we need not take Descartes's words to mean he is writing solely as a physicist, but rather understand that he believed his distinctive contribution to the tradition of philosophical inquiry into the passions was to coordinate psychology very tightly *with physiology*.

After familiarizing ourselves with the whole work and engaging its explanations we notice three things: the physical and psychological particulars are often false and sometimes even comical to our twenty-first-century sensibilities; the explanations he gives of nerve, spirit, and other physiological activities indeed represent an application of the kind of mechanism that he had begun undertaking in the 1630s; and the causation he appeals to seems to work sometimes from the physiological to the psychological, sometimes the reverse. Despite all the shortcomings, Descartes is attempting an account that is recognizably like later scientific approaches. If he talks about spirit flows and motions in the nerves, to and from the pineal gland, we do something similar when we talk of ions flowing across synaptic connections and electrochemical impulses moving along neuronal axons to and from the brain. Descartes is also attempting to connect the psychological features of passions with the various physiological activities that help define them, just as a contemporary neurophysiologist might try to explain different aspects of an appearance (say, horizontal boundaries in vision) according to where the nerve signals are processed in the brain. And, looking backward, one might say that he is trying to give Aristotle's psychophysiology, or more exactly the conceptual topology that presents the soul as the activity of a body divided into organs, a revolutionary new basis. If in other respects Descartes saw himself as refuting Aristotle's philosophy, as practitioner of theoretical psychophysiology he was in essence continuing a kind of research that Aristotle had first defined and started to put to work.

Descartes's mechanistic physiology of the 1630s had been concerned to lay down a broad outline of the physiology of sensation, imagination, memory, and motor activity. Nerves were hollow tubes with a fiber running down the center. When the nerve in a sense organ was stimulated by an external object the resulting motion would be conveyed along the fiber, all the way to the chamber (the center ventricle) in the lower middle of the brain where the pineal gland was suspended. The hollow of the chamber was filled with animal spirits (the liveliest spirits that had been filtered from blood and food). When a nerve motion from a sense-organ nerve arrived at the periphery of the chamber it was translated into spirit pressure directed toward the gland. According to God's institution of basic correlations between motions of the pineal gland and what is experienced (i.e., ideas, as part of the human being's total nature)—an institution that could be partially modified by the person's life history—the thinking thing would see, hear, taste, etc., and then respond in some way (perhaps in an act of will, perhaps automatically). The response would be translated into a pineal gland motion that caused new spirit flows, some of which might communicate with places in the brain where traces of previous experience were preserved, others of which would be communicated to the muscles, through the opening and closing of the ends of the nerve tubes reaching into the spirit chamber around the pineal gland, in order to produce physical motion. This is a system predicated on rapid and efficient response and action—and, apart from the motions of the pineal gland initiated by thinking, it works similarly in human beings and animals.

Despite his practical and theoretical emphasis on imagination in his earliest mathematical and scientific writings as well as in the *Regulae*, in his 1630s productions regarding anatomy and physiology Descartes did not give much attention to how imagination fit into them. Yet, oddly enough, he did retain *images* as a central part of the descriptive and explanatory apparatus of his psychophysiology. It is odd precisely because of the nature and history of mechanistic explanation. Galileo, for instance, had argued that nothing about the process of moving a soft feather tip back and forth across the skin was the feeling called tickling. Sensible qualities like color, sound, smell, and taste do not resemble their actual causes, which are the size, shape, quantity, position, and motion of the parts of bodies. For Descartes, once light stimulates a nerve ending in the eye, that stimulus is translated into nerve motion, which is not an image. If we conceive the same thing happening in tens of thousands of nerves across the retina, what we get is tens of thousands of nerves transmitting their motion, each independent of the others, to the brain. At the periphery of the central brain chamber the arrival of the nerve motion affects the opening of the nerve tubes and changes the pressures and the flows of spirits in the central chamber; when those pressures and flows finally reach the pineal gland they change its position and shape, if only slightly. None of this mechanism involves images per se. Although you might say that an outline of the viewed scene is thereby impressed (quite literally as pressure) on the surface of the pineal gland, there is no immediate reason (without knowing a great deal more about the arrangement of the nerve endings and the fluid mechanics of spirit flows in the chamber) to assume that there is a particularly good or accurate "image" of the scene there. And there is even less reason to speak of images when some other sense than vision is in question.

Descartes nevertheless continued to speak of images. For example, in the *Meditations* Descartes calls what is physically/physiologically formed in the brain a "corporeal image"; he carefully distinguishes it from the idea that is psychologically experienced. He says that the latter derives from the mind's attending to or being directed toward the corporeal image. Clearly he is not invoking the so-called homunculus, as though a little man in the brain directs the little gaze of his little eyes toward the little image formed on the gland. The mind attending to the corporeal image may be an expansion of the schema that he had tentatively introduced in rule 12 of the *Regulae,* when he suggested conceiving the differences between colors as corresponding to differences between patterns of two-dimensional figures. That is, the physics of the cosmos and the physiology of the body lead to the production of some kind of patterned impression on the pineal gland. Descartes does not want simply to say that such a pattern will automatically produce a given color (say red) in the mind, because he frequently invokes the phenomenon of human consciousness in which our mind or attention is elsewhere than on what lies before our noses and eyes—so that we do not notice the red until our attention is called back to our sensory experience. Thus even though he claims that God has instituted certain correspondences between gland motions and ideas as part of our nature, Descartes does not want us to think of this simply as what we might call a stimulus–response or reflex theory. In any case, the stimulus is a complex, imageable pattern, and his deep scientific conviction is that the methodically directed imagination of scientific research can correlate such physical and physiological patterns with patterns of perception. The perceived pattern does not, however, have to be of the same quality as the stimulus pattern—if the pressure pattern on the pineal gland corresponding to red is "cross-hatched," that does not mean that anything at all about perceived red will be cross-hatched.

Perhaps these correlations between impressed patterns and perceived images are uneasily reconciled remnants of an older way of thinking—or perhaps they are positions embraced within the underlying conceptual topology. At any rate, they appear to have encouraged Descartes to look for new possibilities and consequences in the old nerve–and–spirit theory. Some pineal gland activity would be produced directly by thought, but insofar as memory was involved, there would have to be spirit flows communicating from the pineal gland to memory locations. The activity of new imagining would induce other changes in the gland and in the surrounding spirits, not to mention the effects of any continuing acts of sense perceiving coming from the sense organs directed toward the changing vista of the outside world. If there are spirits in all the nerves communicating with all the organs of the body, and a sea of spirits bathing the parts of the brain, there are almost limitless possibilities for eddies, flows, and currents not directly connected with imagination, memory, sensation, and motor activity. Do these have any psychological effects? In the 1640s, in the *Passions*, Descartes gives a clear, affirmative answer.

The *Passions* in fact conceives all the parts of the body in contact with the spirit system as capable of inducing motions and impressions in the spirits. Thus virtually the entire body (and especially certain privileged areas, like those around the heart, liver, and stomach) is connected with the psychologically crucial, central spirit chamber and so can contribute to "spirit flow turbulence"; this turbulence is translated into appearances at the pineal gland in ways less definite and determinate than regular sensations. These appearances are feelings, emotions, passions, stray incipient images of dreams, daydreams, hallucinations, and the like. That is, in the *Passions* Descartes expanded the realm of psychophysiology beyond determinate sensations, images, memories, and motor activities to begin accounting for a fuller range of psychological and physiological phenomena and their essential interaction than he had before.

Already in the 1620s Descartes had conceived the knowing power's relationship to the body as always involving the organ or gland of *phantasia*: imagination is the knowing power at work on the gland, memory the knowing power accessing brain memory locations through the gland, sensation the knowing power extending to the sense organs by way of the nerves that arrive at the gland, and motor activity the result of automatic and deliberate (i.e., produced by the knowing power) impulses issuing from the gland out to the body's muscles. The *Passions* does not abandon this schema, but rather more expressly recognizes that everything involved with sensation, internal sensation (as the medievals called it), and motor activity is part of a single complex system of nerves and spirits. Directed imagination, which can be used for cognitive purposes, is in the first instance a question of will's forming images in the gland; but there is also a kind of incipient imagination—call it para-imagination—that is largely a byproduct of the physics and physiology of motions in the spirits and nerves as they are affected by the other parts of the body. Much of this does not lead to clear and definite images but rather vague and transient ones, and some of the spirit flows do not produce object–images at all but rather establish a background or foreground of *feeling* with greater or lesser duration.

Perhaps the most remarkable thing of all is that Descartes does not treat all of this as physiological and psychological "noise" disturbing the rational processes that are the philosopher–scientist's central concern. Nor does he give in to the impulses of stoicism that would demand the suppression of feeling and passion. In response to the rationalist temptation he countered that the passions of the soul produce a system that naturally directs our minds to objects and holds our attention on them. Wonder is a passion that sensitively responds to what is new in experience and holds our attention long enough for us to begin intelligently dealing with it. Love, hatred, joy, fear, and desire, the other five primitive passions, keep us focused on objects of relevance to actions. All six primitive passions, in their myriad combinations, give rise to other particular passions that further diversify our attention and the character of our lives. Causality in this realm is bidirectional and biplanar: the physiological changes associated with a passion produce the psychological experience of the passion, and psychological reactions tend to prepare, sustain, change, or suppress some of the physiological responses.

This conception provides imagination with a new and unexpected function.[57] Descartes noted that the will does not have a direct effect on physiology: we cannot stop being sad simply by willing it. But the will can form images in a directed way; thus by volition we can choose to saturate our pineal gland with gladdening images that tend to relieve sadness and bring joy. We can think pleasant images and scenarios, we can read amusing books, we can view comedies; the thoughts they produce will, outward from the pineal gland, induce new impulses in the eddies and currents of the animal spirits and counter (if not always overcome) the physiological conditions that have produced the undesirable passion. Of course our intelligence is also at work in this process: we are, according to Descartes, supposed to determine the will to what is best, and it is intelligent perception, the power of comparing things to one another and seeing the appropriate ratio that holds between them (typically with the help of images, of course), that provides the volitions with

[57] New and unexpected in Descartes, at least in the most rationalist interpretations of his philosophy, but less original and unexpected insofar as they were traditional topics of imagination theory.

rationality. But human beings are not just intelligence, not just will, not just imagination, not just memory, not just sensation, not just motor activity, not just passion, emotion, feeling: they are all of these together, in a psychic economy instituted as the whole nature of the human being by God. The thinking part of us has its essence precisely as thinking—though thinking begins with the activity of volition rather than with the passion of perceiving, even intellectually (recall that perception is a passion, not an action)—but our total natural being is precisely as complex as God created it. He could have made us otherwise, of course, for example as pure rational beings. But he did not. We are the way we are, and our task is to live well as the beings he made us.

Thus the *Passions*' accounts regarding interactions of will, imagination, and passion directed toward what intellectual perception determines is best are not just another ruse of a rationalism claiming to be the master faculty in rational passion management. What these complications show instead is that, by the time of his death in 1650, Descartes had begun to come to terms with the entire economy of the psyche in its physiological incarnation. It gave him a way to show in detail that human life is not all about knowledge, that human beings are not created simply to be knowing beings. Perhaps it was just his Jesuit training coming out: the spiritual exercises of Ignatius exercised the imagination and reason in meditations in order to arrive at a resolution of the will to live as best one could in accordance with God's will. The aim of living is to live well. The intellect and all its helps exist not primarily for the sake of proving the existence of the self and God or doing mathematics and science but for keeping body and soul together (literally!) and for always willing, and thus trying to *do*, what the intelligence determines is best among all the appearances the mind–body system shows.

If any proofs are needed, there are two strong supports for this "humanistic" interpretation, one from Descartes's letters, the other from the conclusion to the *Passions*. In discussing the primitive notions Descartes pointed out that soul can be known only by intellect, that extension can be known by intellect but much better by imagination, and that soul and extension together in the incarnate human being are known by sensation. This would seem to settle the matter of their relative value, seeing that sensation has always been the lowest of the faculties of soul. But he then advises his reader (Elisabeth) to spend no more than a few hours a year in metaphysical-intellectual reflection and no more than a few hours a week in scientific and mathematical speculations that occupy intellect and imagination together. These activities are very troublesome and fatiguing, he points out. The vast majority of one's time should be devoted to conversation and other pleasures of the senses (AT III.692–693). If this sounds very uncartesian, perhaps it is because we misjudge Descartes: we think of him as though he were merely a follower of the Descartes of our interpretations—the Descartes that cartesians made of him—rather than a thinker who directed his attention to the sources of things. If we think that Descartes did not follow his own advice insofar as he worked on metaphysics, mathematics, and natural science, perhaps that means only that we do not know enough about the facts of his day-to-day living or notice that he does not seem to have spent much time after the 1644 *Principles* on metaphysics and physics/physiology, nor much time at all on creative mathematics after the 1637 *Geometry*. It is lost to us, but we know that the last work he composed, at the court of Queen Christina of Sweden, was a play, an entertainment (technically, a masque).

Perhaps Descartes meant exactly what he said at the end of the *Passions*. In the very last words of the concluding article 212 he writes:

> As for the rest, the soul can have its separate pleasures; but as for those that are common to it with the body, they depend entirely on the passions, so that the human beings they can move the most are capable of tasting the most sweetness in this life. It is true that here they can also find the most bitterness when they do not know how to employ them well and if fortune is against them. But wisdom is principally useful in this point, that it teaches to make oneself so much a master of them and to manage them with so much address, that the bad things they cause are very supportable, and even that one draws joy from all. (AT XI.488)

The power of imagination, which in the early notes could strike sparks of poetic insight that philosophers could reach only by plodding, had gone through a long period of discipline that culminated in the mathematics of analytic geometry and the cosmological science of universal physics. It seemed to go into eclipse in the works of his metaphysical maturity, but that was more appearance than reality. In the last analysis—an analysis that began in his correspondence with Elisabeth of Bohemia and came to fruition in the *Passions of the Soul*—imagination was recast as the will-directed art of entertaining and managing all the appearances that redound to the human being who is soul and body together. The originals of appearance for the most part come to us through sensation, feeling, and emotion; imagination is the power by which we can "pull back" from the immediacy of the appearances of the moment and play with their possibilities. Through this play of imagination the best (the goal of will) can emerge and appear to us (in intellectual perception) and thus help us live well and experience all the sweetness that the God-given passions afford us.

But this imaginative vision of a good life vanished in the hearts and minds of those who came after; it went into nearly total eclipse. Many of the best of his followers overlooked it in favor of the rationalism they took his philosophy for, and some of those who recognized its consequences took steps that undermined it. In the history of Western thought, imagination was never quite the same. The next chapter will try to explain why and how.

References

Adams, Alison, and A. J. Harper (eds.). 1992. *The emblem in Renaissance and Baroque Europe: Tradition and variety. Selected papers of the Glasgow International Emblem Conference, 13–17 August, 1990*. Leiden: Brill.

Aristotle. 1993. *De Anima*. Ed. D. W. Hamlyn and Christopher John Shields. Oxford: Oxford University Press.

Baker, Gordon, and Katherine J. Morris. 1996. *Descartes' dualism*. London/New York: Routledge.

Brann, Eva T. H. 1991. *The world of the imagination: Sum and substance*. Savage, Md.: Rowman & Littlefield.

Descartes, René. 1964–1976 (cited as AT, with volume number). *Oeuvres*, 2nd ed. Ed. Charles Adam, Paul Tannery, et al. 11 vols. Paris: Vrin.
Descartes, René. 1987. *Abrégé de musique, Compendium musicae*. Trans. Frédéric de Buzon. Paris: Presses Universitaires de France.
Eco, Umberto. 2002. *Art and beauty in the middle ages*, 2nd ed. Trans. Hugh Bredin. New Haven: Yale University Press.
Gouhier, Henri. 1958. *Les Premières pensées de Descartes: Contribution à l'histoire de l'anti-Renaissance*. Paris: Vrin.
Hume, David. 1739–1740. *A treatise of human nature: Being an attempt to introduce the experimental method of reasoning into moral subjects*. 3 vols. London.
Hume, David. 1748. *Philosophical essays concerning human understanding*; 1758 edition retitled *An enquiry concerning human understanding*. London.
Kambouchner, Denis. 1995. *L'homme des passions: Commentaires sur Descartes*. 2 vols. Paris: Albin Michel.
McCracken, Charles. 1983. *Malebranche and British philosophy*. Oxford: Clarendon.
Nolan, Lawrence. 2005. The role of the imagination in rationalist philosophies of mathematics. In *A companion to rationalism*, ed. Alan Nelson, 224–249. Oxford: Blackwell.
Risse, Wilhelm. 1970. *Die Logik der Neuzeit*. 2 vols. Stuttgart-Bad Cannstatt: Friedrich Frommann Verlag.
Rodis-Lewis, Geneviève. 1998. *Descartes: His life and thought*. Trans. Jane Marie Todd. Ithaca: Cornell University Press.
Sepper, Dennis L. 1989. Descartes and the eclipse of imagination, 1618–1630. *Journal of the History of Philosophy* 27: 379–403.
Sepper, Dennis L. 1996. *Descartes's imagination: Proportion, images, and the activity of thinking*. Berkeley/Los Angeles: University of California Press.
Sepper, Dennis L. 2000. The texture of thought: Why Descartes' "Meditationes" is meditational, and why it matters. In *Descartes' natural philosophy*, ed. Stephen Gaukroger, John Schuster, and John Sutton, 736–750. London: Routledge.
Shea, William R. 1991. *The magic of numbers and motion: The scientific career of René Descartes*. Canton: Science History Publications.
Warren, E. W. 1966. Imagination in Plotinus. *The Classical Quarterly*, n.s., 16: 277–285.

Chapter 7
The Cartesian Heritage: Kant and the Conceptual Topology of Imagination and Reason

Descartes's doctrine of imagination, if we might call it that, did not have direct influence on the next generations. The doctrine was hidden from view, more by accidents of publication and nonpublication than by intent. By the time he had worked out the consequences of his early speculations and had learned how to regulate the power of imaginative work (the work of the *ingenium*) in problem solving that employed figurative imaging of a problem's givens and symbolic representation of the figures, productive imagining had become second nature to him. For most of his readers, by contrast, the use of imagination had become inapparent, for several reasons.

It had become inapparent, first of all, because the *Meditations* and the *Discourse* appeared to have placed imagination infinitely below reason among human capacities. Those works had apparently shown human being in its essential form as pure thinking thing capable of turning away from everything connected with the material world in order to focus on purely intellectual ideas (in particular the ego–self and God). Second, the mathematical and scientific works, where Cartesian imagination did its most vigorous work, did not discuss that work as imaginative—though neither were they *totally* silent about the fact. Third, the scientific and even more the mathematical writings were often beyond the capabilities and judgment of readers; if they were read, they were properly appreciated by few. If imagination is supposed to be easy, works like the *Geometry* could hardly be imaginative, or so it seemed. Better to think of them as the achievement of nonimagining reason.

There is a deep—one might even say tragic—historical irony here. Descartes did have readers who fully grasped and affirmed the imaginative character of mathematics and natural science. For instance, there was Nicolas Malebranche (1638–1715), who read Descartes's posthumously published treatise *Man* in 1664, the year he was ordained a priest; it revolutionized his thinking.[1] He immediately began adapting its nerve–spirit–brain account of human psychology to his philosophical

[1]Paragraphs following about Malebranche are adapted from Sepper 2005a, 170–174, with permission of Blackwell Publishing.

D.L. Sepper, *Understanding Imagination: The Reason of Images*, Studies in History and Philosophy of Science 33, DOI 10.1007/978-94-007-6507-8_7,

and theological purposes.[2] The power of beauty or ugliness over people—for example, the effects of stories, works of art, and rhetoric—was due to the fact that psychological and physiological responses are bound together. This led him to develop a social psychology of imagination that even provided a mechanism for the transmission of original sin from Adam and Eve to all their descendants.

Malebranche presented a well-developed elaboration of his theories of imagination in the *Search after Truth*, which he published in two volumes in 1674 and 1675. It is a kind of moral-philosophical anthropology and psychology with an ultimately theological framework. It is divided into six major parts, called "books." The six books are dedicated, respectively, to the senses, the imagination, understanding, the inclinations ("the mind's natural impulses"), the passions, and method. Book 2, on the imagination, is the longest of the six. It is divided into three parts, with constant emphasis on cartesian nerve–and–spirit physiology, though many of the topics (such as a pregnant mother's strong imagination of strawberries producing a strawberry-shaped birthmark in her child) are as old as the history of speculation about imagination. The first part of book 2 considers, on the basis of cartesian nerve–spirit–brain matter physiology, basic phenomena of imagination and their organic causes and concludes with a discussion of ways in which children's imaginations are formed, from the effect of the mother *in utero* to early childhood influences and education. The second part treats the imagination of different categories of person, chiefly according to the typical anatomical and physiological characters of their bodies derived from nature or acquired by association and habit. Briefly considering the imagination of women, men, and the aged, it follows with several chapters about the possible deleterious effects on the imagination of studiousness and typical distortions of imagination among the learned. It concludes with brief discussions of the imaginations of the "effeminate," the "superficial," "persons of authority," and "those who perform experiments." It includes a treatment of how authority over others can be exercised by a physiologically powerful imagination. That anticipates the major concern of the third part, which begins by explaining how the human disposition to imitate primes people to be affected by powerful imaginations and describing the brain and nerve conditions that underlie this susceptibility. The last half of the third part treats characteristics and dangers of strong imagination and some examples, then the strength of imagination of authors (the most dangerous of whom are "the freethinkers" who mock tenets of religion) and considers individually Tertullian, Seneca, and Montaigne. Just before concluding he takes a brief look at the imaginations of those who think they are sorcerers and werewolves.

The brief conclusion to book 2 points out that the first two books show "that all the thoughts the mind had through the body, or through dependence upon the body, are all for the sake of the body; that they are all false or obscure; that they serve only to unite us to sensible goods and to everything that can procure them for us; and that this union involves us in infinite errors and very great miseries," without, however,

[2] *Man*, or *De l'Homme*, from the early 1630s, was planned by Descartes as the concluding part of *The World*.

our feeling most of them as miseries (Malebranche 1997 [1674–1678], 195, bk. 2, pt. 3, ch. 6.2). We develop throughout our lives by a series of unions: with our mothers' bodies (which joins us to concupiscence and sin), after birth with our parents and nurses (whom we imitate), and thereafter with all other human beings we encounter through the rest of our lives. It is in this last phase "that we live by opinions, that we esteem and love everything that is loved and esteemed in the world, despite the remorse of our conscience and the true ideas we have of things." (Malebranche excludes from condemnation the positive effects others can have through the mind, though the next book, book 3, does point out the many ways in which the abstractions of the mind continually deceive us.) "This is how all the thoughts we have as a result of our dependence upon are bodies are completely false, and the more dangerous for our soul as they are useful to our bodies." In more than one respect this third part of book 2 is one of the most influential, though principally negative, early modern contributions to what we might call the sociology of imagination. The very last two sentences of book 2 then point to the "cure" that book 3 will undertake, a cure that points to an even more fundamental "union" of the human soul to another. "Therefore, let us try to deliver ourselves gradually from the illusions of our senses, from the visions of our imaginations, and from the impressions that the imaginations of other men make upon our minds. Let us carefully reject all the confused ideas we have as a result of our dependence upon our bodies, and only admit the clear and evident ideas the mind receives through the union it necessarily has with the divine Word, or with eternal truth and wisdom, as we shall explain in the following book, concerning the understanding or the pure mind." If Descartes in his *Regulae* was wary of saying much about how intellect operates on its own, if his method taught not to reject corporeal ideas as false but to clarify them by subjecting them to rigorous imaginative analysis, if in the *Passions of the Soul* he proposed understanding and appreciating the togetherness of body and soul, Malebranche has no such qualms or concerns. Our destination is eternal, and that means we should reject everything that is not, or no more than tolerate it as long as we must.

Malebranche concisely expresses the book's general concerns in the preface. "The mind's union with the body…infinitely debases man and is today the main cause of all his errors and miseries." "We have such a close tie with our body and depend on it so much that we do well to be apprehensive about not always having distinguished the cacophony with which the body fills the imagination from the pure voice of the truth that speaks to the mind." "The mind must judge all things according to its inner lights, paying no heed to the false and confused testimony of its senses and imagination; and if it examines all the human sciences in the pure light of the truth that illumines it, then assuredly it will scorn practically all of them and will have a higher regard for the science that teaches us what we are than for all the others combined." A chief goal was to "combat several errors and especially those most universally received or those that cause a greater disorder of the mind, and I show that these errors are almost all consequences of the mind's union with the body" (Malebranche 1997 [1674–1678], xxxii, xlii, xl–xli, and xxxix). The disorder is moral as well as cognitive. For example, imagination exaggerates the attractions and minimizes the faults of what we like. We can form many images of what is sensory;

this means that it is easier for most people to see the advantages of riches than of justice. The search for truth, both cognitive and moral, thus had to begin with the elucidation of the causes of error.

Yet Malebranche often vacillates in the *Search* about the value of the lesser soul powers, and there is even a kind of imagination that he treats positively. Strictly speaking, he asserts the unity of the cognitive faculties in a manner reminiscent of Descartes. "Men are capable of sensation and imagination only because they are capable of pure intellection, since the senses and the imagination are inseparable from the mind; yet no one finds anything amiss in treating these two faculties of the soul separately, though they are by nature inseparable" (Malebranche 1997 [1674–1678], 338, bk. 5, Ch. 1). In the conclusion to the first three books he describes the complicated interrelationship between (1) the senses, (2) imagination and memory, and (3) pure mind or understanding. In sense we receive ideas from God mixed with sensation, on the occasion of certain movements taking place in our sense organs in the presence of objects. In imagination and memory we receive "from God ideas mixed with images, which are a kind of weak and languid sensation the mind receives only because of certain traces being produced or aroused in the brain by the flow of spirits." Pure mind or pure understanding occurs when the mind

> receives from God entirely pure ideas of the truth, with no admixture of sensations or images, through its union not with the body but with the Word, or the Wisdom, of God... not in order to know mutable things suited to the preservation of the life of the body, but to enter into immutable truths, which preserve in us the life of the mind. (Malebranche 1997 [1674–1678], 261, bk. 3, conclusion)

Yet, after offering this glimpse of a rationalist's dream—the possession of perfectly disembodied truth in this life—Malebranche warns against expecting too much of pure ideas. Sense and imagination enable us to know the relation of external bodies to our own and are entirely for the benefit of the body. Because they bind us to our bodies and to sensible things, we cannot put our trust in them when our aim is knowledge. "No truth whatever can be clearly discovered through the idea of the senses or the imagination" (Malebranche 1997 [1674–1678], 261, bk. 3, conclusion). But one cannot deduce the relations between bodies (including our own) from pure ideas; without the senses the mind knows bodies and their situations only "in a confused way." So to live well in this world we need to pay close attention to things. The culmination of this necessary use of sense and imagination is scientific understanding.

After having presented the nearly limitless ways in which imagination leads human beings into error in the second of the six books of the *Search after Truth*, Malebranche eventually arrives at pure mathematics and universal cartesian science. He points out that mathematics, especially as practiced by Descartes, is a method of pursuing truth that enables us to avoid error and to see true ideas of things clearly. Descartes, of course, argued that God had created the fundamental truths of mathematics, which included the fundamental truths of physics, both as the foundation of the cosmos and as providing a basic structure for the human experience of the world that is accessible to intellect. Malebranche makes the connection of these truths to God stronger and more intimate. He notes that most people would be incredulous were he to claim that

> in applying itself to these sciences [metaphysics, pure mathematics, and all the universal sciences that determine and contain the particular sciences] the mind applies itself to God in the purest and most perfect way of which it is capable, and that it is in perceiving the intelligible world that these sciences have as their object that God Himself knows and produces the sensible world that bodies depend on for their life as minds depend on the intelligible. (Malebranche 1997 [1674–1678], 367, bk. 5, ch. 5)

This means that mathematical imagining produces a knowledge of the sensible world that knows the sensible things *in God*. Mathematics, even if it is not rationality in its purest form, can thus prepare us for the ultimate task of *seeing ideas in God*. Moreover, in a discussion headed "the imagination's use in preserving the mind's attention," he says of geometry that it

> should be regarded as a kind of universal science that opens the mind, makes it attentive, and gives it the skill to control the imagination and to draw from it all the help it can give; for with the help of geometry the mind controls the imagination, and a controlled imagination sustains the mind's perception and attention. (Malebranche 1997 [1674–1678], 429, bk. 6, pt. 1, ch. 4)[3]

Thus book 6, titled "Method," teaches the essentials of the mathematics-based scientific knowing presented by Descartes in his writings on method, mathematics, and the sciences.

Though he was well informed about the sciences and mathematics of his day, Malebranche was not a professional mathematician. Pascal and Leibniz were, and although they did not emphasize as he had the positive moral force of mathematically disciplined imagination, they both recognized the imaginative character of scientific and mathematical knowing. Blaise Pascal (1623–1662), who knew Descartes personally and professionally, acknowledged this role of imagination almost reluctantly. A generation earlier than Malebranche, in the *Pensées* (*Thoughts*), he dissected imagination as the psychological power that reflects what the human heart is set on, and because of original sin that means it is set on ourselves and our aggrandizement; it flatters our desires, and it magnifies the importance of our little part of the world that we treat as though it were the whole. We are impressed by the color purple because it has been associated with majesty; elaborate images and scenarios put into words by an expert orator sway us according to his (and our) desires; whatever we call to mind we imagine as conducing to our wealth, power, and prosperity.

Descartes's mature conception of the difference between what is very large and what is infinite resonates in Pascal's wager, for example, and similarly points to and tries to educate us about the limits of the imaginative way of conceiving things. Give up on the possibility of an infinite gain and risk infinite punishment for the finitude of an earthly life of pleasure: that is a formulation that tries to make it possible for the ordinary human being to use his imagination to think beyond it. We reckon by imagination the importance of something finite; if we can conceive, beyond that, the

[3]Notice that, unlike Descartes, Malebranche reifies geometry as a science existing apart from the mind that can be used as a tool to control imagination. For Descartes, geometry is precisely a cognitively rigorous imagining of the world.

true infinity of the afterlife, its punishments, and its rewards, we can rationally grasp that it exceeds any amount of imaginable pleasure, wealth, and power that a few-score years can bring.

No mere mention of Gottfried Wilhelm Leibniz (1646–1716) can even begin to do justice to the metaphysical and epistemological background to his thinking and thus to the deeper reasons behind one or another assertion from his writings. If one is aware of his notion that even sense perception is a kind of cognition, one might suspect that he would be less inclined to draw hard and fast boundaries between intellectual powers and sensitive powers (which traditionally include internal senses, one of which was imagination) than Pascal and Malebranche were. As Robert McRae has pointed out, although one can find passages where he appears to be a hard rationalist distinguishing between images and exact ideas, his subtler doctrine is that the mathematical sciences and their exact ideas are products of imagination.[4] In correspondence with Queen Sophie Charlotte of Prussia,[5] Leibniz retains the traditional distinction between external senses and internal ones. The external sensibles perceived by the external senses (for instance color) are very familiar to us, but we do not truly understand them; they are clear, though not distinct. The notions or ideas we attribute to the common sense both *appear* and *are definable*—number and shape, for example. The common sense is not sufficient for conceiving these clearly and distinctly and for building sciences from them, however. For that, we need "something which the senses cannot provide and which the understanding adds to the senses" (Leibniz 1989, 187). The internal sense that allows us to unite the perceptions of the different external senses, so that we can compare (for example) numbers and shapes in color with the number and shapes involved in touch, is imagination. Imagination

> contains both the *notions of the particular senses*, which are *clear but confused*, and the *notions of the common sense*, which are *clear and distinct*. And these clear and distinct ideas, subject to imagination, are the objects of the *mathematical sciences*, namely arithmetic and geometry, which are *pure* mathematical sciences, and the objects of these sciences as they are applied to nature, which make up applied mathematics. (Leibniz 1989, 187–188, emphases in original)

In order to produce demonstrative proof rather than just inductive and observational truth, sense and imagination need the assistance of intelligence. It deals with what is "only *intelligible*, the *object of the understanding alone*; and such is the object of my thought when I think of myself" (Leibniz 1989, 188). Thinking of a

---

[4] See McRae 1995. An example of hard rationalism comes from a work directed against Locke, in dialogue form and not published in his lifetime, *New Essays on Human Understanding*, where the representative of Leibniz's thought remarks "how essential it is to distinguish images from exact ideas, which are composed of definitions" (Leibniz 1981, 137). Earlier in the same work, however, he wrote that "it is an admirable arrangement on the part of nature that we cannot have abstract thoughts which have no need of something sensible, even if it be merely symbols such as the shapes of letters, or sounds; though there is no necessary connection between such arbitrary symbols and such thoughts" (Leibniz 1981, 77).

[5] The letter is published in Leibniz 1989, 186–192.

color and considering that one is thinking a color are "two quite different thoughts," then. My recognition of the "I" and of the right of other beings to say "I" as well allows me to conceive substance; and similarly from considering myself I arrive at other metaphysical notions like "cause, effect, action, similarity, etc., and even those of *logic* and *ethics*. Thus it can be said that there is nothing in the understanding that did not come from the senses, except the understanding itself, or that which understands" (Leibniz 1989, 188). Mathematics, geometry, and universal mathematics "fall under the imagination." They are the "science of imaginable things" or "the science of universal imagination." Universal mathematics deals with "that which falls under the imagination or that which I call the logic of the imagination."[6]

Unfortunately, almost none of these texts were publicly available during his lifetime. Nor were Pascal's aphorisms about imagination, intended for the *Pensées*, until more than a century later. Descartes's understanding of mathematics as essentially a rigorous use of imagination was either neglected or regarded as of minor importance. As a non–mathematician Malebranche, whatever his reputation, had no particular authority with respect to claims about the imaginative nature of mathematics and the sciences and their ability to bring otherwise unruly imagination under control. The one-sentence summary of the longest single part of his magnum opus, dedicated to a comprehensive look at imagination, was that it led to error. The many qualifications of this summary judgment were less likely to be remembered, especially by later rationalists. On the empiricist side, one could conceive mathematics either as the understanding's most accurate treatment of the real, primary qualities of bodies (extension, position, motion, and the like—with Locke as the prime example), or as the imagination's most accurate use of ideas considered apart from reality, that is, as fictional (with Hume).

If we more fully exploit the analogical possibilities of the notion of conceptual topology, we might put it this way. Like mathematical topology in comparison to geometry, a conceptual topology has considerable flexibility in comparison to the specific topographies that hold the field. Topographies are actual maps representing the territory as it has been experienced and explored. In any real topography, some parts are marked as more habitable than others, some are poorly known, and the boundaries between them are murky. The least inhabited territories are terra incognita. Topographical maps are cultivated by authorities who represent the current state of a field and its methods and who transmit these to a new generation. The new generation's task is to preserve and renew the maps, and to improve them where possible. More adventurous thinkers head for the unfamiliar spaces. When one of them returns with strange tales of species and places unlike what is known in the capital, the authorities may try to assure their followers that everything is already under control and that no thoroughgoing remappings are necessary.

Probably most adventurers overestimate the significance of what they have seen, and indeed it can often be absorbed, with modest revisions, into the existing mappings of the world. But a few know that the topography of their teachers is untenable

[6]As translated in McRae 1995, 182; the original is from Leibniz 1903, 348.

and needs to be reconfigured. Rarer still is the adventurer who comes back with not just tales but also with the sketch of a new topography based on the extended possibilities of the topology. Rarest of all is the one who has glimpsed the outline of a previously unrecognized topology. Yet the stories of the adventurers (and another kind of story, that of the archeologists[7]) can have an effect only if they are heard and preserved. If they are lost, or minimized, or suppressed (whether by the adventurer, or by others), it is essentially as though they never existed.

Conceived in a narrow sense, what I have said here and in the previous chapters supports the contention that until at least the later seventeenth century it was a recognized part of the conceptual topographies that imagination has a positive significance and perhaps even plays a central role in knowledge, in particular in some of the best and most articulate knowledge that human beings possess: mathematics and the natural sciences. But the scientific revolutions in the natural sciences progressively detached themselves from established topographies of logic, ontology, metaphysics, and the basic principles and developed organization of knowledge. There came the revolution in mathematics with its new analytic and infinitary processes that concomitantly forced a reconception of nature and it workings. The figure most central to this development was Descartes, precisely insofar as he developed the existing conceptual topology of the human psychological powers to provide new impetus to mathematical and natural scientific understanding. The renewed topology led him early on to recognize metaphysical implications, and finally to revise (in the *Passions*) the topology of human psychology and anthropology more comprehensively. This deep interconnection of fundamental themes began with his exploration of how the human psychological powers can be used for the sake of efficient problem solving. Ultimately this work drove him to a reconception of philosophy, a reconception that qualified him (in the eyes of the future) as the father of modern thought. But the origin and engine of this exploration and reconception was lost to view. Descartes gave way to conventionalized and sedimented cartesianism.

A few who came after glimpsed something of these connections and even clearly recognized (as in the case of Leibniz) that they were working within a reorganized topology. But they were on the other side of the historical divide. Descartes reached the epochal divide, the ridge of the chain of mountains that form the eastern horizon for territories in the west and the western horizon for those in the east, and saw both sides. Pascal, Malebranche, and Leibniz found themselves on the other side of the mathematical divide but still had a sense of where they stood on the continent. But their maps, too, were lost to the future. When Malebranche conceded that imagination was used well and essentially in mathematics and the sciences, it was no longer clear why one should think that, since otherwise the valence of imagination had been turned negative: error-prone, error-inducing, at best fictional or hypothetical.

[7]By "archeologists" here I mean either those who dig deeper into the ground to explore the underlying strata (for instance, the follower who really wants to understand more thoroughly and articulately the master's deepest presuppositions and frameworks), or those who look into the past of the discipline to find out how it was formed.

From the perspective of the new age, the old theories of knowledge had considered imagination as at best *transitional* from sense experience to knowledge. Where in new topographies there ought to have been well-developed theories explaining the positivity of imagination, there was a gap. The gap likely did not seem to be important, however, precisely because imagination had become a matter of secondary importance. In such cases gaps are filled by whatever lies at hand and is not at cross-purposes with one's primary commitments. And thus the age of reason became forgetful of imagination.

7.1 How Imagination Got Misplaced, Part 1: The Way of Ideas

In contemporary historiography of philosophy, the early modern philosophers, both empiricists and rationalists, are often characterized as following the "way of ideas."[8] This provides a unifying perspective on what are often seen as dichotomous traditions based in conflicting theories of knowledge and its sources. The concept implies the epistemological point of view that has been prevalent since Kant[9] while emphasizing that the concept of idea underlies both empiricist and rationalist traditions. It amounts to a recognition by historians that there are many possibilities to be explored when one is committed to the centrality of ideas in thinking. In that sense it implicitly acknowledges that *idea* is part of a conceptual topology that ranges beyond epistemology and epistemological topics to psychology, anthropology, logic, language, ethics, and metaphysics—with the possibility or even probability that what we conventionally understand as empiricism and rationalism mark distinctive topographies within the same topology.

If nothing else, Descartes's use of "idea" is at the origin of the conceptual topology of modern philosophy. Historians of philosophy continue to discuss the sources of this use and the paths by which they influenced Descartes. If the present study has anything to say about questions of sources and paths, it is that one must not forget an almost casual remark in his replies to Hobbes. He adopted the term "idea" because it was already used of the forms of divine perception. Of course God does not possess a perceptive faculty like human beings. Ideas are like divine phantasms—but of course God does not have phantasms because he does not have common sensation, imagination, or memory. Descartes knew enough Augustine (and, by extension, Plato) that he could easily have explicated the analogy further. The divine ideas are the exemplars according to which God creates all things. The use of the

[8] The term was introduced into wide circulation by Yolton 1956.

[9] Interpreting the history of modern philosophy as essentially epistemological has severe shortcomings that are increasingly well understood. See, for example, Loeb 1981 and the respective Cambridge Histories of Philosophy for the seventeenth and the eighteenth centuries (Garber and Ayers 1998; Haakonssen 2006). Nevertheless, there is still an almost overwhelming tendency, especially in introductions to the early modern period, to make the narrative thread chiefly epistemological, with rationalism and empiricism playing the leading roles.

word suggests something not merely static but also actively formative. Hobbes, however, thought of ideas as fixed units of thought.

I am less interested in looking from Descartes to the past of "idea" than to its future. Let us consider, then, one of the most influential books of the second half of the seventeenth century, *Logic or the Art of Thinking*, known more familiarly as the *Port–Royal Logic*. It is a classic not just in the rationalist tradition but also for the entire modern period. It was first published, anonymously, in 1662. Its authors were Jansenists—followers of Cornelius Jansen (1585–1638), bishop of Ypres in Flanders—who practiced Jansen's strict, Augustinian regime at Port–Royal, a former convent about 20 miles west of Paris. The principal author of the book was the theologian and philosopher Antoine Arnauld (1612–1694), assisted by Pierre Nicole (1625–1695). Arnauld had been one of the earliest correspondent–critics of Descartes's *Meditations*—his was perhaps the only criticism that Descartes truly esteemed—and by the time of the *Logic* he was considered a leading representative of cartesian philosophy. The *Logic* was one of the chief instruments through which basic cartesian notions became part of the very element of modern philosophy.

The introduction starts by defining logic as "the art of directing reason to a knowledge of things for the instruction of both ourselves and others"; "this art consists in the reflections that human beings have made on the four principal operations of their spirit, *conceiving*, *judging*, *reasoning*, and *ordering*" (Arnauld and Nicole 1992 [1662], 30). The first three operations had been accepted for centuries as basic "matters" of logic[10]; the last, ordering (also called method), was a more modern, cartesian concern. All four, the *Logic* says, begin with and depend on the first, conception.

> One calls *conceiving* the simple view that we have of things that present themselves to our spirit, as when we represent to ourselves a sun, an earth, a tree, a circle, a square, thought, being, without forming any express judgment about it; and the form by which we represent these things to ourselves is called *idea*. (Arnauld and Nicole 1992 [1662], 30)

Ideas are thus the most basic units of logic; they are what we use to represent things to ourselves before we come to judgments. The basic presentations in consciousness are portrayed as singular: a sun, a tree, a being, a circle; conceiving the singular thing is viewing it in its presentation to consciousness; the idea is the (repeatable) form that *represents* the thing and that we *use* for such representing.

When we expressly put two of these ideas together in an affirmation or a negation we make a judgment ("Horses are mammals," "Horses are not rabbits"); at the next level, reasoning, we conclude new judgments from several judgments already given (from "Horses are mammals" and "Mammals are animals" we conclude that "Horses are animals"); and finally it is by method, by understanding the forms and ways of the previous three levels, that we "arrange the various ideas, judgments, and reasonings we have on a certain subject" in order to understand. Logic, then, is the general theory of forming, using, and organizing ideas in order to *know*, and the ideas are, by definition, the basic representations of things when they are viewed or conceived by the mind.

[10]The corresponding "form" is provided by the logic of terms, categorical propositions, and syllogisms.

The section that follows takes pains to show that ideas are not to be confused with images. By "images" are meant solely the *corporeal* forms of things that we perceive when we imagine (or sense, or remember). They are a subset of ideas, not the whole. In making this distinction the *Logic* follows the track laid down in Descartes's *Meditations* by citing the example of the chiliagon, which we can clearly and distinctly understand but not imagine. The *Logic* then proceeds to defend the thesis that ideas are not (necessarily) images by criticizing the theories of two materialist critics of Descartes, Thomas Hobbes (1588–1679) and Pierre Gassendi (1592–1655). The reason for Arnauld and Nicole insisting on the difference was precisely that materialists and empiricists like Hobbes and Gassendi equated the two, so that the entire process of thinking would be nothing other than having and controlling sequences of images. In the very first chapter of the first part of the *Logic*, Arnauld and Nicole do everything they can to forestall such theories—especially those of "the Englishman." They combat them as "very absurd and as contrary to religion as to true philosophy" (Arnauld and Nicole 1992 [1662], 38).

For the early modern empiricists, the senses receive (or produce, in response to a physical motion, particle, or impulse) images of things—Locke will later call them "ideas"—and the mind thinks by comparing, classifying, and altering the images and their sequences. Hobbes noted that, unlike brutes, human beings have the capacity to remove themselves from any current sequence of images and to enter a different one through memory, imagination, and language. For empiricists in general, understanding is what allows one a certain freedom with respect to any sequence of images that currently occupies the mind. The ability to stop one and begin another is due to understanding's ability to take notice of and indicate—for Hobbes and for Locke, by using marks, words, or signs—resemblances and differences of various kinds in the images. Hobbes had a maximally mechanical conception of this power: images are carried by sequences of vibrations of the nerves; by marking images in one sequence with the same mark used for similar images in another, the mind can more easily and freely move between different image sequences; furthermore, this creates the possibility of thinking in terms just of these marks or names, insofar as their reference to images allows us to note equality and differences.[11] Locke went so far as to allow that noticing and keeping track of all the complex differences and identities of ideas even *required* names and signs (Locke 1690, bk. 3, ch. 9, sect. 21).

One might conclude that empiricism was marked by the tendency to understand thinking as occurring first and foremost as a kind of appearance or presence at the level of the sensory or the quasisensory (a term used to expressly include the internal senses like memory and imagination alongside the external senses); the power of understanding, which appears to reside elsewhere than on that level but may not itself constitute an integrated level, is what permits us to organize, direct, and negotiate the level of quasisensory presence. Even if the empiricists, unlike

---

[11] For Hobbes, in particular, reasoning is the kind of thinking that calculates using words, the words being themselves aural or visual images that refer to other images. Words like "God," however, literally refer to nothing—as Hobbes says to Descartes, it is only a name. Arnauld and Nicole take special pains to attack and ridicule this claim (Arnauld and Nicole 1992 [1662], 35–37).

rationalists, were reluctant to discuss the (superior) ontological status of understanding, through the theory of signs and language they had a surrogate for the higher kind of intellectual idea to which rationalists appealed in their theories.

Rationalist thinkers did not deny that there are image–ideas, but they repudiated the notion that there is nothing more than these and that knowledge is just a special way of having image sequences. Rationalists like Descartes conceived of thinking as having more than a single level, and consciousness as being intrinsically capable of concentrating on levels at will or of bringing two levels into a compound view in which one tries to see the "higher" through the "lower" or vice versa. They offered in general a more fully articulated account of the different levels of mental experience—which means that image–ideas would be differently articulated by their relationships both to one another and to the nonimage–ideas at play in higher levels of the human soul. Images had differently nuanced roles by virtue of the typical rationalist affirmation of the biplanarity of thinking. Images present themselves at one level, judgments and reasonings about them at another, with the possibility of a simultaneous presence of the levels.[12]

Both empiricism and rationalism, at least as they are conceived generically, comfortably use the same term, "idea," for the contents of consciousness, despite the fact that they disagree about the character and kinds of ideas. There is, however, at least one major problem intrinsic to the way–of–ideas approach. It highlights the *objects* in mental acts more than the *acts* themselves or the human *powers* expressed in the acts. With imagination, this makes almost inevitable defining it as the psychic viewing of an image–object.

The image–as–object is in practice and theory far more congenial to empiricist thinkers than to rationalists. This helps account for one of the ironies of the historiography of early modern imagination. The paradigm of imagining is taken to be having and holding in mind a sensory object–appearance without the presence of the object itself. Rationalists, however, are almost by definition regarded as ultimately dismissive, or at least neglectful, of this kind of experience. So it is the empiricists who draw detailed attention in histories of imagination, and the rationalists are relatively neglected or misrepresented.[13] Empiricism takes "idea" as a

[12] It is no accident, in this light, that the modern approach to aesthetics—which analyzed sensation according to its simple elements, combinations, and relations and presented this analysis as the basis of a science—originated in rationalist circles (in German speaking lands). At the level of sensation, memory, and imagination the image–appearances had relations to one another that were relatively independent of the level of rationality.

[13] Lesser and moral rationalists are more likely to appear in histories than the great rationalists. A standard account of the rise of Romantic imagination, Engell 1981, identifies the major line of development before Kant as proceeding through Hobbes, Locke, the Cambridge Platonists, Shaftesbury, Hume, and Tetens; the Cambridge Platonists and Shaftesbury are the only arguably rationalist figures among them (there is a brief glance at Leibniz because of his notion of active force). Kearney, after a chapter devoted to the Middle Ages, takes two sections in a single chapter to make quick nods to mysticism in the Renaissance, Leibniz, and the three empiricists Hobbes, Locke, and Hume, before reaching Kant; and Leibniz is treated as representative of the "Cartesian hostility to imagination" that "was shared by such rationalist philosophers of the seventeenth century as Leibniz, Spinoza, and Malebranche" (Kearney 1988, 162).

well-defined unit of thought, whose provenance in the first instance is ordinary sense experience. This means, again in the first instance, that "idea" is more or less equivalent to "unit–phantasm" or "unit–image." Once the ideas are drawn into the mind by sensation, what happens next? For the most part they are subjected to mind or soul operations (for example, the power of understanding classifies and manipulates them according to various kinds of resemblances, contiguities, and causes), and these operations in turn give birth to other ideas (principally ideas of the operations of the understanding, like thinking, doubting, conjecturing, and the like). In this sense, early modern empiricism shows a strong tendency to conceive thought as the mind's observing and dealing with unit–ideas. That any given idea is a unit does not preclude breaking it down into parts, and that in fact facilitates treating it as something that can be combined with others—and that in turn facilitates conceiving imagination as a process of assembling things (possibly unprecedented things) out of components. Perhaps that made it inevitable that empiricism would settle upon one of the oldest ways of portraying imagination as essentially reproductive and unoriginal, even when it hits upon new assemblages of idea–units. The mind becomes thereby a mix of passive and active picturing, of imagining, though typically *not* under that name: receiving the ideas through the senses is their passive source, comparing the new to the old is active but usually mechanical, and recalling, dividing, and reassembling the ideas produces hybrids.

There is little room in this conception for intrinsically biplanar fields of imagining, however—for example, Descartes's habit from very early on of taking what he had experienced by sense and reconceiving it in the form of a geometric figure, used as a simplified way of viewing the original. The greatest and most influential rationalists of the seventeenth century all retained the sense of biplanarity in thinking (shifting backgrounds and foregrounds) that had been characteristic of Plato and Aristotle.[14] One of the most basic features of the founding conceptual topology of psychological experience, it lent the theories of mind the rationalist thinkers developed a potential for internal dynamism and for a dimension of depth. Those theories allowed for the shifting focus of attention from one level of experience to another and back.

The account I have given so far in this section is more an indication of some tendencies of the early modern philosophy conceptual topographies than a history of the period. As Locke, Berkeley, and Hume's examples show, the quasimechanism of Hobbes's portrayal of thinking as nothing more than sequences of picture–images and name–images was not intrinsic to empiricism (in particular its insistence that the name "God" has no reference), but it was certainly one of the topological possibilities. No other major rationalist thinker accepted Malebranche's notion that at the point where human beings ascend to knowledge in a full sense they are *seeing the ideas in God*, but it is an intrinsic possibility of the biplanar/multiplanar character of rationalism: one developed level of psychological apprehension projected upon another, higher level.

---

[14]To the usual trio of Descartes, Spinoza, and Leibniz one should add Pascal and Malebranche, and one might even allow Berkeley to jump the empiricist ship he is ordinarily made to travel on.

As a reminder, and to bring a little more definiteness to the vocabulary of topographies and topologies, one might say something like the following. Topologies are the most general descriptions and articulations of fields of concern. They present a field's basic objects (which do not have to be substances or qualities but could be, for instance, impulses or vector forces or anything else that can be imagined), the objects' characteristics, their configurations, the places they occupy, and their differential tendencies in those places. Biplanarity in psychology presumably commences with sensory experience, but biplanarity as a topological feature does not by itself determine the constitution and/or independence of the field of sensation (it could, for example, have sublevels drawn up into an overarching level, as is the case in the relationship between proper sensation and common sensation in Aristotle or modular theories of mind). Nor does it determine *a priori* what the higher or lower levels upon which projection takes place will be like. Most versions of empiricism retain at least a vestigial level (viz., understanding) roughly correspondent to Plato's *dianoia*, but it is far less subject to attempts at thorough description and characterization. In Hobbes and Locke understanding seems to be the place of names and nominal essences; in Hume, it virtually disappears insofar as everything is absorbed into the laws of association of image–ideas. According to the Appendix of the *Treatise of Human Nature*, even the ego or self is nothing more than an association—albeit the association of *all* the appearances that the self appears to "have." This is, as it were, Hume's attempt to set the parameter for that *to which* psychic events appear—a fundamental topological feature of almost any psychological theory—to zero.[15]

To continue using the concept of parameter[16]: At a certain level of consideration, the basic topology of Plato and Aristotle does not determine whether the primary source or stimulus of knowledge is sensation or common sensation or memory or imagination or discursive intellect (*dianoia*) or noetic intellect (*noûs*). To choose one or another of these is to move from a family of possibilities to a determinate one. When the determinations are sufficient to constitute a defensible theory—which means at the least that it must take some credible position with respect to matters that are considered to be part of the theory's field—it can be called a topography. Since most of those thinkers widely regarded to be in the first or second rank of the history of philosophy would likely have theories of such kind, we can call them

[15]Even though setting the parameter to zero gets rid of the I–thing or I–substance, it does not annihilate the *phenomenon*, the *appearance* of the I. What is required in the first instance, then, is a careful redescription of everything that was formerly regarded as the actions and passions of the self. From the perspective of a topology, if a post-Humean generation stops referring to the self or ego, the question becomes whether the topology has truly been altered (or even abandoned) or whether the theme is still implicitly there but ignored. Many problems are temporarily "solved" by pushing them into a corner and then refusing to look there.

[16]I am applying a relatively strict figurative use of the mathematical concept "parameter." It is a value in a solution that must have a definite value but that can be set *ad libitem*. Indicating the parameter as such means that there will be a family of possible solutions rather than a unique solution.

topographers. Note that this does not simply establish a terminological equivalence between "theory" and "topography." Insofar as historians of philosophy, whether casual or serious, are concerned, their focus tends to be theories rather than topographies. A topography is a topology that has been parametrized by specifying and phenomenalizing basic elements and features of the topology. The topography is further articulated by explicit concepts and theoretical assertions, and as such becomes a well-defined field of concern.[17]

In the senses of these terms as I have just described them, history of philosophy is inadequate when it is considered primarily or solely as history of theories. And the more theories are conceived as propositional (without attention to the *proposita*)[18] and the propositions are conceived in a merely logical sense, the less one understands of the theory—because what makes for understanding is the thoroughgoing attempt to deploy and place the concepts and propositions of the theory in the field of play. Propositionalism would be, to use a contemporary philosophical analogy, the equivalent of an Alpha Centauran coming upon the remnants of earth civilization a million years hence and trying to understand football and the society that spawned it from three randomly selected chapters of the rulebook of NCAA football.

Indeed, those who argue that (almost) everything is text often do not draw the proper consequences from the claim.[19] A text is a thing woven, and since weaving is for the most part orderly it is possible to follow the order. But the text is woven on and around a pattern or template that includes what it is supposed to be about, its objects. Even if there is no theoretical limit to how far weaving can be taken, real weaving (even if the real thing woven is a text) fills out the limits of its frame and the weaving template—its field; and if the first woven thing is woven over in a larger frame, there can still be traces of the first frame, and literally myriad crossings of the threads will still be present. Those crossings establish a matrix, and the things of the world fit somewhere—sometimes uneasily—in the weave. The weave of perception might be woven around the weave of natural things, the weave of concepts around the perceptions, the weave of words around the conceiving of the mind, etc. What this gives rise to, obvious to anyone who has attended to real weaving, is complex texture and specific patterning. Appealing, in a more strictly regulated use of the notion "text," to the density and amplitude that real weaving develops in its texturing, pattern, and extent, we can conclude that the elaboration of a phenomenal field through basic concepts that articulate a basic experience leads to a field that *is* context and *has* texture.

[17] A topography is then roughly equivalent to what Thomas Kuhn called a paradigm.

[18] The parenthesis is a reminder of the origin of "proposition": a *propositio* is in the first instance what has been placed before (*pro–posita*) the mind. "Proposition" is a reification of this phenomenon that, when taken absolutely (according to *praescissio*), tends to falsify it.

[19] I must beg, in the paragraph following, more than a little indulgence for the postmodern idiom.

7.2 How Imagination Got Misplaced, Part 2: The Psychological Reification of the Idea

The intrication through weaving of text and texture words is not just a play on words. It is meant as a challenge to our basic ways in the early twenty-first century of conceiving what human (and for that matter also animal) psychology is and does. The philosophical impulse almost invariably is to simplify and to isolate, to purify concepts, to educe propositions and principles. But nothing is known or explained until they are put in their place, that is, until they are put to work in their appropriate fields and in the ways of human experiencing. An isolated notion of perception makes sense methodologically, and against the background of an already fairly well-developed theory of psychology it may make sense factually as well. The notion might well be as illuminating and accurate as one likes when applied to the case of the very simplest animals. But in animals with memory and imagination it must be wondered whether any perception is not interwoven (texturally intricated) with them, perhaps in a manner that in the last analysis is not simply a weaving but a fusing.[20] Certainly one can legitimately wonder—and at the end of this chapter we will wonder—whether imagination can be disentangled from conceptuality.

It may seem ironic to mention Descartes here, since he seems to be one of the greatest historical offenders against the intrication I have just invoked. Beginning in the next paragraph I will appeal to him substantively with respect to how he conceived the *process* of thinking. But it is also possible to appeal to him methodologically. Of the four rules of method in part 2 of the *Discourse*, I call the fourth the neglected rule. The rules are: accept nothing as true that is not clearly and distinctly perceived; divide a problem into as many parts as needed or possible; from the analyzed parts look for an order that will allow one to put (and weave?) parts together with the aim of building back to wholes, in particular a whole solution generated from the original wholes of the problem; and, last, to make a thorough review to make sure that nothing has been left out. The fourth injunction is often interpreted as simply meaning "double-check your answer," for example (to use an algebraic example) by actually plugging into the original equations what your solution process says is the answer, doing the calculations, and showing that it works. But projected against the background of the *Regulae*, what the fourth rule means is that we cannot say simply that we have worked out an answer, so we are done. We have to work our way back through the whole treatment of the problem to see that we have not oversimplified, that we did not leave out dimensions and facts that in the last analysis matter, that we did not leave out a level of thinking or being crucial to the problem solution. This, I take it, is also one of the points that he is making in rule 8, when he points out that we cannot solve the problem of the refraction of light just by deriving a mathematical curve for notional lenses. We cannot pronounce the problem solved until we actually take into account the physical realm, the field of material interaction of things that could give rise to such a mathematical equation.

[20]This is a theme that appears in Saussure. See Sect. 8.7, below.

All praise to the man or woman who comes up with the sine law of refraction, but it is not an answer until we know how and why it works physically. A mathematical model is never adequate by itself, except in pure mathematics. It is perhaps no accident that, after discussing these things in rule 8, Descartes says that the most difficult and honorable example of the problem is provided by the human psychological faculties. Given what we have already researched in this book, what that means is that unless we understand human psychology as a whole, we don't understand the parts, no matter the accuracy of our observations, the precision of our experiments, the articulation of our equations, or concepts, or propositions, or logic. When we apply knowledge in particular we must also apply it with completeness. An equation describing a curve is not really grasped or understood unless it is seen actually—I am tempted here to use the medieval expression "in act" or to appeal to the corresponding Aristotelian notion of "being at work"—as the equation of that curve and no other. Whether or not Descartes always or in the long run thought exactly this is beside the point. It is a consequence of the topography of his philosophizing, and it is an intrinsic possibility implied by occupying and traversing the topology of the field of human psychology.

If one looks to the usage of the concept "idea" after Descartes, one notices an oddity: it is regularly used with the terms clear, distinct, obscure, confused. Descartes himself does this rarely, although in rather conspicuous spots—a conspicuousness that undoubtedly has something to do with the initial implausibility of claiming that Descartes's usage is different. But both before and after Descartes's use of the term "clear and distinct idea" in Meditations 3 and 6—that is, in his published and unpublished writings before and after 1641, including the *Discourse*, the *Principles*, the *Passions*, and the rest of the *Meditations* itself—Descartes used these adjectives of psychological *acts*, or adverbially with the verbs of those acts. One sees, perceives, understands, etc., clearly and distinctly; one has a clear and distinct intuition, perception, understanding, and so forth. The first-approximation way of expressing this is that "clear and distinct idea" is a reification of a clear and distinct act of perceiving,[21] whether it is sensory or imaginative or intellectual. One finds evidence of this reification in Arnauld and Nicole's *Port–Royal Logic*, in Spinoza, in Leibniz, and even more thoroughgoingly in the lesser rationalist followers of the way of ideas. Thus it is not just empiricists who treated ideas as fixed, basic units of thought with their own qualities and features; the tendency was also at work already in post-Descartes rationalism, and at certain moments in Descartes himself.

Before the *Discourse* of 1637, Descartes almost always used *idea* in the sense of corporeal idea, phantasm, or image. The earliest specifically datable occurrence of *idée* (French) or *idea* (Latin) in the large and comprehensive sense that includes

[21] I have used the term *reification* previously in this book. In this note I want to suggest, but without argument, that reification is a basic topological function that arises from the marking of the field of appearance. Wherever one sets down a landmark to highlight a feature of the field, among those who are expert in the field there is a tendency gradually to appeal to the landmark (the word) without more than a gesture toward the feature and its context. The beginning of reification is a familiarity that breeds contempt.

both what is intelligible and what is sensible occurs in correspondence exactly contemporary with the preparation of the *Discourse* for publication (in a letter of March 1637, probably to Silhon). Part 4 of the *Discourse* also uses *idée* in this large sense, though where Descartes presents his physical theory (in part 5 of the *Discourse* and the accompanying scientific essay *Dioptrics*) he still uses it as equivalent to "phantasm/corporeal idea." The Latin *idea* first occurs in the large sense in correspondence of 1640, roughly contemporaneous with his completion of the *Meditations*.[22]

Descartes actually provided a formal definition of "idea" in the second set of responses to objections to the *Meditations*. "By the name *idea* I understand that form of any thought whatever, through the immediate perception of which I am conscious of this same thought."[23] In the first instance this cannot mean a Lockean idea: for example, not a well-defined complex of the simple idea of blue with the simple idea of square. If the principle attribute of the thinking thing is a kind of thought–substratum, which would be like matter with respect to form, the idea is the momentary formation of that substratum. I use "formation" not in the sense of containing this or that form but as the *total* formation of that thought–substratum. Consciousness *as a whole* is shaped, and the form proper is the *whole shaping form* of the moment corresponding to the *whole conscious appearance*, however many levels of consciousness (sensory, imaginative, memorative, intellectual, volitional, etc.) occur. In the first instance we cannot know whether there is any definitive analysis or decomposition of this total thought formation into elements, and whether any analysis we arrive at makes real distinctions or merely distinctions of reason. When I am angrily plotting revenge against someone for real and imagined wrongs, or when I am inquiringly wondering whether it is possible to produce a circle having twice the area of a given square, the idea is not a static, simply picturable event, thing, or shape, devoid of any affect or emotion: it is the entirety of the momentary formation of the experiencing ego, including the anger or the wonder, the hope for success, the imagined scenarios of the personal offenses given or previous ways of dividing the geometric figure, etc. There is no apriori way of fully specifying this formation for another, nor even of duplicating it again in the future for oneself.

That last point opens the way to despair: if the idea is the formal structuring of the totality of consciousness, and that totality is not necessarily repeatable for anyone, not even myself, then we end with a paraphrase of a famous Heraclitean saying: You can't dip your thought into the same idea twice. And knowledge would not be possible. This despair is premature, however. For if it is possible that we can and do have ideas that are too complicated for us ever to have again, that does not

[22]In my search, I have used the CD-ROM version of *Œuvres Complètes de René Descartes*, produced by the Connaught Descartes Project (André Gombay assisted by Calvin Normore, Randal Keen and Rod Watkins) and distributed by the InteLex Corporation in their Past Masters series.

[23]"*Ideae* nomine intelligo cujuslibet cogitationis formam illam, per cujus immediatam perceptionem ipsius ejusdem cogitationis conscius sum" (AT VII.160); "par le nom d'idée, j'entends cette forme de chacune de nos pensées, par la perception immédiate de laquelle nous avons connaissance de ces mêmes pensées" (AT IX.124).

mean that all ideas are that way. Indeed, from the *Regulae* through the *Meditations*, Descartes was teaching nothing if not how to direct the mind so that it could start with unwieldy complex thoughts and gradually clarify and simplify them, to the point of achieving things simple enough to be knowable and repeatable.

Let us attend more particularly to what Descartes says about "idea" in the larger sense in two letters, one from just before the appearance of the *Discourse*, the other from the period of the *Meditations*. In the March 1637 letter to Silhon I mentioned three paragraphs ago, Descartes asks whether he had done enough to prove that the existence of God and the human soul are the most evident and certain of truths and that the soul has nothing corporeal about it.

> One even, by stopping for a sufficiently long time on this meditation, acquires little by little a knowledge that is very clear, and I dare say intuitive, of intellectual nature in general, the idea of which, being considered without limitation, is that which represents God to us, and limited, is that of an angel or of a human soul. (AT I.353–354)

To Mersenne, discussing in July 1641 a Hobbesian-style objection to the very idea of God, he says the following:

> I do not call by the name idea simply the images that are depicted in the [organ of] phantasy; on the contrary, I do not call them by this name insofar as they are in the corporeal phantasy; but I call generally by the name idea everything that is in our spirit when we conceive a thing, in whatever manner we conceive it. (AT III.392–393)[24]

Neither passage allows us to conceive an idea as something we passively have or possess. In both passages we start with something very general: a generic *thing* in the 1641 letter, an *intellectual nature in general* in the other. In each case something more must happen; the thing or intellectual nature as such is only indeterminate, not yet a fully developed and clarified idea. In 1641 we must *conceive the (generic) thing* in some manner or other—and *all* that is in our spirit *as we conceive* the thing *in some manner or other* is exactly what the idea is. In 1637 we start with an intellectual nature in general; that turn of phrase suggests that the general must become more particular. How? A developed idea is a specific form in which the intellectual nature in general offers itself to our consideration in a determinate way. We must take note of the initial idea and contemplate or consider it now this way, now that. In one instance we do this in the unlimited mode and have as a result the idea of God; in the other instance we consider it in a limited mode with the result being the idea of an angel, on the one hand, or the idea of a human soul, on the other. Even the unlimited mode is not presented as the first thing that comes to mind when we

[24]The first passage in French: "Même en s'arrêtant assez longtemps sur cette méditation, on acquiert peu à peu une connaissance très claire, et si j'ose ainsi parler intuitive, de la nature intellectuelle en général, l'idée de laquelle, étant considérée sans limitation, est celle qui nous représente Dieu, et limitée, est celle d'un ange ou d'une âme humaine." The second: "[J]e n'appelle pas simplement du nom d'idée les images qui sont dépeintes en la fantaisie; au contraire, je ne les appelle point de ce nom, en tant qu'elles sont dans la fantaisie corporelle; mais j'appelle généralement du nom d'idée tout ce qui est dans notre esprit, lorsque nous concevons une chose, de quelque manière que nous la concevions."

follow the instruction "insert intellectual nature for consideration"; we have to work our way to contemplating this in the proper way, a task set first of all by the *context* and our *concerns*. As we consider and reconsider we gradually modulate and modify the idea, so that it changes; the *tout ce qui est dan notre esprit* (everything that is in our spirit) progressively *develops* as we methodically regard the nature in question. We focus here and there, we vary what we are considering, we shift our concerns, we change the context, we intricate and disintricate the elements of our thought, we simplify until we see clearly and distinctly (or not, as the case may be). Some of these thoughts would be, strictly speaking, unrepeatable, but many of them we can manage to produce again or communicate to others. What the two passages mean, then, is that Descartes conceives ideas as total mind–forms that have an intrinsic dynamism of possible development.[25] The task of thinking is to develop and refine them actually so that they finally become repeatable forms of knowledge.

Clear and distinct (or perspicuous) seeing, therefore, would occur not of an unmediated thing or nature but only as the result of a way of inquiry about it. If the idea one has is the current total form of one's thinking substance—the total formation of the *res cogitans*, which is shown to be a *res* by the fact that it can have a limitless number of *cogitationes* that invest it—then it is evident that only in rare and special circumstances can an idea appear simply, clearly, and distinctly, much less in isolation.[26]

Now we can better understand how an idea might be correctly designated as clear and distinct. It is most proper, in the first instance, to use the various terms regarding clarity, distinctness, perspicuity, and the like as adverbs and adjectives with, respectively, verbs and nouns of perception: I clearly and distinctly *see* or *perceive*, I have a clear and distinct *perception*. A clear and distinct perception is an accomplishment. There are few objects or natures, ideal or real, that, when put before the mind's eye, would immediately reveal themselves as clear and distinct. As soon as we summon an object to mind we certainly do have an idea, but it is not the idea of the object and the object alone; rather, it is the form of the whole awareness that is in our spirit at that moment. Only if the thought has undergone sufficient analysis, thinking, and rethinking can it reach the point of being clearly and distinctly perceivable. The "intellectual nature in general" that Descartes mentions in the 1637 letter to Silhon is not yet clear and distinct when it first enters the mind; it takes on distinctness (and greater clarity) as one works at considering the idea in as many germane variations as we can manage; especially relevant in this specific case are the infinite mode and two finite modes. This produces distinctions that may allow the idea of intellectual nature to be perceived both clearly and distinctly—though whether it is perceived in that way depends on the total formation and preparation of the mind at that moment.

[25]Thus we also have an answer to the often-posed question of why Descartes called his work of 1641 *Meditations*: meditation is hard thought–work that starts very far from anything determinate but slowly and progressively develops thought to the point where we can perceive with knowledge. To overlook this is to understand Descartes as someone other than who he was.

[26]The idea as it is used in the empiricist tradition *easily* lends itself to isolation in consciousness—altogether too easily.

An inappropriate mood, a failure to clarify and distinguish certain aspects of an intellectual nature, now or previously, or a premature approach to the nature by way of an inappropriate question or background concern might all block proper clarity and distinctness. But when you finally manage to see it clearly and distinctly you *easily* notice it as such. Because of the work of simplification, clarification, and distinction you have done, the total idea now shines out in its significance and differentiates itself from the previous, unsuccessful versions that preceded.

One might therefore go one step further in the elucidation of the clear and the distinct: it is not ideas that are intrinsically clear and distinct, nor is it simply perceiving that is, but rather the clarified and distinguished seeing comes about because the forms of consciousness (those are by definition ideas) have been made to stand out and shine in their situated significance. The situation is important; for example, in Meditation 2 it is doubting that allows the self to stand out. It is not self as a mere concept, however, but as actively thinking being. That actively thinking self has been there all along, but it does not shine out until one clears away all the sensations, the memories, the images, and the busyness of ordinary thinking. At the end of the process one has "intellectual nature" in a simple but present and active form, and it is made to stand out in an indefeasible way by activating it in the precise format of the activity of conscious doubting, one that tries to doubt its consciousness of itself right out of existence—and fails, and fails perspicuously.

If there is a basic coherence in Descartes's progression from the *Regulae* to the *Meditations*, the position he expresses about ideas in the letters to Mersenne and Silhon we have looked at helps us conceive it more clearly. When he says that imagination is not a corporeal depiction on the gland (not a brain state, in modern parlance), he is in effect taking issue with the long-held notion that imagination was in essence the forming and having an image in an organ or in a brain ventricle. The corporeal forming and having may well be indispensable for imagining, but they fall short of it: the imagining is in being conscious of the formation and the configuration, and only the knowing power can do that. Moreover—and here one needs to apply the lessons of the *Meditations* speculatively to the methods of the *Discourse*, *The World*, and the *Regulae*—imagining is much less having a single impression in the corporeal organ that the knowing power steadily views than it is the nimble activity of the knowing power as it moves from one image to another. The movement of the knowing power transforms a present image, step-by-step or even continuously, into new images. The knowing power sees one image as representative of another or in terms of another, it sees it as a known or an unknown, it sees the very same image now as a line in the sand, now as a geometric line segment, now as the answer to the question of what the length of the side of a square must be in order for it to have double the area of a given square. Looking at a right-angled triangle is not to recognize the Pythagorean theorem, though one cannot prove the theorem without the triangle and relating the steps of the proof to the figure. An image of a dog imprinted on a part of the brain might naturally activate brain processes that cause memory locations where images of other dogs are preserved, or the image of a dog might by the same type of processes become joined to a memory–appearance of the dog's owner. But only *intellect* can see, and *will* affirm, that this appearance is of a dog, and that the relationship of ownership holds between that dog (not its appearance) and that man (not his appearance). Thought of the self may lead to the imprint in the brain organ of one's physical appearance, but no gaping at the physical appearance amounts to the self unless intellect recognizes itself as intimately connected with that appearing body. Like the proverbial shark, the knowing power keeps knowing only by staying in motion.

To put these reflections into a formula: the conceptual topography of Descartes insists that, if having an image in consciousness is marked by the term "imagination," then imagination is of limited value. But as soon as we talk of the various ways that a consciously perceived image (regardless of its corporeal substratum) can be taken, manipulated, and recontextualized, then we are talking not about imagination but about imagining. Imagining is the activity of the intellect or knowing power with respect to the dynamic field variability and field projection possibilities of images. The active power is clearly and distinctly different from the organic forming and holding of an impression, and it is even clearly and distinctly different (in its specific manner) from the conscious registration of corporeal appearances. The activity that knows it is changing and directing the corporeal appearances could not know itself as clearly and distinctly differentiable from those appearances if it did not have this directive experience of imagining. Imagination is what the knowing power does with images and the imageable.

7.3 The Rationalist Loss of Confidence in Imagining, and the Rise of Aesthetics

For religious reasons as much as for philosophical ones, Pascal and Malebranche were suspicious of imagination's indifference and even opposition to truth, its preference for frippery and fashion, and its subornation of reason. They nevertheless affirmed that, in mathematics and the sciences, its use was essential, even constitutive. Leibniz, as I noted in Sect. 7.1, above, agreed about the mathematical and scientific uses. His general theory of knowledge was simultaneously more nuanced and less stratified, that is, divided into levels, than that of most of his contemporaries and predecessors, so he was inclined to think of the progression from sense perception to intellection as a continuum rather than a step-wise climb.[27] For example, he considered sense perception to be not simply a passive registration of what is there in front of us or a response to a stimulus. In accordance with his metaphysics of the monad, every substance (one of which is the human soul) is active and also expressive of the whole universe. Only one monad (God) is truly infinite, so that every other monad is expressive of the universe in a limited way. Space does not exist in itself, it is an appearance, though a true one—a true phenomenon, as he calls it—in which every monad (except God) shows itself as having place. Space is an unlimited appearance of the totality of monads, and the perceptive acts of each

[27] Paragraphs following about Leibniz are adapted from Sepper 2005b, 330–334, with permission of Blackwell Publishing.

monad are productively true expressions of this whole. Because all monads—now we need to focus just on the human ones—are limited, their perceptions cannot manage to produce the limitless richness that God has established. Yet in a sense every act of perception by a monad reflects the totality, so all the appearances in perception are simplified expressions of more complex truths. Although this is not perception in its conventional sense, it is a constantly productive act of consciousness, and because there is a continuity between perception and intellection that passes through imagination one might without distortion call this kind of intellectual activity *imaginative perception*.

Perhaps Leibniz's most original contribution to the theory of imagination is his theory of signs. His *universal characteristic* helps make clearer in what sense he thinks the true phenomena of space and appearances can express something about deeper reality. The universal characteristic was the symbolic writing system that he worked on throughout his life and hoped to develop to the point that it could express, beyond the limits of natural languages, the structural relations of complex truths. As such it is an ancestor of modern symbolic logic. It can also be seen as a radicalization of Descartes's method (Leibniz actually had access to Descartes's *Regulae* and his early notebooks in the 1670s). Descartes had represented problem data by figures and ultimately by line segments and plane figures, then used letter symbols to stand for lengths and areas so they could be incorporated into algebraic equations for the sake of calculation, then a retranslation back into geometrical figures.

An early brief dialogue Leibniz composed in August 1677 gives some insight into the relationship between this universal characteristic and imagination. The speaker (labeled A) representing Leibniz's position points out that although there can be thoughts without words, there can be no thoughts without signs or characters:

> if characters can be applied to reasoning, there must be some complex arrangement, some order which agrees with things, an order, if not in individual words (though that would be better), then at least in their conjunction and inflection. And a corresponding variegated order can avoid the difficulty. For though the characters are arbitrary, their use and connection have something that is not arbitrary, namely, a certain correspondence [*proportio*] between characters and things, and certain relations among different characters expressing the same things. (Leibniz 1989, 271)

Whatever the words or characters one chooses, then, as long as the characters refer to the same elements (or at least as long as one can produce an equivalent of those elements in alternative format), and as long as the structural relationships expressed are the same, it does not matter precisely what characters one uses, for the (possible) truths will remain unchanged. Leibniz's speaker A affirms that "the most useful of characters" are those that maintain "a certain similarity" to the object.[28]

---

[28]Leibniz does mention here the notion of a "blind symbol," one that has no resemblance to the original and in that sense is not an image. Its value is not as a representation but as a token in calculation. Anything more than its bare serviceability as a placeholder would be a hindrance to the efficiency of the calculation. Its emptiness is not a virtue per se, and it serves for knowledge only when the calculation is done and the symbols are retranslated into what they stand for. It should almost go without saying that Leibniz is treating both the characteristic and the objects to which it applies as subsisting in structured fields of relationships (that the characteristic can therefore express).

About figures in geometry he says that they "must be regarded as characters, for a circle drawn on paper is not a true circle, nor is it necessary that it be, but it is sufficient that it be taken by us for a circle" (Leibniz 1989, 271).

Leibniz understood the universal characteristic as an extension of what we talked about in connection with Descartes: an extended imagination, one that uses images to stand for structures and relations as much as for objects, and that then marks those images with symbols that allow a clear, distinct, and manipulable representation of those relations and the elements they relate. For Aristotle's "no thinking without phantasms" Leibniz substitutes "no thinking without signs." The signs most conducive to thinking clearly and distinctly are those that image the fixed properties of relations. Thus insofar as we use an imagined circle to think about the properties of circle we are no longer simply using a weakened sensation, but instead a *schematized appearance* that preserves, in its resemblances, relations that hold in the "true circle." This schematized appearance is not a rational concept, either; it is, to be exact, an image. Perhaps not until Charles Sanders Peirce and Ferdinand Saussure, that is, in the late nineteenth and early twentieth centuries, did anyone press further the signitive use, the sign use, of imagination. Logic itself, insofar as it is expressive of sign relations, might turn out to be a radical extension of imagination—radical in that is goes to the very *root* of imagination. But that is a question left for later.[29]

After Leibniz, European rationalism became progressively more scholastic—that is, it developed, in the setting of university education, into a conventionalized, abstract form. A key figure in this development was Christian Wolff (1679–1754), professor of mathematics and philosophy at Halle and Marburg in Germany.[30] Wolff wrote many treatises and compendia, in German and in Latin, very few of which have been translated into English. For the purposes of understanding imagination, the most important works are the *German Metaphysics* (1720)[31] and, in Latin, *Empirical Psychology* (*Psychologia empirica*, 1732) and *Rational Psychology* (*Psychologia rationalis*, 1734). All are very tightly organized. They begin with

[29] I omit discussing Spinoza's conception of imagination. Although at first glance it is easy to see him as a critic of imagination, it is only of the misuse of imagination that he is critical. Imagination is included in the first of the three kinds of knowledge he defines in the *Ethics*. Because he thinks there are in human beings two infinite attributes isomorphic to one another, thought and extension, the appearance of the extended in thought is not illusory but a field of sensory appearances (images) in the field of thinking. Just as much as Descartes—not least because on this question he was a careful reader of Descartes's works—he believes that imaginative thinking is *necessary* for adequately understanding extension. In addition, Spinoza radicalizes Descartes's initial attempts (in the *Passions of the Soul*) to include the affects of passions, feeling, and emotions in the same conceptual topology as imagination. See Sepper 2005b, 323–329.

[30] Paragraphs following about Wolff are adapted from Sepper 2005b, 334–340, with permission of Blackwell Publishing.

[31] "*German Metaphysics*" ("*Deutsche Metaphysik*" in German) is the conventional designation for the 1720 work, the actual title of which translates as "Rational thoughts about God, the world, and the human soul, also all things whatsoever." It is the most compact presentation of Wolff's metaphysical thinking. Beginning with the 1727 edition it was expanded by adding as a second part the explanatory *Remarks* (*Anmerckungen*) on the *German Metaphysics* that he had published separately in 1724.

basic concepts and principles, and then carefully introduce new phenomena and concepts by closely linking them to what has previously been discussed, where possible by logical deduction. In general, Wolff's approach brings a rationalist's aims to an empiricist's awareness of the richness of sensory, memorative, and imaginative experience.

Wolff's account of images and imagination is conventional but also rigorously set into the context of sensation and judged according to clarity and distinctness. Images are obscurer than sensations; when I imagine a person I have seen before I can present the figure, size, and position very nicely, but "the colors remain almost entirely absent, and everything becomes nearly black" (§237).[32] When they appear without the simultaneous operation of the senses, for example as in dreams, these images are much clearer, though never as clear as in sensation itself. Images take their origin from the senses by association and similarity. Whenever our senses present us with what "has something in common with a sensation we had at another time," the latter comes before our minds. When part of the present sensation shares something in common with just a part of a past sensation, "the entire past one comes forward again." The same holds true of images we have had or produced in the past. They are constantly shifting by virtue of this association and reassociation based on resemblance and part-sharing in sensations and images. In dreams these changes occur by leaps because the successive images are not well grounded in one another. But the imagination involves not just things that we have already thought: "we can also present to ourselves what we have never sensed before. We experience this in geometry, when we present to ourselves the drawing of a curved line of a kind we have never before seen, also when following this we draw the same line on paper and thereby bring it to sensation for the first time" (§241).

This marks a decisive and unanticipated turn in Wolff's presentation. In the following paragraphs he explicates this *productive and inventive* power of imagination. His chief examples are from the arts, which are hardly so much as mentioned by earlier rationalist philosophers. What for Wolff justifies including the arts is the analogy to the imaginative function that was crucial for Descartes, Leibniz, and even Malebranche: its constitution of mathematics. The first "manner" of this imaginative production of what has never been experienced before depends on our ability to divide what we have already experienced into parts, then recompose the parts as we please, to arrive at a composite being like a mermaid, a winged angel, or the gods of the pagans. Wolff calls this "the power of feigning" (*die Kraft zu erdichten*; in the Latin works this becomes *facultas fingendi*). Through it we often produce something that is not literally possible, what he calls an "empty image." This first use of the power is often not under our control, because, as we engage in feigning, the free association of images can carry us far beyond our original intention. This explains "the images of painters, sculptors, and other artists who bring wild adventures to the market" and other bizarre creations. But there is a second "manner" of producing things never before seen. In it the imagination "employs the principle

[32]References are to the *German Metaphysics*, by section number.

of sufficient reason, and brings forth images in which there is truth." Wolff provides three examples: "the image under which a sculptor presents to himself a statue and into which he has brought everything beautiful he has seen in the human species," after engaging in much research and effort; the aforementioned geometrical curve never before seen or drawn; and finally the image of a building "which an architect presents to himself in thoughts according to the rules of the art of building" (§§242–245).

Wolff goes into detail, here and even more in the *Empirical Psychology*, about how architects engage in this second manner of imaginative production. First, they gather pictures and plans of already existing buildings as well as plans for others that have not been constructed. Then, in designing a new building, they examine these pictures one after another, in the same way that "the imagination brings forth, one after another, things that have a relationship to the thing we are thinking about. What pleases them they bring together afterward in a new design plan" (§246). The role of the principle of sufficient reason—a crucial notion in Leibniz, which states that everything that is or is possible must have reasons or causes sufficient to produce it precisely as it is—is to bring to bear not just the rules that all architects follow but also to give the individual building an "appropriate ground of perfection." When, on the other hand, the architect instead follows the first manner, the manner of feigning with its free association of images, he ends up introducing all sorts of errors and imperfections into the design.

Wolff's discussion here bears some resemblance to theories of artistic production of the late seventeenth and early eighteenth centuries that had been influenced by rationalism. In them imagination was involved as serving the subordinate function of adorning or decorating the portrayed object in a manner that gives it some distinctiveness and specificity, according to the taste, skill, and experience of the artist. This of course connects the rationalist period to earlier ones, going all the way back to Greek antiquity, when the purpose of art was above all to present good models according to good proportion.[33]

Wolff, like Descartes, recognizes that imagination allows itself to be exercised and increased (§§262–263). Like Descartes and Leibniz, Wolff (himself a professional mathematician) conceives mathematics as an imaginative function. The distinct use of imagination requires attention, the mind's power that allows for reflection by directing our attention now to one aspect or part of what we imagine, now to another. This enables us to distinguish things more carefully; and noting the similarities and differences of things allows us to achieve the presentations of species and genera, "which one is accustomed to call actual concepts and that are the ground of universal knowledge" (§273). The progressive refinement and combination of these ultimately leads to scientific knowing. Immediately after explaining how attention to sensations and images produces concepts Wolff introduces the faculty of understanding: the power of distinctly presenting what is possible. He points out that, by themselves, the senses and imagination can at most achieve clear

[33] See Becq 1984 and Eco 2002.

presentations; if you bring in the understanding they can be made distinct. Understanding, strictly speaking, has its sufficient reason in the distinction and naming of the presentations of sense and imagination. This conception of understanding that is a methodically controlled work with images is reminiscent of what we find, in various forms and under various terms, in Descartes and Leibniz.

But this is the point in Wolff's account where the conventionally understood "rationalist" separation of intellect from sense and imagination begins—a sharp separation that is in fact uncharacteristic of his great rationalist predecessors. Wolff distinguishes between distinct knowledge by means of pure understanding, and indistinct knowledge with impure understanding. "Understanding is separated from the senses and the imagination whenever we have fully distinct knowledge." In discussing this conclusion he dismisses the claim of "those who pretend the pure understanding is an empty image of mathematicians," that is, who think that mathematicians in talking of such a thing are describing an empty, contentless, or vacuous thing. Such people do not understand the difference between pure and impure understanding, he says. But then, in a puzzling turnaround, he immediately concedes that the understanding is *never* entirely pure (§282).

The reasons for Wolff's indecision become progressively more apparent. Wolff's account of understanding culminates in the making of judgments. Judgments, of course, put two concepts into propositional relationship, and they require conceptual distinctions made in experience that are tracked and registered by distinct signs or words. When we attend to a glowing iron rod we have a concept of it, says Wolff. But to turn that experience into a judgment we need to distinguish the glowing from the iron rod as two separate things, and then at the same time we need to see these two things as joined together. So for a judgment we really need three concepts: the thing or substance, its property, and their being united (§§289–290). Precisely here Wolff introduces words as a type of sign that tracks such distinctions: "words are nothing but signs of thoughts" (§291). Thus understanding is the power of presenting in words what is possible. Properly speaking words do not stand for the individual thoughts we have but rather for their kinds, their species and genera. As Wolff examines the different grammatical categories of words he notes in particular that because the imagination presents things clearly but not distinctly it leads to the tendency of human beings to substantialize—we might say "reify"—the properties of things. Thus, for example, the virtues that are proper to the soul are, in allegorical drama, represented as individual characters.

The discussion of words leads to a further conclusion relevant to imagination. Either we present to ourselves things themselves, or we present them through words or other signs. Signs allow for the exact tracking of things in "figurative knowledge"; it is opposed to "intuiting knowledge," which presents the thing itself or its image (§316). The words and signs are class names based on our having already compared and contrasted the various relevant sensations and images; they are the marks of what we have made distinct in our imaginative experience. Thus the figurative kind of knowing, knowing in propositions of natural language or other systems of signs, is distinct; by contrast, intuitive knowing can be no better than clear. There is, nevertheless, the danger in figurative knowing of its being contentless or empty.

Words really indicate nothing definite or distinct without our remembering that they "indicate a certain thing of which we have had a concept, that is, in remembrance of the intuiting knowledge" (§323)—so figurative knowledge based on language does not have by itself any certainty or clarity. Oddly enough, Wolff immediately qualifies this by arguing that it is conceivable that there is a better kind of figurative understanding:

> It is possible that clarity and distinctness can also be brought into figurative knowledge, and that it even can, as it were, place before the eyes what is to be met with in a thing, and through which one distinguishes it from others, in a manner that if, following this, composite signs that are indifferent to the concepts are held up against one another, one can also see from it the relation of the things to one another. (§324)

The example Wolff gives of this kind of presentation in signs, one that can bypass the concepts and directly mirror the relations of the thing, is modern algebra. He does immediately acknowledge a problem: there has been so little success in achieving this algebra up until now, and so few people understand what has been done, that the technique of perfect figurative knowledge has hardly been invented yet. He mentions specifically that Leibniz's universal characteristic intended, but did not achieve, this kind of figurative knowledge (the concept of which seems to evoke Leibniz's "blind symbols"). Thus Wolff once again makes a strong assertion of the independence of real knowledge from sense and imagination but then immediately qualifies and even contradicts it. It is possible, it exists, but no one has achieved it.

Progressing toward the ultimate faculty of reason, Wolff advises us to use common experience and the refined experience of deliberate experiments to note and name all the differences in our sensations and in the corresponding changes in our soul, and to "name them with their right names, so that we do not mix imaginings and preformed opinions with experience" (§§325–326). After showing that experience is expressed in judgment–propositions, and that higher knowledge is a knowledge of these propositions according to the canons of logic, he presents the culminating human faculty: reason, *Vernunft*. Reason allows us to rise to a knowledge that does not even need to take its propositions from experience—though once again we read that most knowledge, including the fundamental knowledge of nature, cannot escape from resorting to experiential propositions. Yet about mathematics he asserts an important difference. "In arithmetic and geometry, and similarly in algebra, we have samples of refined reason [or *purified reason, lautere Vernunft*]: for here all the conclusions proceed from distinct concepts and reasons that are separated from the senses" (§382). Presumably, then, they are also separated from imagination. Wolff does not qualify this conclusion, even though it rests uneasily alongside what he has said earlier.

Wolff's method of rational derivation of higher level concepts tends to reify them. They are treated as objects or things rather than signs. Signs refer to other phenomena; images are like signs in that they always have a reference, to whatever they are images of. The more that concepts are absolutized as existing per se, the easier it is to conceive the life of the mind as taking place in complete abstraction from sense, imagination, memory, and even the signitive function of mind. As always, the history of the psychological powers turns out to be enmeshed with the ontology of their direct and indirect objects. What is surprising for our more immediate purposes is that Wolff's attempt to put images and imagination into their place is so halting and so conspicuously inconclusive. But that did not prevent the next generation from taking the matter as settled. Students obviously do not always learn from their masters what they should.

7.4 Aesthetics, Genius, and the Ordinary Mystery of Sensation: The Need for an Expansive Concept of Imagination

By the first third of the eighteenth century the status of imagination was undergoing a shift. Under the inspiration of the scholasticized rationalism that was settling into the universities with Wolff and his followers there was an impulse to make even the study of sensibility, both external and internal, into more of a science. This provided an accommodation between the different approaches in Britain and on the Continent: even if people disagreed about what the status of reason and its insights was, they could agree that the ideas or images assimilated through sensation had to be reviewed and organized by reason or understanding. Thinkers who agreed with Hume's suspicion that the ego is nothing more than the entire series of impressions and ideas could put all the more emphasis on grasping the relations that hold between images.

If the products of sense and imagination were too various to provide matter for a science in the strict sense, it appeared that they could be organized according to empirically discovered principles that would allow some understanding and a more articulated use. This underlay the emergence of early eighteenth-century aesthetics—that is, of a science of the sensible and imaginable under that name, which previously had been taken simply as the Greek word for "things pertaining to sense," "sensible things." Although Alexander Baumgarten's mid-century *Aesthetica* (1750–1758) is the best known, not least for its influence on Immanuel Kant, it was actually the culmination of nearly two generations of work along similar lines. Its presupposition was that there is a dense enough network of relations among sensibles that permits them to be organized into a relatively stable structure or appearance–field. The stability did not, however, derive primarily from things but from appearances. This aesthetics was about the qualities and relationships of sensory appearances, not about the things that underlie those appearances. By investigating the kinds and relations of the sensory qualities the science also sought to understand the typical effects those appearances have on human beings (perhaps also on animals, at least those having similar sense powers). It would then be up to artists, engineers, and other practitioners to incorporate this knowledge into their artistic and technical works and to learn how to arrange appearances in ways that would be maximally pleasing or achieve other effects or combinations of effects. In the course of their work, the practitioner would expand the understanding of the imaginative science of aesthetics.

There were also changes in the public sensibility for art that affected the conceptual topography of imagination. Before 1700 or so, both artists and nonartists could largely agree on the objects most fit for artistic portrayal, but only the artist knew the elements and materials well enough to judge how they were to be employed. Although the artist and the nonartist shared certain conventions and symbolism—purple as a sign of royalty, for example, or the accoutrements that would signal the identity of a Christian saint or a goddess of ancient mythology (in this respect an artist might need to acquire a great deal of learning that the audience of his patron–connoisseurs already possessed)—the artist's skill and the associated familiarity with artistic materials played the major role in the actual execution of the work. The plastic arts in particular had about them certain basic representative purposes, but the existence of principles of idealization and ideal objects would never allow realistic portrayal to be the only or even the primary standard of excellence. Art was about copying a rationally justifiable, not a material or immediately visible, ideal.

Once the sensible realm is understood more scientifically the position of the audience and its expectations change. Insofar as the relevant scientific knowledge is widely accessible, those familiar with it will become a much broader part of the audience, and they will insist that the artworks they see or hear conform with what they (presume to) know. Thus the audience can easily become more demanding about the proper use of sensory images and about their coherence in the overall sensory impression made by a work.[34] Besides presenting appropriate objects, the works now had also to observe a more elaborate lawfulness that was not simply left to the artist's taste. Indeed, the taste of both artist and audience had more than ever to be educated by the laws and principles of science and quasiscience. This of course does not eliminate the rights of taste, but it does begin to transform them.

With the increasing scientific knowledge about the anatomy and physiology of sense organs and nerves and breakthroughs like Newton's correlation of the different hues of color with degrees of refractibility of light by prisms, there was an increased expectation of how far scientific knowing could reach into areas where it had formerly been out of place or inconceivable. The distinction between theoretical and practical knowledge began to shift in favor of the former, and practical knowledge was left to deal with what was not yet accounted for (but probably would eventually be taken over by science) or with what was quite unaccountable (if there were things beyond the reach of the sciences). Such a conceptual shift brings with it a shift in valuation. If there are unaccountable things, they cannot be set into a ratio

[34] Here one should mention the modern theory of the primary and secondary sense qualities. In brief, what until the seventeenth century had, following Aristotle, been called common sensibles (space, time, location, motion, duration) became known as primary sense qualities, whereas the proper sensibles (color, tone, aroma, and the like) came to be called secondary sense qualities. Even more important, the primary sense qualities were understood to be real (that is, the appearances of being extended, moving, relative positions, etc., *are* as they *appear*—barring dreams, hallucinations, and the like), whereas the secondary qualities were considered to be not in the object but results of human physiology and psychology and imputed by the mind to objects. It is likely that this distinction further encouraged the notion that sensible ideas or qualities could be studied in a purely aesthetic sense.

or proportion with anything else; they are, to take the term quite literally, irrational. If nevertheless there are people who have a special gift for dealing with or mastering such things, there is no accounting for this gift. It would have to do with their personal talent or genius. But this shifts the meaning of genius. It still has to do with personal, often inborn characteristics, but it is no longer considered part of the natural variability of human talents—for example, that some artists have a genius for drawing, others for color, others for grinding pigments, etc. All of these are matters that have an expected range. The new notion of genius is about the unaccountable and thus the unexpected. The more comprehensive that genius is, the more unexpected, even amazing, it will be. The artist who has a genius not just for one or another element of painting but who produces compositions that escape any accounting begins to open a conceptual space for genius as understood by romanticism. And that means the romantic concept was already emerging under pressure from the changing conceptual topology of the psychological powers in the eighteenth century.[35]

7.5 Kant's Response to the Challenge: Transcendental Psychology

Many stories have been told about what stirred a professor of philosophy at the University of Königsberg in East Prussia, trained in mid-eighteenth-century rationalism, to awake from his dogmatic slumber and to become the originator of the critical philosophy, or transcendental idealism. However he came to his discovery, Immanuel Kant (1724–1804) questioned the possibility of metaphysical knowledge, of a science of metaphysics as it had been practiced before him. A major part of that questioning derived from his fundamental reconception of the basic human cognitive functions and their relationship to one another.

To put it rather crudely, Kant argued that all metaphysics before him claimed either positively that human beings were attuned to things external to them in a way that allowed them to know those things just as they are, or negatively and skeptically that this kind of knowledge was impossible. Invoking the Copernican Revolution in astronomical cosmology, he proposed to turn things around: rather than *assume*, or *deny*, that we have one or several powers that grasp external things as they are, we

[35]I am not trying here to restore any simpleminded apriori constructs of an autonomous intellectual history, as though it is concepts and only concepts that rule the world. The point is this: if there is a certain field or level of human experience and activity that we are accustomed to experiencing and describing using certain concepts, and if the circumstances of the field begin to change, even without any other associated changes the traditional concepts will undergo pressures that alter their application to that field. When the understanding of what the sciences can explain shifts, so does the practical and technical knowledge concerned with those things. It is always easier to see why and how after the fact, but often it is possible to anticipate the effect of the pressures while they are still in progress.

instead *argue* that the things we know are shaped to our powers of experience. To put it even more starkly: we do not see and know things as they are but rather experience things only in forms that our own powers make possible, so that what we judge the being of things to be is intimately tied to human psychology.

Every philosopher worth the name has, of course, some notion of what is, how we experience and come to know things, and what the relationship between the two is. In Plato's *Republic* the good images itself in the different levels of forms, mathematical knowables, sensible things, and images of sensible things, and the human being has powers that allow some apprehension of the things of each level and their relationship. Aristotle argues that what makes particular things what they are is essences informing matter; the senses are capable of apprehending the sensible form–qualities (phantasms) of those things, and intellect is capable of recognizing in the phantasms the intelligible forms of the corresponding things—and intelligible forms are what essences become in human intellect. For Descartes (of the *Meditations*), even if we cannot, to begin with, know that what appears to consciousness is in reality the way it appears, we can nevertheless become experienced in the appearances as such; and once we have proved the infinite existence of God, recognized that He is perfectly good and therefore no deceiver, and concluded from this at least some of our powers must show us things that are, we can reason further to the certainty of everything else that we see clearly and distinctly (for example, to the reality of spatially constituted matter and its laws of motion). Of these three, Descartes felt the need to *prove* that we do know what we know, whereas the others developed theories of the congruence of being and knowing without any particular proof that the congruence really exists. Kant's critical philosophy attempts to show why no such congruence can ever be known to exist in any fundamental sense, much less proved; and yet, he affirms, we have scientific knowledge of the world, and it is true.

The clauses of the last sentence are not contradictory. What we know scientifically is things *as they appear to us*. Furthermore, things as they appear to us (the totality of which Kant says is what we call *nature*) are not things apart from us as the Forms are for Plato or real space is for Descartes. (If they were, the question would have to be repeated: how then, do we know this new kind of thing apart from us as it really is.) A thing–as–it–appears–to–us is not a thing apart or a thing–in–itself (as Kant designated it): it is a phenomenon, an appearance. It is a showing, and our knowledge is a knowledge of those showings. And it is a showing to us—we are the witnesses, no one else—so we are already united to the thing–as–it–appears, precisely insofar as it is an appearance to us.

In summarizing some of these results in the *Critique of Pure Reason* (1781, 2nd ed. 1787), Kant remarks that his philosophy is both an empirical realism and a transcendental idealism. When we take the objects of experience as they present themselves to be the objects of scientific investigation and try to explain them according to our ordinary scientific categories we take our empirical experience as reality. But when we ask about the source of our most fundamental scientific categories and how we can know them to be true we cannot point to any "things" that present themselves in the way that the objects of experience do. We are trying to explain the fundamental scientific ideas we use in a way that has to transcend our ordinary experience.

Kant frequently uses the adjective "transcendental" of his investigation. "Transcendental" means something quite distinct from "transcendent," and a careful observance of the distinction is crucial. "Transcendent" means "beyond (possible) experience"; that is, if X is transcendent, when we try to think of X not only do we not have an experience of the object that the name X intends to designate, it is also *impossible* for us (as we are presently constituted) to experience the object. A God who is in a heaven that we can experience only in an afterlife is transcendent, as perhaps are Plato's Forms, at least in typical interpretations. Kant adds to this traditional roster the "thing–in–itself," a term that we might be tempted to use as we accumulate experiences of something in the realm of appearances (trees, for instance) and begin wondering whether they exist as we experience them, or whether our experience corresponds to anything foundationally real at all.[36] The thing–in–itself is the thing as it exists in itself, apart from all human faculties, powers, and knowing; but since we cannot experience anything at all apart from these (these apprehensive and reasoning powers are, after all, prerequisites and participants of all our experience), it names what totally transcends our experience, now and forever. The "transcendental," on the other hand, is that which lies at the limits of knowability and experienceability, given our faculties and powers.

Early in the First Critique Kant argues that a good investigator needs to get to know his tools before investigation begins; it seems strange to him that no one before had thought to investigate the competencies and limits of the most basic instruments of science, the human powers of experiencing and knowing. However unfair or even false this is as an accusation against all previous philosophers—it is clear from rule 8 of the *Regulae*, for example, that this is exactly what Descartes thought he was doing—it seems to be quite justifiable to demand some preliminary accounting of these powers, what they do, and what sorts of limits they have. Kant calls this transcendental psychology. The term "transcendental" distinguishes it from the varieties of rational psychologies and empirical psychologies that had flourished up to his day and emphasizes that this is a psychological inquiry at the very limits of the possibility of investigation. Empirical psychology engages in a careful observation of human acts, omissions, words, and behaviors, from the ridiculous to the sublime. Rational psychology attempts, from the standpoint of metaphysics, to provide the fundamental justification for the kinds of things and concepts that psychology deals with: the soul as the organizing unity of the human being, thought as its basic attribute, the division into fundamental powers like the nutritive, the sensitive, and the cognitive. Kant has no qualms about empirical

[36]We perhaps should remind ourselves that it is not just hyperbolically doubting philosophers who ask such questions. The search for the thing–in–itself can begin whenever we see that there are underlying elements and overarching principles that constitute things of appearance. If a tree is a community of cells, if cells are communities of organic compounds, if organic compounds are groups of molecules, if molecules are aggregations of atoms, if atoms are composed of subatomic particles, etc., and if we reach a level where nothing any longer properly appears (even with the assistance of electron microscopes, telescopes, or other technical devices), we have approached and perhaps reached a level where we are going to begin talking about a thing–in–itself.

psychology insofar as it identifies phenomena of interest, observes them, and tries to explain them. The problem with rational psychology is not that something like it exists but that until now it has been conducted as a fundamental branch of dogmatic metaphysics. The First Critique is directed against the traditional claims of metaphysics. What it puts in place of dogmatic metaphysics is an inquiry that asks this question: granted that in experience we appear to have certain basic psychological powers and experience of certain basic kinds, what is the *minimal* set of psychological concepts and principles that can explain the fundamental structures of that experience? This is an inquiry into the fundamental conditions of the possibility of (the kind of) experience that human beings have—which is another way of saying it is a transcendental inquiry.[37]

So what are the minimal concepts and principles that account for human experience? That is a result of the inquiry rather than a beginning postulate, of course, but a basic articulation goes like this. The most basic division is twofold. Human beings have, to begin with, a part of their experience that appears largely to be passively received through the senses, in sensibility. They also have a more active power, the power of thinking about what they have received in sensibility; this power can be called understanding.

Put this way, the division seems unexceptionable, philosophically bland, something that no responsible philosopher or researcher could fundamentally disagree with. That is one of its virtues. Kant's aim is not to *prove* this division, since it is something that more or less shows itself to anyone who pays attention to experience. With our contemporary sensibilities we probably would want to undertake at this very point an investigation—perhaps in the spirit of Husserl's phenomenology—of precisely how this division shows itself in all conceivable circumstances. That is, we would want to justify, articulate, and ramify the very notions of *sensibility* and *understanding*. Kant performs no public phenomenological investigation, but one is also compelled to think that he must have done it privately—if not according to contemporary phenomenological method at least according to the rigorous demands of Kantian philosophy. What the First Critique presents and develops is, however, an account of certain basic features of human experience that amplifies our understanding of the chief psychological powers but that also requires us to acknowledge their complicated interrelationship, even entanglement. The single word that most directly and most accurately expresses Kant's conception of this interrelationship is *synthesis*. And, as he says in the First Critique, synthesis is a function of the imagination in its transcendental use (see, e.g., A118).[38]

---

[37]In this sense, the attempts since the mid-nineteenth century to interpret Kant as fundamentally antipsychologistic, in the sense explained in Chap. 2, above, were misbegotten. The critical philosophy is directed against the excesses of rational psychology, not against the need for some basic human psychological structure that underlies the very possibility of knowing.

[38]The First Critique, *The Critique of Pure Reason*, is cited using the traditional designations A plus number for the pages of the 1781 first edition and B plus number for the 1787 second edition (both, when the passage is found in both editions). Quotations of the First Critique are drawn from the Pluhar translation (Kant 1996 [1781, 1787]).

It is common to explain the First Critique as a response to Hume, who Kant said awoke him from his dogmatic slumber: in particular a response to Hume's critique of causality. Kant supposedly recognized that it threatened to undermine any real notion of scientific knowledge and developed his transcendental philosophy to counter the threat. Unfortunately this may be false historically.[39] It is also too advanced a position from which to see what Kant had in mind. His insight is much more basic: it is that if we think about our ordinary experience of the world as it is commonly conceived—though here I would immediately add, commonly conceived *in the early modern period*—we will see that there is a kind of structure to that experience that cannot be explained at all by interpreting sensibility simply as a passive reception of what is given. Causality is indeed one of the structures of our experience, but it is built up on a much more basic nonpassive structure that is a necessary condition for something like causality to appear.

Empiricists and rationalists did not typically disagree about the passivity of sense perception and the activity of understanding.[40] Let us focus on sense perception. Ordinary experience of the world is the result of sensible ideas or images entering by way of the sense organs; if an idea is sensible one cannot have it unless it has entered by the senses. Though Locke and others made a distinction between primary and secondary qualities, this did not affect the fact that one does not have them till one has gotten them. The former—space, position, motion, shape, and the like—were veridical, that is, they appeared more or less as they really were. The latter—color, flavor, sound, the feeling of tickling when a feather brushes one's skin, and so forth—did not resemble their causes, and in that sense they were false or at least not accurate representations of what caused them.

Shortly we will take up the special problem posed by space and time. But for a moment I want to transport us to the post-Kantian worlds of evolution by natural selection. Sense organs as they exist today were not the intended goal of evolution. To focus just on vision: the eye of the human being evolved ultimately from the fact that a living thing through some genetic accident acquired a few light-sensitive cells that improved its ability to survive and reproduce. That original light-detecting function may not even have been conscious. If you think of the way that a person touching a hot stove draws away his hand before feeling the pain, by analogy you can understand how light-sensitive cells might work similarly: say that a predator casts a shadow that sets off a cascade of neuronal events that make the threatened animal start running in the other direction. It would be false to say that the cells or

[39] For an argument contesting the usual claims while emphasizing the positive influence of Hume on Kant, see Hatfield 2001.

[40] This is not to imply unanimity, nor that their conceptions of the kind and source of activity or passivity (in particular of understanding) were the same, but that they did not fundamentally disagree on these basic topological features. Yet I would argue, for example, that Hume's interpretation of understanding in the *Treatise* is ultimately more reactive than spontaneous, and that Leibniz's mature conception of perception is not simply passive. That might to some degree explain why both Leibniz and Hume stimulated Kant to thought about the conceptual topology of experience.

the animal has *detected* a predator, or a threat, or even a shadow. If one insists on the fact of consciousness, at best one might say that the animal has perceived a shadowing event—but we would have to be careful to eliminate any conceptual perception that an incautiously anthropomorphic description might imply. The ultimate point is that the original light-sensitive cells would not be dedicated to perceiving any *thing* in the world, only to eliciting behavior by stimulus and response.

By multiplying the number of light-sensitive cells we do not automatically get an eye, of course, even if we reason that an animal having tens of thousands of them together might have consciousness of the *outline* of a shadowy figure. The neurophysiological "wiring" and operation would need to be quite sophisticated in order to accomplish this. All the nerve paths from the eye to the brain would have to keep the same basic orientation to one another, or else they would simply scramble the result. If we consider further that in the human eye in ordinary daylight there are enormous numbers of photons falling on each rod and cone receptor, faster than the chemical reactions could track, it is amazing that the eye, and more generally the visual system, can make sense of it all so that we perceive a world of relatively stable appearances.

There is the rub. The seventeenth and eighteenth centuries did not have evolutionary theory, rods and cones, or photons, but they possessed sufficient anatomical, physiological, and optical knowledge to understand the physical complexity underlying the phenomena of sensation.[41] Of course physics and physiology do not translate directly into appearance. The complexity does suggest, nevertheless, that there must be organizing principles that allow for the translation of organ input into sense perception that has everyday clarity and distinctness. We cannot just plop down a ripe Gala apple in front of any eye and expect that the animal having the eye will see exactly what is there, ovoid shape, red and yellow striations, glistening of the waxen coating, and so forth. (Indeed, the more we know about the variety of eye structures in the insect and animal realms the more unlikely becomes our expectation that any animal eye will simply "see what is there.") Moreover, Leibniz's theory of sense perception as true phenomenon also forced the issue of the difference between the fine-grained complexity of things and the accuracy of what we experience macroscopically. In principle, thought Leibniz, the process of sensation involves innumerable tiny perceptions (*petites perceptions*) that are not perceived in the mass. He pointed out that the sound of sea waves is a complex sum of the sounds of all the tiny wavelets that compose it; sensation does not discriminate the wavelets but rather creates a kind of simplified mass sound that is nevertheless a true phenomenon, the structure of which does not contradict its imperceptibly fine-grained cause. In addition, he argued that space does not exist per se but is instead an ontological and epistemological product of the totality of monads that allows them to be positioned with respect to one another, in another true phenomenon.

[41] Descartes, for example, portrayed the retina as packed with circular nerve endings (e.g., Descartes 1964–1976, AT VI.146). It is curious that Locke, who had studied medicine, announces at the outset of the *Essay* (Locke 1690, bk. 1, ch. 1, §2) that he will not discuss at all any of the scientific questions connected with sensation. It is precisely those questions that give weight and substance to theories built on them—something that has been true since Aristotle first decided to think about body organs to understand the powers of soul.

It is in the first substantive portion of the First Critique, in the "Transcendental Aesthetic," that Kant addresses the special character of space (and also time) and ultimately avoids both the metaphysically-produced phenomenal space of Leibniz and the absolute space of Newton. Yet his immediate object is to argue that space and time are unlike color, flavor, sound, and other sensible appearances, and so cannot be accounted for in the same way. This is not, in the first instance, surprising: the distinction corresponds to Aristotle's pitting common sensibles against proper and the early modern contrast of primary and secondary qualities. The critical philosopher, on the other hand, is interested first of all in the basic conditions of the possibility of our experience. All of our sensation is localized spatially and temporally; to have a sensory experience of any kind, space and time must already be present. To use an analogy: putting to the side certain varieties of avant-garde theater, for there to be a play, no matter who the characters and what the action, there has to be a staging place and an appropriate stretch of time. Without the stage, none of these can show themselves at all; without the time, there is only a frozen, momentary tableau. Kant, unlike Locke, does not appeal to child development in discussing the origins of ideas (*Vorstellungen* in Kant's terminology, singular *Vorstellung*, the German for Latin *repraesentatio*), but a thought experiment with a newborn helps clarify the difference. What happens when a newborn opens his eyes? For Locke, it would *get* its first idea of space and its first idea of time. For Kant, space and time would both *be constituted* as they always are experienced.[42]

This is not to say that empiricism is totally helpless in the face of accounting for space and time—philosophical positions of long standing always have a ground. The first time you see red you certainly do see it, though because you do not have the concept "red" you cannot experience it *as* red. With the second seeing of red you are in a different position: the retained image of the first seeing serves as a point of reference, and so you are on your way to a full-blown concept. There seems to be no difference with space and time: you get the appearance of space and time the first time they occur, though you have nothing to compare them to. Still, they have a special status. For one thing, they never appear simply as themselves, without anything else combined with them, and every other idea or experience type requires the presence of space and time. They are experienced by the newborn along with the white or blue or green or pink of the medical staff's clothing, the loud clapping sound of the nurse performing the Apgar test, the acrid odor of disinfectant, the pain from the slapped backside. They are part of *every* experience; we are primed for them in a way unlike that for any other sensory experience.

There is more, however. It would be wrong to deny that our experience of time and space can be refined, like any other experience can be. Our concept of red improves as we encounter more shades of it. Even a newborn destined for

[42] Since neither Kant nor Locke uses the example, presenting it is speculative. The infant could not, of course, tell us which happened! It would not be a decisive objection against Kant to argue that the infant will in any case not at first have a well-developed and completely reliable sense of space and time. There only needs to be a before and an after for time, and a copresented center and periphery, right and left, up and down, near and far for space. See also the next two paragraphs.

mathematical greatness does not in his first moment of vision clearly and distinctly see three-dimensional cartesian space extended indefinitely and isotropically in all directions. Yet once again there is a difference. In the conceptual topography of empiricism, experiencing crimson does not guarantee we can conceive burgundy or the red of a desert sunset, much less teal, chartreuse, lilac, or ultramarine. Space when it is first experienced appears in a way that is essentially homologous, virtually identical, with all future encounters. It may not be indefinitely extended but it *is* extended, it extends a little in every direction and has a sense of "aroundness" (here and hereby, there and thereby). Time will have its duration, its just–before and just–after, even if it is not yet measured by a watch or an atomic clock. All other sensibilia have to be learned slowly and progressively; with space and time, the learning curve is nearly vertical.

For Kant, the presence of spatiality and temporality is primitively and fundamentally given in sensible consciousness. They are given because they are there as the most basic forms of sensible consciousness; they are cogenerated with sense. Therefore there are no puzzles to be solved about whether space and time appear the way they are. Space and time are not things to which consciousness must correspond if it is to have a true idea of them. The only space is the one that is at the very foundation of our consciousness of other things, one that is produced as a basic condition of the mind's sense–orientation. Similarly, time is never all there as a thing, but it extends backwards and forwards indefinitely, and at will we can take into account larger or smaller expanses of it in continuity with the present (for instance, when we think back to imagine the founding of civilization or ahead to the sun's going supernova). There is a fundamental lawfulness of our experience according to which the framework of temporality and spatiality is produced to embrace everything else that can appear.

This "production" of space and time is not voluntary, of course, at least not in the sense that we might choose it or not. It is *connatural* with us; it is the deepest part of our nature.[43] The production of space and of time is one of the most basic, most primitive functions of the human being: to produce, or one might say project, a three-dimensional space that is indefinitely extendable, and to produce–project time and its passage. Before any other *particular* sensation, perception, imagination, or memory can take place or any conceptual marking can be made, there must be this original projection of a "theater of operations" for sense. For Kant, this is what it means for the forms of space and time to be not *a posteriori* but rather *a priori*. Quite obviously, because we are born infants, we human beings cannot identify or talk about space and time before we have acquired language, and we cannot refer to them in any way at all (just like any other kind of experience) until after we have experienced them—*posterior* to experience. But every other sensory experience we have comes contingently, whenever it happens to come and not before. We might

[43] Since, as it turns out, space and time are produced by transcendental imagination as the synthesis of the manifold of sensibility, I am anticipating a passage from the *Critique of Pure Reason* that says the transcendental functions of imagination are "a secret art residing in the depths of the human soul, an art whose true stratagems we shall hardly ever divine from nature and lay bare before ourselves" (A141/B180–181).

be 60 years old before we see the shade of color that everyone is calling "metallic Pacific mist." Our sensory and conceptual relationships to such experiences are *a posteriori*. But our relationship to space and time is immediate: for everyone, always and everywhere, who has human experience, space and time are there from the beginning; they do not depend on our having previous experience, they are *a priori*. And, finally to return to the subject matter of this book, they are, according to Kant, a result of imagination in its transcendental function. In its transcendental function imagination synthesizes and thus provides unity to the so-called *manifold of sensibility*. Without this function the manifold would be many without unity, more or less a blooming, buzzing confusion—perhaps, to take just a visual example, like the chaotic pixelation, each pixel rapidly changing color, of a malfunctioning computer monitor. The first synthesis—there are multiple syntheses due to imagination—is the one that spatializes the manifold (in what Kant calls "outer sense") and that temporally sequences the perceiving subject's experience (in what he calls "inner sense"). This and all the other syntheses are, furthermore, syntheses according to rules. Because the syntheses follow rules, they are regular (no pun intended, though *regula* is Latin for "rule"), knowable, and predictable, at least for a being that has the power of understanding according to concepts. What is special about the fundamental syntheses of the manifold of sensibility (which start with space and time but proceed to a synthesis according to concepts as well) is that *they build understandability right into the appearances*.

These rules are not voluntary, so that one might obey them or not, as one willed; they are more like algorithms that are followed whenever there is an appropriate input. In that sense they are due to the spontaneous activity of the experiencing subject on what is passively received through the senses. In principle one might speculate that different kinds of organisms (other than human) have different basic syntheses of sensibility, ones that (in nonrational animals) are organized but not thought in any way. One might even speculate (and Kant at least gestures in this direction) that other kinds of thinking being could exist that had a radically different source and organization of sensibility.[44]

Before we go on to the higher syntheses of transcendental imagination, we must mention that this transcendental production–projection of space and time makes possible an *a priori* science of mathematics. Geometry is in essence the science that results from exploring, articulating, and knowing the features of the purely formed intuition of space, and arithmetic is at least in part the result of exploring, articulating, and knowing the features of the purely formed intuition of time. Thus imagination in its productive use comes fully into its own in the realm of intuition in a way that might have pleased Descartes, insofar as it gives further and in fact deeper

[44] We can of course *know* nothing about this, but we can speculate about it in analogy to the relation of sensibility to rationality in us. For example, there are scenarios in which science-fiction authors devise intelligent extragalactic beings whose sense organs are based on the chemistry of silicon rather than carbon. And of course intelligent robots might be designed to have "senses" that respond to and organize far different kinds of inputs (say X-rays rather than the spectrum of visible light) than human organs do.

support for his conception of a dynamic, figure-producing geometry that can be tracked algebraically. Kant is very insistent on the fact that the imagination must produce particular figures (for example a straight line) progressively, which is to say through time. It is by the synthetic-extensive powers of the imagination that we do this. In the space that is itself already an original and foundational intuition of imagination we can conceive particular figures according to rules. These figures have to be "inscribed" in the original space. If we draw a penciled line on a sheet of paper we can depend on graphite's adherence to the paper to preserve what we have already drawn as we further extend the line, but when we draw a line in our imagination we must constantly reiterate what we have already drawn as we conceive other parts of the line to be produced, and when we try to hold a geometric figure in mind we need to constantly renew the production of the figure. Even when we use a pencil, we must constantly attend to what we have drawn already and extend it accordingly: however slight the effort there is in making sure we draw it straight, we are constantly conceiving the next moment of extension of the already-accomplished extension that began when we touched pencil to paper. We can use a computer analogy here: what has been produced on the screen of imagination must constantly be reproduced or refreshed. With computer monitors we in fact talk of a "refresh rate," the number of times per second that the screen is rescanned by an electron beam (or by other means), which has to be done even to maintain a static image, much less to produce a moving one. Without the constant refresh operation the image would quickly disappear.[45]

For Kant the synthetic work of the imagination is the source of mathematics. If many of his predecessors and contemporaries were struck by the instability and unpredictability of imagination, for Kant both the foundational projection of space and the constantly refreshed production of geometric figures were governed by absolutely regular principles or rules of the transcendental use of imagination, and it is precisely these rules and their being intrinsic or inherent to the human mind that makes possible the certainty and reliability of mathematics. Even the fact that we cannot always follow these rules in perfect strictness does not impair their importance. If a drawing on paper or an imagined figure deviates from the rule, then that only demonstrates that the rule is more reliable than the various exemplifications of the rule in different media. (This also amounts to an anticipation of the schematism of imagination; see below.) The truths of mathematics provide the elemental facts (*factum* means "made thing") about what is at the foundation of all sensibility and receptivity, or more precisely at the foundation of sensible intuition of objects *within* space and time. The nature of space and the truths of mathematics, whether exemplified in abstraction from actual sensation or not, is drawn from no other source than the fundamental regulating principles of the mind itself. The most basic

[45]There is also the fact that, in Meditation 3 (AT VII.49), Descartes explained time as God's continual re–creation of the world. Thus a divine refresh operation sustains the world and produces all its movements. There is a similar speculation in an essay of Newton's, "De gravitatione et aequipondio fluidorum," which he wrote in the early 1660s but did not publish during his lifetime; see Tamny 1979.

transcendental functions of imagination provide the *where* and *when* for particular sensed things and their relationships. This holds for sensation, for memory, and for imagination in all the conventional senses of those terms.

The transcendental structure of spatiality and temporality produces sufficient stability and articulability so that "secondary" rules and laws can generate further, more particular structures in it, in an orderly and reproducible way that allows for scientific knowing. The fact that we can "do mathematics" is a result of the secondary articulation of the founding institution of space and time. There is more, however. The very same structures that give rise to space, time, and the further articulation of space and time are incorporated into our actual sense experience. Again, Kant uses the term *synthesis* for successive acts of mind that take up in a certain unity the results of earlier acts and further unite with them later ones.[46] Previous syntheses are incorporated into successive ones. When the senses are passively stimulated, imagination in its first transcendental function spontaneously produces space and time and brings the sensibles into this "theater." That brings us to something like a level of consciousness involving awareness of colors, sounds, aromas, pressures, positions in the field of sense, and the like. The imagination has thus brought "unity to the manifold of sensibility."[47] At this level there is already a geometry to the world, especially with respect to vision: the colors, the areas of lightness and dark, can already have quite determinate shapes. There is a parallel spontaneous synthesis in the thinking of a mathematician. His mathematical work is not the result of a passive stimulation by sense, but rather for the most part a spontaneous production of geometrical figures "drawn" to the mathematician's purposes. The corresponding degree of synthesis would be less a (say) mathematical plane already filled with determinate figures than the local but extendable part of the experienced plane with emergent lines and shapes.

In the Western conceptual topology of imagination, Kant is the first to systematically explicate imagination as essentially and fundamentally productive *before* it is reproductive.[48] More precisely, for Kant the reproductive power of imagination is

[46] There is a risk of distortion in assuming that Kant conceives of synthesis as an empirically successive process, in which first there is level-one synthesis, taken up into level-two synthesis, taken up into level-three synthesis, etc. One reason that he replaced the 1781 "Transcendental Deduction" with a completely new one in 1787 was that the 1781 approach seemed too empirical/sequential.

[47] Kant scants on examples, so it is more than a little speculative to associate specific experiences with the different syntheses. Short of this first synthesis of the unity of the manifold, one might imagine that all one could be conscious of would be short bursts of sensation, like a blinding light, a poke in the ribs that wakes one from a dreamless sleep, or the like. Moreover, any actual synthesis involves more than the pure syntheses of *possible* experience that are the precise concern of the First Critique. A mathematical example: the space of geometry is a formal intuition, but any geometer actually doing geometry already has a far more particularly determined space in mind than that of formal intuition of space all by itself. Space and time as *pure* formal intuitions cannot be perceived at all (A166/B207).

[48] This is not to say that there are not partial anticipations of this: for example, in Descartes, Spinoza, and Leibniz, imagination's role in constituting the place of space and mathematics, or, in the passage from Descartes's *Compendium musicae* discussed in Sect. 6.2, its role in producing sense's musical perception. But these are not counterexamples to the claim of Kant's originality, because none is all at once systematic, fundamental, and essential.

inconceivable without something more basic: that there is organized and intuitable presentation of appearances in the first place. Without the organizing capabilities of imagination the physical and physiological events to which our sense organs are subjected would yield chaos rather than perception, a phantasmagoria of transient light flashes, aural buzzes and snaps, fleeting pains, whiffs of unidentifiable smells, and snatches of indeterminate flavors.

If Descartes's dynamic understanding of a generative mathematics based on rules, albeit divinely created rules, makes clear that Kant's conception of mathematics is not entirely original, it does not minimize Kant's philosophical creativity or his insight into mathematics. Descartes looked at the fields of ancient mathematical sciences and taught himself to traverse them differently than the ancients had; he thus brought dynamism to the concepts of mathematics and more generally to the ways of producing and moving objects in the spatial field. A largely static ancient topology of mathematics was turned active. Kant did not need to read Descartes closely or have access to his posthumously published writings to see this, because the dynamic conceptual topology Descartes had innovated became the substrate of seventeenth- and eighteenth-century mathematics and mathematical physics. It is likely that Kant recognized the same topological features that Descartes had by looking at the current state of mathematics and the sciences and thinking the matter through for himself. That is, more than a century later he thought about the same kind of objects and practices situated in the same kind of background field articulated according to most of the same concepts.[49]

7.6 How Previous Philosophy Failed to See the Syntheses of Imagination

Sensibility is of course filled with changeable content. That almost goes without saying: to be of use informing us about what is going on in the world around us, it has to be in tune with contingency, with the constantly varying and even accidental character of the world open to sense. My senses are ready to see, hear, smell, taste, and feel whatever presents itself, even if no particular thing is there to be sensed. Yet, as Kant conceives it, even more fundamentally ready is the spatiality and temporality that provide a matrix with which these particular qualities of sensing will be placed. Space and time constitute a field of possibilities that preexists all particularities and contingencies. For there to be the wafting aroma of yellow primroses, there has to be place and temporal sequence.

[49]Notice the advantage of thinking about this in terms of conceptual topologies rather than influences, paradigms, conjectures and refutations, purely conceptual theories (lacking the topology of the underlying field or substratum), and all the varieties of relativisms. It is possible over time to think the same objects against the same backgrounds and using the same concepts congruently in a topology—mathematicians and scientists do it all the time—and thereby to gradually develop and change the topology without overthrowing it.

But presumably primitive animals have a sensibility with basic temporal and spatial character. What makes the human experience of the world different? It is the power of thinking, more particularly the power of understanding. If this sounds unsurprisingly conventional, that is all to the good. Kant's conception of the internal structure and constitution of basic human capacities may be revolutionary, but to begin with he wants to specify fundamental conditions without which human experience is not possible. There is what we receive from the world through the senses more or less passively, and there is what we actively bring to our inquiry into what we see: sensibility on the one hand, understanding on the other. So far in considering the manifold of sensibility, we have mentioned the (unconsciously) active synthesis by imagination of intuitable space and time. Before we can passively receive the information from the world that allows us to put together yellow, sweet smell, distinctive shape, etc., into a primrose, our organism constitutes a where and a when for it.

For the most part, those aspects of the things of the world that we receive passively are what for Aristotle were the proper sensibles and what for Locke were the secondary qualities of sense: colors, odors, sounds, tactile feeling, flavor. Explicitly (in Aristotle) or implicitly (in Locke) these sensible qualities coming from different sense organs had to be coordinated with one another in order for human beings to have an experience of things. What Kant was doing in the transcendental aesthetics of space and time was arguing that Aristotle's common sensation (which allowed for the emergence of sense qualities not perceived or at least not clearly perceived in individual senses) or Locke's primary qualities of sense were not simply noted and registered by sensibility but were transcendentally produced for it as prerequisite for any particular sensing. All the things that the individual sense organs deliver to the organism are, for Kant, passively received and then placed in the intuitable space and time of transcendental imagination. They are not, as such, knowable *a priori*. That is, we have no access to them apart from having actually experienced them. We cannot, for example, "see" metallic Pacific mist even in our imaginations until we have first seen it in sensation. (The not insignificant possible exception is whoever invented and named it.) Although at birth we cannot philosophize about the transcendental functions of imagination, they are already operative in us, and in our experience we are actually already producing the rule-generated, repeatable certainties of space and time, right from the very beginning.

For Aristotle the phantasms constituted by proper (e.g., color and aroma) and common (e.g., place, movement, and time) sensibles could be thought, in the sense that the intellect or noetic power could grasp the intelligible forms in them. For Locke, once the ideas of the primary and secondary qualities of sense were received they could be surveyed, compared, contrasted, and classified by understanding. The difference, of course, was that for Aristotle the intelligibility was present in the thing's phantasm (culminating perhaps in intelligible essences), whereas Locke's understanding contingently identified and associated the ideas of its experience under nominal essences (essences in name only, always subject to reassociation and revision). Kant's theory in the first instance resembles Aristotle's. We in fact see in the things of experience intelligibilities. But the transcendental aesthetics provides the model for how we are to understand the source of the intelligibility of experience: it is the structuring rules of our mind that build the intelligibility into what is sensed. The basic concepts we use to think them come not from the ontological essences that make them what they are in the world of nature (Aristotelian) nor from the associations we consciously make of the sense ideas in understanding (Lockean). They come instead from the syntheses in and of our own minds: our minds are grasping their own rule-bound though initially unconscious work in organizing the manifold of sensibility in ways that are reliably knowable.

Early in the first or A[50] edition's version of the "Deduction of the Pure Concepts of Understanding," the Transcendental Deduction—in the subsection titled "I. On the Synthesis of Apprehension in Intuition"—Kant presents a possible obstacle to our understanding of passivity and spontaneity by his remark that "any presentation as contained in one instant can never be anything but absolute unity" (A99). It appears to imply that there is an organized synthesis before what I have called the first synthesis, which includes both spatial and temporal unity. To interpret it this way is to get things backwards, however. The correct Kantian response is to point out that synthesis is an act of spontaneity, and that the "absolute unity" of an instant's presentation is passive. How do we make sense of that? First, the only way we can arrive at the notion of an instant's presentation is through an analysis of a synthesis that has already taken place. At the start of the subsection Kant notes that all syntheses—in the immediate context that would include any synthesis of "outer sense," for example any snapshot image of sensation of the external world—are subject to "the formal condition of inner sense, i.e., to time. In time they must one and all be ordered, connected, and brought into relations. This is a general comment that must be presupposed throughout what follows" (A99). Therefore the absolute unity of the instant's presentation is factitious or artificial: we cannot assume that what we arrive at by analysis exists as a real thing. Without the synthesis of time, we cannot have such an absolute unity as an actual presentation. Even if we succeed in imagining a passively perceiving organism that is capable of "sensing" such a unit of appearance at a given moment, we must realize that at the next moment that unit would be totally effaced by the next appearance, and then by the next, and the next, and the next, like the images on a television screen as we rapidly zap or channel surf through the available offerings. Even if each momentary image is "clear and distinct," all this would mean at best is that the passively receiving sense apparatus per se produces a sharply articulated output that does not cohere with anything else. Without the continuous synthesis of each moment's output with the previous moment's and the next moment's there would be no objects at all for the organism to perceive, much less to remember. Therefore this organism could have no experience in the proper sense, since Kant defines experience as the result of intuitions (which are themselves synthesized as unities) that are further unified by concepts—that is, what has been sensed has in addition been understood.

[50] Although there have been fierce debates over the consistency of the A (1781) and B (1787) editions precisely with respect to the role of imagination, all that is necessary for our purposes is that in one or the other edition Kant recognized the possibilities of the conceptual topology of imagination discussed here. For a persuasive and exquisitely detailed argument that the two editions are for the most part consistent, see Longuenesse 1998.

Unlike Descartes, what we know of Kant does not allow us to reconstruct much about his own practice of imagining. But it is hard to conceive that he would have assigned to imagination transcendental functions if he had not given considerable attention to ways in which the theories of his day fell short of adequately accounting for sensation and imagination. His acknowledging the unity of the momentary appearance of sense is an indication of his clear insight into these shortcomings.

A basic presupposition of Locke and Hume, but also of many rationalists, was that our original sense impressions are relatively independent and completely contingent data units, meaning that they have no intrinsic connections before the understanding (in Locke) or before the principles of association (in Hume) begin working on them. That Kant intentionally undercut this assumption is clear from something that he wrote just two paragraphs before our last quotation:

> If each singular presentation were entirely foreign to—isolated from, as it were—every other presentation and separated from it, then there would never arise anything like cognition; for cognition is a whole consisting of compared and connected presentations. Hence when I ascribe to sense a synopsis, because sense always contains a manifold in its intuition, then to this synopsis there always corresponds a synthesis; and thus *receptivity* can make cognition possible only when combined with *spontaneity*. Now, this spontaneity is the basis of a threefold synthesis that necessarily occurs in all cognition: viz., the synthesis of the *apprehension* of presentations that are modifications of the mind in intuition; the synthesis of the *reproduction* of these presentations in imagination; and the synthesis of their *recognition* in the concept. Now, these three syntheses guide us to three subjective sources of cognition that make possible the understanding itself and, through it, all experience, which is an empirical product of the understanding. (A97–98)

Let us focus on the first half of this quotation. Kant's first sentence is a devastating criticism of an oversight empiricists commit: they treat consciousness as though it were nothing more than a neutral container with unconnected bits floating about. But before there is any input there is already a synthesis that establishes this place of presentation, the place where the appearances already belong together. He calls this initial synthesis *synopsis*, which etymologically means, roughly, "seeing together." One can't have even the sense of a manifold (of something that is many–fold) without there being some kind of synthesis beforehand. This and all other syntheses are a result of on the one hand spontaneity and on the other what is given in receptivity (and, for higher syntheses, what is given by the previous syntheses).

Spontaneity in Kant's transcendental psychology has recently drawn considerable scholarly attention.[51] In the first instance we might be tempted to think of synthesis as being an automatism, machinelike: there is input, the synopsizing machine starts its work, there is output. This is not an entirely illicit conception with respect to synopsis. In a sense all we need in order to "synopsize" is to open our eyes. Even so, if we open our eyes but our attention is otherwise absorbed (say we are sitting in a busy restaurant, trying to solve Fermat's last theorem) it might be hard to say that the synthesis of synopsis is fully effective, since we may notice nothing at all of what is going on around us. Even to see what's going on we need a certain minimum of voluntary attention: we must to some degree want to see, enough to take a look.

[51] See, for example, Pippin 1997.

Synopsis will work at some basic level if we are healthy and our eyes are open, but the visual input also has to be spontaneously acted upon so that it can be emplaced by another synthesis.

The problem with fully mechanizing synthesis is that it would make it not spontaneous but a mere response (respontaneity, to coin a term); it would be reactive rather than active. Kant understands the experience we actually have as requiring, in the last analysis, a transcendental psychology that includes both receptivity and spontaneity (A50/B74).[52] The psyche both receives and acts, and it needs what it receives before it can act in any determinate way whatsoever. In the most accurate sense, we must say that there are two sources of cognition, the first being the receptivity of presentations or of impressions—which a paragraph later he calls *sensibility*—the other being our ability to know objects through these presentations or "the spontaneity of concepts" (A50/B74)—which a paragraph later he calls *understanding*. But sensibility, too, requires a kind of spontaneity: the production of space and time and the inception in this spatial and temporal place of the categories of appearance. If these are not the products of direct willing, they are spontaneously produced—by us as living organisms—as the basic field of possible appearance.

Spontaneity is ultimately the source of freedom, but it is not simply identical with freedom, much less with spontaneous free choice. In the First Critique it is roughly analogous to an aspect of soul or *psuchē* recognizable from Aristotle's theories. In the first instance soul is the first actuality of an organized body having life. It is the base level of living activity from which all other more specified activities spring; and those other activities are all the ones by which the animal or plant does more than just survive, and by which the organism can flourish in performing all its specific activities. Animals and plants spontaneously live, move, and respond to the world, one might say, but that re–sponse is itself an affirmation of a spontaneity that is situated, the spontaneity of living out their natures. Kant does not consider plants and animals in his conception of spontaneity, but the notion does seem to indicate a kind of living–out–into–the–world similar to Aristotle's. And since for Aristotle the most noble human power is intellect, and intellect in the highest sense is always active, the relation to Kant would be all the nearer. This is a relation not of influence but of a common conceptual topology of mind and soul.

[52]This assertion occurs in the very first sentences of the "Transcendental Logic." Apart from prefatory material and the introduction, the First Critique as a whole consists of the "Transcendental Doctrine of Elements" and the very much shorter "Transcendental Doctrine of Method"; the former is divided into two parts, the "Transcendental Aesthetic" and the "Transcendental Logic," with the "Transcendental Analytic" being the first substantive portion of the "Transcendental Logic." This makes the statement as conspicuous and central as possible. The spontaneity of reason will play an even more prominent role in the next philosophical generation, beginning with Fichte's understanding of the ego as a spontaneous and outflowing *drive* (*Trieb*). With respect to the quarrel between psychologism and antipsychologism, one might say that Kant insists that the minimum of "rational psychology" required to give an account of how we know is the distinction between receptivity and spontaneity (and any powers of mind necessary to support the distinction).

7.7 The Higher Syntheses

Space and time are spontaneously synthesized in the most basic transcendental function of imagination as pure intuitables that provide a place for the "matter" received by the senses. The *final* outcome of this "placing together" (a plausible translation of the Greek *synthesis*) is a thoroughgoingly unified manifold of sensibility, which is the surveyable total field of our sensory experience. But this last sentence conceals more than one synthesis, as is evidenced by both the A and the B versions of the Transcendental Deduction. The A version spells out that after the synopsis (the synthesis of apprehension in intuition) come the synthesis of reproduction in imagination and the synthesis of recognition in the concept. Apprehension in intuition starts with space–time and articulates a sensory array therein; reproduction in imagination involves the ability to produce the same sensory array again; recognition in the concept means that the reproducible form of intuition is gathered up under a concept so that it becomes thinkable. This is a sequence we have witnessed before, in Aristotle's conception of the different levels of animal soul: the animal that is merely responsive to sensory stimulus, the animal that in addition has memory, and the animal that is able to see forms in phantasms by intellect. Indeed, one might easily conclude that Kant was simply understanding as progressive syntheses what medieval Aristotelians had assigned to different internal senses and organized in the brain: common sensation, imagination–memory, and the cogitative.[53] But that would miss a central point: for Kant, the syntheses beyond the pure intuition of space and time presuppose that the spontaneity of the categories acts upon already synthesized sensibility. That is what is truly distinctive of Kant in comparison to virtually all his predecessors: even very basic levels of the experienceable world are *formed* by concepts in/of one's own mind.

The synthesis of apprehension begins with taking up what is presented in the synoptically viewable manifold that has been presented to intuition (and in that sense has already affected the mind). The synthesis of reproduction in the imagination repeats and refreshes, but also simplifies and varies, what has been apprehended.[54] The synthesis of recognition then connects this new phantasm (for that is what it is, in the vocabulary of the older tradition) to the concept that allows it to be identified. These syntheses might take place all in the blink of an eye, but they are at least conceptually distinguishable.

[53] The cogitative, Thomas Aquinas's particular reason, assigned names and concepts to phantasms; it was where intellection met sensation. See Sect. 5.3, above.

[54] Why add "repeats and refreshes"? Because that is what Kant describes as happening with imagination in the production of something as simple as a line in imagination: at every moment it must confirm what it has represented and refresh it, to renew it in space and extend it temporally. Why add "simplifies and varies"? Because in comparison to the manifold of sensibility, imagining simplifies the background and often the object as well; and the potential for alteration, change, and variation are of the essence, even in lending the least attention to simple things like a triangle or line. It is only at this level that, for empiricists, the association of ideas can begin.

Despite the fact that imagination is named only in the middle synthesis, the synthesis of reproduction, it is clear that all three of these syntheses are due to imagination in its transcendental use. In the explication of the synthesis of reproduction Kant refers back to the synthesis of apprehension as itself involving the transcendental power of imagination. The synthesis of recognition in its turn is connected to the imagination in the sense that the argument anticipates a notion that he introduces later (after the "Deduction of the Categories"), at the beginning of the second book of the "Transcendental Analytic": the forbiddingly named "schematism of the pure concepts of the understanding." Described as another function of the imagination in its transcendental use, it performs functions that were in medieval thought attributed to reproductive and productive imagining or some other internal sensation like the cogitative power, through which a well-prepared phantasm received its conceptual designation. All these imaginative syntheses further undercut the assumptions of empiricism and of conventionalized rationalism: imagination is at the beginning, the middle, and the end of ordinary *and* scientific experience and understanding.

Where imagination is no longer fundamentally at work is in the transcendental unity of apperception, in the constitution of the ego (except insofar as the ego is temporal); where imagination no longer has a place at all is in pure reason, when it thinks concepts without the presentations of the manifold of sensibility (except, once again, insofar as the temporal synthesis of the empirically experienced ego might require imagination).[55] The mark of the transcendental unity of apperception is that everything I think can have *prepended* to it the words "I think that...." It is a kind of Kantian affirmation of Hume's claim that he could not find a *thing* in consciousness that corresponded to the self, but only a complex series of impressions and images. What Kant is saying is that the ego is precisely not a thing or substance, but rather the consciousness that attends all the impressions and images and successively unifies them as the ego's own experience.

In effect, to contradict in paraphrase the Aristotelian slogan, Kant believes that the thinking of pure reason is not of images and does not require imagination, except insofar as this pure mental activity uses concepts that have an origin in experience and insofar as the imagination's synthesis of time in inner sense is required for thinking to observe the sequence of pure reasoning. Trying to reify this kind of experience as an ego–substance existing outside experience would be to posit a thing–in–itself. The ego is not a thing, since things appear only as part of external nature, that is, in the outer-sense spatial synthesis of the manifold of sensibility. The thinking in inner sense can, at best, ascend to a kind of conceptual thinking of possibilities (that is, thinking concepts in a way not violating the principle of noncontradiction) formed in analogy to our thinking of things presented in space and time but that one knows can never be exhibited in real existence

[55] Kant may have underestimated here the implications of his own thinking. The temporal qualifications of the transcendental unity of apperception and of pure reason are not adequately addressed, and they give purchase to Martin Heidegger's conviction that imaginative synthesis is the common root of sensibility and understanding. See Heidegger 1929, 37 (§6) and 160–161 (§31).

(that is, not in outer sense). This kind of thinking Kant calls regulative. It is a thinking of pure conceptual possibilities, not of understanding. Understanding, by contrast, is the use of reason with respect to the manifold of sensibility in accordance with the pure concepts of the understanding. Those pure concepts are synthesized into the unity of the manifold of sensibility by the schematism of understanding in its transcendental use.

Schematism in fact also involves a certain automatism. Kant identifies schematism as yet one more function of the imagination in its transcendental use; but he also takes pains to explain that a *schema* (plural *schemata*) is not an image. A schema is a cognitive connection between a presented image and a corresponding concept as well as between a concept and its possible presentations in various images. Schematism is the mind–power that accounts for the formation of schemata. Schemata are not static forms but generative rules.

Kant's examples of schemata (triangle, dog, and fiveness) are somewhat misleading. The *Critique of Pure Reason* aims to explain the pure transcendental cognitive functions of the human mind, but of course examples almost always muddy the issue, because to be describable they need to refer to particular contingencies that have nothing to do with *pure* transcendental functioning, prior to all experience. The essential function of schemata is to implement the pure concepts of the understanding in sensibility. That is, they incorporate (synthesize) pure conceptual forms into the intuitions of sensibility.

The pure concepts of the understanding, also called the categories, are presented in four groups of three: categories of quantity (unity; plurality; allness), of quality (reality; negation; limitation), of relation (inherence and subsistence, or substance and accidence; causality and dependence, or cause and effect; community, or interaction between agent and patient), and of modality (possibility–impossibility; existence–nonexistence; necessity–contingency). Although Kant's source for this constellation of categories is uncertain, his basic argument is based on the implicit consensus of philosophers about the fundamental conditions of logical assertion. These are all categories that are essential for identifying terms and making judgments with them, that is, for taking terms or concepts A and B and saying "A is B" or "A is not B," with consideration of all the various fundamental quantities (e.g., all or some of a kind), qualities (e.g., affirmative or negative forms), relations (e.g., this substance has such and such an attribute), and modalities (e.g., necessity or possibility) in such judgments. The transcendental question for Kant is where we get the right to apply these fundamental logical categories to experience. Once again he argues that they are *a priori*, rules in accordance with which our experience is fundamentally structured, in yet another synthetic function of imagination in its transcendental use.

Just as space and time, the pure intuitions of sensibility, have to be already constituted in the most basic synthesis of the manifold of sensibility in order for there to be any kind of more particular sensation or sense experience, so, too, do the pure concepts of the understanding have to be implemented/schematized in the manifold so that particular things can be constituted there in various kinds with various quantities, qualities, and relations. Any and every empirical concept, like squirrel or tree or brown or bushy, presupposes that our field of experience, the manifold of sensibility, has already been schematized by elemental unifications that enable us to individuate and identify "things" that display characteristics "belonging" to them, things that stand to one another not just in spatial and temporal relation but also in conceptual relation, etc. From Kant's perspective Aristotelian induction explains how we get *empirical* concepts but not how we get the most basic concepts of all. Aristotelian induction *presupposes* a synthesizing activity: as toddlers we see one squirrel, another squirrel, a third, etc., and at some point we realize they are all the same thing, and from that point we are able to say truly, "Squirrel!"

Aristotle had a theory of fundamental categories, of course: the most basic things that are are substances, which have qualities, quantities, relations, etc.; and he had a theory that human beings by their nature can come to know these things precisely as substances, qualities, etc. But, from Kant's perspective, that is to *presuppose* that we human beings by nature are constituted to know things in their natures, not to explain that fact or even to recognize it as such. That amounts to a dogmatic metaphysics that claims to know things–in–themselves, a metaphysics that is no longer, in Kant's eyes, justifiable. It has been undermined not just by skepticism but also by science.[56] His critical philosophy demands, negatively, that henceforth we acknowledge the inadequacy of all such dogmatic claims (not just Aristotelian ones), and, positively, that henceforth we need to give an accounting of the basis of our experience like that of transcendental aesthetics and transcendental logic.

The categories are built into the manifold of sensibility by progressive synthesis. That is, at a certain low level of synthesis, say in the A edition's synoptic apprehension, what the manifold of sensible appearance shows visually is an articulation into areas of distinguishable color and color boundaries. To some degree there is already categorial effect at this level. Consider the categories of quantity. For there to be distinguishable (though perhaps not yet cognitively distinct) colors and boundaries there must be a certain overall *unity* to the color field, several subsidiary unities (a *plurality*) in the individual patches of distinct colors, and an *allness* (the field as a whole field, a whole of color and color boundaries). Quality is implicit as well (each color is *real* insofar as it is currently showing in the manifold, each color excludes or *negates* other colors where it shows itself, and each area of color is demarcated from the others by a *limit*). If at this level there are no substances and accidents per se (within a field just of color as color we, or the animal consciousnesses we are at this level, do not ascribe these colors to any *things*) and certainly no cause and effect (the first two of the three forms of relation), there is nevertheless an implicit *community* of colors in the current field, as well as in the fact of coloration per se, in which the colors potentially and actually interact (say in color contrast effects). But it requires further levels of imaginative synthesis to enhance these already implicitly categorial features so that they might more fully emerge as explicit.

[56] Undermined by the science available to Kant and even more by the science available to us. Our sense organs do not simply "take in the world" as it is but gather a bewildering number and variety of "data points" (e.g., quadrillions of photons impinging on tens of millions of receptors in the eye) that then get "synthesized" in the appearances we see, hear, touch, etc. See Sect. 7.5, above.

Schematism is the general name for the way(s) in which the pure concepts of the understanding are implemented in human experience. Unlike the previous paragraph, Kant gives no details about the initial emergence of categories in the appearances of the manifold. In fact, he describes not categorial schemata but contingent, empirical ones like those of triangle, dog, and the number five. The schema of triangle enables us both (a) to see a triangular shape and associate it with the concept triangle and (b) to think triangle and produce a triangle or a triangular shape. Similarly for the schema of dog. Whether we see a Pekinese or a greyhound, or anything "in between," by virtue of the schema we can think and say "dog"; moreover, when we start thinking about dogs, we are in a state of readiness to begin to portray, with a pencil or mentally, images of dogs.[57] Kant is careful to point out that he does not mean that the schema has us produce a perfectly rendered portrait of any particular dog. That is certainly a possibility, of course, but it is the moment of our *beginning* to produce an image in light of the concept that is crucial. Even a very rough sketch of a dog-looking thing counts as a product of the schema. Thus Kant captures here in explicit form one of the crucial characteristics we found in imagining back in Chap. 2: its incipience, its inceptive character. In the very act of beginning to imagine something that has any conceptual delimitation at all, we are employing a schema.

The example of the number five does in fact at least point in the direction of the pure transcendental use of schemata. Here is the crucial passage:

> A schema is, in itself, always only a product of the imagination. Yet, because here the imagination's synthesis aims not at an individual intuition but at unity in the determination of sensibility, a schema must be distinguished from an image. Thus if I put five dots after one another, like this, ….. , then this result is an image of the number five. Suppose, on the other hand, that I only think a number as such, which might then be five or a hundred. Then my thought is more the presentation of a method for presenting—in accordance with a certain concept—a multitude (e.g., a thousand) in an image, than this image itself. Indeed, in the case of a thousand I could hardly survey that image and compare it with the concept. Now, this presentation of a universal procedure of the imagination for providing a concept with its image I call the schema for that concept. (A140/B179–180)

It is this kind of passage that justifies our saying that Kant, for all his radical reorientations of the traditional conceptual topology of imagination, did not totally detach himself from it. (If he had, it is not clear that he could have allowed himself to talk about imagination in any sense at all.) Imagination does not simply work (individual) intuitable images into the manifold of sensibility; it provides the rule–schema along which the mind moves from conceptual cognition to images and back. The schemata of quantity in their pure use allow us to relevantly "*unitize*" things of our experience. We can then repeat the units to allow the establishment of a *plurality* of units, and we can unite some of the pluralities into new plural units (five units taken together constitute a new unit we call five). *All* the things we generate according to this fundamental rule of quantity belong to the same quantitative

[57]The examples implicitly show that there will be further, more particular schemata: Pekinese, greyhounds, right triangles, scalene triangles, etc.

field. Because these schemata are flexible rules, it does not matter whether what we are dealing with is as concrete as traffic cones, as geometrical as points or line segments or sides of n–gons, or as ultimately abstract as "unit" pure and simple.[58]

One can take schematism as literally or idealistically as one likes: that is its virtue. As with Plato, it allows one to implement forms in mathematical, physical, and imaginative formats, but without the commitment to an ontology of ideal forms. In view of Aristotle, it justifies an even more complicated and aggressive version of the thesis that there is no thinking without images and that intellect sees forms in things. As with Descartes, it attributes an inner dynamism to imagination, which is constantly formative of appearances with potential cognitive relevance. The notion of schema can even work within fairly traditional versions of faculty psychology without committing itself to an introspective science or ascribing to human psychology powers that are evidenced solely in the private inwardness of individual consciousness. Even the "I think that..." of the transcendental unity of apperception is there for everyone around to hear as an utterance. The schema accounts for our ability at a moment's notice to begin figuring things out—that is, to think current presentations in terms of other presentations, and to move freely between different tiers of experience (at a minimum that of sensibility, that of the pure forms of intuitions of mathematics, that of natural causality, and that of human conceptuality—which is not even yet to include human ethical practice, the articulation of purpose or purposiveness, and the experience of the realm of beauty). Presentation (*Vorstellung*) at the sensible level is already impregnated with the presentations of concepts, and the conceptual is always capable of being exemplified in images. Images, whatever other content they may have as being quasimaterial, are intelligibly structured. Thus, with the doctrine of the schemata of empirical and even more of *a priori* concepts, we reach an historical extreme in one of the basic possibilities of the conceptual topology of imagination. Through schematism, the understanding imaginatively forms sensibility precisely so that it is "markable," nameable, and describable by concepts.

Kant is perhaps the first to think out the occluded-occulted tradition of imagination in its fullest consequentiality.[59] He brought the incipient, appearance-provoking,

---

[58]The schemata of quantity as Kant explains them in the quotation above can, as transcendental functions of imagination, be expanded into a criticism of Descartes's exclusion of imagination from the clear and distinct understanding of a chiliagon. Kant's example, proceeding from the schema of five to the schema of number capable of producing, in principle, any determinate number whatsoever, shows that schemata can be more and less abstract, but that as functions of transcendental imagination they are as necessary for producing an image of a thousand (or a thousand-sided figure) as for an image of five (or a five-sided, or even a three-sided, figure). An imaginative schema is implicit in a chiliagon or an n–gon, even when it is not drawn. Thus Descartes can show that there is a difference between imagining/conceiving a triangle and imagining/conceiving a chiliagon, but the difference does not mean that conceiving a n–gon can completely prescind from or leave behind the functioning of imagination.

[59]The "perhaps" leaves open, for example, the possibility that someone like Fichte thought out the consequentiality with even greater ruthlessness. But Fichte considered himself a follower of Kant, though more radically Kantian than Kant.

intelligible *dynamism of imagining* to a high point that was not exceeded even by the romantics, who adopted (but also routinized) Kant's specific conceptual topography. He presents a multiplicity of fields (starting with the manifold of sensibility) produced by synthesis that match the variety of levels on Plato's divided line, he goes beyond Aristotle's articulation of fields of sensible/imaginable qualities structured by contraries with his theory of transcendental and empirical schemata that give conceptual structure to images. In addition, he explains the thoroughgoing temporality of human existence more satisfactorily than Aristotle (for whom time is a measure of change, with human beings the principal measurers). More fully and satisfactorily than Descartes and his successors he implements the fundamental mathematical character of nature. Since nature is the totality of subjective appearances to human beings, and those appearances are all synthesized as a temporal and spatial unity of the manifold even before any particular sense quality or sense object is produced, there is a thoroughgoingly mathematical (and scientifically physical) character to the world.

Kant expected that future scientific and philosophical research would further explore and refine the implications of transcendental philosophy. Scientific understanding was, after all, the explanatory articulation of the manifold of sensibility understood as the totality of nature—in accordance with the pure intuitions of space and time, with the pure concepts and the pure principles and schemata of the understanding, and with the concepts and principles of an *a priori* physics (of the kind that he began elucidating in one of his last works, *The Metaphysical Foundations of Natural Science*). But he also knew that the sciences would never be able to jump over the shadow of the appearances to get back behind them. They might speculate about some region behind the appearances, of course, but to cross over the boundary of sensibility puts one in the realm of thinking without a net of evidence, thinking about no–things in particular with only the vague analogy of concepts that are originally and rightfully directed toward the explication of the manifold of intuition. Sciences that make the claim to encounter things–in–themselves end up not as science but as speculative metaphysics: just as dogmatic as the most dogmatic systems that Kant believed he had forever banished to the wastebin of useless but instructive efforts of human inquiry. In human beings, what keeps experience grounded is not reason, which ultimately pushes us into questions we cannot answer, but imagination in its transcendental functioning.

7.8 Aesthetics, Ethics, and the Limits of Kantian Imagination

In the introduction to the Third Critique, the *Critique of Judgment* (1790), Kant explains that the three critiques are founded on a basic division of human powers. Theoretical reason is considered in the First Critique; in particular the work teaches how sense experience provides content for judgments that count as knowledge about objects. The Second Critique is about practical reason and how we can bring order to desire by universalizing our judgments of duty (expressed in maxims) as rational commands. The Third Critique treats of the human power of *feeling pleasure and displeasure*. Pleasure and displeasure seem to be imprescindibly subjective and impossible to encompass in universal judgments, much less *a priori* ones. But by focusing on purposiveness in experience, Kant shows that there are special ways in which pleasure and displeasure can be involved with universal claims that are directed toward objects taken aesthetically—even if those claims cannot achieve objectivity pure and simple.

As we have seen, in the First Critique imagination is essential both to unifying the manifold of sensibility under the intuitable forms of space and time and to schematizing the very possibility of objects according to the pure concepts of the understanding. In the Second Critique Kant keeps imagination at bay, because sensibility in any form would detract (Kant thinks) from the *a priori* universality of the commands of reason. But in the Third Critique imagination plays a central role in achieving aesthetic experience that is stable enough to take on a universality in form and a regulated pleasure that, if not objective, is nevertheless disinterested. The key to this universality with respect to aesthetic judgment is purposiveness.

Purpose, having and pursuing a goal in the proper sense, occurs only in the realm of freedom: that is, in the practical activity that is the field of the Second Critique. In studying I purpose to learn; in speaking to another I purpose to be understood; in assisting someone in need I purpose to carry out a maxim governing my ethical action. Ancient and medieval science erred by thinking that the realm of nature has purposes in this sense. A stone does not fall toward the center of the earth because that is its natural goal, it is because of gravity, a mechanical cause—and mechanical science is governed by physical law, not elective purposes. Yet there is no doubt that some complex natural phenomena display a structure that suggests something like purpose: for example, in many activities of living things there appears to be a structure of the *in–order–to*. A plant takes in water and nutrients through its roots in order to grow; a spider spins its web in order to capture flies; an elephant makes a trumpeting sound in order to signal its mate. The appearance of purpose even where there may be none Kant calls *purposiveness*. The purposiveness of nature presents itself in experience in two ways, aesthetically and logically. The logical presentation occurs "on an objective basis as the harmony of the form of the object with the possibility of the thing itself according to a prior concept of the thing that contains the basis of that form" (Akad. 5: 192)[60]; this logical purposiveness is the concern of the second half of the Third Critique and encompasses examples from nature like those mentioned in this paragraph.

Properly speaking, I cannot know that any living thing has a purpose, at least not in the sense that Aristotelians designated "final cause." As a rational being I know that I pursue purposes; in the realm of nature, on the other hand, I expect to find

[60]Quotations of the Third Critique are drawn from Kant 1987 [1790]; page numbers are given according to volume 5 of the Akademie edition of Kant's works (Kant 1900 ff., indicated henceforth as "Akad."), numbers that are given in the margins of Kant 1987 [1790]. Square-bracketed phrases are the translator's suggestions.

mechanical causes having to do with matter, motion, and objective forces. Yet I also recognize—this is an important motivating theme in the second half of the Third Critique—that attempts to explain life *totally* by means of mechanical causation fail, and are likely to continue to do so (see Akad. 5: 400). This leads me, as a rational being, to suspect that something like purpose is possible in nature, even if it is purpose as narrow as self-preservation, maintaining or cultivating one's being as the kind of thing one is. This feeling that there are such purposes is, properly speaking, *purposiveness*. It has a plausible rationale but not a definite natural reason or cause. Indeed, if it had such a definite natural reason or cause in mechanical scientific explanation, it would not be a purpose at all.

The first half of the Third Critique deals with purposiveness in aesthetic presentation. It can be differentiated from the logical form in that it is not primarily oriented to an object presented to us as an object of understanding. Aesthetic purposiveness is "on a merely subjective basis: as the harmony of the form of the object (the form that is [manifested] in the *apprehension* (*apprehensio*) of the object prior to any concept) with the cognitive powers—i.e., the harmony required in general to unite an intuition with concepts so as to produce a cognition" (Akad. 5: 192).[61] That is, the appearing form of the thing in our organized and schematized sensibility *seems* to "fit" with the powers of knowing, as though it were purposefully made for them, but it is not (at least not in the first instance) a question of placing the thing in a cognitive category. Rather, the way that sense, imagination, and understanding usually interact produces a sense or feeling of harmony.

The principle of purposiveness assures that my experience of nature will be an experience of nature as a whole related to the totality of my powers. In transcendental philosophy, this means it is not a merely contingent or merely empirical psychological quirk. It is, rather, built into the very structure of our experience from the beginning. If there are purposes in a nature that is not thoroughgoingly purposeful (for example, in a nature that is otherwise mechanical), we would have to arrive at the concept of that purpose, and its verification, through a process of induction. Induction requires a few successful examples that are verifiably unified by a concept. If squirrels, trees, and eyes seem to us to have purpose or to be purposes, we have the dual problem of not being able to reconcile this aspect of appearance with mechanical science and not having any directly self-exhibiting concept of purpose to justify it. Yet our mind is primed to search constantly for unities in the manifold of appearance. If the human mind is constituted to seek unity, that means that the human mind is from the start governed by the principle of purposiveness (or even by a purpose, though what that is is unknown to us)—even if purpose is not a category fundamentally schematized in the manifold.

The logical and aesthetic aspects of purposiveness typically occur together. They are, however, distinct in what they refer to and how they refer to it, and it is the

[61] Notice that the last clause appears to evoke schematism, transcendental imagination's production of a link between concept and intuitable image/appearance, and suggests by "harmony" a special kind of unity and relation not simply reducible to the categories of quantity (unity, multiplicity, allness) or relation (substance and accidents, cause and effect, and interaction).

outcome of the process that determines whether we are dealing with something belonging to the understanding or with a matter of properly aesthetic judgment. Kant takes the example of the senses presenting things outside me (say, a forest in spring). Focusing only on the space I perceive: the space in which the things I see are displayed is a subjective feature of the presentation (because space is produced by the transcendental imagination, not by the forest's extent in space "in itself"); and in general I cannot assume that the forms of presentation in sense tell me what things are in themselves. But even though this presentation of space is subjective, space is a necessary element in our cognition of natural things as appearances. Thus on the one hand space is presented in a merely subjective way that is not binding on the *nature* of things, yet it is also associated with the necessary conditions of *knowledge* of the *appearances* of things and consequently has an *objective* reference as well (that is, in the appearance of the object called *forest*). Kant then points out that there is a subjective feature of presentation "which cannot at all become an element of cognition": the pleasure or displeasure connected with the presentation. The pleasure or displeasure experienced is not like space in the preceding example, for there is no pleasure or displeasure to be ascribed to the forest per se (although perceiving the forest can produce or cause pleasure in you or me, given the right subjective conditions).

Such experiences, Kant says, "refer the presentation not to the object but solely to the subject; and the pleasure cannot express anything other than the object's being commensurate with the cognitive powers that are, and insofar as they are, brought into play when we judge reflectively, and hence [expresses] merely a subjective formal purposiveness of the object" (189–190). What this means is that all the powers of mind that are called into action when we attempt to understand something are also primed to shape the subjective apprehension of the scene: they are potentiated and ready to go, they are implicitly but not expressly in action. The pleasure that we may eventually feel in coming to terms with an object is an experience of the adequacy and appropriateness of our powers to the appearances. It gives us a sense of "fitness": a fitness of our powers to the object's appearance *merely as appearance*.

Even more intriguing is what Kant says immediately thereafter. It is a landmark passage in the history of theories of imagination:

> For this apprehension of forms by the imagination could never occur if reflective judgment did not compare them, even if unintentionally, at least with its ability [in general] to refer intuitions to concepts. Now if in this comparison a given presentation unintentionally brings the imagination (the power of *a priori* intuitions) into harmony with the understanding (the power of concepts), and this harmony arouses a feeling of pleasure, then the object must thereupon be regarded as purposive for the reflective power of judgment. A judgment of this sort is an aesthetic judgment about the object's purposiveness; it is not based on any concept we have of the object, nor does it provide such a concept. When the form of an object (rather than what is material in its presentation, viz., in sensation) is judged in mere reflection on it (without regard to a concept that is to be acquired from it) to be the basis of a pleasure in such an object's presentation, then the presentation of this object is also judged to be connected necessarily with this pleasure, and hence connected with it not merely for the subject apprehending this form but in general for everyone who judges [it]. The object is then called beautiful, and our ability to judge by such a pleasure (and hence also with universal validity) is called taste. (Akad. 5: 190)

These compact remarks contain the core of Kant's understanding of aesthetic judgment, beauty, and taste. A detailed explication would take us far beyond the limits of imagination. The nub, however, is this. In the context of our efforts to understand the natural world (which Kant takes up in the second half of the Third Critique), things presented to our awareness can produce in us a pleasure that is not individual and contingent pleasure but something that all human beings are capable of feeling because of the *general* conformity of their faculties to the appearances of nature. We can come to *understand* things by virtue of the transcendental principles that organize nature according to schematism (which connects concepts to image–forms), and we feel–and–"know" this before the understanding has even commenced its work. (The feeling and knowing come together without at first being clearly distinguishable, and the knowledge is not objective but a feeling of knowledge or possible knowledge—therefore the expression feel–and–"know" takes the hyphenated and ironic quotemarked form I have given it.) Before the understanding commences its work, however, this feeling can be only a subjective *anticipation* of what will be fulfilled ultimately (and later) in a scientific or conceptual understanding of what presents itself. When the moment of understanding arrives we take a different pleasure, in the *accomplishment* of scientific judgment, though Kant has little to say about it.

Something different is happening, however, when we concern ourselves only with the appearance of what has presented itself to our sensible intuition (the concern of the first half of the Third Critique). I will speak here chiefly of art, although one of the most remarkable things about Kant's explanations, implicit in what I have just discussed with respect to the forest, is that for the most part they appeal not to works of art but to the appearances of natural things and natural vistas. *Art* is not made in general to illustrate or exemplify *scientific knowledge*; analogically, we have no reason to think that natural things exist and appear to us only for the sake of being understood scientifically. In the production of art what counts is that the artist is trying to express in the sensible realm supersensible ideas, sometimes of ultimate human meaning, and often in a form that is not consciously apprehended by the artist herself. There is actual purpose in works of art, of course, at least when judged from the artist's expressed intentions. The viewer (or hearer), in the presence of the appearance–form of the artist's object, undergoes an experience similar to the kind Kant describes in the case of viewing natural things in anticipation of understanding them. But the logic of aesthetic experience is not a logic of understanding or of concepts. The aesthetic logic cannot in fact be expressed conceptually (and if it can, so much the worse for the beauty of the so-called work of art or the beauty of the natural vista). As with the presentation of sense in the context of science, there is possibly an anticipatory pleasure, but the full pleasure of the aesthetic experience as such is a result of a *play* of the faculties, of an interplay or interaction that is not intrinsically governed by the purposes of science, much less of morality. If this playful interaction achieves a harmony or balance, that is due not to the individual subjectivity of the viewer of the art or vista but to fundamental characteristics of the transcendental psychology of human beings as such.

At the transcendental level there is something shared by all human beings, since we all have sensibility, imagination, and understanding. Thus there is in principle the possibility of a universal communication of one's mental state in the presentation of something to sensibility, a *communicability* that is *not conceptual.* That is, the judgment form "A is B" can be filled with a content—the beauty of a brilliant crimson sunset—that is humanly shareable but is not like the judgments of either science or morality. It is important to recognize how paradoxical this is, at least at first glance: all judgments unite concepts, but here a judgment is universal in its claim (in the sense that you and everyone else should agree with, or at least see the point of, my claim that the sunset is beautiful) but nonconceptual, because no concept fully captures the experience I am having and the associated pleasure.

Later, Kant gives a fuller explanation of the transcendental basis of this universal communicability of feeling:

> If, then, we are to think that the judgment about this universal communicability of the presentation has a merely subjective determining basis, i.e., one that does not involve a concept of the object, then this basis can be nothing other than the mental state that we find in the relation between the presentational powers insofar as they refer a given presentation to *cognition in general.* (Akad. 5: 217)

When this happens, the cognitive powers brought into play by this presentation are in free play, because no determinate concept restricts them to a particular rule of cognition. Hence the mental state in this presentation must be a feeling, accompanying the given presentation, of a free play of the presentational powers directed to cognition in general. Now if a presentation by which an object is given is, in general, to become cognition, we need *imagination* for combining the manifold of intuition and *understanding* for providing the concept that unites the presentation. This state of *free play* of the cognitive powers that accompanies the presentation by which an object is given must somehow be universally communicable. In cognition, it is a *conceptual* determination of the object given in presentation that produces a "harmony"—the concept or the conceptual judgment—that holds for every subject.

Kant's point is not that we reduce artworks or striking natural vistas to concepts—we do not—but rather that our typical aspiration to understand sets our imagination and our understanding into a state of trying to make conceptual sense of what is being seen. There is a sense or anticipation of meaningfulness that comes first, which spurs the interactive play of the faculties not just in those with artistic temperaments but in all human beings. Although this interplay is not expressible in a conceptual judgment, it strives to communicate itself in a judgment of taste, in particular in the judgment that something is beautiful. This judgment is universal in its form, but unlike scientific judgments it is not objective but subjective; yet it is subjective in a manner that is truly about an "object" (about the intuitable form of the appearance) rather than just my feeling about it. As the result of the interplay of imagination and understanding, the judgment *in general* is the same for all human beings. Thus it provides a genuinely universal basis for the assessment of aesthetic qualities, like beauty, derived from this interplay.[62]

[62] In his analysis of taste and genius later in the Third Critique, Kant argues further that the *sensus communis* is the ground of the communicability of aesthetic judgment, to be treated later in this section.

Much further on, in section 50 of the *Critique of Judgment*, Kant writes that the imagination produces a wealth of thought, and spirit animates the mind. "What this principle uses to animate the soul, the material it employs for this, is what imparts to the mental powers a purposive momentum, i.e., imparts to them a play which is such that it sustains itself on its own and even strengthens the powers for such play" (Akad. 5: 313). Notice the synonymy that is established. A purposive momentum in the mental powers is a play that sustains itself unto itself, and it strengthens the powers that play and even the vigor of the play itself. An analogy might be taken from what happens when political powers are divided, for example into executive, administrative, judicial, and legislative powers. Powers divided are powers augmented. When there is only one authorized user of power, many possible expressions of power will lie fallow; but when there are several users, the power that is not put to use will often be coopted by one of the other users. Exercise of powers becomes interactive and competitive, and the total expression of power increases.[63]

Intellect or understanding per se is indifferent to beauty. Sensibility is busy providing matter for intuition in space and time according to schemata. Feeling is occupied with pleasures and pains. Understanding tries to survey the appearances and works to bring them under some kind of unity. Imagination synthesizes; as it shuttles back and forth between the other powers it spurs them on to different considerations, surprising appearances, new pleasures and pains, so that their mutual indifference now becomes a search for an equilibrium. What one gets, in the case of beauty or any similar predicate,[64] is a harmony in/of all the powers, one that would not be possible or even conceivable without the interplay. Harmony is an equilibrium point—call it a solution to the problem of synthesis—that is not proper to any of the powers individually, only to all working together. Finding equilibrium requires a manifold activity of trying to achieve balance.

---

[63] This is borrowed from Arendt 1970, esp. 40–45. It is perhaps not accidental that Arendt was a student of the political relevance of Kant's theory of imagination; see Arendt 1982.

[64] Kant treats one other aesthetic predicate in the Third Critique, that of the sublime, the awe-inducing—examples would be our aesthetic response to vistas of the Grand Canyon or Niagara Falls—which ends not in the harmony of the faculties but in a recognition of an ultimate disproportion between our imagination and reason. In the face of what causes a feeling of unease or threat because it exceeds the limits of our imagination, the unlimited capacity of reason allows us to see ourselves placed as rational beings above and beyond such sensible or imaginable threats. It is significant that both Kant and Descartes came to recognize the true scope of human reason in contrast with the large but limited capacities of imagination, though Kant drew an aesthetic consequence that he understood as a symbol of the moral status of the human being, whereas it led Descartes in the direction of the metaphysical distinction between thinking and extension. In both thinkers, the mind makes its discovery by shuttling between the plane of the imagining power as a whole and what presents itself as exceeding it.

There is no reason to assume that the aesthetic predicates in Kant stop at two. When one looks at certain art works, one gets an overall impression of rest; with others, a sense of dynamism. "This work is restful" and "That work is dynamic" would thus be aesthetic judgments with universal claim based on a feeling of an equilibrium in the play of the faculties. This means that the Third Critique justifies not only judgments of beauty and sublimity but also extended, aesthetically interpretative discourse. See Makkreel 1990.

Earlier, in section 40 of the Third Critique, Kant suggests that there might well be a common ground for human taste in *sensus communis*. He does not mean the medieval *sensus communis*, which was one of the internal senses alongside imagination, memory, and the basic estimation of good and bad, although what he says about it does not entirely exclude these. Kant's common sense is the developed and educated power of the human sensitive reaction to the appearances, things, and forms that the world presents; it thus includes something of what people mean by horse sense, although it is intended as a public, aesthetically formal sense. Here is what he says:

> we must [here] take *sensus communis* to mean the idea of a sense *shared* [by all of us], i.e., a power to judge that in reflecting takes account (*a priori*), in our thought, of everyone else's way of presenting [something], in order *as it were* to compare our own judgment with human reason in general and thus escape the illusion that arises from the ease of mistaking subjective and private conditions for objective ones, an illusion that would have a prejudicial influence on the judgment. Now we do this as follows: we compare our judgment not so much with the actual as rather with the merely possible judgments of others, and [thus] put ourselves in the position of everyone else, merely by abstracting from the limitations that [may] happen to attach to our own judging; and this in turn we accomplish by leaving out as much as possible whatever is matter, i.e., sensation, in the presentational state, and by paying attention solely to the formal features of our presentation or of our presentational state. (Akad. 5: 293–294)

Thus educated taste lets go of the merely sensory in favor of the formal. Color, savor, aroma, feeling, and tone are released in favor of the formal alone. In visual art he is leaving us with a black-and-white world, it seems, unless black and white themselves belong too much to the merely material realm. Perhaps then it is a world that is blind, dumb, and flavorless.

Were these not Kant's words, I would be tempted to call them obtuse. They are at least a disappointment. The disappointment is connected as well with his understanding of the possible judgments of others, as he explains later in section 40. In presenting the maxim that we should think from the standpoint of others, he says that we can bypass the *actual* judgment of others in favor of their *possible* judgments. But this would mean that already in our *conception* of *another's* judgment we abstract from the particular details of anyone else's mind in order to recognize the purely formal features of our own presentational states. We are looking for the single unifying form that should be the skeleton of all correct judgment about the object in question; the judgment of another is helpful merely in displaying incorrectness and one-sidedness in our own judgment. If everyone followed this method, presumably each would converge upon the single formal truth behind the variety of presentations. One might well question whether this is a method calculated to acquaint us with the possible judgments of others or rather to do away with having to make a good faith effort at trying to think the way they do. It could be the pretext of a philosopher who does not care to approach the unfamiliar too closely and who prefers the formality of concepts to the messiness of the manifold of sensibility.[65]

[65] For a further discussion, see Sepper 2013.

Such defects in Kant's theory of taste also affect his theory of artistic genius. The eighteenth century witnessed a prolonged development of the notion of genius, as the craftsmanly conception of the arts gradually gave way to the notion of unexpected artistic creation. The upshot was that the artist came to be conceived as someone whose imagination simply could not be accounted for.[66] Kant ratifies and philosophically seals this conception by saying that nature prescribes its rules to the rest of us, whereas the genius prescribes rules to nature. In the last analysis genius "explains" what the genius artist produces by asserting that it is something that cannot be explained by rules. At what level the genius does this—whether in the formlessness of sensory matter, or in forms beyond those of transcendental imagination and understanding—Kant does not explain. He asserts further, however, that the rules prescribed to nature by the rule-prescribing genius cannot be expressed as rules; therefore they cannot be taught. One does not have to be an antiromantic to be appalled by this Kantian dictum. The history of the arts is full of episodes of creativity that schooled the future, and even if some works of genius do beggar both our imagination and our understanding, we learn from them nevertheless, even if not in the form of rules. Artistic styles are often born at the hands of genius, and those who come afterward are always able to learn a great deal about those styles[67]—although perhaps, when all is said and done, there must always be an element in the greatest works of art that escapes analysis. That might be the truth Kant was getting at. Yet truly great works of art are few, even in the oeuvre of the greatest artists, so it is only rarely that they prescribe to nature in a way that cannot be taught. One thing is clear, however: attempts to understand the psychological effects of works of art are better approximated by artists' rules of practice ("rules of thumb") than by the dumbfounded reaction of a nonartist staring at a great work. The rules of thumb are based on an incipient recognition of forms; they teach us to deal imaginatively with what we have not reduced, and perhaps cannot reduce, to concepts. In that sense they are the measuring stick that, when exceeded, lets us recognize genius.

In the 1790s the German poet, statesman, and sage Johann Wolfgang von Goethe, having already for more than a decade undertaken scientific and technical studies with respect to mineralogy, botany, animal morphology, and color, was inspired in his methodological reflections by Kant, and especially the Third Critique. Particularly in the theory of color, he had recognized that prevailing conceptions of science ignored sensory qualities. His research showed, however, that the visible qualities of hue reveal a network of formal relationships between the colors (for instance, what we call complementary and contrast colors, in both the physical and the physiological senses), and he sought to devise experimental techniques and methods that would highlight such formal and relational properties. Here the much-controverted question of whether he understood Newton's theory can be set aside, although for the purposes of what he was trying to show one has to say that he understood Newton well enough, and that Newton did not explain or in many cases even notice these particular phenomena.

---

[66] See Sect. 7.4, above.

[67] See Sects. 3.7 and 3.9, above.

Despite the fact that his philosopher friends told him his understanding of Kant was a "strange analogue" of the transcendental philosopher's thought, he understood well enough the conceptual topology that underlay Kant's claims about purposiveness in the experience of nature (including its aesthetic aspect) and the interplay of the faculties as they reckon with an orderly but unexplained appearance. One of the fundamental organizing principles of Goethe's approach, as he came to see later, was that of polarity. But that was in an important sense a rediscovery of the Aristotelian topology of the senses: the qualities of sense range proportionally across a gamut that stretches from one extreme to another, from one contrary to its opposite, from one pole to the other pole. This is a type of formal relationship that cannot obtain without the qualitative appearance of the qualities. Talking about contrast colors without reference to the "matter" of color is talk by the blind.[68]

In his methodological reflections Goethe recognized that thinking and experiencing things like others did in fact require an educated *sensus communis*, but unlike Kant he saw that the education had to be sensory and imaginative as well as formal–intellectual. He conceived already in the early 1790s, from his study of the history of several sciences, that there were different styles of *cognitive imagination*. When faced with a phenomenon, there were some whose first inclination was to think genetically, in terms of how it had been generated in time; some were analytic, inclined immediately to divide it into parts; there were those who looked for mathematical structures, others who searched for material principles; those who would seek the phenomenon in all its more complex forms, versus those who immediately look for the simplest form. That is, each person's approach to understanding natural things would be marked differentially by styles of questioning that direct the mind to take its very next step in one of many possible directions. Goethe called these *Vorstellungsarten*, presentation–types, and his mature history and philosophy of science is predicated on their variety. Unfortunately the polemic that broke out over his rejection of Newton's color theory obscured Goethe's discovery of the constellation, even system, of ways of cognitive imagining. A rising tide of positivism led researchers to claim that they were guided only by the facts. Presumably they had learned to think in the place of others in the precise, unidirectional manner Kant had recommended.

For Kant, intuition of the manifold of sensibility requires both affection–receptivity and spontaneity. Imagination in its transcendental use unifies the manifold of

---

[68] A sign of Kant's failure to see that his fundamental principles might give rise to forms inalienable from qualities is the poverty of his "anticipations of perception," the second group of "principles of the pure understanding." These principles are, in essence, *implementing schemata* for the pure concepts of the understanding; thus they are a function of imagination in one of its transcendental uses. The anticipations are universal features of sensory perceptions due to the schematism. The only anticipation Kant identified was that of intensity, which is simply a principle of less and more, down to zero and indefinitely upward in intensity (as with pain and brightness). But the pure concepts of the understanding, with their dialectically productive dualities and triplicities (the first two of each group give rise dialectically to the third; see the First Critique, B110–111), ought to have provided resources for conceiving as well that qualities can be structured by contraries. By recognizing and developing this notion, Goethe was more faithful to Kant's approach than Kant himself.

sensibility spatially and temporally and then schematizes and weaves into the manifold the concepts and the principles of the understanding, so that we experience an object-filled world of articulated sense qualities and relations and not just random sensations. But there is the rub: what is properly sensible in the manifold is what is passively taken up by imagination and understanding, so color and other Lockean secondary qualities/Aristotelian primary sensibles are no more than the bare matter that the spontaneous powers mold into experience. Color fills in the blanks of the geometric shapes; imagination paints by the numbers—or, rather, does not paint the colors at all, since they are due solely to the passive receptivity of sensibility. It is mere "matter," and as we know from the long history of philosophy, matter always takes a back seat to form. But as the early chapters of this book argue, and the Aristotelian treatment of primary sensibility as structured by contraries in a substratum suggests, there can be and in fact there are formal principles that are intrinsically embodied in the substratum–matter of sensation.[69]

The problem for Kant appears to be something like the following. The basic powers that he identifies have to be kept distinct from one another, in some fairly strong sense of "distinct." There has to be a real difference between reason, imagination in its productive uses, imagination in its merely reproductive uses, desire, and feeling. For example, in its pure uses, reason must be separable from imagining and desiring, so that it does not turn out to be merely a hidden function of these others. In its use as understanding, it may govern the work of the imagination (through the pure concepts of the understanding), but it must not be tainted by the contingencies of reproductive imagining (especially when that reproduction involves sensations).[70] With imagination, there has to be a

[69] I mentioned in the preceding note that the principle of the understanding called the "anticipations of perception" does impose the formal feature of intensiveness on qualities; thus colors should be brighter or darker, more or less matte, and so forth. But after Newton irrevocably displaced the Aristotelian theory that color was a dynamic mixture of light and dark, there was no longer a plausible way to account for differences in chroma or hue (to use technical terms from color science to identify the feature that makes color appear as specific colors) as a difference in intensity. Pale green and intense green with the same chroma can be differentiated according to the presence of more or less whiteness, but a yellowish green and a bluish green, much less yellow and blue, cannot be differentiated in the same quantitative way.

[70] This is true insofar as the schematism of contingently acquired concepts, like triangles and dogs, partially governs any reproduction of actual triangles drawn in the sand, or of dogs playing in it. But this does not contaminate the pure uses. First, these two concepts are produced in accordance with but not deduced from the pure concepts, so the particularity in them (as matter) is distinguishable from the universality (as form). Second, schematism correlates the concept with images, but only with the formally effective aspects of the images. What allows me to recognize a dog as dog is not its color or aroma but its shape and the configuration of its organs and body parts; and when I reproduce the figure of a dog, color and aroma are ordinarily not necessary. If they are, they once again are not essentially bound to the concept. Of course dogs cannot naturally be green or smell like lilacs, so even here there is potentially the need to redraw the boundaries between the faculties. But perhaps we should be given pause by this consideration: many animals, including dogs, can identify and distinguish their own kind by smell. It is not inconceivable that similar phenomena can occur with (some) human beings. This is another sign that Kant overlooked a matter of potential scientific and philosophical importance.

clear separation of the functions of productive and reproductive imagination, especially when the latter is conceived apart from schematism and therefore more in line with traditional kinds of image reproduction.

As we mentioned earlier, in the Third Critique's introduction Kant explained that the three Critiques each dealt with a separate fundamental human power: thinking, desire, and feeling. What makes thinking cognitive is form, what brings desire under control is form, and what makes feeling universalizable enough to make an aesthetic claim is form. We have only mentioned a rather conspicuous fact about imagination in the three Critiques: that it is almost entirely absent from the second, the *Critique of Practical Reason*. As he explains in its preface, both the First and the Second Critiques are about pure reason: pure reason in its theoretical use and pure reason in its practical or moral use, respectively. The first teaches how the matter of sense and the intelligible structuring by transcendental imagination give rise to propositions that embody knowledge, the second how the ability to find maxims that guide our actions leads to a univeralizability culminating in absolute commands of reason (the categorical imperative). We have propositions in the declarative mode in the one case and propositions in the imperative mode in the other.

There is much more to Kantian ethics than the categorical imperative, of course, but the advantage of the imperative is that it isolates precisely what makes an action right: not pleasure, not utility, not doing good for ourselves or our friends, not obeying the command of a parent or a leader or a God, but conformity to a duty that we legislate for ourselves and simultaneously recognize as an obligation for all rational beings whatsoever. There is no doubt that this lends a certain formalism to Kant's ethics. If someone shows up at your door determined to kill you or your brother, you have no *right* to lie. The categorical imperative allows no exceptions. In a way there is something rationally beautiful about this rigor, and Kant's point is that once you start making exceptions there may be no limiting them. Yet one might wonder, in a quite literal sense, whether this does not betoken a certain poverty of moral and political imagination.

I do not intend in this case to accuse Kant of obtuseness.[71] He was lucky that he did not have to experience twentieth-century totalitarianisms and the ethical and political crises they produced. Moreover, one cannot simply assert that a more educated *moral imagination* overlooked by Kant would automatically save us from wrong. Yet there is a point in the Second Critique where the imagination might have naturally fit, a position analogous to that in the First and Third Critiques. Kant explicitly rules out imagination at that point and instead substitutes what he calls

---

[71] I acknowledge that in the past several paragraphs I have been somewhat unfair to Kant. My criticisms have nevertheless followed genuine structural vectors implicit in his transcendental psychology. Only by examining the topography and tendencies more carefully is it possible to justify Kant's position or, alternatively, to find some other possibility that takes one beyond Kant in a truly Kantian way.

Typik, typics or typology, which, he says, has nothing sensible or imaginative about it. As far as he was concerned, the sensible realm, and the imagination insofar as it occupies itself with sensible appearance, can only confuse the issues moral reason is concerned with.

Is there any way for imagination to reenter these considerations in a Kantian way? If a friend comes to us desperate for help and the situation he explains seems impossibly confused, we have to engage a kind of practical imagination to conceive the various scenarios that might show us a way forward. Moreover, the whole effort of rising to the categorical imperative seems to require a complex analogue of *schematism*: one is looking for a concept that allows one to move from a real-world situation to a command in universal terms by way of an intermediate degree of generalization, the maxim. Where do maxims come from? There is a dearth of reflection about this in Kant's writings, yet he sees choosing good maxims as the best guarantor of ethical action. Is it not possible that arriving at a maxim is precisely the work of imagination in yet another transcendental function? Transcendental imagination "fits" the manifold of sensibility to understanding by way of schematizing the understanding's foundational concepts and principles. Why cannot something very much like it be at work in fitting our practical actions to the complex setting of the natural-social world?[72]

The test applied in universalization is a strictly logical one—contradiction—but otherwise there are many places for imagining to enter. The aim that one is pursuing in the action is one such place. Sometimes we just respond to a situation with a kind of automatism, so we hardly think or reflect on it; but especially when we are considering an action prospectively we often engage imagination. This is crucial if we are to arrive at a *real* action. Suppose we have the impulse to give a gift. What is the maxim? Maybe there is not one to begin with, only a generous impulse, and the only way to find one is to consider further what it is we want and how we conceive fulfilling the action. Wanting to do something nice for a person might lead us to consider giving a gift, and it seems universalizable in an appropriate way: "Always be nice to others." Without any further qualification, that might have us giving gifts to everyone at every waking moment of our lives! More to the point is that before making such judgments we have to think about specifics: about what the person might like, about what would make the gift distinctively ours, about what we can afford, about when we can make a shopping trip and where, etc. In fact doing an insufficient job of thinking and imagining toward concreteness might well lead to premature or inappropriate universalizations and questionable moral action.

These are the very types of consideration that are the substance of ethical action in Aristotle. The reason that Aristotle does not present a *typology* of ethics, a set of types of moral command, is that in his view there is no such thing. One might put it in a dichotomy that would stipulate a refined use of basic moral/

[72] Johnson 1993, 68–74, echoes scholarly criticism of Kant's matter–form dichotomy and goes on to show ways in which his "typics" *requires* metaphor and imagination.

ethical concepts: *morality* is expressible in commands, thus essentially propositional, and located in an abstract realm that strips away all particulars, whereas *ethics* is intrinsically a matter of assessment of situations, and situations are placements in a field of factors interrelated in particular ways that must all be considered. Indeed, the basic ethical fact for Aristotle is not acts but virtues, the *multiple dispositions* to *various kinds* of acts typical in a *polis*/city for citizens who have been raised and educated to live in that *polis*. This does not amount to "ethical relativism," even if a good act indeed has to be related to many different conditioning factors. The *polis* has shaped citizens' tendencies to pleasure and pain into desires and aversions that help them avoid extremes in the pursuit of comprehensive happiness in that specific *polis*. Such a citizen, should he find himself in another city, would be at a disadvantage, but by already having learned to negotiate moderation in pleasures, pains, desires, and aversions of all types he can adapt his current states of virtues to the practices of the new city. There is even a basis here for arriving at certain principles that all citizens of all cities could agree to. But the fact is that even in one's home city concrete thinking—imagination—is needed in order to act ethically. It is only by seeing one's act *in situ* that one can exercise virtue.

In one formulation of the categorical imperative, we must affirm that universalizing the maxim of our action is compatible with our being a member of the community of all rational beings. Strictly speaking, that is not just formal but utopian—nowhere, to be literal about it. Yet Kant, too, has a conception of virtue ethics, and although our possession of reason allows us to see that we are superior to all the contingencies of the world and its material basis, it also ultimately makes us recognize that we are finite beings who have to live here and now. It leads, for instance, to his claim in *Anthropology from a Pragmatic Standpoint* that the best kind of activity in human life is having a delightful conversation over dinner (Akad. 7: 278). But perhaps in the last analysis Kant did not see all the forms of reason's self-subversion that are implicit in its infinite aspiration. That may sound harsh, but in fact the transcendental dialectic of the First Critique was precisely an extended analysis, over hundreds of pages, of the ways in which reason in its cognitive use inevitably drives us to questions that cannot be answered—and, at least until Kant himself discovered critical philosophy, answering them dogmatically was the easiest and most common thing to do. The philosophical temptation to conceive ourselves as part of the community of all rational beings, and to do this full of confidence that there may be a Divine Being who guarantees the ultimate coherence of this rational community with the manifold of sensibility—and with the further confidence that we legislate for ourselves precisely as that Being legislates for us and even for Himself—is a noble temptation, but a temptation for all that. And perhaps it is what German Idealism and Romanticism succumbed to in taking up the conceptual topology of imagination and reason Kant had brought to only imperfect fulfillment.

References[73]

Arendt, Hannah. 1970. *On violence*. New York: Harvest Books.

Arendt, Hannah. 1982. *Lectures on Kant's political philosophy*. Chicago: University of Chicago Press.

Arnauld, Antoine, and Pierre Nicole. 1992 [1662]. *La logique ou l'art de penser*. Ed. Charles Jourdain. Paris: Gallimard.

Baumgarten, Alexander. 1750–1758. *Aesthetica*. 2 vols. Frankfurt/Oder.

Becq, Annie. 1984. *Genèse de l'esthétique française moderne. De la Raison classique à l'Imagination créatrice (1680–1814)*. 2 vols. Pisa: Pacini.

Descartes, René. 1964–1976 (cited as AT, with volume number). *Oeuvres*, 2nd ed. Ed. Charles Adam, Paul Tannery, et al. 11 vols. Paris: Vrin.

Eco, Umberto. 2002. *Art and beauty in the middle ages*, 2nd ed. Trans. Hugh Bredin. New Haven: Yale University Press.

Engell, James. 1981. *The creative imagination: Enlightenment to romanticism*. Cambridge: Harvard University Press.

Garber, Daniel, and Michael Ayers (eds.). 1998. *The Cambridge history of seventeenth-century philosophy*. 2 vols. Cambridge: Cambridge University Press.

Haakonssen, Knud. 2006. *The Cambridge history of eighteenth-century philosophy*. 2 vols. Cambridge: Cambridge University Press.

Hatfield, Gary. 2001. The "Prolegomena" and the "Critiques of pure reason". In *Kant und die Berliner Aufklärung: Akten des IX. Internationalen Kant-Kongresses*, ed. Volker Gerhardt, Rolf-Peter Horstmann, and Ralph Schumacher, 1, 185–208. Berlin: de Gruyter.

Heidegger, Martin. 1929. *Kant und das Problem der Metaphysik*. Bonn: Friedrich Cohen. English edition: Heidegger, Martin. 1997. *Kant and the problem of metaphysics*, 5th ed. Trans. Richard Taft. Bloomington: Indiana University Press.

Johnson, Mark. 1993. *Moral imagination: Implications of cognitive science for ethics*. Chicago: University of Chicago Press.

Kant, Immanuel. 1900 ff. *Kant's Gesammelte Schriften* (known as the Akademie Ausgabe, "Akad."). 29 vols. to date. Königlich Preußische Akademie der Wissenschaften. Berlin: de Gruyter (before 1922 Reimer).

Kant, Immanuel. 1987 [1790]. *Critique of judgment*. Trans. Werner S. Pluhar. Indianapolis/Cambridge: Hackett.

Kant, Immanuel. 1996 [1781, 1787]. *Critique of pure reason: Unified edition*. Trans. Werner S. Pluhar. Indianapolis/Cambridge: Hackett.

Kearney, Richard. 1988. *The wake of imagination*. New York: Routledge.

Leibniz, Gottfried Wilhelm. 1903. *Opuscules et fragments inédits de Leibniz*. Ed. Louis Couturat. Paris: Presses Universitaires de France.

Leibniz, Gottfried Wilhelm. 1981. *New essays on human understanding*. Trans. P. Remnant and J. Bennett. Cambridge: Cambridge University Press.

Leibniz, Gottfried Wilhelm. 1989. *Philosophical essays*. Trans. Roger Ariew and Dan Garber. Indianapolis: Hackett.

Locke, John. 1690. *An essay concerning human understanding*. London.

Loeb, Louis E. 1981. *From Descartes to Hume: Continental metaphysics and the development of modern philosophy*. Ithaca: Cornell University Press.

[73] If there is a second, square-bracketed date, it indicates the year the work first appeared in its original language.

Longuenesse, Béatrice. 1998. *Kant and the capacity to judge: Sensibility and discursivity in the transcendental analytic of the "Critique of pure reason"*. Trans. Charles T. Wolfe. Princeton: Princeton University Press.
Makkreel, Rudolf A. 1990. *Imagination and interpretation in Kant: The hermeneutical import of the "Critique of judgment"*. Chicago: University of Chicago Press.
Malebranche, Nicolas. 1997 [1674–1678]. *The search after truth. Elucidations of "The search after truth"*. Trans. Thomas M. Lennon and Paul J. Olscamp. Cambridge: Cambridge University Press.
McRae, Robert. 1995. The theory of knowledge. In *The Cambridge companion to Leibniz*, ed. Nicholas Jolley, 176–198. Cambridge: Cambridge University Press.
Pippin, Robert B. 1997. Kant on the spontaneity of mind. In *Idealism as modernism: Hegelian variations*, 29–55. Cambridge: Cambridge University Press.
Sepper, Dennis L. 2005a. Cartesian imaginations: The method and passions of imagining. In *A companion to rationalism*, ed. Alan Nelson, 156–176. Oxford: Blackwell.
Sepper, Dennis L. 2005b. Spinoza, Leibniz, and the rationalist reconceptions of imagination. In *A companion to rationalism*, ed. Alan Nelson, 322–342. Oxford: Blackwell.
Sepper, Dennis L. 2013. An obligation to cognitive pluralism? In *Kant und die Philosophie in weltbürgerlicher Absicht. Akten des XI. Kant-Kongresses 2010*, ed. Bacin Stefano et al. Berlin: de Gruyter.
Tamny, Martin. 1979. Newton, creation, and perception. *Isis* 70: 48–58.
Wolff, Christian. 1720. *Vernünfftige Gedancken von Gott, der Welt und der Seele des Menschen, auch allen Dingen überhaupt*. Halle a.d. Saale.
Wolff, Christian. 1732. *Psychologia empirica*. Frankfurt/Leipzig.
Wolff, Christian. 1734. *Psychologia rationalis*. Frankfurt/Leipzig.
Yolton, John W. 1956. *Locke and the way of ideas*. Clarendon Press/Cumberlege: Oxford/London.

Chapter 8
After Kant: Appropriating the Conceptual Topology of Imagination

Uttering a word is like striking a note on the keyboard of imagination.[1] *(Wittgenstein 1953, §6)*

In *Kant and the Problem of Metaphysics* (1929), Martin Heidegger contends that Kant lost his nerve. The B edition (1787) of the *Critique of Pure Reason* reduced the role A (1781) had assigned to imagination and subordinated it to the understanding. Although recent scholarship has argued that the status of imagination does not substantially change between editions,[2] Heidegger still has a point. The most striking revision occurred precisely in the transcendental deduction of the categories, the section that in A anatomized and described the functions of imagination. Version B says less about imagination and more about the functions of understanding.

Heidegger points further to a remark Kant makes about a hypothetical *common root of sensibility and understanding*. Heidegger argues that this common root is imagination, but that acknowledging this would have thrown Kant's division of human powers into abyssal turmoil. Kant supposedly drew back from this abyss—not least because it would have undermined the sharp line separating the passive givenness of sensibility from the free spontaneity of understanding and reason.[3]

Heidegger had specific reasons for interpreting the First Critique as he did. One of the essential insights that led Kant to his critical philosophy was that neither empiricism nor rationalism gave a plausible account of the coherence of the experience of time. They either took time sequence as a brute given in need of no explanation or invoked some grand metaphysical thesis. Although time is one of Kant's two

[1] Das Aussprechen eines Wortes ist gleichsam ein Anschlagen einer Taste auf dem Vorstellungsklavier.

[2] Most persuasively in Longuenesse 1998.

[3] See Heidegger 1929, 37 (§6) and 160–161 (§31). The passage in Kant is at A15/B29 (references to the First Critique will take this form of citing pages in the A edition followed by pages in the B edition, unless the passage in question occurs in just one of them).

D.L. Sepper, *Understanding Imagination: The Reason of Images*, Studies in History and Philosophy of Science 33, DOI 10.1007/978-94-007-6507-8_8,

pure intuitions, it takes more than 150 pages for him to get around to explaining its schematism (with reference to the triad of the categories of relation), which produces the thoroughgoing time–connectedness of human experience.[4] Schematism is, of course, one of the transcendental functions of imagination. For Heidegger in the late 1920s, in the immediate aftermath of *Being and Time* (1927), the transcendental temporality of Kantian imagination might well have looked like an incomplete anticipation of his own philosophical aims. Heidegger claimed that after *Being and Time* it was necessary to think being within the horizon of time, rather than (with the entirety of the rest of the Western tradition) time within the horizon of being. As he had asserted in a lecture course before *Being and Time*, human being (more accurately, *Dasein*) *is* time.[5] Any future fundamental ontology needed to begin with this.

The accuracy and plausibility of Heidegger's interpretation of Kant will doubtless continue to be the object of debate. But where the conceptual topology of imagination is concerned, Heidegger's approach is on the mark. It asks why Kant's topography[6] set certain topological parameters as it did regarding sensibility, imagination, understanding, and reason. It makes evident that Kant's conceptual topography of imagination was an attempt to draw sharp boundaries within the more encompassing conceptual topology of human psychology. Most later nineteenth-century developments were, in one way or another, responses to Kant.

8.1 The Idealist-Romantic Appropriation of Infinite Imagination in Art

Kant found repellent the little of German Idealism and Romanticism that he lived to see. Fichte, for example, who claimed he was the true heir of Kantian philosophy, argued in the first *Wissenschaftslehre*[7] that human beings possess

[4]This happens chiefly in the sections devoted to pure principles of the understanding. The supreme principle of *analytic* judgments is not intrinsically temporal (A152–153/B192). It is the supreme principle of *synthetic* judgments that introduces the unique "sum total that contains all our presentations: viz., inner sense, and its apriori form, time" (A155/B194). The schematic implementation of time commences with the axioms of intuition and culminates in the three analogies of experience (see especially the A edition's general statement of the principle of analogy, A176–177).

[5]The course was held in the summer semester of 1925 and first published as volume 20 of the Gesamtausgabe of Heidegger's works. See Heidegger 1979, 267 and 442.

[6]A reminder: "conceptual topography" indicates a particular way of interpreting the possibilities opened by an underlying conceptual topology. See Sect. 3.6, above.

[7]Fichte regularly lectured on *Wissenschaftslehre* or "Doctrine of Science" over two decades. The most historically influential was the first version of 1794 along with the two introductions (1794 and 1798; they are included in Fichte 1970 [1794]). But the last word about the meaning and scope of imagination in Fichte has not been written.

intellectual intuition. Kant had expressly asserted that for human beings there was only one kind of direct seeing of things, *sensible* intuition—that is, the intuition of unities *in the manifold of sensation*. Claims to a direct seeing of concepts or other intelligible things—intellectual intuition—were the major source of the fallacies, paralogisms, and antinomies of traditional metaphysics.[8] Intellectual intuition, according to Kant, implies that the knowing of a thing is immediately united with, indeed is the same as, the known thing. In any being that genuinely had intellectual intuition, the intuition would in effect *generate* the being that it knew: that is, having the purely intellectual appearance of the thing in consciousness would be a *creation* of the thing. Romanticism enthusiastically embraced and expanded this quasicreative power.[9]

Romanticism is a story for another day, but if Kant's transcendental imagination was prologue, Fichte wrote the introduction by affirming the creativity of intellectual intuition. Kant had contested our right to treat the ego as a substance, but he nevertheless allowed the notion to be used regulatively. That is, treating the ego as substance*like*, though not as an actual substance, allowed one to lend to *self* a certain *hypothetical* unity that could assist thinking about human being. The hypothesis could be rendered innocuous by constantly reminding oneself and others that it was not a concept proper (because it introduced unity that exceeded the manifold of sensibility). Fichte recognized that the conceptual topography of Kant's transcendental psychology treated the self as a dynamic system of synthetic activity, and he therefore believed that there was a topological way of getting around Kant's prohibition against taking the ego as a substance or thing in itself. It was not a thing but a *Tathandlung*, a semi-redundant word meaning "deed–action." The ego or self was not a thing but an intensive action: self-generating synthetic activity. Thus the self was, contrary to what Kant had suggested, creative: it was constantly producing *itself* in its very thinking activity. That would mean that the standard of what "intellectual intuition" meant for Kant was met: a knowing that in its knowing produced the knower as it was known.

Fichte acknowledged that this went further than Kant, but he believed that here one needed to be more Kantian than Kant—because, as the following quotation indicates, intellectual intuition was implied by Kant's *apperception*, the final act of synthesis of understanding with sensibility, corresponding to the "I think" that accompanies all our experience. "Since the *Wissenschaftslehre* derives the entire concept of being only from the form of sensibility, it follows that, for it, all being is necessarily sensible being....The intellectual intuition of which the *Wissenschaftslehre* speaks is not directed toward any sort of being

[8]Whether Kant ever noticed that he in essence was reaffirming Aristotle's principle that there is "no thinking without phantasms," at least for thinking about the natural realm, is uncertain.

[9]There is no lack of secondary literature on Romanticism and its sources, although the historical scope is rarely sufficiently ample and the philosophical analysis rarely deep. For a standard account, see Engell 1981. For a literary-critical approach that is philosophically incisive and acutely attentive to Romanticism's use and abuse of Fichte, see Walser 1981, 11–75.

whatsoever; instead it is directed at an acting—and this is something Kant does not even mention (except, perhaps, under the name 'pure apperception')."[10]

Fichte went further yet. Able to claim, once again, a Kantian justification in reason's unconditional aspiration to unify all its objects into a whole, Fichte conceived ego as an outward drive [*Trieb*] that constantly reached limits, rebounded from those limits, and then, recommencing, moved outward once more. The limit of the outward drive was the *world*. This limit was the limit of the ego's own activity of synthesizing, and thus amounted to the not–ego or not–I. The world, experienced as different from the ego, is constituted precisely as the ego experiences it. That is, it is constituted in accordance with all the pure intuitions and pure categories Kant had already identified as part of the transcendental synthesis of imagination and understanding. The ego, once it recognized that this world was not a thing in itself but its own product, would start its outward drive anew in order to press beyond the world as it currently appears, until the drive encountered new obstacles in a new version of the not–I. This led to a practical solution to what some people found discouraging about Kant, that we could never know the world as it is. For Fichte, the ego, singly and in community, could through its ethical and social activity constantly transform the world in accordance with its emergent conceptions. In practice (and in technical and artistic making), one remade the world to conform to one's ideas. Ethics, technics, and politics were thus the *practical* solution to the Kantian "paradox" of knowledge—a "solution" that never came to an end because the drive of infinite aspiration never stops.

Fichte exhibited no Kantian shyness about imagination's part in this infinite practical-productive drive. It was the vehicle for ever new and creative uses of imagination in its transcendental function. Imagination is what again and again reconstitutes the limits of the I's experience (and thus remakes the I) by producing new versions of the not–I, the world. In the first *Wissenschaftslehre* Fichte explicates imagination as the productive function that puts–in–place both the *content* and the *forms* of intuition. This content–and–form begins with the outward, not–I–constituting drive that originates with the I/ego. The back-and-forth movement between I and not–I is self-bounding and creative. In the back-and-forth movement—called *Schweben* by Fichte, a hovering, oscillating movement—the drive produces situations in which, as the resulting tension between the I and the not–I plays itself out, *new* appearances emerge. These are images, and they are experienced precisely as such (Fichte 1970 [1794], 149–150). Thus Fichte made formal and central to his philosophy the theme of imagination as productively incipient: imagination thinkingly and productively presents in a more determinate appearance what it has already experienced in the tension of its hovering–oscillation between the poles of what it is trying to conceive.

Imagination as incipience of appearance is a theme we pointed to in Chap. 2 as native to the conceptual topology of imagination almost since its beginnings.

[10]See Fichte 1994, 56. Fichte distinguished between the ordinary empirical sense of self (corresponding to Kant's inner sense of temporality) and the philosophically educated transcendental experience of self as active and productive.

It has rarely had more than an uncertain, noncentral, even spectral presence in the topographies of imagination held by individual philosophers. Few thinkers have been comfortable with autonomous imagination. They always tend to subordinate it to something more stable: reason, sensation, even memory.[11]

Whether he was aware of it or not, Fichte was also resurrecting a theme from Aristotle: imagination is a motion, a motion that at crucial nodes of human experience gives rise to a reappearance of its original likeness, recontextualized and sometimes newly developed. This motion is open-ended, with no single goal (for Fichte, and also to a significant extent for Aristotle). For Fichte, every time the imagination conceives a new appearance it becomes possible, and often mandatory, to implement it in the world. Implementing it changes the world, and that guarantees that the drive of the ego will oscillatingly hover between I and not–I differently than before, until new images–appearances emerge. This happens again and again and again. In this sense, Fichte began the Romanticizing process of infinitizing imagination per se (as opposed to infinitizing other psychological powers, like the will in Descartes or reason in Kant).

Fichte also emphasized more strongly than Kant the role of imagination in works of art. Whereas Kant noted that such works can symbolize ideas of reason, Fichte understood artworks as the embodiment of imaginative activity itself.[12] This kind of understanding is more familiar to English-speaking audiences through Coleridge's extensive adaptation in his *Biographia Literaria* (1817) of ideas and passages from the *System of Transcendental Idealism* (1800) by another German Idealist, F. W. J. Schelling. Schelling's *System* was in turn heavily indebted to Fichte, not least in the sections on art that conclude the book. Schelling argues there that works of art are the preeminent way in which transcendental imagination produces the appearance of the unprecedentedly new, of a new object–appearance

[11] This tendency to subordinate imagination to more reliable powers is the engine of Castoriadis's critical history of the imagination. In a more constructive vein, while conceding that imagination must "lean on" nature, he argues that without imagination our rational explanations fail to achieve the total determination of natural things and events, and that not even natural causality and rationality together can account for human and social reality. Imagination, by leaning on nature, brings the human and social world into full being, and it always remains creatively open to the future. Thus politics, ethics, religion, language, and even the institution of science are all part of the "imaginary institution of society." His account of the schemas of *legein* and *teukhein* (speaking–conceiving and making) are perhaps the most rigorous extension of Kant's schemata since Kant himself. See the introduction to Chap. 9, below, and Castoriadis 1987 [1975]. Another notable attempt to develop schematism in a new, more comprehensive direction is Johnson 1987.

[12] See the Second Critique's discussion of symbols, the typics of practical reason, and imagination in "On the Typic of the Pure Practical Power of Judgment" (Kant 1900 ff., 5: 67–71—henceforth Kant 1900 ff. will be referred to as "Akad.," and the translations of passages from the Second and Third Critiques will be drawn from Kant 2002 [1788] and Kant 1987 [1790], respectively); also section 59 of the Third Critique, "On Beauty as the Symbol of Morality" (Akad. 5: 351–354). Fichte's theory of artworks, "On the Spirit and the Letter in Philosophy," dates from 1795, the year after the first *Wissenschaftlehre*; see Fichte 1984 [1795].

from which new philosophical thought can begin. Later, more popularized forms of romanticism made this theory of imagination's infinitization and creativity a standard, and increasingly trivialized, cultural trope.

Perhaps the best nineteenth-century analysis of imagination's activity in art making is found in the art criticism of Charles Baudelaire. Baudelaire presented an analysis of the use and abuse of imagination in painting that is sometimes taken as typically romantic (for instance, with his characterization of imagination as "the queen of the faculties").[13] Baudelaire is too subtle an observer, writer, and thinker to be typical, however, and labeling him as "romantic" or "idealist" is profoundly misleading. If he attacks some of the tropes of past theories of artistic production ("copy nature!"), it is not for the benefit of romanticized artistic genius. As one might expect of a poet (and that unfortunately one cannot expect of the professional philosopher), his descriptions, images, and analogies are exquisitely measured and considered. Although he was not a practitioner of painting, he knew it well both as critic and as friendly visitor to the studios of contemporary painters. The argument he makes can easily be adapted to the other arts.

Baudelaire's fullest development of his theory of artistic imagination is contained in *The Salon of 1859*. He begins with the observations that (1) nature is ugly, (2) it is only the complacent who say art is about copying nature, and (3) imagination is properly speaking an asset of the dissatisfied. Although he remarks that imagination "is positively akin to infinity," in context it is clear that the kinship can be remote as well as near.

The first observation may seem merely provocative, but to the author of *The Flowers of Evil* the ascription of beauty to nature as such is facile. The admonition to "copy nature" does not rise to the level of the problems the artist or even any human being faces. First, no one grasps the whole of nature, and it is ridiculous to think that copying a part automatically leads to beauty or other aesthetic values. Second, human beings are natural, too. Yet not all human beings see nature in the same way, and that, too, is natural. Even if the command to "copy nature" could take account of these things, Baudelaire argues that artists do not typically begin with a well-formed seeing of nature that they simply transcribe into their medium. When that happens, it leads to merely formulaic art. Baudelaire highlights instead the slow, progressive activity of the artist, moving back and forth from thinking and looking to first implementation (painting, in the particular examples he is discussing) and then back from the first implementation to looking and thinking. The process can begin with something evanescent, like a feeling or a dream. Even as such, it is already formed, with a form that has an implicit situation as part of a world and that the artist has to realize in a medium. This world–form is gradually worked out in the art's medium, with the artist calling on all his skill, knowledge, and memory as well as the historical practices, styles, and traditions he has acquired in order to elaborate—often through many drafts and sketches—a fully formed piece.

[13] See, e.g., Casey 2000, 177.

> A good painting, faithful and equal to the dream that conceived it, must be produced like a world. Just as creation, as we view it, is the result of several creations whose preceding ones are always completed by the next, so a harmoniously conducted painting consists in a series of superposed paintings, each new layer lending more reality to the dream and raising it one degree closer to perfection.[14]

Baudelaire's use of the trope of world–creation is carefully controlled. If the painting begins with a dream, it is a well-formed dream with an inner dynamism. If this is creation, it is creation *ex creatione*, out of a prior creation, and not *ex nihilo*, out of nothing. If there is any *copying* of an idea going on here, it is of something that is only fully evident when the work is finished—and thus it is not copying at all. But every layer laid down along the way is a moment in progress toward the realization of the original idea's formative unity. From the complex of anticipations in the human soul there is a projection onto the plane of the canvas—into the material of the medium—of an incomplete determinacy; by means of a quasi-Fichtean process of "hovering–oscillating" recursion, the anticipations in the soul become better organized as the artist immerses himself in the earlier stages and superimposes on the plane a more perfect determination.

Is Baudelaire's artist a Kantian genius? For Kant, genius is a force of nature and does not merely copy an original. The artistic genius does not follow rules but prescribes them. That formulation does not work for Baudelaire's conception, however. To begin with, he does not use the rhetoric of genius; moreover, in implicit partial rebuke to Kant, he says that it is lesser, school-forming spirits who conceive what the great artist does in terms of rules. According to Kant, geniuses speak primarily to other geniuses. The genius uses the elements of nature analogically and metaphorically to go beyond natural significations, and concepts to symbolize ideas of reason. It is unlikely that Baudelaire could agree. To begin with, he does not conceive geniuses apart from the spirit of craft (they are first of all masters of a craft and thus know the work of geniuses and journeymen alike), nor does he have the artist ascending incomprehensibly to the ethereal realms of symbolized pure ideas. Baudelaire is too strongly grounded in the necessities, particularities, and stages of artistic practice for that.

Borrowing from his friend Eugène Delacroix, Baudelaire remarks that "painters who obey their imagination seek in the dictionary the elements which suit their conception." The dictionary trope signifies that the imagination calls on ready-made meanings and devices, many determined by convention, that are (first) selected and (second) modified according to the demands of the conception of the whole work. The formative idea of the work is of primary importance. It accretes to itself and transforms what it needs for implementation. Baudelaire sees this dictionary of the elements of art as a social knowledge that is intrinsically defective insofar as it is routinized. In this he takes a step that Kant probably would have been reluctant to

---

[14]Baudelaire's implicit analogy is between the successive phases of realization and the 6 days of creation in *Genesis*. My analysis is heavily indebted to Frey 1996, esp. 61–98. This translation of the Baudelaire passage appears on p. 72 of Frey; the original is from the section "The government of imagination" of *The Salon of 1859*.

follow. It is a threat to the very concept of progressive enlightenment insofar as it calls into question the adequacy for use of already expressed truths, even those (especially those!) expressed in purely logical format (like definitions). For Baudelaire, truth is native to active seeing, thinking, remembering, and imagining; truth, to be expressed in abstract and communal form, must undergo one or several degrees of alienation. Enlightenment, by contrast, understands itself as *secured* in social institutions like the university and in communal practices and works like encyclopedias and dictionaries.

Baudelaire's criticism of the established "dictionary" of art can easily spill over into a critique of socially approved techniques, symbols, and meanings. (It also helps explain why he called imagination the "asset of the dissatisfied.") This is not to be chalked up simply to Kant's being a proponent of bourgeois culture, whereas Baudelaire is a *flâneur*–critic of it. The theme of reason's infinity that Kant so strongly affirmed in the topology of human psychology and that the Romantics shifted to imagination had by Baudelaire's day come to be accompanied by a question mark.[15] For Baudelaire, even if imagination is *akin* to infinity, it is naturally situated in the here and now. Imagination is not the absolute emperor of the human faculties and of art, nor even the king: it is the queen. It does not make something out of nothing but invigorates and activates what is at hand: *creatio ex creatione*!

Although for Kant the transcendental principles of nature more or less guarantee that nature is the same for everyone, by virtue of the ethical autonomy of reason human beings are not completely natural. Baudelaire, by contrast, considers the human being as *fully* natural, and each human being (potentially) sees the world (both the natural and the moral world) differently. There is not one way of human seeing, there are many. Thus the concept *nature* cannot be simply abstracted from the manifold ways of *seeing nature*. Seeing is not just a question of the imagination's transcendental unification of the manifold, it is also a matter of the tendencies of historical groups and even of individuals to follow distinctive approaches to the limitless range of nature's ways of showing itself—approaches that, however, can be individually named and explained even when they are owed to genius.[16] The infinite aspirations of human being are always and everywhere expressed and instantiated in finite ways. The present reality one lives is intuited according to one's experience and situation. In thought, that experience can be limitlessly varied and repositioned against innumerable possible backgrounds, but the limitlessness is more virtual than real insofar as the variations themselves fall into patterns and styles.

Baudelaire's theory explicates and develops the notion of imagination as fundamentally placing and re–placing. Whether that would be, for Kant, a source of pride

[15]This would be the place to reflect on another mid-nineteenth-century phenomenon, the philosophy and the imaginative practice of another urban *flâneur*–critic of bourgeois society, Søren Kierkegaard.

[16]Johann Wolfgang von Goethe extended the idea of the transcendental functions of imagination to these more particularized approaches in his scientific research and historiography. See Sepper 2009.

or humility, arrogance or despair, joy or anxiety, is not something that the Critiques' presentation of the topology of human cognitive powers can entirely determine. Kant was more concerned than Baudelaire to keep theoretical, practical, and aesthetic matters distinct from one another and to isolate them in separate realms (cognition, desire, and feeling). Perhaps it was less theoretical consistency than an acknowledgment of the complexity of human existence that led him in the Second and Third Critiques to discuss what he called *intellectual feelings*, like respect for the law and the awe felt before the sublime. Whether the very concept of intellectual feeling is consistent with Kant's strict boundary-drawing between powers is doubtful.[17] But sometimes inconsistency is evidence of ruthless honesty in a thinker. Kant himself, in the Third Critique, pointed the way to a more interactional understanding of such phenomena by explaining both the beautiful and the sublime as universal–predicates–accompanied–by–feeling produced by the interplay of the psychological powers. Even if there is a certain anticipation of this approach in his early modern predecessors, especially Spinoza (specifically with the intellectual love of God), it is to Kant's credit that he gave these phenomena a prominence that made the inadequacy of certain of his accounts more conspicuous.

At any rate, once such difficult phenomena were broached, the need for a better understanding of how they comport with established Kantian and non-Kantian theories became evident, and the topological vistas they opened within the theory of human psychology become irresistible. For later Idealism, Romanticism, and post-Romanticism to have ignored them would have betokened not philosophical discipline but an inordinate degree of intellectual unresponsiveness to the topology of transcendental psychology.

8.2 Tendencies of the Post-Kantian Topology

What paths of response to imagination are still topologically open after Kant? Is the topology of imagination and, more generally, human psychology still relevant or defensible? That is, even if it serves as a useful historiographic tool, has it become outmoded by virtue of better ways of "saving the phenomena"?

Radical antipsychologism is not an option, not least because it is not concerned at all with phenomenality, with the specific ways of appearing of psychological

---

[17]In the Second Critique Kant begins his discussion of respect for the moral law by remarking that "the moral law as determining basis of the will, by infringing all our inclinations, must bring about a feeling that can be called pain; and here we have, then, the first and perhaps also the only case where we have been able to determine *a priori* from concepts the relation of a cognition (here a cognition of pure practical reason) to the feeling of pleasure or displeasure" (Akad. 5: 73). At Akad. 5: 117, however, Kant notes parenthetically that "an intellectual feeling would be a contradiction." The Third Critique's investigation of the intellectual feeling of the sublime takes place from sect. 24 to sect. 29. The sublime is specifically defined as a (disproportionate) relationship between imagination and reason, and Kant arrives in sect. 26 at a definition: "Sublime is what even to be able to think proves that the mind has a power surpassing any standard of sense" (Akad. 5: 250).

phenomena. This is not to say that certain questions posed by antipsychologism cannot contribute to the effort to preserve the rights of the phenomena. Consider, for example, the conventionalized notion of imagination as forming and holding before the mind's eye a quasivisual experience. Antipsychologism sharply rejects the idea of an introspective intuition modeled on visual perception. This same rejection was one basis of our agreement, in Sects. 2.2, 2.3, and 2.4, above, with French philosopher Alain's assessment of imagination and memory's inadequacies when it comes to matters like recalling the appearance of the Parisian Panthéon. But we also pointed out that even Alain granted that there were "flash appearances" to consciousness: appearances insufficient to support an accurate column count but nevertheless justifying the claim that something senselike appears—and that this is so even if there are still important distinctions to be made between imagining and remembering.[18]

A better response than the total rejection of the analogy between perception and imagination/memory would be to proliferate examples in order to see how our first-approximation claims can be appropriately differentiated and refined. Few people can count the columns of the remembered Panthéon; few can represent to themselves a heptagon (regular or not), much less a chiliagon. But if asked, many could form in imagination the general appearance of a schematic portico having six, seven, or eight columns, and even in the medium of the purely imagined they would begin to notice distinctions of appearance (for example, how increasing the number of columns affects the overall proportion of a façade, and how having an odd number of columns gives a less stable appearance than an even number). In reflecting about the internal angles of a regular heptagon or chiliagon, most people will probably take advantage of what Kant called a schema and sketch for themselves—privately in mind, or publicly on paper—figures that will help them to think more determinately. Just because they do not meet the definitional standards of mathematicians (for whom lines have no width) does not mean that they are useless or irrelevant but only that they are an imperfect realization of the cognitive goal (and even mathematicians resort to figuration as they pursue their inquiries). A singer who has been practicing a song a capella while driving to the studio is not wrong in believing that she was imagining *some* accompaniment, even if it was not as fully determinate as she felt it to be. Although she might realize that the tone of the studio piano has a brightness she had not imagined, she nevertheless can stop rehearsal and explain or even demonstrate to the pianist that her till now only imagined dynamics would provide greater propulsive force than the dynamics she has just heard. To deny all privacy and inwardness to imagining is to turn a basic truth (that *I* cannot hear her imagined accompaniment) into a minor form of obtuse nihilism.

[18]There is, moreover, another point to be made: that with enough columns, even direct sense perception can be inadequate for counting them (for example, counting the columns visible around the base of the exterior of the Panthéon's dome). Not even perception translates immediately into accurate conception—which reinforces the need to *distinguish* cooperating psychological powers in every mental act rather than *proscribe* their participation.

A more productive, if somewhat darker, path followed by thinkers responsive to the interplay of the human psychological powers can be traced from Fichte and Schelling to the psychoanalysis of Freud. Although this path is hardly unknown to historians, its philosophical roots have been insufficiently explored. Cornelius Castoriadis, who intended but never completed a history of imagination up through Freud, makes the point that it was not under the rubric of imagination (*Einbildungskraft*) but rather fantasy (*Phantasie*) that Freud thought about the phenomenon, primarily in the form of symbol- and symptom-producing functions in dreams, daydreams, and repression.[19] For Freud, these functions culminated in his theory of the Oedipus complex. The Oedipus complex—according to which the son wants to do away with the father so as to have the mother all to himself—is a syndrome of image–symbols and drives, a complex psychosexual schematism that shapes and colors the cognitive and affective relationship between mother, son, and father. Particular, and thus already imaged, childhood experiences are synthesized and affectively charged (cathected) with emotional values that simultaneously express and mask (or suppress) relational conflicts, tensions, misunderstandings, and anxieties. According to Freud, this naturally plays out in the psyche of each person (or each male?) as a drama that is paradigmatically expressed in the myth of Oedipus, who killed his father and married his mother.

This is no mere empiricist associationism—it is passionate, naturally determined association with a vengeance (quite literally!). If there are relatively few philosophers and scientists today who take Freud's theories seriously, that need not deter us from appreciating how his theory of fantasy exploited and explored (however dogmatically and speculatively) the topology of imagination, reason, and desire and reconceived the usual notion of their relation. To a Kantian, Freud would seem to be almost a parody of the play of the faculties that Kant introduced in the Third Critique, where this play accounts for the experience of aesthetic phenomena and the corresponding universal attribution of aesthetic predicates to objects that set loose the interplay of faculties in the first place. "Almost a parody" is, however, a reminder that the parody has some basis in the original. The parodic interprets what it parodies as merely one among many possibilities. The original through alternative possibilities becomes a field of play, and there the parodist plays out another possibility of the field and thereby also allows us to step back and see the field as such. The step back is necessary precisely to see the field. It sets up the basic biplanar situation that in Sect. 3.8, above, we described as characteristic of imagination. Imagination can both immerse itself in the field as such, and it can remove itself from the field far enough and long enough to see what it has done there as one possibility among many.

Kant adumbrated the biplanar play of imagination but inadequately explored its planar placements. He adumbrated it in the sense that the transcendental function of imagination establishes or constitutes the pure intuitions of space and time as the where–and–when of all (possible) objects and events of human experience. This original constitution is not simply an association of discrete idea–units into

[19] See Castoriadis 1997, esp. 175–181. On *Phantasie* in Freud, see Laplanche and Pontalis 1968.

a networked relationship. Before there are any units to associate, the place of any future networking exists already as the where–and–when. The where–and–when as such is not conspicuous because our attention is focused initially on all the "things" that take place there, each of which is a determinate possibility not just of time and space but also of matter and physical law and forms of appearance. But the mind, at certain appropriate moments, can step back from its ordinary focus and attend instead to the continuum of space and time that is the basis of pure geometry and arithmetic and of the dynamic possibilities of physics. This stepping back is a crucial step toward cognition, but it is founded precisely by imagination as *transcendental*. That is the first act of transcendental imagination. In the second act, imagination transcendentally realizes (gives concrete form to) and presents in appearance the things cognition will come to know in a double function. The first and more original function of the second act embeds the fundamental categories in the field of space–time by implementing (i.e., fulfilling) the schemata of the understanding; its subsequent function schematizes the field according to the multitude of empirical schematisms. The schematisms are a back and forth movement of imagination, which can start from the concept and go to the corresponding appearances or start from appearances to home in on corresponding concepts. The concepts themselves constitute a more discretely established but nevertheless still interactive field—one might say, without too much fear of punning, a more *schematic* field than that of the sensory appearances—which corresponds formally, though not materially, to the presentations in the dynamic space–time field of the manifold of sensibility.

To put it as simply as possible: Kant articulated human imagination as fundamentally transcendental, constitutively immersing us in the fundamental forms of the manifold of sensibility on the one hand and raising us up to the corresponding concepts of the understanding on the other. Transcendental imagination allows us to be immersed *and* to stand apart—and, to some degree, to do both at the same time. By coming more consciously into possession of the schemata, we can do things like construct mathematics and hypothesize dynamic physical laws. These are different biplanar modes of encountering the field of the manifold of sensibility through the mediums of mathematical fields and physical fields and fields of empirical schemata. To paraphrase in Kantian terms the traditional Aristotelian slogan: there can be no human thinking without the fields of transcendental imagination.

But Kant no more than commenced the investigation of the dynamics of the conceptual topology of imagination and fundamental human psychology; in that sense he only adumbrated what I have just explained. Even as adumbration it was inadequate. Kant (unlike later Idealists, especially Fichte) failed to draw the conclusion that, just as there is the field of physics that builds on the field of space–and–time, so too there might be further field–communities built upon both of these. These supervenient fields could also be conceptually marked and articulated. There might be many such fields, especially if one found ways (not countenanced by the formalism of Kant) to understand the "matter" of appearances (which Kant regarded as passively given and thus intrinsically unknowable) as infused with form (which is conceptual and ideal)—for instance, as one finds can organize and represent qualitative characteristics of color and sound in mathematical and geometrical displays (like color solids that display all combinations of hue, saturation, and brightness). The Third Critique makes at best just a few tentative gestures in this direction, pregnant but underexploited.

Even feeling and desire might be articulated in an analogous way. This is an insight that goes back at least to Aristotle's conception of virtue as the habituation of feeling to moderate desire and aversion, which was developed in different ways by early modern empiricists and rationalists. More recently, Freud and his followers recognized that all "parts of the soul" have to be articulated in order to become articulate. Even unexpected and "irrational" subjective explosions (or implosions) of affect are often anticipated and expressed in symbolic image–forms that are "readable" if an analyst knows enough about the experience of the subject–patient. In the middle of the twentieth century, Jacques Lacan produced a striking attempt to refound Freud's thought along these lines. One of the remarkable things about it is that it returned to a very old topic in the history of imagination, one as old as Plato's insistence that *logoi* are iconic: that speech is a form of image. One of Lacan's most famous dicta was, "The unconscious is structured like a language." More generally, one can say that all the basic powers of soul or mind are structured like language or at least invite such structuring. If the linguistic structure that supervenes on natural powers is dynamic, then the soul would constantly be infused and structured by images of speech.[20]

Considering in detail how, for Lacan, the soul is structured like a language is beyond the scope of the present book. It is nevertheless possible to use a well-known Lacanian example, the mirror image of the infant, to indicate how basic the *linguistic organization of images* is to Lacan's dynamics of soul.[21] Lacan describes a moment in the infant's (still largely *in–fans*, nonspeaking) early development, when the mother puts the baby on her lap in front of a mirror. Up to that point the baby's experience of itself is volatile, poorly organized, and inconsistent. Simply seeing what shows in the mirror is only a condition of what happens when the

---

[20]The step from the unconscious structuring to the structuring of the entire soul is short, in particular since in all Freudian topographies of the soul the infant is almost totally dominated by the as yet undisciplined id or unconscious. Any higher powers must be, quite literally, the result of partial structurings of the unconscious, and over time the unconscious as such is shaped by the experiences of the subject. When in the main text I say that there is a linguistic structure supervening on the natural, I am looking beyond Freudianism to include a more Aristotelian conception, in which the soul is the fundamental actuality of the organized body. The virtues or excellences are structurings by experience of the natural pleasure/pain and desire/aversion one feels, and thus they supervene on natural structures. Similarly and more basically, to train an infant to use a spoon or a cup is to restructure natural powers.

[21]Lacan eventually conceived the structuring of the soul according to three major aspects, the Imaginary, the Symbolic, and the Real. The brief explanation I give here of the mirror experience does not observe the precision and complexity of Lacan's presentation, but it does, I hope, give an at least initial insight into the entanglement of the Imaginary and the Symbolic. Lacan's original paper on the mirror stage, presented but interrupted (by the session chair, Freud's student and biographer Ernest Jones) at the 1936 International Congress of Psychoanalysis, was lost. The most accessible later consideration is the paper Lacan delivered at the 1949 International Congress of Psychoanalysis; see Lacan 2006.

mother calls the baby's attention to its image–appearance in the mirror and names it: "That's you, that's Harry! There's your nose, and your mouth, and your ears, and your fingers, and your toes…." This moment is pregnant with consequence. The disorganized experience of self acquires a point of focus. One might in fact have to say it acquires several coincident points of focus. The baby sees a baby, but now that baby is not just any baby but itself. The baby sees itself as having a structured appearance, divided into many parts but nevertheless structured as a distinctive whole. It identifies itself with that appearance: it now has a self-image. That self-identification is, furthermore, now marked by a name, and all the "baby parts" that the mother identifies are marked as parts of self. The occasion of the self-identification and the naming is the mother's imposition, and thus it is not only an intrinsically familial event but also, through the use of the language that marks the mother as part of already existing society, the infant's initiation into that society and into accepting the society's way of identifying and organizing experience. To paraphrase Aristotle's definition of soul with a Lacanian twist: the self has been linguistically constituted as a first actuality of an incipiently organized body having social life.

This mirror experience certainly has a mythic aspect; in that respect it is not unlike the Oedipus complex. The plausibility of the Oedipus complex has always been enmeshed in both cultural and ontogenetic difficulties. It privileges the modern European anomaly of the nuclear family; it presupposes that the young child experiences a "primal scene" that might never actually have happened; it appears to impose a dogmatically monolithic view of familial relations; and it leaves obscure the psychodrama of the female. Even if each of these difficulties can be answered, there is still a gap between the theory and the possibility of its universal application. Lacan's mirror stage, by contrast, is far more strongly connected to manifest stages of child development, and also more basically and originally constitutive of the human personality as ego–in–a–community. Whether the speaking partner in the event is mother, father, sibling, or grandparent does not radically change its significance. Nor does the event have to be compressed into a single, well-defined moment of time; it can be a discovery stretched out over days, weeks, or months. Although in Lacan's telling it appears to depend on the existence of mirrors, thus on a historical contingency, it is actually only dependent on the young child's encountering some phenomenon of mirroring, if only in polished metal or the surface reflection of water.[22]

---

[22]Of course the wide availability of large, high-quality mirrors in a domestic setting makes it far more likely that the scenario will be played out as described. This and related historical and cultural contingencies raise important questions about the sequence of acquiring linguistic competencies as well as about whether there is a truly decisive *aha!*–moment in the acquisition of the ego function. Unlike with the Oedipus complex, however, they are not likely to evoke simple incredulity about the general plausibility of the theory. Such problems in fact invite further development of the theory (e.g., might historical, technological, and cultural differences affect the ways in which personhood and the I–function are lived out in various cultures) rather than dogmatic reassertion.

In the later development of the theory Lacan insisted that the mirror stage was not so much a contingent event in the life of each child as a fundamental condition of being human that is constantly repeated. Its aspect as condition is expressed in the entwinement of the Real, the Symbolic, and the Imaginary.[23] In three sentences: the Imaginary, instanced in the mirror stage by the mirror's image–reflection of self, presents a fundamental integration–in–image of the infant's experience of self; s/he gains from the image–reflection here-and-now a unifying focus for all prior and subsequent phenomena of the self. But this insight is not fully formed or constituted except in the Symbolic, a term that includes the social sanction (in every sense of "sanction") of what has shown itself by means of the mirror; the word gives definition and identity to the Imaginary as part of a community project. What the Real expresses is more elusive, precisely because it is that which the Imaginary and the Symbolic both indicate but cannot encompass, neither individually nor together; it is that in which the self-interested project of the now constituted life of the self will take further substance, and in that sense the here-and-now is only *beginning* to be formed.

For our purposes it will have to suffice if we point out two relevant themes. One is that Lacan has in effect seized upon the topological opportunity presented by Hume's thematic of the passionate self as the indirect object of imagining.[24] In the *Treatise of Human Nature*, Hume had argued that every mental action is undertaken in light of an interest of passion and desire, thus in light of what I call an indirect object (alternatively, one might name it, borrowing from Latin grammar, the *dative of interest*). If in the appendix to the *Treatise* he acknowledged his failure to find the ego or self as anything other than the entire series of impressions and images that constitutes mental life, perhaps that means that he overlooked the significance of his earlier discovery: always present and at least partially directing the train of impressions and images is the nonautonomous but interested (and self-interested) subject that to a large degree defines itself indirectly by its objects. To concentrate overmuch on the impressions and images that populate consciousness is to overlook this other point of focus. When consciousness focuses on its objects, it is typically unaware of itself.[25] By claiming that the mirror stage is not just a single event in the first year or so of life but a condition that constantly reoccurs and reconditions the human being's life, Lacan is trying to reconceive human psychology so that we can see the ego in constant formation and reformation, in its private experience (imaginary), in its social being (symbolic), and in its working out a place in the medium of living (the real).[26]

I do not wish to argue that Lacan's theory of the mirror stage is an ultimate and true development of the conceptual topology of imagination, nor that the brief

[23]The entwinement was expressed emblematically in the Borromean knot.

[24]See Sect. 6.9, n. 53, above.

[25]Lenses and lens systems have two points of focus, in two different planes: one is at the object, the other where the (subject's) eye is positioned to see the object clearly. This is of course one of the themes that Lacan exploited in his 1950s presentations of the mirror stage.

[26]This reiterated movement between objects and ego can be traced back to Fichte's *Ich/Nicht–Ich* dialectic.

interpretation I have given is even close to the last word. Nevertheless, the mirror stage offers an especially pregnant confirmation of the topology's reality[27] and of possibilities for its development. It integrates several of the basic themes that have been at the heart of this book. Imagination appears in the first instance as expressed in a loose association of independent image–units of sensory experience but is superseded by becoming organized into a differentiated imaginal unity. Imagination occurs not just in a single organized field or plane, however, but between planes, and thus it exemplifies, indeed generates, the biplanarity (even the multiplanarity) of human experience. Imagination becomes human not simply as the reproduction and transient association and reassociation of image–units but by virtue of marking and naming features in fields, fields that can subsequently be recognized in their own essential unity and from which new images can be elicited.

The association that really counts for humans is not so much that of the elements in a field (for instance, the play of colors) as it is the coherencies between the elements that come fully to appearance by experiencing one field in terms of another (for instance, understanding the play of colors as exhibiting harmonies grasped in the first instance by analogy to musical harmony).[28] Imagination comes most richly into its own through the recognition, differentiation, and development of the possibilities of a plane according to the markings that have significance in a second field (for example, when geometrical lines are viewed by marking their positions with respect to one another and translating those markings into the idiom of algebra). This biplanar reassociation is expressible and thus shareable: that means that the imagining is both individual and social. Imagination constitutes the experienced world as a coherent, traversable realm unifying and differentiating elements, characteristics, and functions that show themselves as variable and combinable. Imagination thus becomes the principle according to which the human being can range through planes from the private to the public, the singularly individual to the universally social, the possible to the actual, the merely planned to the effectively realized, the quasireal to the definitely fictional. If this is true, it would require a radical revision of our average-everyday understanding of human psychology, epistemology, logic, ethics, and politics—that is, of human existence. The realm of imagination would be as wide as all possible ways of cross-sectioning planes and fields from real and possible worlds.

8.3 The Signitive Placement of Imagination

Given recent attempts to rehabilitate the importance of affect for understanding human perceptual and cognitive functions, Lacan's reworking of the Humean theme of the indirect object of imagination could have a more than casual interest

[27] Reality not as *res* but as the condition of the possibility of any institution/constitution of the psyche.

[28] The point is not that interpreting color according to musical concepts is true or thoroughgoingly useful, but that the very fact of trying to correlate one field with another helps one to make more careful discriminations and distinctions. All the better, however, if the correlation is perfect!

for contemporary researchers. (The association with Hume might offer a less dubious point of entry for those who fear that the French psychoanalyst lies too far outside the contemporary mainstream.)[29] The conceptual topology of imagination has historically been deployed in topographies that treat it as a middle place, between sense perception and knowledge. In Lacan's theory of the mirror stage, however, imagination is both (re)cognitive and affective. It is both medium (the remnant–trove of loosely organized experience) and origin (of new activity when that trove is better organized). This duality of imagination as medium *and* origin leads to the second topological theme of Lacan's theory I wish to highlight: that the imaginary, the potentiated trove of the experiences that come to us through (in the first instance) the senses, is potentiated to renewed activity primarily by language. The consequence is that the human imaginary (the supply of organized, available imagery) is constituted, or at least coconstituted, by language. But language, although by its nature social, is not just other-directed. Language is about both self and other. Without this duality, the imaginary could not be the medium of *radically* human imagination—that is, of imagination *rooted* in the nature of being human.

Although in the last third of the twentieth century Paul Ricoeur was the leading voice for a non-Lacanian interpretation of imagination as essentially linguistic, the theme is, as we have seen, as old as Plato, for whom the primary form of making icons/images was *logoi*/speaking.[30] Aristotle took no more than a step toward developing this association, most famously at the outset of *On Interpretation*, where he says that the *pathēmata* or things suffered by the soul have spoken words as symbols and the spoken words have as symbols written marks (the *pathēmata* of the soul in turn correspond to the things of the world human beings have to do with). The historical persistence and influence of this work in and beyond antiquity determined the basic topology of signs and language—a topology that was over the next two millennia more often varied or inflected than criticized or rejected.[31]

The traditional "obviousness" of this topology probably impeded exploration in depth of its features and consequences. The ancient rhetoricians and the Stoics, to be sure, further elaborated the conception of the sign understood as an indicator

[29]There is no doubt that Lacan was inclined to speculation and, especially in his later work, obscure. And there is, in addition, the worry that at moments he deliberately acted the provocateur—or, to put it less flatteringly, as Noam Chomsky did, that he was a bit of a charlatan. I would be the last to argue against the importance of trust in the sciences, but it is too easy an out for inquirers to make the accusation of charlatanry against someone and thereby to be done with whatever he said. The nihilistic rejectionism that I brought up in Chap. 1 has many forms. Moreover, it is worth more than just a moment's reflection to consider that a charlatan's success depends crucially on producing at least a simulacrum of truth according to imaginative possibility.

[30]For an introduction to Ricoeur's semantic approach to imagination, see Ricoeur 1978.

[31]The topology of the sign relation was not invented by Aristotle, of course. It was already well established in the Hippocratic medical tradition, before which came the use of signs in divination. See Manetti 1993, ch. 1–3. Whether one should make a great deal out of the distinction that Aristotle draws in the passage between symbol (in noun form) and signing or signifying (in verb form) has been a matter of intensive debate since antiquity; see Kretzmann 1974.

helping effect the transit between mind and thing. I indicate something I have mentally experienced, to myself and even more to others, by a sign. This, too, had a kind of obviousness about it that discouraged tarrying long over the problems that the sign function raises.

In Platonic portrayal, the *logoi* are images that intermediate between (a) the world situation being described by the words and (b) the meanings that originate with the forms.[32] On the other hand, it seems that in Plato's dialogues the interlocutors ordinarily get at the forms by considering single words rather than a complete *logos* or sentence—for example, by asking about the nature and meaning of "justice" or "the good." Words became objects of inquiry in their own right, with the proximate goal of defining them. But this also made the inquiry communal, insofar as language is communal and the typical Platonic-Socratic inquiry is undertaken by a group of people asking how they all use the words and what they think of when they do. Words or signs are thus manifestations of the imaging of meaning that stands between (and joins) the person who utters them and the things they indicate and reflect; they also stand between and join the person using them and other persons considering them and what they indicate. Moreover, as rhetoricians were fond of pointing out, signs are often a means of indicating not things but other signs.[33] Thus signs are positioned as both means and ends.

Even without a detailed re-investigation of all the Platonic and pre-Platonic sources, however, it is easy to see that one of the ambiguities in the understanding of words and signs is their *place*, place understood both absolutely and figuratively. Words are never (or rarely) the end absolutely. Typically they are a medium that we do not reflect on (since in order to bespeak the things we talk about they have to be largely transparent). When we attend specifically to what we say and write, however, we have to shift the plane of our attention primarily to how the words stand with respect to one another. Primarily, but not exclusively: for even when we attend to the

[32] Unfortunately (or deliberately), in the *Republic* Plato's Socrates does not discuss the ontological status of words. One way of putting the question would be to ask where (and whether) words should be positioned along the divided line of Book VI. If they are images they would seem to belong on the "lowest part," along with reflections in water and mirrors. Insofar as they correlate with concepts, however, they seem to belong on the "highest part," that of the ideas. Plato appears to leave the question to us; to answer it we would of course need to take into account other dialogues, chief among them the *Phaedrus* and the *Cratylus* (without forgetting Socrates's autobiographical reflections in the *Phaedo*, where he describes turning away from Anaxogoras's philosophy to an examination of *logoi*). I do not believe for a moment that Plato was unaware of the question. I suspect that the answer would have to reflect the fact that Books VI and VII of the *Republic* present the cosmos as organized by what I called (in Chap. 3, above) "ontological imaging." That is, language is a human participation in the ontological imaging that, emanating from the good itself, joins all the parts of the line and all the parts of the cosmos into a well-ordered, representative whole.

[33] See, for instance, the theory presented by Augustine in *De magistro* (On the teacher). For Augustine, the role of the sign was incomplete without acknowledgment of the ultimate *verbum*: the mind of God, Christ, the second person of the Trinity.

words we cannot completely close off our minds to what they are about. Just as with images, we find ourselves in a situation that is not merely planar but biplanar. In fact with language more than with images we need to entertain the possibility of going beyond biplanarity: there is the plane (1) of the things we are dealing with, (2) of our thought, concerns, feelings, and hopes, (3) of the language we use to name, describe, and discuss the other two planes, and (4) of the audience.

Aristotle's elemental description of the linguistic situation in *On Interpretation* exhibits multiplanarity: the phantasms in the soul are in the middle, the world–things and two planes of symbols, spoken and written, are extremes. As with the rhetorical and Stoic development, the signs (at one extreme of the four-place sequence world thing/soul/speakable sign/written sign) point us back toward the things of the world. In his description of the generative sequence Aristotle only gestures toward explaining how the soul takes things in from the world—he says (at 16a8–10) only that one needs to look to another work for that account, presumably *On the Soul*—and thus leaves readers to wonder how the passively experienced appearances in the soul become susceptible of symbolization. *On the Soul* does not, of course, say anything directly about language, apart from a reflection on animal sound-making which distinguishes voice from mere sound. Voice requires soul: "the voice is the striking against the so-called windpipe of the air that has been breathed in, by the action of the soul in these parts," and "it is necessary for the part that causes the striking to have soul in it and some sort of imagination with it, since the voice is some sort of sound that is capable of carrying a meaning [*sēmantikos gar dē tis psophos*]" (see II.8, 420b28–33).[34]

Besides Plato's understanding of *logoi* as images, Aristotle's characterization of voice as semantic (insofar as it is a sound along with or produced with imagining) and his conception of intellection's constant requirement of images compelled later Aristotelian thinkers to at least touch upon language's connection to the "abstraction" of intelligibility from phantasms. Unfortunately the passages in *On the Soul* give no further instruction about the matter. The just-quoted passage from II.8, along with the even more difficult discussions of images and imagination in III.3 and III.7–8, contributed to the development in medieval Aristotelianism of the doctrine of the inner word. That is, it was the intellective power in the presence of the phantasms prepared by the psychological workings of inner sensation that gave rise both to the apprehension of intelligible form and the formation of the proto-word that precedes public utterance. Topologically one could make this relationship very close indeed or try to create a space for different phases; but medieval thinkers could not simply ignore it—particulary since "word" was one of the names for Jesus Christ, second person of the Trinity.[35]

---

[34]The distinction is generic because it applies to animals as well as human beings, and in human beings would not distinguish words from whistles or cries of pain. For further discussion, see the next paragraph, below, and Sect. 5.10, above.

[35]Recall that "abstraction" is a medieval term that has only limited justification in Aristotle's Greek. On the "inner word" in medieval adaptations, see Lonergan 1997 (a book first published in 1967 and based on articles published in the 1940s).

As we noted in Chaps. 6 and 7, above, early modern philosophers recognized that there was a relationship between imagination and naming, but they differed about its strength and character. Hobbes argued that thinking is reckoning in names; names mark similarities between phantasms or ideas, and as such they allow us to move mentally from one sequence of phantasms to another. Without this we would have an existence dominated by whatever train of phantasms currently occupied our attention—chiefly the phantasms of perception, but also those of memory, dream, or hallucination. Descartes and Locke both appear in the first instance to hold that our thought is a *private mental language* complete in itself, though in need of words or other marks once there is an intention to communicate. Of course Descartes does not, like empiricists, think that the only source of ideas is sense, so that a fundamental part of that mental language is the cognitive power's ability to perceive natures and to note, represent, and expressly mark resemblances, differences, and quantities of difference (and in the first instance he understood his mathematics as a kind of representative imagining of those quantities of difference, and the algebraic marking of geometric figures as a formulaic imagining of the geometric relations). Even Locke differentiates between the ideas that have their source directly in sensation and those that we discover from the mind's operation with those sense ideas; more significantly, he remarks that thinking works differently, according to names, once we have given names to ideas.

Kant took a crucial step, but only a step, toward (re)establishing an even stronger link between imagination and language. In the first instance the schematism of the pure concepts of understanding joins images to concepts, without mention or intervention of words. A mere sketch of a dog will put us in mind of the concept "dog," and thinking the concept readies us to portray, in a real or a mental place, some degree of dog image. For living, breathing human beings, what puts them in mind of a concept is for the most part language. Where does language fit in schematism?

Schematism is, of course, a work of the imagination in its transcendental functioning, and it bypasses the need for invoking a psychological process of abstraction.[36] Like the pure intuitions, the pure concepts, and the pure principles, it is native to human experience and understanding as such, so it is universal in a way that naming in particular languages is not. Even before we speak a word, much less get put in front of Lacan's mirror, our minds are dynamically structured to put images and concepts together. But even if Kant did not have much to say about language per se (and virtually nothing about a possible connection between schematism and words), schematism provides the natural *topos* for joining the intuition of the world in sense to its linguistic articulation in any quasi-Kantian philosophy of language. In its most intrinsically transcendental function, schematism does not join a thing–concept to sensory and imaginary depictions of the thing (dog to dog appearances, triangle to drawn triangle) but the pure concepts of the understanding

[36] One cannot simply call schematism a theory of abstraction because it is a two-way street: the way from image to concept is the same as the way from concept to image. Schematism might therefore more properly be called a theory of abstraction–and–concretion.

to the manifold of sensibility. That is, (1) unity/plurality/allness, (2) reality/negation/limitation, (3) inherence–subsistence/causality–dependence/community, and (4) possibility–impossibility/existence–nonexistence/necessity–contingency are expressed in the appearances of the manifold by means of schemata, in the direction *concept to image*, and the appearances of the manifold are brought to the threshold of conceptual unity in the direction *image to concept*. These mental structures do not correspond to words naming things but to linguistically-eligible or -susceptible functions relating concept to possible imaging. It is through these transcendental functions of imagination that we can conceive the very possibilities of beings apart from actual existence, and it is typically in language that we express these possibilities. But where the word emerges in the functional process is unclear. Is it in essence an image of the concept, and thus at the image–end of the process? Or is it closer to the concept, by its nature conceptual rather than imagistic?

The exploration of the relationship between language and imagination begins to develop ever richer resources in post-Kantian thought. But perhaps it is still surprising that it took more than a century after Kant for language to emerge as a fully philosophical subject matter in its own right. It was only in the last decades of the eighteenth century, contemporary with Kant's critical philosophy, that the study of language began to aspire to scientific status, especially as researchers discovered and methodically applied principles of comparative morphology; it was similarly late in Kant's career that important philosophical questions were being asked about language's anthropological origins, for instance in the posthumous publication of Rousseau's *Essay on the Origin of Languages* (1781) and in Herder's *Treatise on the Origin of Language* (1772; Herder had been a student of Kant's). German Idealism played a significant role in nineteenth-century developments, for example in Wilhelm von Humboldt's efforts to understand the characteristics that contributed to the specific spirit or genius of each language. But none of these specifically addressed the questions of transcendental psychology that had concerned Kant.

Of the major Idealists it was Georg Friedrich Wilhelm Hegel (1770–1831) who attended most specifically to the psychological processes underlying language. In the *Encyclopedia of Philosophic Sciences*,[37] most particularly in the *Philosophy of Spirit* (the third of the *Encyclopedia*'s three major divisions; it follows the *Logic* and the *Philosophy of Nature*), he expressly acknowledges Aristotle as perhaps the only other

[37] Hegel supervised three editions in his lifetime: 1817, 1827, and 1830. Later editions are typically based on Hegel 1840–1845, which published the three parts of the *Encyclopedia—Logic* (1840), *Philosophy of Nature* (1842), and *Philosophy of Spirit* (1845)—as separate volumes. Hegel 1840–1845 arose as follows. In lecture Hegel read to his students from the book's sections, which were typically a paragraph or two, and commented on them. After his death, students collated notes they had taken during the lectures and added them, in reduced typeface, to the appropriate sections of the text of Hegel's third edition. These notes, called *Zusätze* (plural of *Zusatz*, "addition") obviously have lesser authority than the main text, though the editorial care of his students and the clearly high standard of their collective note-taking make them illuminating sources. Since the 577 sections of the *Enzyklopädie* are individually numbered, I will cite according to these section (§) numbers, with "*Zusatz*" if the passage is from the student-added notes. All translations are my own.

thinker to have made genuine progress (in *On the Soul* and his other "psychological" writings) toward the goals of the *Philosophy of Spirit*.[38] The *Philosophy of Spirit*, for its part, traces the dialectical development of spirit (*Geist*) from its first emergence in the animal *homo sapiens* to its culmination in "absolute knowing"; accordingly, it is divided into the three parts "Subjective Spirit" (which progressively develops the consciousness of the individual human being as such), "Objective Spirit" (which presents the social and institutional realizations of human spirit), and "Absolute Spirit." "Subjective Spirit" in turn has three parts: "Anthropology: The Soul" (about the differentiation of the human species from animal existence), "Phenomenology of Spirit: Consciousness" (about reason as emergent from the development of consciousness and self-consciousness), and "Psychology: Spirit."

"Psychology: Spirit" is itself divided in three, into theoretical spirit, practical spirit, and free spirit. It is in the first of these, theoretical spirit, that Hegel traces the development of human psychology from sense intuition to thinking by way of *Vorstellung*. Generically this should be rendered as "representation," especially insofar as, unlike Kant (for whom "presentation" is a better rendering of *Vorstellung*), Hegel excludes from its semantic range the original presentation in *Anschauung*, the intuition–view of sense perception proper. After one had had the original presentation of *Anschauung*, however, one was in the realm of *Vorstellung*, representation. From that point forward Hegel developed his own updated version of the internal senses[39]: two types of memory (*Erinnerung* and *Gedächtnis*) preceding and following imagination (*Einbildungskraft*). With memory and imagination as the places of *Vorstellung*, it would be faithful to Aristotle to translate the term as "phantasm–having."

In view of the context we have established, what is most interesting is that Hegel's discussion of imagination culminates in the sign, and the second kind of memory (*Gedächtnis*) has to do chiefly with words. There the sign becomes the name, or rather the synthesis of name and meaning as permanent and universal; then it turns into the vehicle of reproductive memory, which "has and recognizes the thing in the name, and with the thing [has and recognizes] the name, without intuition and image" (§462); and finally it takes the very last step in the development of representation/*Vorstellung* by turning into an activity of thought without any separation of intending intelligence from intended meaning. Accordingly, in the third and

[38] The *Philosophy of Spirit* begins with §377 and ends with §577. At its outset (§378, thus just the second section of this third part of the *Enzyklopädie*) he says this: "The books of *Aristotle* on the soul, along with his treatises on its special aspects and circumstances, are consequently still the choicest or [even] only work of speculative interest about this subject. The essential purpose of a philosophy of spirit can only be this, to introduce the concept once again into the knowledge of the spirit, and thus also to open up once more the meaning of those Aristotelian books." In a note on *De anima* from 1820, Hegel described the task of a modern philosophy of spirit this way: "in all this it comes down to translating it into our (admittedly more cultivated) way of thinking" (quoted, in German, in Ferrarin 2001, 234).

[39] "Internal senses" doctrines, of course, refer to the late-ancient and medieval theories of mind powers situated between the external senses and the intellect that had been inspired by Aristotle's psychological writings.

last part of theoretical spirit, Hegel treats thinking proper (in four sections: its three stages are understanding, judgment, and reason) and thus concludes "Theoretical Spirit," the first division of "Psychology: Spirit," after which he turns immediately to the second division, "Practical Spirit."

This is, of course, no more than an outline of Hegel's trajectory from representation to thinking. As such it is not, and cannot be, perspicuous to anyone unfamiliar with the *Philosophy of Spirit*. Yet, despite the fact that even many professional philosophers experience Hegel's thought as alien territory, it is nevertheless possible to make a few crucial points about how it decisively inflects the conceptual topology of imagination in the direction of language.[40]

Right at the outset of the development of *intuition* (see §446; intuition/*Anschauung* is the first part of theoretical spirit, preceding representation/*Vorstellung*), Hegel explains that the proper object of consciousness is stuff (*Stoff*),[41] which takes up a position opposite consciousness as something relatively or merely other. Once we have advanced to spirit, however, stuff is given "the rational determination of being the other" of consciousness. Stuff is thus turned into something that the spirit expressly takes possession of inwardly and that is formed and organized by intelligence's focused attention; the object is received, and as received it has become an acquisition of spirit, indeed a *part* of spirit, although its being does not yet become *identified* with intelligence. Stuff's being is recognized as distinct from that of the intelligent subject, in that (in a Kantian moment) it is looked upon *in space and time*; that is, the experience of intuiting requires that the inwardized, focused experience be seen as elsewhere than in the self-conscious intuiter/perceiver: as being in space and time.

If Hegel, in general at least, holds that ontogeny recapitulates phylogeny, we can take this as describing two things: the (perhaps rapid) temporal development of the intuiting process as the intelligent subject becomes more experienced, and also a long-term, almost evolutionary development of intuiting in higher forms of animal life. Every stage of a process, whether long-term or short-term, bears in itself the elements or moments of later developments, though whether those developments will take place *here* and *now* is not inevitable. As you quickly look around a room or experience some fleeting aroma you do not fully and determinately take an inward hold on anything with sharply focused attention, but as soon as something catches your attention (as we say) or the aroma becomes determinate the development Hegel has just

[40]In the following brief discussion it is important to keep in mind that Hegel's dialectical philosophizing is based on the underlying unity of apparent differences that are overcome by the progress of spirit, whether individual or collective spirit. Oppositions and differences start out appearing stark, almost dichotomous, but as they become more familiar one begins to glimpse ways in which they are united and eventually brings about a unity by thinking and living with the differences. This is true whether one is dealing with what is subjective, what is objective, or with the relationship between the subject and object.

[41]I prefer the colloquial English cognate "stuff" to the more usual "matter" or "material," in part because of the diverse denotative range of German *Stoff*: cloth, fabric, material, matter, solid, stuff, subject matter, (chemical) substance, substrate, tissue. In the first instance what Hegel is indicating is that consciousness faces a world filled with all sorts of things—stuff.

described accomplishes itself: a visible thing becomes *what* you see *in your seeing*, an odor becomes *what* you smell *in your smell sensing*. The more decidedly you take hold of the experience, the more sharply defined a position it takes in space and time beyond yourself, as a thing out there that you have taken hold of; but it also takes up an ever more sharply determined position as one thing among the many that your particular sense powers can actually sense, as one position in your relevant plane or field of sensation. That is ontogenetic, a matter of an individual's experiential maturation. But we can also say that Hegel believes that in animals there must be some that are incapable of this full development. Something of these two moments of the human being's inward appropriation of an external perception can occur in any animal with a certain degree of sensory development; but the degree and the distinctness (or distinguishability) of the two moments is different in different animals. Animals that have a well-developed sense of space and some experience of the continuity of time undoubtedly are able to place a thing they are focusing on now with respect to others they have focused on before and in anticipation of future others; those that have a duller sensibility may for the short term be aware of a before and after, but it passes away quickly. The dullest just pursue or run away or are indifferent.

The influence of Kant with respect to space and time as products of human consciousness—of transcendental imagination, to be exact—is clear enough. But there is also an Aristotelian influence, which one might expect given Hegel's conception of *On the Soul* as an adumbration of his own aims. The moment in which intuition is inwardized Hegel calls *Erinnerung*, and it is the first stage of representation/*Vorstellung* proper. *Erinnerung* is used generically in German for memory, but etymologically it suggests an inwardizing movement. Hegel takes pains to emphasize exactly this. He does not name Aristotle, but he is clearly alluding to Aristotle's definition of both sensation and *phantasia* as forms of movement (with the movement of *phantasia* being a continuation of the movement of sensation). For Hegel, it is precisely in this inward movement that the initial constitution and preservation of the form of the intuition occurs; this takes place as *Bild*, image (§452). More expressly than Aristotle, he tries to show a detailed progression through the higher powers of sensation, memory, and imagination of implicit formal principles that become explicit by virtue of the progression; as Aristotle said, what is in intellect was first in sensation, and intellect grasps the form in the phantasm–image. Thus a good, explicative translation of Hegel's *Erinnerung* is "inwardizing image memory."

The notion that seeing–as–intuiting involves a field or subfield of our experience has roots in both Aristotle and Kant. There is, first, the intuition of space and time and the inner acquisition of a field of space–time that the intuiting subject can itself (re)project. But there is also the "stuff that spirit is": that is, Hegel argues that what spirit appropriates becomes *its very own* stuff and possession. That stuff is not completely independent atoms or units of experience but the recognition of sensation as having a formal principle capable of shaping and articulating all the relevant sensations as part of a field that the self can project; Hegel thus implicitly develops principles of comparability that underlie all the sensations. Intuiting progressively develops and deepens its stuff by formally diversifying its realm of experience.

In this section of the *Philosophy of Spirit* intuiting takes place *as* self-conscious spirit rather than at the earlier dialectical levels of mere consciousness or the even more primitive one of anthropological feeling. As a result, intuition and the inwardization of intuition are not just feeling related to outward otherness, but also more specifically developed sensory articulation of rational form. Intuition is therefore a kind of rational feeling. This, too, can be understood as an elaboration and radicalization of what Aristotle had understood as the (topological) space of sensation oriented by contrary extremes between which individual sensations are positioned.[42] It is a culmination of the conceptual topology of the phantasmal field that originates in sensation and that is the basis for the regeneration and differentiation of the original appearances in the inward senses of memory and imagination.

Despite this culmination, however, Hegel does not take great advantage of it. The most obvious reason would be that he was not pursuing an investigation of the external and internal senses per se, but using them only as the dialectical origin for a rapid transition to ever higher and more encompassing powers of spirit. Unfortunately, in this dialectical transition Hegel fails to overcome the conventions of tradition and thus misapprehends the relationship between imagination and reason.

The distorting weight of tradition is betrayed by his image for the inwardized image memory: the *unconscious shaft* of image memory (§453). This image of the shaft (or *mineshaft*, which I shall use henceforth) in fact appears much earlier in the work. In the "Anthropology" division, in anticipating the equivalents of *Vorstellung* and *Gedächtnis* at the level of quasi-animal feeling (§403), Hegel had said that "every individual is an endless richness of sensation determinations, representations, facts of knowledge, thoughts, etc.; but *I* am still a wholly *simple thing*—a mineshaft without determination, in which all this is preserved without existing." Each image or other mental fact or unit exists there in random storage. In the first instance the animal *homo sapiens* is a repository in which the image content of experience is preserved in an indiscriminate way. Although at certain points in the later *Vorstellung* section there is a more nuanced development, Hegel continues to use the figure of the mineshaft and its darkness, and he reinforces the sense in which each image is a disconnected unit floating about randomly in that dark shaft. It remains that way until intelligence pulls an image out and associates it with other images (in §455, the second paragraph after he reintroduces the mineshaft figure).

This is surprising because Hegel is the philosopher par excellence of finding the conceptual/ideal at every level of consciousness and existence. A premiss of dialectic is that there is formal truth at every level that is preserved and taken up (*aufgehoben*) in the next level. In the analysis of intuition there is no shortage of structure, but his representation of the collective storage of memories in inwardizing image memory loses all trace of this. In this way, if in no other, Hegel joins the list of thinkers confirming Castoriadis's dictum that even philosophers who

[42] Hegel does not mention Aristotle's conception of contrarieties in sensation. If he had, is there any doubt that he would have been able to develop and ramify them dialectically? There are already hints of this in his anthropological discussion of sensation in §401 and its *Zusatz*.

have the most profound insights into matters of imagination fumble them away in the elaboration of their thought. I would add: when push comes to shove, they tend to ascribe anything of true worth to reason. That is certainly the case with Hegel, for whom the goal in "Theoretical Spirit" is to "discover" the autonomy of intelligence.

In the sections immediately following inwardizing memory, sections that treat the three different states of imagination (reproductive, fantastic–productive, and sign–making),[43] Hegel shows how the intelligence makes ever more abstract use of images and how the content of the representations becomes correspondingly less concrete. Image–representations are typically a sensory and concrete kind of representation; other representations, by contrast, have concepts and ideas as content. This difference is not, however, an all-or-nothing affair. Emblematic of this is his explicit criticism of conventional notions of association, which for his philosophical taste are too concrete insofar as they take the unit–character of the original images of sensation as a permanent feature of subsequent images. Instead of the rather crude empiricist idea of similar images piling up or "falling upon one another" (§455), he emphasizes their gradual subsumption under universals. Again he resorts to an imagistic portrayal to illustrate this. The crude conception risks being nothing more than pure accident and conceptlessness, he says, unless there is something like a force of attraction between similar images, "which would simultaneously be the negative power of rubbing away on one another what is still unlike in them." (All images of squirrels, to give an example, must be drawn to one another in a way that eliminates what is merely particular in each.) Such a positive force of attraction that produces the negative result of grinding away dissimilarities is the intelligence, "the I identical with itself, which through its inwardizing memory immediately gives them universality and *subsumes* the individual [act of] intuiting under the image that has already been made inner."

But *is* this a process driven by intelligence's autonomy? If the force of attraction is, as Hegel says, the intelligence, then it is not in the things. Hegel's process therefore does not allow for the possibility that intelligence might *discover* structures in fields and subfields of the inwardized intuitions. Instead, he uses the smoothed-out images to pave the entrance ramp to the highway of abstraction. Getting rid of the rough edges of dissimilarity lets the images be subsumed under an abstract, intelligence-derived template. The individual images, as things turn

[43] In the *Zusatz* to §455, imagination is described as "*above all* the *determining* of images" (emphasis in original). Hegel lists the three fundamental kinds of *Einbildungskraft* as (1) reproductive imagination, (2) fantasy (productive imagination expanded upon as symbolizing, allegorizing, and poetizing imagination), and (3) sign-making fantasy (productive imagination that reaches the verge of abstract thinking); see §§455–457, at αα, ββ, and γγ. The first and second make a standard modern distinction—the first in fact appears to be Hegel's creative adaptation to his philosophy of Kant's schemata—whereas the third is Hegel's true innovation. Especially in light of how he develops sign–making into words and then into thinking in the sections devoted to the third stage of representation, he anticipates to some degree Saussure's ontology of language–signs.

out, have no intrinsic relationship to one another, no more than they have in the mineshaft of images he described earlier. The force behind the image–rubbing is somehow directed by intelligence.

Of course, some such cognitive force would be necessary to assure that individual squirrel images rub against other squirrel images, hamster images against hamster images, star images against star images, color images against color images, and so forth. But insofar as Hegel is to any significant degree following the path traced by Aristotle, this would make animal imagination unintelligible. If the force that allows an image to come into contact with images of the same kind is in intelligence rather than the things or their phantasms, then imagination could not serve the purposeful activity of animals; the images would have to associate in some way that Hegel does not even begin to articulate. But if, as one ought to expect with Hegel, something is carried forward from earlier levels of the dialectic that is their truth on the next level, one should expect that something of the sensory image, the remembered image, or the poetic image ought to remain even once intelligence has done its work.

Perhaps, then, Hegel ought simply to have declared that the conceptual realm achieved by dialectic must not be tainted by mere sense images or by analogies based on sense images (which latter are produced by the symbolizing and allegorizing versions of imagination). Instead, he employs verbal imagery and figures to progressively banish the sensory (that is, the image) character of images in favor of the abstract and the conceptual. Yet he offers nothing to warrant that the image–character of the remaining appearances can be totally eliminated. All the rubbing of image against image guarantees is a less distinctly defined image, not the elimination of the imaginal—although clearly the hope is that once the distinctiveness of individual images is removed, the *typical* form that each shares will become clearer.

There is an important positive point to be made here about Hegel's development of the conceptual topology of imagining. Etymologically, the German word for "image" or "picture," *Bild*, suggests "structure" even more than anything visual. Kant was the first philosopher of historical stature to emphasize the express structuring power of imagination, most articulately of all in the schemata of the pure concepts of the understanding. But given the fact that, even before it deploys the schemata, transcendental imagination establishes the structure of space for all external experience, Kant's way of thinking about imagination tends strongly to the visual. Hegel is one of the few philosophers of imagination whose principles allow him to avoid being trapped by the visual image. The formative or structuring character of imagination, both ontologically and psychologically, is far more important to him than the appearances produced. In the dialectical development of the ideal from the concrete–material, what counts is the ever more distinct appearance of the formal, not as something static but as a dynamically formative principle of the things generated. If we have seen the desire for the emergence of the formal/ideal in all the thinkers we have considered at length, it reaches new heights in Hegel. But for all the detailed attention Hegel tries to give to what happens to images in order to make form apparent, the emergence of the ideal, of patterns, and of forms that he attributes to reason is still deeply mysterious.

8.4 If Signification Is Imaginative, Can Reason Leave Imagination Behind?

At the end of the intuition section of "Theoretical Spirit" Hegel introduced *Erinnerung*, inwardizing image memory. At the end of the imagination section he introduces *Gedächtnis*, which also can be translated as "memory," though if the word *Erinnerung* is suggestive of the original inwardizing of memory, *Gedächtnis* suggests a further development. For Hegel it matters that *Gedächtnis* is built on *gedacht*, the past participle of *denken*: it is memory subsequent to the having–been–thought of images. Intelligence thinks images as it takes possession of them in ways that *signify* universals and culminate in *naming* the forms of the images. The memory of *Gedächtnis* is thus a retention of the signs and names that indicate this past thinking. It is name-using, sign memory.

What is at first glance very odd about the sign memory section is that Hegel's account of the progressive abstraction of the sign from what is concretely given goes so far as to eliminate meaning and reference. To oversimplify only a little: after claiming that the arbitrariness of making signs and names reflects the spontaneous autonomy of the intelligence, Hegel points out that a word repeated mechanically over and over gradually loses meaning and reference. Why, of all the possible directions to take in examining language, did Hegel choose to focus on this? The discussion does not get even as far as the verb, much less other parts of speech, and about syntax it is totally silent.

Perhaps this can be explained largely by the goal of this part ("Psychology") of the *Philosophy of Spirit*. It is an account of the powers of spirit as they ascend from a merely receptive dependence on the material and animal world to the pure self-possession and self-activity of an ego, a soul–self, a *psuchē*, a spirit that has realized itself as pure rational intelligence. It is an account that is purely individual; that is, there is hardly a mention in the emergence of the sign of the sociality of language, the fact that it is acquired by each individual being taken up into the community of adult speakers who are already in possession of a fully developed language. It appears, in the first instance, as a modernized version of Aristotle's *On the Soul* account of the progression from sensation, through imagination, to the noetic powers. Capacities that human beings share with animals—sensation and imagination and purposeful image–reckoning—lay the basis upon which arises the highest function human beings are capable of, thought.

I say this "appears" to be a modernized version of Aristotle because Hegel fails to be Aristotelian here in two decisive respects: he cannot accept the dictum that there is no thinking without phantasms, nor that intellect's grasp of forms requires the presence of phantasms in which those forms are grasped. Thus Hegel is not genuinely Aristotelian at all! For Aristotle, the intellect grasps the form in the phantasm; it sees the phantasm and the (kind of) thing and situation (a light at night moving across the plain) from which the phantasm arises in the light of this form. That brings one, in effect, to the threshold of predication, and thus it goes beyond the phenomenon of isolated words referring to isolated images (even after rubbing!). Similarly Hegel abrogates the rights of Kantian schematism. If Kant grants that there is a kind of pure reasoning that rises above the "limits" of the realm of sensibility, it is neither theory nor knowledge but rather the power of a rational being to direct its actions according to autonomously generated and affirmed universals—practical, not theoretical thinking.

Moreover, one might even contend that in at least one crucial respect Hegel's drive to evacuate thought of sensory content violates the principles of his own dialectic. Dialectic is supposed to preserve and raise up to a higher level the truth already contained in an initially posited claim, and to leave behind any falsity; it does this precisely by considering the initial position in light of a second, contextualizing position.[44] The kind of imagination that establishes signs, called signitive imagination, involves abstraction that leaves behind the sensory. Rhetorically one can then put the question this way: is the color and bushiness of a red squirrel's tail a falsity that is left behind by scientific understanding? Is the spatiality of a geometrical figure left behind (as something merely quasisensory) in accomplished geometrical science? Is the reference of algebraic formulas to curves in cartesian space a falsity that is left behind in some kind of higher, purer mathematical understanding?

There is something perverse in the claim that the algebra of analytic geometry leaves behind geometry; more generally, there is a perversity in any claim that when insights into the structure of an appearance and its field are formulated more abstractly, the appearances are thereby fully transcended in reference and meaning (whether or not what is left behind is subsequently called false).[45]

[44]Since the misrepresentation of Hegel's dialectic as thesis–antithesis–synthesis is still so widespread, I insert the following note: Dialectic understood according to this thetic triad is more Fichtean than Hegelian; it has been handed down to the present chiefly by Marxist tradition. Marx may have acquired it from lectures given in Berlin in the mid-1830s by Heinrich Moritz Chalybäus. Hegel occasionally mentions the thetic triad, but when he does so he is critical of it, because he finds it to be an unworthy, merely mechanical understanding of dialectic. On the myth of the Hegelian triad, see Mueller 1958. More genuinely Hegelian is the illustrative image of the dialectical process Hegel gives in the preface to the *Encyclopedia* (§13): dialectic is a process of breakthrough from an initial circle to a containing circle, and then to yet another circle from the perspective of which the two previous circles are seen as a unity. Thus I have qualified the second position by the term "contextualizing." In light of the present book's arguments in behalf of *conceptual topology*, one might easily argue that this way of understanding dialectic as the re–positioning of circles makes Hegel the philosopher of conceptual topology par excellence. But in his insistence on liberating thought from the falsity of appearance, he falsifies imagination and also reason's relation to imagination, and in a sense he even becomes unHegelian—that is, he falls short of the greatness of his profoundest thinking.

[45]"Perversity" should be taken here literally as well as figuratively. Literally it indicates a thoroughgoing turn away from X (here, the sensory–imaginative) toward its opposite Y (the purely abstract–conceptual). In the literal sense, calling Hegel's attempt to leave the sensory behind "perverse" is merely factual. In the figurative sense what is "perverse" is ordinarily thought to have something of the immoral about it. The figurative use I intend is ethical insofar as it has to do with the ethos of philosophy and the ethos of inquiry. What consequences are there in denying the importance, relevance, or even being of something that one wants to leave behind in the course of an inquiry? Is the nihilism implied by such denial merely a logical matter, or is it ethical and ontological as well?

Although extreme in terms of how it is expressed, the Hegelian claim is an instance of the more traditional philosophical belief that, in some way, reason *does* completely transcend the realm of the sensory. Here one needs an argument rather than mere assertion or dependence on tradition.[46] Hegel, at least elsewhere in the *Encyclopedia*, recognized that the mechanically applied abstractness of traditional logical forms was a bar to reason rather than its vehicle (accordingly he distinguished *Verstand*, understanding, from *Vernunft*, reason); thus he begins the entire project of the *Encyclopedia* with a new logic that, wherever and however one starts, will generate its own totality. That is, out of its inner dynamism and capacity it always produces wholeness, the totality of things taken in the largest conceivable sense of "totality" and "things," including their appearances. Hegel's virtue is that, in comparison, almost every other thinker merely *postulates* that logic transcends concrete reality.

In this book I can do little more than put a question mark after conventional claims about the rational status of logic. It is one thing to say that rational understanding is not the simple mind–possession and –contemplation of a sensory image as such (as, for example, Descartes says), quite another to say that rational understanding takes place without images of any kind, and possibly even without reference to the imaginable in any residual sense. As I have argued throughout this book, imagination is *both* abstractive *and* concretional. If that is the case, then it is at least plausible that what has traditionally been called abstraction is as imaginative as it is rational. Conceiving one's yard as a trapezoid is an abstraction by imagination, as is treating the trapezoidal area as an algebraically calculable quantity measuring the size of the yard. The biplanarity of imagination implies that whenever one "abstracts" from an original, one is temporarily stepping out of the originating plane and leaving behind (though not overcoming as false!) features in the originating plane; but at the same time one is projecting features from the original plane into a more formal, simplified, and schematized aspect in a second plane. "More formal, simplified, and schematized" does not, however, mean devoid of all appearance. Wherever there is appearance, there is a medium or matter, a field–stuff, be it mental or real, that takes on now this, now that appearance; where there is form, there is that which is formed.[47] To say that we leave behind and forget the originating plane is a kind of Alzheimer's disease of reason. Although, when the mind moves in the abstractive direction, this second plane is "less concrete" than the first, it is still an imaging plane and has its typical appearances, even if they are not of the same type as the

---

[46] As a counterexample, recall that, strictly speaking, Aristotle's "no thinking without phantasms" means that even the most perfectly noetic understanding cannot entirely escape the human condition of having phantasms.

[47] This last clause identifies the locus of the problem. The crucial question is whether there is a level of form that is perfectly free of matter. Almost all nonmaterialist metaphysical systems say yes, and they try to get there by a kind of extrapolative argument. What this book has done is to reinforce the question mark after the question; it demands that the philosophers justify rather than merely invoke the right to step beyond the process of extrapolation and actually reach the purely formal realm.

original appearances. Thus the plane or field of geometric abstraction in which one conceives the trapezoidal shape of one's yard is two-dimensional, no matter how complex the topography of the actual yard; it is not inexorably temporal in the way real yards are (as grass grows and leaves fall), it is marked by virtual boundaries that do not exist in reality (the property line), it does not have hillocks and mailboxes, etc. The algebraic formula for calculating the area of the trapezoid does not have intrinsic spatiality, but the formula still has to be arrayed spatially, symbolically, and sequentially on a page or a screen, and though it does not have an intrinsic temporality it is implemented step by step; more to the point, since for every image we expect that it should be in at least one respect like the original, the symbols correspond to the distinct lengths of the sides of the trapezoid and are relationally greater or lesser corresponding to the length and shortness of the sides. Even a logical formula retains certain minimal features and distinctions of the natural language propositions (and elements of propositions) they represent. Just because something does not have colors (or aroma, or temporality, or any other property that can be sensed) does not mean it cannot be an image. Moreover, although what imagination or intellect does with images (even when they are sensuous) is not in the first instance itself sensuous, exactly what it can do with the image depends precisely on the image's image–character and aspects of its materiality as image, whether that materiality is maximal or minimal.

What I have been implicitly exercising here is a different "sensibility" for what happens in abstract thinking than is conventional. It is a sensibility that derives in part from the fundamental possibilities of the conceptual topology of the human psyche, possibilities that I contended earlier were realized to some significant degree in Descartes's philosophy, in Aristotle's, in Plato's. To focus only on the first: Even if the mature Descartes thought that a kind of pure noetic thinking was possible for human beings, it could be achieved only *after* prolonged imaginative activity, and it was not a state of thinking that could be maintained for very long. Indeed, in his conception of mathematics, Descartes constantly affirms that we run the risk of making mistakes and not even thinking about anything at all if we forget to portray the imaginable as concretely as we can manage. Human beings have the tendency to lose track of where their thought has gotten to, and it takes every expedient of constant attention and method to overcome the defects to which this makes us subject. If I wish to take even more seriously than he the dictum "no thinking without figures and images," it is in part to rigorously follow out the consequentiality of the way of thinking he established.

Unfortunately—or perhaps not so unfortunately—these are questions that cannot be settled in a few chapters, much less by dictate or dictum. They require thought, and thoughtful attention to the thinking of those thoughts in all respects. What is unfortunate is that the legacy of antipsychologism has alienated us from our sense of ourselves as psychological beings. To put it simply, we have not only lost the habit of using old psychological terms, we have rejected them without anything to replace them. The reasons for rejecting or at least deemphasizing them were perhaps valid within their originally motivating field—for example, to forestall making appeals to an inadequately substantiated introspection in psychological explanation.

But the spread of antipsychologism took on a life of its own; it was extended and universalized without adequate supporting argument. Overtaken by the conceptual and methodological momentum of this movement of thought, philosophers and psychologists failed to attend to the topological features of psychological experiences and raced right past them. Thus they deprived themselves not only of some possibly viable explanatory tools but also of terms needed for a first-approximation description of basic psychological events. In the process they fell back on an even less justifiable, more spectral formalism than that of the traditions they rejected.

It would be foolish to think that all we need to do now is reinstate some single, authoritative old way of speaking about the psyche. One thing this book shows is that, without our making the effort to see things whole, there is no single, privileged past example. But it also shows that a few of the old ways of speaking reflect a profound and more adequate way of thinking about the matters they attempt to narrate, and that if we are to have any hope of significant discoveries and rediscoveries in the same field, we have to hear and learn their lessons.

8.5 Wittgenstein and the Imaginative Supports of *logos*

Fichte and Hegel represent different outcomes from a Kantian starting point. In Fichte imagination reaches its (recent) apogee. It creates new appearances in thought where previously there was a tension or a gap. The merely imagined could then be implemented and embodied in the world by corresponding human activity, which in turn produces a new situation in which new tensions and gaps appear so that the process of new imagining begins again. Schelling made this process both objective (in nature) and subjective (in spirit), and Romanticism eagerly accepted, then assumed, and finally took for granted this legacy of creative nature and creative artistic genius. More invoked than studied or understood,[48] imagination took on an almost wholly positive, creative valence in everyday parlance, until it could no longer bear the weight of the expectations that had been placed on it. Imagination had become virtually identified with the novel in art; by the same token it took up a position (particularly in the psychology of genius) in starkest opposition to ordinary reason and bourgeois culture. That opposition ultimately brought imagination into discredit. Viewed as essentially untethered fantasy, it became in the course of the nineteenth century a countercultural but unserious option in an age of scientific, often positivist, reason. To assert that there is no thinking without images was to taint thought.

Hegel, of course, rationalized imagination by arguing that it culminated in nonsensuous thought. But if Hegel's *Encyclopedia* taught that the image was superseded and overcome by the sign and the word, the immediate effect on the

[48] Robert Kugelmann has pointed out to me that in Boring (1950), the second edition of Edwin G. Boring's classic *History of Experimental Psychology*, the index entry for "imagination" lists a single occurrence. This is a sign of the lack of interest in it as a topic for scientific investigation.

study of language reinforced tendencies in comparative linguistics that had been prospering since Sir William Jones (1746–1794) proposed that Sanskrit or a Sanskrit-like language was the ancestor of other Indo-European languages. The proper object of this linguistics was the language and the language group. The history of the changes of individual words and structures revealed lawlike transformations that over long time periods differentiated languages from one another. The scientific linguistics that developed from Idealism, particularly in the work of Wilhelm von Humboldt, treated these languages as organisms with distinctive, gradually developing characteristics. Languages were understood as the chief organ of the culture of a people, expressive of a governing spirit or genius, the evidentiary basis for which was chiefly literary and artistic. If Hegel ascended through empirical evidence and the functions of individual consciousness to arrive at the universal, Idealist linguistics abstracted from individual, empirical usage to focus on the particularized expression of universal culture that each language represented. By the second half of the nineteenth century there were, however, countervailing trends that tried to rise above historicist approaches and to adopt empirical methods like those of the natural sciences. The Neogrammarians, for instance, aimed to register the basic facts of current language use by representing spoken utterances in a universal phonetic alphabet and applying inductive methods to arrive at verified generalizations. The role of language in the individual psyche (or rather in the psyche of the idealized individual) was largely ignored, and possible connections with imagination all the more.

Yet this narrative about the beginning of the "linguistic turn" becomes more problematic if one looks carefully at three giants of the study of language at the end of the nineteenth century and the beginning of the twentieth. Charles Sanders Peirce, Ferdinand de Saussure, and Ludwig Wittgenstein would each ordinarily be considered as to some degree antipsychologistic. Peircean semiotics presents language as a system of sign use with an objective taxonomy and a dynamic pragmatics. Saussurean structuralism understands language as a differential system of signs conceived as the arbitrary fusion of sound and meaning, a system that can be studied apart from the vagaries of individual human consciousness or historical change. Wittgenstein, whether early or late, dismisses the notion that there is anything decisive for philosophy and science in what is said to go on "in the mind"; early he understands language as the vehicle for what can be said about the world, late he presents it as a game (*Sprachspiel*) played out within the practices of ways of life.

These things are all true, as far as they go, but the question is whether they go far enough and look to the right places. Wittgenstein's interest in imagination, for example, was lifelong and complex. If certain parts of his work show a behaviorist inclination, that does not mean that he is determined to deny the existence of images; rather, he argues that they cannot accomplish the epistemological functions they were usually assigned. When in *On Certainty* he contests the theory that, if one *knows,* there must be some special kind of thought (say, an idea with a mark of clarity and distinctness), he is arguing not that we never have clear and distinct images but rather that knowing is something other than a private experience of a special, inward mind–presence.

Wittgenstein's antipsychologism did not preclude a lively interest in and even a positive appreciation of images and imagination, an interest that in many of his writings is quite striking. For example, a reader paging through the *Philosophical Investigations* will come across figures demonstrating perceptual paradoxes, like the famous duck/rabbit. The drawing as drawing is quite determinate, but that determinacy does not fully determine what we take it for. The fact is that we do take it in one way or another, and we cannot take it in more than one way at a time. What we see depends in part on our experience: the first time we see the duck/rabbit, for example, we may see only the duck, so that the next time we see it we will probably see the duck again. With some figures a person might not, on the first try, see either of the expected figures. But if a friend gives us some guidance in how to look ("you see the duck's bill, right? now think of it as a pair of rabbit's ears"), or if quite unexpectedly we see the alternative, it is easy in future to choose to see it first one way, then the other. Wittgenstein calls this kind of phenomenon *noticing an aspect* (Wittgenstein 1953, 193e) and a *change of aspect* (196e).

In the same aphorism in which he discusses the duck/rabbit Wittgenstein denies that when we sit down to eat we *take* the cutlery *as* cutlery (German *halten…für*; see xi, 195e). This does not mean that we do not take the duck/rabbit now as a duck, now as a rabbit, but rather that the "taking as" locution really only has a place in the language game of everyday description when we are aware of alternatives or after we have produced an alternative. It is the average, everyday philosopher, the one who likes to tell us what we have been doing (or should have been doing) all along without knowing it, who universalizes a situation so that it represents all: who says, for instance, that not just with the duck/rabbit but with everything we encounter in the world we must *take it as* one way or another. According to Wittgenstein, however, what typically happens is that we simply see what we see, whether a duck or a rabbit or a duck/rabbit, whether a knife–fork–spoon or cutlery.[49] It would make perfect Wittgensteinian sense, on the other hand, for airline passengers to ask a flight attendant whether he really expected them to take the flimsy, plastic "antiterrorism" knives and forks provided with meals as cutlery. The question legitimately arises precisely because, in many of the ordinary actions of the language game called dining, plastic utensils do not successfully play their expected role. The knife bends too much to cut meat, the spoon softens in hot soup, the tines of the fork break trying to spear an undercooked carrot.

[49] I have seen the duck/rabbit figure so often that, when I turn the page of a book and come upon an instance, I see it immediately not as a duck or a rabbit but as "the duck/rabbit." As a result, one has to be cautious about overgeneralizing Wittgenstein's point that we do not take these implements on the table as cutlery. There may well be a biographical moment in the life of virtually every diner in which he learns to take knife, fork, and spoon together all at once as cutlery—certainly a cutlery saleman looks at them this way—but that does not mean that henceforth when he comes to table he ascertains first a knife, a fork, a spoon, and then adds to this a mental operation of taking the group as cutlery. That kind of explanation is psychologistic and false. It is false not because there is no psychological activity but rather because the explanation manufactures a scenario corresponding not to what happens but to what the explainer thinks the explanation should look like. Psychologism is often made-up psychology. But antipsychologism cannot guarantee that just by virtuously opposing psychologism it avoids being made up itself.

One must recall that antipsychologism in its originating sense does not deny psychological life and functions but rather insists that scientific knowledge becomes inexplicable when it is simply reduced to a particular kind and sequence of such functions.[50] Antipsychologism does not have to deny the existence of human psychology, only inappropriate invocations of human psychology. Wittgenstein would deny that the person who understands the theorem that angles opposite equal sides in a triangle are equal must have a particular (marked or unmarked) figure in the imagination, although for some people understanding of the theorem might very well always involve having such a figure in mind. What understanding is depends on the practices of the language game and the moves in it made by individual persons. In a geometry class it would involve being able to produce the figure on paper, or a chalkboard, as one developed a step-by-step proof. In an engineering class understanding the theorem would probably not involve anything so elaborate; just remembering where to find it would do. At a certain level of theoretical algebra one must be able actually to *derive* the formula for the two solutions to a quadratic equation; in calculus class all one needs to do is *remember* the formula, and in a practical applications class one might only need to be able to remind oneself by looking it up. Thus "understanding" does not have a unique or standard meaning to which all usages must refer, but a local usage relative to the language game being played. Across the different language games there may be some *family resemblances* in the ways that a word is used, but only rarely will it be possible to specify them rigorously.

Wittgenstein thinks that philosophers exceed their reach either by universalizing the practice of a single language game as though it applies to all or by arriving at some pretended essence common to all language games. Philosophy is not, despite its centuries-long pretensions, the language game of all language games. This does not mean, however, that people must confine themselves to a single language game. Quite the contrary: people constantly shift from one game to another. Although Wittgenstein does not provide much guidance about how to draw boundaries between games—one might in fact expect that he would look upon the idea of drawing boundaries between games as a remnant of the kind of philosophizing he repudiates—it does not seem too difficult to arrive at reasonable first-approximation divisions. At work one might be engaged in a language game in which the acts and words one uses aim at making and selling a product; if one practices a profession like medicine, law, or ministry it has typical places, activities, and terms that anyone in the field would be familiar with. If, at noon, one goes to a health club to exercise, one shifts not just place but also activities and words.[51] At home one moves into another language game

[50] A similar criticism could be made, of course, of cognitive and neurobiological theories that reduce psychology to a sequence of (nervous system) functions.

[51] The games can cross boundaries, of course. When the doctor jumps off the elliptical machine to assist the weightlifter whose knees buckled during a power lift, he is (figuratively) jumping back into the way of life of his medical practice.

where the significant places are kitchen, dining room, den, bedroom, recreation room, and yard. In the recreation room one might do and say at least a few things that would be in place at the health club, and over dinner one will perform many of the same actions that one would at a business lunch. But at home the manner of speaking at table may be relaxed in a fashion unacceptable in a formal setting. If at a business lunch one spoke as if to one's children, or if at home one talked deals, everyone would notice the breach of the rules, of the accepted and expected way of life.

Negotiating the differences in the language–games is, in effect, negotiating shifts in imaginative fields. In terms of philosophical grammar, the analysis of our ordinary uses of language, this means that the more language–games we "play" the more easily we can negotiate transitions from one place to another, literally and figuratively. Another thing we learn is how to recontextualize what is part of one language–game in another. This would amount to a deliberate, biplanar employment of imagination to reconfigure something that plays a role in one plane so that it can play an analogous role in another. It is precisely the linguistic marking that fixes the thing in its original "field of play" so that we can see how to project it into a new one.

No grand unified theory of mental life is needed to apply these Wittgensteinian lessons. In Sects. 3.7 and 3.8, above, we discussed the imaginative character of playing and practicing basketball. The rules of the game, the painted lines on the field/court of play, the typical actions players can take, etc., provide "real-world" support for the work and gestures of imagination. For example, the point guard does not see a painted line on the floor and then interpret it as the time–line he must cross in 8 or 10 s, nor does see first a rectangle of red enamel, then an out-of-bounds area: he just sees it (peripherally) and avoids bouncing the ball there. So, too, in the fine arts, where oil, pigment, gesso, canvas, etc., are the material supports for the artist's imagination, the artist is at home in her workspace and does not (ordinarily) have to switch between thinking that a yellow pigment is a rare earth and that it serves well for painting a sun. But both player and artist can perform sensitive and cognitive shifts where they are appropriate and serve his or her work. In a flash the array of bodies or paint before his/her eyes shifts from being mere background to focused opportunity and back again. Rules, markings, typical configurations, routinized actions are all familiar and all subject to variation according to circumstances—and not everything (and perhaps precious little) is planned or well anticipated. Much of what goes on would be publicly and behaviorally accessible. But not everything. The superior point guard, being positioned where he is, sees something, personally and to begin with privately, that no one else, not even his point guard coach on the sidelines, can yet see. The artist can pick up a color that her kibitzing fellow artists tell her is wrong, but then they marvel at how it works once it is on the canvas. The precious little that is imagined against the publicly accessible background is crucial to success, to the beauty of the work or the game.

This glance at the placement activity of imagination in the later Wittgenstein in fact prepares us to notice how imagination functions even more basically in his early work, in particular in the *Tractatus Logico-philosophicus*.[52] The *Tractatus*, as is well known, presents a picture–theory of language. Elementary propositions indicate the basic situation of the most basic things of the world that we can talk of; from these elementary propositions more complex propositions and meanings can be derived. The function of logic is to organize the elementary propositions into more complex forms; these complex forms can be determined with respect to their truth or falsity—that is, their truth values can be determined—as a strict, quasimathematical function of the truth or falsity of the elementary propositions out of which they are built. Logical truth presupposes, or rather is indifferent to, ontological or substantial truth, the actual truth or falsity of the elementary propositions and of the complex propositions built up out of the determinately true or false elementary propositions. Logic breaks down the structure of a complex proposition so that the logician can determine the cases when it is true and the cases when it is false. It considers all possibilities of assigning true or false values to the elementary components of the complex whole; if the actual truth or falsity of each component is known, so is the truth or falsity of the whole. Tautologies and contradictions, propositions that are always true and always false, respectively, are so by virtue of how they are logically put together. Given clarity about propositional structure, logic is an ideal instrument for sciences that must construct and assess generalizations about empirical facts. The most basic empirical facts are expressed in the elementary propositions; all other knowledge is the result of their composition. Anything that cannot fit this scheme is unknowable. The world of this logical scheme is the world knowable by science.

Wittgenstein did not provide a metaphysical theory of this world; rather, he described the logical structure of any world that consists of basic, expressible facts that can be combined according to logical rules. He placed ethics and other mysterious things in an inexpressible realm, where language speaks nonsense but to which you might ascend by using the ladder described at the end of the *Tractatus*. To say anything in language, whether ordinary language or the disciplined usages that strictly conform to scientific demands, is to speak about the logically knowable world. To appreciate properly the mysterious matters of metaphysics, ethics, and similar concerns, however, one needs to pull the ladder up after oneself (ascending the ladder, it seems, is making your way through the

[52] Wittgenstein 1922. This was a dual-language edition, with German and English on facing pages. The German version had been published a year earlier in *Annalen der Naturphilosophie*. The work is divided into short sections—some only one sentence long—that are enumerated decimally (thus section 1 has a subsection 1.1, which has three subsubsections, 1.11, 1.12, and 1.13, followed by subsection 1.2 and its single subsubsection, 1.21). I shall cite the work using these decimal numbers. I will often modify the English.

Tractatus and achieving the insights it contains) and thereafter hold one's peace. "Truth" in such matters—the word has to be put in ironic quote marks because it is technically an equivocation—cannot be said; it can only be shown and seen.[53]

Wittgenstein freely admitted that in the *Tractatus* he was trying, before he pulled up his own ladder, to say things that in literal truth could only be shown and seen. This saying of the inexpressible includes what he says in the first several sections of the *Tractatus*, where he describes how the basic facts of the world get expressed. This is where the "picture–theory" of the world enters. Propositions are portrayals or pictures of the world—or rather of the possibilities of the world, since logically any proposition might be either true or false. Propositions are a picture–language. In the German of the *Tractatus* Wittgenstein uses *Bild* for "picture." It can equally well be translated "image," and it suggests something formed, a pattern or structure.[54] Thus language, in the form of propositions, logically images world–structures.

Wittgenstein's theory in the *Tractatus* is nothing if not an account of how the logical structure of language derives from structure implicit in propositions as images of the world. He composed the work as a sequence of seven major propositions, each of which (except for the seventh) is elaborated in numbered subpropositions.[55] About halfway through section 4 he begins to treat propositions as truth–functions, an approach that continues, after announcing the general form of the truth function, until about halfway through section 6, where he begins to contrast the scientific and "philosophical" uses of language. What precedes, section 1 to the middle of section 4, is preliminary, in the sense that it discusses what is prerequisite for the truth–functionality of propositions (or, to use a less philosophically overinterpreted rendering of the German *Satz, Sätze:* the truth–functionality of *sentences*). Proposition 1 says that "the world is everything that is the case"; proposition 2 that "what is the case, the fact, is the subsistence of relations of things"; proposition 3 that "the logical image [*Bild*] of the facts is the thought"; and proposition 4 that "the thought is the meaningful sentence."[56]

---

[53]The early positivist exploitation of the *Tractatus* referred to anything said about what was at the top of the ladder as nonsense, *Unsinn*, but their use of the word had a quite different valence from Wittgenstein's. From the logical/scientific/linguistically expressible standpoint of Wittgenstein, any attempt to bespeak these things does not make sense, because language is about and directed to the scientifically describable world. The distaste that reverberates in the exploiters' usage has little in common with Wittgenstein's; it is an emotion-laden accusation rather than the ascertainment of a basic incapacity of language (in particular of scientific language). Whether metaphysical and ethical "things" "exist" is beyond the capacity of language to express and scientists (and philosophers) to judge; only human beings can decide, according to a different and perhaps more basic experience than that of the logically expressed world.

[54]Grimms *Wörterbuch* says that the word was originally applied to statues, then extended to paintings and drawings and appearances more generally.

[55]From this point onward I will refer to the major proposition N at the start of the Nth part as "proposition N," and to the entirety of the Nth part as "section N."

[56]My renderings diverge from that of translations that, till recently, have been standard—with the *Tractatus* having entered the public domain new translations are proliferating—but my choices are lexically and contextually every bit as justified as the standards. I will retain the conventional "fact" for *Tatsache*, because there are not many alternatives and because it is relatively unproblematic as long as one avoids the word "fact" for rendering other terms (e.g., as when *Sachverhalt* gets turned into "elementary fact," "atomic fact").

Wittgenstein begins to lay out the corresponding theory of picturing/imaging/(re)presenting in the second section of the *Tractatus*, after asserting that the world is divided up into facts. Proposition 2, again, states that "what is the case, the fact, is the subsistence [*Bestehen*, usually translated "existence"] of relations of things [*Sachverhalten*, often rendered by the problematic "atomic *or* elementary facts"]."[57] This assertion is one of those items that properly speaking the work ought merely to show rather than try to speak; it is the basis of the *implicit* metaphysics of the truth-functional logic later presented. The connection between sentences and facts rests on how things enter into relations with one another and the human mind's capacity for recognizing, or rather projecting,[58] the form of the fact–situation into the form of the sentence.

The fundamental insight into the shared form of facts and sentences may have been occasioned by an article about a Paris court trial that Wittgenstein read, to which he refers in a notebook entry of September 1914. It told how a traffic accident was represented in court by an artificial scenario employing toy cars, toy trucks, and figurines. He realized that the relation of a sentence to a fact of the world is like that of such a scenario to the situation it portrays.[59] The sentence pictures/images the world; it shows us how things stand. Moreover, just as the scenario might or might not portray things the way they were, so too does the sentence. "The image presents what it presents, independent of its truth or falsity, through the form of the depiction"; "what the image presents is its sense" (*Tractatus*, 2.22 and 2.221). Taken as a picture, the sentence is a possible presentation of a situation, and its meaningfulness lies precisely in this

[57] My divergence from the standard translations is simply a question of not getting ahead of ourselves (much less ahead of Wittgenstein!) by overinterpreting at the outset. *Tatsachen*, facts, are the way the world falls into parts; the facts can be ultimately analyzed down into elementary facts, beneath which there is no further possibility of analysis, but complex facts are facts, too. "Existence" would be unproblematic for *Dasein* or *Existenz* but is too tendentious for *Bestehen*. In older philosophical parlance this would be better translated by "subsistence," which can be used of the existence of both substances and accidents of substance and tends to suggest persistence in existence, as *Bestehen* in fact does. As for *Sachverhalt*, pl. *Sachverhalten:* it should be out of the question to translate it with any version of "fact," since it is historical accident that English *fact* corresponds to German Tatsache, deed–thing, and that *Sachverhalt* contains the stem *Sach–* that also appears in *Tatsache*. Doing otherwise is to confuse matters. Wittgenstein immediately establishes a basic synonymy of *Sache* with *Gegenstand* and *Ding*, which should only reinforce the decision to render *Sache*, unless circumstances demand otherwise, with the generic English "thing." *Verhalt–* appears also in *Verhältnis*, which variously can mean "proportion," "circumstance," "condition," "relation(ship)"; and *verhalten* as reflexive verb indicates acting or behaving, conducting or comporting (oneself). "State *or* states of affairs" would thus be a somewhat anodyne but acceptable alternative.

[58] See the end of the next paragraph for a justification of "projecting."

[59] Wittgenstein may have been thinking of this kind of model in 2.0272 of the *Tractatus*, and in 3.1431 when he says that the "sense of the sentence" (regarding "tables, chairs, books") is expressed by the reciprocal spatial situation of these spatial objects. The Paris court scenario is mentioned in the notebook entry dated 29 Sept. 1914; see Wittgenstein 1998, 7; also the discussion in McManus 2006. For a strong philosophical-historical presentation of Wittgenstein's theory of the unity of proposition and image throughout his career, see Perrin 2007.

possibility. "In the sentence the thought expresses itself as sensorily perceptible," as Wittgenstein says in 3.1, then immediately elaborates this way: "We use the sensorily perceptible sign (sound– or writing–sign etc.) of the sentence as a projection of the possible thing–situation. The projection–method is the thinking of the sentence–sense" (3.11). The sentence is no mere mechanical assemblage or mixture of words—"just as the musical theme [is] no mixture of tones" (3.141)—but is rather something that is articulated, an articulated whole. This is therefore a sophisticated theory of the imaginative character, the *Bildcharakter*, of language as projective between fields—even if it largely takes for granted (which is not to say eliminates) psychological imagining. One might even quite easily conceive the Plato of the *Republic* saying something very similar.

Although any further attempt to interpret particulars of the early sections of the *Tractatus* would draw us into many controverted questions, two parallels to Kant are worth mentioning. Kant had shown what the fundamental conditions were for even the possibility of articulating objects in the manifold of sensibility. By virtue of the pure intuitions of space and time and the schematizing of the pure concepts and principles of the understanding, what appears to the senses falls out into things with properties and relationships. By accepting the basic forms of general logic as universally valid he appeared to obviate explicating the conditions of the very possibility of logic and its use. Wittgenstein fills this gap by asking an analogous question: what are the fundamental conditions of the possibility of any sentence at all about the world? Without addressing the ultimate metaphysical question about what the things are that exist in the most basic sense, Wittgenstein provides a minimalist account of how the world must be so that modern propositional and predicate logic can express truths about things in it. If there are to be true basic sentences (atomic or elementary facts), then there must be elemental things (for things in this sense he uses in section 2 the words *Ding* and *Gegenstand* rather than *Sache*) that are capable of entering into all possible relations of things (*Sachverhalte*).[60] Just as for Kant the manifold of sensibility is thoroughgoingly articulated so that it appears in a manner that is conceptually knowable, for Wittgenstein the world must be divided into things (*Dinge* and *Gegenstände*) that can enter into the relations (*Sachverhalte*) that constitute the facts that are expressed in elementary sentences and that can thereafter enter into complex sentences that are true or false according to truth functionality.

Wittgenstein does not, of course, attribute this division/articulation of the world to transcendental imagination. Yet we must not be hasty in judging what this implies about imagination. How the world is, apart from all possible sentence expression, is beyond human understanding, just as for Kant we cannot have any knowledge of things in themselves. Yet what the human being experiences *as* the world is the sentence-expressible totality of the facts. As Wittgenstein says in 1.13, "the facts in logical space are the world." This notion of logical space is an analogue to Kant's

[60] This is intensively dealt with in propositions 2.0121–2.0124, which are comments on 2.012: "In logic nothing is accidental: if the thing [*Ding*] *can* occur in the relation of things, then the possibility of the relation of things must already be prejudged in the thing."

manifold of sensibility. It is the realm of the sentence–expressible, which means the sentence–picturable or –imageable, understood as a totality. That totality is not just an assemblage of things arbitrarily placed one next to the other, but (as we have already pointed out regarding 3.141) an articulated whole. The things in this whole are part of a manifold, a field, a space for the logically conceivable—which is to say the logically imaginable. "Every thing is, as it were, in a space of possible relations of things. I can think this space for myself as empty, but not the thing without the space" (2.013). In the immediately following commentary of 2.0131 he remarks that a speck in the visual field does not have to be red, "but it must have a color: it has, so to speak, color space around it," and similarly for the qualities of sound and touch. Not only is there a common field for all possible things in logical space, there are also subfields articulated by qualities, qualities that are proper to those fields and therefore to the very conditions of the possibility of experience of that kind. Conceiving possibilities is a kind of logical imagining in logical space and its subspaces.

Even at this early point in his career Wittgenstein resisted making states of mind an explicit part of the process of expressing the world and one's knowledge about it. Kant, of course, had coordinated by schematism all concepts of understanding with the manifold of sensibility, which had to be synthesized and unified in a thoroughgoing way by the transcendental imagination. The richness of the world is all in mind, and our science is an expression of its *formal* properties. Wittgenstein left the richness of the world to the world in the sense that he did not speculate about precisely how and in what terms the mind reflects the world. All that he needed for his purposes was that the mind somehow gets a basic but nevertheless highly articulated picture of the facts. The human mind or the mind–with–body is a projector or transducer of the states of the world into sentences. It performs a conversion of format from one medium to the other; it is a black-box device that translates the situation of things into statements. We have seen this structure repeatedly in this book, for example in Plato's levels of being and knowing, which projectively image one another, or in Descartes's isomorphism of geometrical figures with their possible transformations, on the one hand, and algebraic equations with their possible manipulations, on the other.

For Wittgenstein, logic depends on this basic picturing function of mind. Anything more complex about the world does not require psychological synthesis but logical complexion and analysis according to the rules of modern propositional and predicate logic. In a sense, then, Wittgenstein's so-called antipsychologism is the result of his coming upon the logical process *after* the mind has performed one of its most basic functions, its logic-founding function. There is no need to deny that there is mental content, because, at the point where logic takes over, the facts have already been pictured in publicly shareable form. The specifically mental can then be left behind, because it has already accomplished its works in the process of projecting-transducing the world situation into language, into sentences that can be classed as true or false. The basic function of the philosopher then is to present the logic of the world that has been experienced and transduced into the forms of complex propositions; he is expert in the ifs, ands, and nots. He points out and corrects the sloppiness that accompanies our ordinary ways of speaking, which mislead us into making extraordinary and unsupportable claims about what we know, do, and can accomplish.

In effect, the *Tractatus* is a theory of imagination, of *Einbildungskraft*, albeit an eccentric one—literally off-center. Perhaps it would be even more accurate to call it a totally extroverted theory of imagining, totally turned outside the mind. Just as with Kant, the science and knowledge we build up is knowledge of a world, but it is not necessarily a world that exists for beings other than we are. The forms of logic apply to particular elementary statements that are *made* by human beings, in both literal and figurative senses; thus the human way of seeing and expressing things cannot be totally factored out. But there is no need to appeal to any internal or psychological conception of reason, understanding, imagination, sensibility, desire, feeling, etc., at least not for logical or scientific purposes.[61] When we try to speak about such things, we end up distorting what is knowable and try to express "things" that are not the things language is able to express.

Even when Wittgenstein abandoned the path of the *Tractatus* he maintained the exteriorization of the mind and imagination. The *Tractatus* was formally oriented toward the things of the world that are ultimately expressible in language. The notion of language games, by contrast, attempts to deal with things where they are, in the world of ordinary places, events, and speaking. Imagining is still embedded in the world: the very sense of game/play in language–game, *Sprach*spiel, suggests the imaginative element in our way of engaging with and expressing the possibilities of the ways of life we engage. This is one of the most promising ways in which the legacy of Kant's theory of imagination has been developed, by radicalization of its logical functions. Imagination has no stronger support structure and expression than the logic of everything we can say in and about the world.

8.6 Semiotics: Thinking the Signification of Sequenced Phantasms

Wittgenstein's image theory of language is not the only way in which the linguistic character of imagination and the imaginative character of language were developed in philosophy's linguistic turn. In Sect. 8.2, above, we noted Lacan's exploration of the imaginary as a languagelike structuring of the unconscious psyche. This tendency in his thinking was strengthened when in the 1950s he became more fully aware of structuralist thought and interpreted the formation of the images of fantasy as the structuring of affect. Images thus became absorbed into a semiotics of

[61] Like Kant's, Wittgenstein's basic theory of imagination is transcendental. One consequence is that later quasi-Wittgensteinian theories of imagining that conceive it as an attitude (such as believing or supposing) toward a proposition are talking about something peripheral, something other than and secondary to a more primordial imagining that propositions reflect in their very constitution.

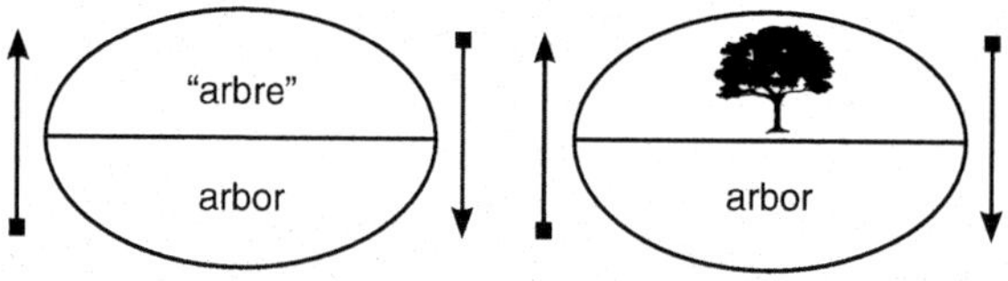

Fig. 8.1 Saussure's ovals, representing signs

passion, emotion, and feeling. Although this passional turn is an innovation, the approach to images as signs was already foreshadowed in the origins of semiotics and structuralism.

A first approximation to the recent history of sign theories would point to the contrast between the triplicity of the Peircean sign and the duality of the Saussurean. Saussurean structuralism considers the sign to be a fusion of sound or signal (the phonological or phonetic aspect of the sign)[62] with psychological signification or meaning. The version of the *Course in General Linguistics* published by Saussure's students in 1916 represents signs by ovals with a dividing bar inside separating an idea from a sound (see Fig. 8.1, after Saussure 1916, 99). Taking one example: above the bar is, say, the idea associated with French "arbre" or English "tree"—alternatively, an image of a tree—and below the bar the sounded form of Latin *arbor*. The sign is not the sound made when we say *arbor* but rather the entire oval; the part above the bar is the *signified* (the signified aspect of the sign, the mental representation of the tree), the part below is the *signifier* (the sign's signifying aspect, the phonological signal /tree/, /arbre/, /arbor/).[63] In the first instance, then, the sign is a fusion of an idealike or imagelike element and a sound. For Saussure, the sign is neither just the signified meaning nor just the sound but *always* the two together. Moreover, signs do not exist in splendid isolation but as a system of positions in a network of differentiations of both signifieds and signifiers. In fact "positions" is a bit misleading insofar as it suggests that words/signs are posits, positive entities. They are, strictly speaking, differences. Especially in later versions of structuralism (for example, Claude Lévi-Strauss's anthropology), this differentiation is presented as constantly and dynamically effective in language, myth, stories, and other cultural productions.

When all is said and done, however, there appears to be something static and very conventional (in the sense of traditional) about the Saussurean sign. It does serve to identify words as the basic elements of language and to make them nodes in a

[62] Saussure's use of "phonology" basically corresponds to contemporary "phonetics." Both terms refer to meaningful sound in language. Contemporary linguists use "phonology" to refer to the principles of sound production governing a specific language, whereas the general theory of linguistic sound production is called "phonetics." Thus the French *u* is not part of English phonology, although it is part of phonetics.

[63] By placing the word between forward-leaning slashes I am referring to the phonetic transcription of the word. I am not actually giving a transcription, however, but merely indicating that one should think of just the *sound* of the word placed between the slashes. Notice that the idea of the tree fused with the sound /tree/ is a different sign (a sign of English) than the idea of tree fused with the sounds /arbre/ or /arbor/ (signs of French and Latin, respectively).

complexly differentiated structural network of signs—this network being the proper object of the structuralist approach. Yet such a theory might be understood as simply formalizing and systematizing the tradition that sees language as a slapping of labels on objects, or rather on object–types in the mind. Even once it is made clear that the Saussurean sign is determined more by a dynamic of negation than of posits, there still seems to be a peculiar stolidity to the overall conception. As governing structure it is meant as a principle of possible activities, but it does not look active per se. It looks like a magnificent but bloodless edifice of formal structure—thus, in a sense, merely a linguistic rationalism. Moreover, the sign–orientation of structuralism appears to leave out of account or take for granted the importance of matters like reference and syntax.

Although the concern about accounting for syntax might also be raised with respect to Charles Sanders Peirce, his theory by contrast appears to give full scope from the outset to both individual and social sign use and to the dynamism of thinking. His conception of the sign has a triple aspect rather than a dual, and in the eyes of many scientists and philosophers of language this is already a decisive superiority over Saussure. According to Peirce, all thinking is sign–thinking, semiosis (or semeiosis, to use Peirce's preferred spelling). There is probably more than an echo here of Aristotle's dictum that there is no thinking without images. In addition—an addition not at all Aristotelian in spirit—there is no end to semiosis. Each sign leads to others.

Every sign use involves a triplicity. First, there is an intended object; second, there is a first-approximation representation of that object; third, there is a more developed form, a second-approximation representation of the object by way of the first-approximation representation. None of these exists in abstraction (or, rather, in prescission) from the others. The intended object is not a thing in itself, not a real-world thing as it would exist apart from human consciousness, but the thing–we–have–been–dealing–with. It is what is indicated by the current sign, and in that respect it is relatively past; as relatively past it can already be taken as an incipient sign of the object. The current sign in the first instance points back to the object it intends, but if the mind goes further—and the mind always goes further—it can do so only by way of another sign that develops the current sign and that, through the first sign, points to the intended object.

In one of the earliest presentations of the theory Peirce identified the three aspects as the *object*, the *representamen*, and the *interpretant*. In the first instance one might explain this by saying that an object of the mind is an object grasped in an already formed experience, and that therein consists the mind's first signification of the object. In that sense there is no absolutely first or primitive experience for human beings, nor does this conception presuppose the immediate mental presence of the object in perception, image, or idea without signitive character. Experience takes place as a signitive and signifying flow of already–having–experienced many things. If we turn from regarding the represented object to the representing of the object we come upon the representamen. If we see a dog (the object) and then think or say /dog/ (the representamen) we nicely make this distinction. But our thought does not stop with the sign representing the object; it goes further in a more developed form that in effect interprets the first representing sign. The dog is a Schnauzer. "Schnauzer" is the interpretant, which now in its turn can become for the mind a new representamen that has as its object the original representamen of which it was the interpretant; and it still refers, though more remotely, to the originally intended object, which was in fact an object–sign.

In first approximation one can cast this in the idiom of empiricism, with the empiricist "idea" or "image" being replaced by "sign." When the idea–sign first enters awareness (in empiricism, chiefly through the senses) it is an appearance that refers to an original in the world, and in that sense it is already a sign. Our thinking, in empiricism, is idea–thinking, of course, but in any particular sequence of thought the next idea refers back to what precedes it and is thus another sign. By grouping and distinguishing present and past ideas according to resemblances, contiguities, and the like, our new ideas develop a complex signitive relationship to the prior ideas—and even more complex when we arrive at the second-order ideas that we know as words. In this way our ideas about the objects of our attention take on a complex structure of signification. Idea–signs always lead to other idea–signs—this would be the empiricist idiom for Peirce's infinite semiosis—all in the train of trying to conceive a thing that we are acquainted with in the first place by virtue of an idea–sign. Even if one is determined to be psychologistic, introducing the signitive function attenuates the dependence on immediate psychic presence. Every representamen points both backward to the object and ahead to further interpretants. The being of the sign is not so much in the psychological present as in the movement of consciousness; yet because it is a sign it is already abstracted in the first degree from one's personal consciousness, potentially communicable to others, and thus quasipublic.

One of the first and most basic applications Peirce makes of this threefold conception is to the assertoric proposition, A is B. In this proposition–form, the term A is a representamen, a representative sign, because it stands for and refers to a thing or a kind, which is the object of the representamen. Peirce was extraordinarily well read in medieval philosophy and especially in suppositional logic, and this claim about "A" is not far removed from the medieval conception that the subject of the sentence stands or is put (is *supposed*) for a substance. The predicate B is conceived as an explication of the subject and the substance it stands for.[64] The specifically Peircean difference is, in the first instance, that all three—the object, the representamen, and the interpretant—are regarded as signs.

Peirce's thought always retained an important empiricist and even realist strain. He had no doubt that our experience is both about the world and its things and about the thoughts that intend those things. This attitude he shared with many other philosophers (like the medievals) who had been strongly influenced by Aristotle. But

[64] The chief medieval influences on Peirce's sign theory were John Duns Scotus and John of St. Thomas (also known as Jean Poinsot). In Scotus' realism the sign has a more direct relationship to the real-world thing than would be true, say, of the Dominican (Thomist) tradition. Given the quasi-Aristotelian background of medieval philosophy in general, the theory was still explained using phantasms. See section 10 ("Mind and Semeiotic") in Burch 2010, and Dumont 1965.

one might say that, for Peirce, to regard the object as though it is an ontological substance that could be apprehended without sign–mediation is a falsification of what is involved in experience. The very act of conceiving or intending an object is a signitive act, an *indexical* act of directing attention to it; there is no such thing as a primitive apprehension of a substance apart from signs, since for every human act of attention there is a long autobiographical prehistory of signitive relationship to things in the world and to things like the one we are currently intending. As he argues in the early "Questions concerning Certain Faculties Claimed for Man" (1868a), there is no cognition that is not determined by previous cognitions, and there is no thinking without signs. "Every thought must be interpreted in another."[65]

In the slightly earlier "On a New List of Categories,"[66] Peirce used very basic notions of psychological acts to explain the emergence of concepts and symbols from the starting point of the Kantian manifold of sensation. He creatively employs both Aristotelian-scholastic and Kantian terminology, makes a nod in the direction of the evidence of the relevant phenomena, invokes the results of psychology, and encourages further psychological research. As such the "New List" might be taken as an immature presentation of Peirce's views, still too caught up in an old-fashioned and highly questionable psychologizing approach to phenomena of mind and language. Yet precisely as a very early work it gives insight into the network of the traditional problems and concepts that were his starting point. They make even clearer the manner in which Peirce's conception of semiosis developed from psychological theories (whether empirical-rational like the Aristotelian-scholastic or transcendental-psychological like the Kantian) in which imagination played the central role.

After pointing out that the "New List" is based upon "the theory already established, that the function of conceptions is to reduce the manifold of sensuous impressions to unity" (§1), he points out that the "theory gives rise to a conception of gradation among those conceptions which are universal. For one such conception may unite the manifold of sense and yet another may be required to unite the conception and the manifold to which it is applied; and so on" (§2). This is both Kantian and more than Kantian. It is clear enough in Kant that there are different levels of synthesis and unification, the more basic ones prerequisite for more complex ones, in particular for the ultimate synthesis that allows one to say "I think that X" for all thoughts X—the transcendental unity of apperception. It is also more or less clear that the *Critique of Pure Reason* gave an account in principle only of the most basic syntheses, and that it left to future philosophers the resolution of questions about how we move, for example, from the overall unity of experience to isolating a particular thing in attention, to giving it a common specific name, to identifying its genus, then placing it under other universals. In "New List" Peirce's focal interest is this large but relatively underexplored middle realm, after or "above" the original

[65] From the last sentence under Question 5 of Peirce 1868a.

[66] Peirce 1867. It is divided into 15 "sections," which range in length from one or two sentences (sections 1 and 2, respectively) to about three pages (section 15). I will cite it by section (§) number.

synthesis of the manifold of sensibility by transcendental imagination and before or "below" the final synthesis in the transcendental unity of apperception.

Peirce's new beginning (in §3) teases out the first transition from the manifold of sense to specific concepts that unify it. "The present, in general" is the universal that is closest to sense. This conception of the present in general "is nothing but the general recognition of what is contained in attention," and "the act of attention has no connotation at all, but is the pure denotative power of the mind," so the conception of the present in general correlatively "has no connotation, and therefore no proper unity." He immediately substitutes the word "IT" for "the present in general" and says that philosophical language renders it by the word *substance* "in one of its meanings." This indexical IT is prerequisite for all subsequent mind acts, like comparison, discrimination, and abstraction, and although what is abstracted is attributed to or predicated of it, the IT cannot be predicated of anything—a formulation that is reminiscent of Aristotle's definition of substance in the primary sense, that it can be neither predicated nor part of anything else.[67]

This passage of the "New List" also shows some similarities to Hegel's discussion of the "here and now" in the *Phenomenology of Spirit.* Yet Hegel's method was dialectical, to show that, although "here" and "now" are used with the intention of pointing to what is most concrete and immediate, they are in fact universal, since they apply indifferently to all places and all times. By contrast, Peirce emphasizes that the present (taken generally) is universal from the outset. The background common to both approaches is, of course, the Kantian synthesis of the manifold of sensibility. Like Hegel, though developmentally rather than dialectically, Peirce is trying to derive and explicate the most primitive kinds of concepts that present themselves in a most basic human situation. Like phenomenologists more generally, he does this by appealing to basic acts of mind. Instead of the (chiefly later) phenomenologists' *intention* he uses *attention*, "the power which directs the mind to an object." This "pure denotative power of the mind" is what renders and delivers an object without predicates, without even a name. One might say, in anticipation of the full-blown theory of semeiotic, that it is a pure indication, where indication is one of the three basic kinds of sign.

But Peirce's interpretation of the present diverges from Kant. Kant thought that one of the unprecedented accomplishments of his critical philosophy was to elucidate space and time as the two pure intuitions of sensibility. Space and time as pure intuitions were a presentation dependent on the most basic of the syntheses performed by imagination in its transcendental use. Although one might argue that the synthesis of space and time by imagination in its transcendental use establishes the place and time within which everything more determinate occurs, and in that sense is prerequisite for the here and now, Kant was quite clearly not attempting to give in the *Critique of Pure Reason* or elsewhere a phenomenological description of our experience of the present in general.

Indeed, one of the most unKantian things about Peirce's analysis is its eventuation in *substance*, when for Kant the first result was the incipient field of

[67]In *Categories*, ch. 5, 2a11–12.

mathematically determinable space and time. Kant did not, moreover, derive his first, elemental concepts from a phenomenological analysis of attention or anything similar (although some early followers of Kant tried to arrive at his results by way of an analysis of consciousness). Peirce, by contrast, seems to be aiming for an *induction* of the most primitive concepts from an analysis of the basic encounter of mind with its world.[68]

If Peirce is right that attention is largely denotative, picking out something without extraneous associations, attention nevertheless works by presenting the present as that in which attention can be further focused upon an object–with–background. He thereby arrives at further concepts derived from experience by a process that in the "New List" he variously calls *abstraction* and *prescission* (§5). That is to graft an old philosophical doctrine (we acquire concepts from phantasms by abstraction) onto Kant's position, which does not rely on an abstractive power in any conventional sense. Peirce takes for granted a Kantian background, but in an unKantian way. This means that he has an uneasy relationship with the field in which a Kantian understands images to have their place. Not coincidentally, this erases *images* in favor of *objects of attention*, which is to say that they first enter the mind in the format of signifying and signified signs rather than of image.[69] By his further invocation of the concept *reference*, Peirce avoids any distinction between the thing in the world and the thing as it has appeared to mind. The locus of attention is the proposition's referential relationship—its signitive relationship—to the present–being that has been isolated by attention. What had been commonly understood as the relationship between a phantasm and conceptual thought thus comes to be conceived as progressive semiosis.

I raise the diversity of possible interpretations and antecedents not in order to emphasize that Peirce had incurred debts to other theories but rather to argue that Peirce was tackling a set of issues rooted in some of the most basic historical conceptions of logic, mind, and being, and that he was using diverse resources to tease out a new emphasis and new insights about their tendencies. What the fourth section of the "New List" announces, for example, is one of the oldest themes of all in Western thought: "The unity to which the understanding reduces impressions is the unity of a proposition." Like Aristotelians, Kantians, and Hegelians of all stripes (not to mention virtually every other philosophical school), for Peirce both the proximate and the ultimate goal of philosophizing is to express what sensory experience presents in (true) propositions. This unity is the connection of the predicate with the subject, he says, and that is the work of "the copula, or the conception of *being*" that completes the reduction of the manifold of sense to unity. He parses this being in an

[68] See especially §6 of "New List." For discussions of the reception of Kant by his early followers, see Beiser 2002.

[69] This is not to assert that signs cannot be images or have typical properties of images like resemblance to the things they image. Moreover, since signs are presentations or appearances they have no less an image character than do any other determinate presentations to consciousness. Nevertheless, it is clear that Peirce thinks that the sign is intrinsically mind-directed and mind-directing (i.e., signifying), not entirely dissimilar from the way the great philosophers of imagination understand the image not as a thing but as a dynamic position in a topological field.

alternative: the copula "means either *actually is* or *would be*, as in the two propositions, 'There *is* no griffin,' and 'A griffin *is* a winged quadruped.'" This is a passage where one might expect imagination to play some role. It also seems clear that the second conception of griffin—the imaginative "would be" kind of being that implicitly refers to "everything we know" about griffins from myth and fiction—has to be implicated in the first kind, where the reality of griffins would be denied.

If these two uses imply that "being" or "is" has different meanings, one way of interpreting that would be to differentiate them by content. Peirce denies that the conception of being (which is the copula) has any content. Yet it is also clear that the mind must appropriately judge the circumstances in which the two different uses of "is" occur and then match the proposition's usage to the situation and some meaning paradigm (a correspondence of world–thing and concept), since he also remarks that "the conception of being contains only that junction of predicate to subject wherein these two verbs agree." Still, what isness means exactly remains rather obscure. Each "is" (one in the sense "actually is," the other in the sense "would be") has a different meaning, so one expects the real meaning of the copula in predication to be the intersection or commonality between the two. Yet one might, alternatively, simply assert that they have an originally common meaning that is modally inflected. The junction between predicate and subject would thus be assessed by the mind as, say, actual or possible (this would correspond to the schematism of Kant's three modal categories, possibility, actuality, and necessity) and the meaning inflected accordingly. That is, whereas Peirce is saying that "is" is the sign of an agreement between the subject–sign and the predicate–sign, a more Kantian alternative might be to say that it is the mark of the mind's successfully unifying the appearance in the manifold of sensibility under the presentation of the schematism of the subject–concept and likewise the schematism of the predicate–concept, in the mode of actuality when existence is in question, in the mode of possibility otherwise—which would be a Kantian way of expressing (through schematism, a transcendental function of imagination) what an empiricist would explain simply in terms of a succession of images.

Of course Peirce is trying to trace out a third way while also reckoning with Kant's famous assertion that being is not a predicate (§4 in fact ends with a suggestion in this direction, that being is inapplicable to a subject taken simply). The possible dollar, or rather (in Prussia) thaler, does not differ from the real thaler, argued Kant, as far as anything conceptually predicable is concerned. The difference is that the real thaler is presented in the manifold of sense, whereas the possible thaler is not.[70] More strictly speaking, the possible thaler (in a novel, for example, or as the profit on a stock purchase I might have made but did not) is only imaginative, even

[70] Whether Kant can literally mean what he says is problematic, however, or at least requires a more refined discrimination of basic cases, since the thaler produced by imagination in its ordinary productive sense (for example, in a story as it is being planned by a novelist) "exists" in the manifold of sensibility for him, even if it is only imaginatively. This does not in itself infirm Kant's larger point, however: that the conceptual content of any subject is independent of every possible realization, even imaginative ones (which would support his contention that being cannot be a predicate). It would, however, require a nontrivial discussion of whether the manifold of sensibility has different modes in sensation, memory, imagination, and the like.

imaginary. As imaginative it belongs to the manifold of sensibility,[71] although as a fiction, spontaneously produced by the fiction's author rather than in a real presentation passively received in sense.

Peirce does not follow a comparable course of interpretation, however. Although he begins with and occasionally recurs to empirical psychology,[72] and although the analysis of the "New List" depends on a theory of mental activity, especially abstraction and prescission, his constant goal is to achieve *logical* results. He is often coy with respect to psychological matters, but the understanding and classification of acts of the mind is essential to his theory. Semiotics for Peirce is in essence an extension of logic, which means that it is a matter of form rather than content; yet as based on the sign, the forms of semiotics always retain the vestigial forms of appearances that refer to other appearances, which for him is the only way in which the mind takes or grasps its experiences. This becomes somewhat reminiscent of Descartes, for whom "conception" was quite literally the mind's intuitive *grasp* of a current appearance (of an object, with "object" taken in the widest possible sense), which grasping points to a different mind–place than the place of the current appearance.

Perhaps most revealing of the degree to which Peirce depends on acts of the mind is his analysis of the example "The stove is black" and his discussions of how mind abstracts/prescinds quality from substance, refers to grounds, makes correlations, etc. He offers his first detailed example of attentive focus in §4, and its development through the next several sections gives further evidence that he is in effect supplanting the traditional phantasm. Given that we say that the stove is black, he asserts, "the stove is the substance, from which its blackness has not been differentiated, and the is [i.e., the assertion of being], while it leaves the substance just as it was seen, explains its confusedness, by the application to it of *blackness* as a predicate." The first question to ask is what Peirce means by the "confusedness." In terms of attention one might need to explain it this way. Attention in the first instance does not predicate, it isolates and views. Attention ranges over the manifold of sensibility and settles upon something. In that moment of settling it acts in a purely denotative fashion, without connotation–predication, so that it provides us with little more than "the present [thing], in general." But there is, virtually in the very same moment, the attention to the stove, or rather the stove–in–its–blackness. Predicating blackness of the stove in its isness (whether one's native language expressly uses the present tense of "be" or not) as it were blurs the focus on the original apprehension of the stove–substance by dividing the original unity.[73] The "is" of the predication–sign

---

[71] As problematic as it sounds, this assertion has to be right, insofar as Kant offers no alternative "locus" where the imagined as opposed to the real-world thaler has its place. See the previous note.

[72] With the basic phenomenology of attention in §3, for example, or in §8, where he explicitly invokes empirical psychology's discovery that we know qualities only by contrast and similitude.

[73] It is as though the plane of the subject and the plane of the predicate *separate* so that the point of visual focus is no longer in either of those planes. Hegel, of course, also remarked a kind of confused focus in every nonidentical proposition "A is B" (that is, where $A \neq B$): the mind, beginning with A, must move to something else, B, following upon which it must (re)establish the unity of the two together.

"The stove is black" produces no more and no less than the degree of consonance between the subject and the predicate that is the mind's reference of the (complex of) signs to thing/substance.

The oddity, of course, is that it looks as though Peirce is saying that the moment we predicate anything of something we lose focus on the something, so that any proposition, any logos with the minimum discursiveness of "A is B," represents a confusion. But this is at most a temporary problem, and in fact it is the engine that drives the limitlessness of the semiotic process. By focusing on "stove" (or any other specific or generic substance) we have gone well beyond the present in general, which Peirce has said has no connotations at all. The mind we are describing, one that can call a stove a stove, is already experienced in arriving at specific universals and their signs (stove, burner, oven, rack, heating, cooking, etc.). This mind has learned how, after the pure denotation of its attention, to carry out a further development, the naming of the denoted, attended-to thing. In Peirce's sign theory, to name is to establish or to reaffirm a regular (that is, lawlike) relation between the sign and the thing it names. Moreover, beyond the naming of the denoted thing there must be some further signitive development to arrive at the specific universal "stove."[74] So the "confusedness" Peirce mentions seems to be something that occurs indefinitely often in individual human experience, with a first occurrence that is shrouded in the mists of individual biography. The only way confusion can be clarified is by asking questions and seeking more information.

In these passages of the "New List" Peirce is silent about the original "naming" of the focal point of consciousness and about the further developments that lead us to say /stove/. But his immediate concern is a different development, that of the quality of blackness, beyond the fact that it at first confuses the substance–denotation "stove." This is not to accuse him of any omission, however. What he is doing is giving a kind of description of the phenomenological emergence of more complicated *conceptual*—and that is to say also *signitive*—relations in experience, relations that, retrospectively, can be applied even to the first emergence of a new kind–concept. Taking for granted the kinds of unification of the manifold of sensibility that Kant had attributed to the *pure* concepts of experience, Peirce is in effect offering an explication of how we move toward ever more complex Kantian schemata for contingent experience. The Kantian schemata in general, of course, account for how concepts are reliably connected with images.[75] In a larger sense, Peirce is also giving an account of how induction works. Induction is a staple of empiricist thought going back to Aristotle. For the latter, induction is the (natural) process of soul by which human beings cognitively ascend from sense experience to conceptualizing what has been sensed. The fuzzy, long-tailed, reddish focus of repeated acts

[74] The point being that a certain thing in the world may be indicated by "George" or "Tappan" (that is, by names), but that is not to designate or know its *kind*.

[75] With this remark I am scarcely more than gesturing toward what requires detailed exegesis. Recall, for instance, that although in the first instance Kant understands the schemata as implementing the pure concepts of the understanding (like unity and substance), the examples he actually gives are for schemata of "dog" and "triangle"; see Sect. 7.7, above. Umberto Eco has explored at length the affinity between Peircean semiosis and Kantian schematism in chapter 2 of Eco 2000.

of attention comes to be called "squirrel." Peirce makes more explicit than Aristotle, and far more explicit than Kant, that there is a crucial division and new synthesis by mind needed before we can predicate qualities of substances in a truly signifying manner.

It is no accident, then, that Peirce immediately turns to the mental act of abstraction, which he also calls prescission (§5).[76] "Prescission" is a term we discussed in Sect. 5.13, above. If the process by which we induce universals from particulars, from phantasms/images, is traditionally called abstraction, prescission is regarded as a special case of abstraction that takes the concept that has been abstracted[77] and treats it absolutely. Thus after abstracting the matter and the form of a physical thing—which matter and form, at least in an Aristotelian conceptual universe, exist only in co–relation—we can go on to talk of form in and of itself and matter in and of itself. We see a gold ring and a silver ring, and we talk not only of the rounded shape of the gold or silver but of roundness in itself, and alternatively we talk of the gold and the silver as though they were purely shapeable stuff and thus neglect the fact that, at every moment, any mass of gold or silver we are talking about must already possess some specific shape. Prescission goes a significant step further. To take the case of body and soul: Aristotle's definition of soul determines it as an activity of body, and that counts as an abstraction. But many people proceed to treat soul as though it can exist apart from body, and animal and plant bodies as though they can exist apart from soul. They even ask questions that treat each as complete in itself, like "How is it possible for soul to unite with body?" That is prescission. We argued earlier, of course, that *abstraction*, the Latinate rendering of Greek *aphairesis*, is not understood in a genuinely Aristotelian sense when it is made equivalent to the human recognition of forms in things (or in the phantasms of things) that are present to the soul; in particular, Aristotle used *aphairesis* almost exclusively for the active power by which one treats a mathematical space independent of the actual place of the things of nature. In this sense (and as we argued earlier), Aristotle's *aphairesis* of mathematical space from natural space corresponds more to the later notion of prescission than to what most people have called abstraction; and his conception of the recognition of forms in natural things is not aphairetic at all. For Aristotle we recognize, we see, forms *in* things and images—forms that are actually there. But we can conceive mathematical space apart from the space of natural things, and in doing mathematics we develop a science of space while disregarding the fact that this space does not exist per se but only as an aspect of nature.

Peirce in fact says that what he calls, equivalently, prescission or abstraction takes something we experience as joined to others things and then *neglects* the other things. This would seem to be a quite natural and plausible extension of

[76] I will continue using the spelling I introduced in Sect. 5.13. Peirce uses "prescision."

[77] Recall that it was later interpreters of Aristotle who introduced the process "abstraction." Aristotle himself calls the process of arriving at this kind of universal *epagogē*, induction. The infant who learns to say "cat" of some animals and "squirrel" or "dog" of others, and later to say "mammal" and "nonfish" of all three, has recognized different intelligibilities in the phantasms presented by sense through induction.

attention's ability to change focus. In "the stove is black" we announce a unity that confuses the original focal point, but by subsequent attention—by the continuation of semiosis, which is infinite in the sense that it never stops in a way that is totally definitive—we abstract or prescind the black from the stove, and then we develop black as a quality, which can happen only by the process of comparison and distinction with other colors prescinded in their turn from their underlying substances. Peirce thereby constitutes a semiotic field that takes on a relative (that is, prescinded) independence from the original field out of which it originated, where it was first noticed. This is another instance verifying one of the recurrent themes of this book: that among very many truly original/originating figures in the history of theories of images and imagination, the foundation of imagination is the emergence and emplacement of the (sensory) qualities we experience in a common field. Peirce, of course, resolutely interprets this phenomenon signitively, semiotically. That is, on the one hand he implicitly argues that there is something other than arbitrary association going on when we compare and contrast black to white, red, green, yellow, etc.; but on the other hand he gives us no direct access to the underlying field that is thereby exploited except *through* the signs. The field exists only as it has been signitively experienced (this means that it is what I have been calling a conceptual topography, a thoroughly labeled and interpreted topology; and insofar as his sign theory is trying to elicit the underlying structure of such topographies, it aims to become a founding conceptual topology). In this sense he affirms (with most modern empiricists) that the object can be reactivated only in the forms already experienced. Yet far more strongly implied in his theory than in empiricism is the understanding that one reactivation can lead to any and every other, and to new activations when one begins inquiring anew. It is a consequence of semiosis and its limitlessness that signification takes the measure of a *field* of experience, and the field is in essence constituted by the complex intersignifications of all the relevant signs.

Peirce's triplet of object–representamen–interpretant can be applied to thinking that is imagistic as well as to propositional thinking. We can take an example that would be congenial to Descartes. A point can stand for a material particle (the point is the representamen, the particle is the object), and thinking does not simply end with the point but rather (re)commences with it. The point is developed as having certain coordinates by measuring its distance from coordinate axes, or it is conceived as being in motion and thus tracing a path. Both would be interpretants of the representamen, and the fact that there can be different interpretants makes clear that thought need not proceed from the representamen to some unique interpretant. The interpretant, in its turn, can be viewed as a representamen with respect to the previous representamen, which thereby becomes the object of this next moment of thought, and the second representamen will have its own further interpretants. That semiosis can continue indefinitely in this way should be obvious. It is true that every "line of thought" factually comes to an end, but there is no absolute necessity in the ending.

This section has not aimed at offering a comprehensive view of Peirce's semiotics. Its goal has been to show the relevance of the *origins* of Peirce's theorizing to the conceptual topology of imagination. The immediate background against which Peirce made his inquiries was a blend of two topographies of that topology: the medieval Aristotelian tradition of the inward senses and the Kantian understanding of concepts as serving to focus on and unify the manifold of sensibility. The questions he brought to this setting were about the dynamism of mind, in particular about how the mind moves in the field of experience. This is a question that was rather less prominent in Aristotle and Kant than in modern philosophy in general. Hobbes, for example, had asked how our thoughts can free themselves from their current stream of phantasms; the answer was by finding the way to the remnants of previous streams by means of the mind's ability to mark similarities and differences with words. Descartes had sought to counter the tendency of the mind to move randomly among sensory, memorative, and imaginative images by regularizing the use of its simplest cognitive functions, intuition and deduction, and placing all objects of attention in imaginatively constituted series according to participation in natures. Fichte and Hegel had emphasized the progressive growth and self-formation of the mind's drive to experience the world in ever ampler totality. One experience leads to another, because experience is saturated with an implicit logic destined for development, and all "conclusions" will eventually be drawn and implemented.

For Peirce, thought is always on the move. Yet, although like Hobbes and Descartes he acknowledged the restlessness of the human mind and soul (for instance with his insistence on the constant irritations of doubt that accompany our experience), he followed the later moderns in emplacing the chief motive force of this movement in the thoughts themselves, but now understood as signs. Signs have implicit motion in two directions: intentionally in the direction of an object that they signify and developmentally in the direction of what one is going to do or say next with respect to the object one has signified. Signs are thus vectors: not just meanings (meaning–positions) but significations with directive tendencies.

Hume had called express attention to the fact that whatever images we are attending to, they always have some at least indirect reference to the passions. Well before Freud, with scientific rather than therapeutic (much less depth-psychological) intention, Peirce expanded the topological possibilities of the resulting "vectors of interest" and positioned them within the signs themselves, within the basic mind–function of signifying. This meant also that the mind exercised an even more subtle and more comprehensive mobility than Descartes had imagined with his biplanar mathematics. But like Descartes's conception, it was grounded in a basic fact: that attention has to find an initial object, a first point of focus, and once it has that it can concentrate on, "take," or "grasp" the object in many different fields and move accordingly. The signitive function is constantly creating differential distances from its objects and constituting temporary and permanent planes of signifying to accommodate them.[78]

[78] A scientific discipline is constituted by various planes in which theory, laboratory practice, and applications work with the typified objects of the discipline. These are routinizations of the pragmatic signification function, which is capable of routinization in planes or fields precisely because it is constantly, ad hoc, taking a distance from objects and then returning to them.

Perhaps no one has analyzed, named, and classified more rigorously than Peirce the multifarious ways in which the mind's signifying works. If signification is not strictly imaginative, it nevertheless generalizes and universalizes the basic imaginative function of entertaining a partially abstracted concrete presentation in a representative way, following out possibilities of the presentation's intrinsic development, and then projecting them back to the originally represented object or onward to a differently constituted abstract concretion. But the question is now whether any hesitation we have in calling this Peircean activity imaginative is because it is not truly imaginative, or instead because Peirce tried to avoid calling it what it was. At least part of the answer depends on the significance of his antipsychologism. Peirce was wary of relying on distinct mental faculties and clear mental presence as having explanatory value, and at least methodologically he denies recourse to them. Whether he was entitled to the denials other than in a carefully qualified sense is doubtful. Precisely insofar as his semiosis tried to avoid the explanatory weaknesses of psychological attribution it had to reconceive and re-express the basic acts of mind.

A further comparison with Descartes is revealing here—not the stock Descartes of the commonplace historiographical tradition, the one whom Peirce tried to savage in "Some Consequences of Four Incapacities" (Peirce 1868b)—but the historical Descartes who was looking for a way to explicate the image–use of mathematics and who liberated a cognitive dynamism with more than superficial similarity to Peirce's conception. In his early writings Descartes, wondering about the step-by-step procedures of geometric proof, asked the question of how we get from the current state of the figure or image to the next. It was relatively a matter of indifference to him whether this figure was mentally present or figured in the world (e.g., on paper). But as someone who had devoted no little effort to training his mathematical imagination and following the imaginative practice of Ignatian spiritual exercises, he would have found it curious that anyone might want to deny its existence or importance. The crucial element in making progress from one step to the next was focusing on the different aspects of the figure, noticing and marking what was known and unknown, and transliterating all this information into symbols that could be algebraically correlated in proportions and equations. This was how the human powers (not faculties as modules) of intuition and deduction could be used to overcome weaknesses of the faculties of sense, memory, imagination, and intellect. It was equally necessary, however, to recognize that this kind of work leads somewhere only if it is contextualized in a problem; the desire to solve a problem is precisely what focuses the mind so that it can do this problem-solving work by working on, changing, and interrelating the representations. None of this work can take place, however, without the ability to focus at exactly the right moment on exactly the right (i.e., the revealing) aspect of the figure and the network of mobile relations one has marked. In this respect, the major difference between this and Descartes's later approach is that the later approach welcomed and encouraged putting the images into continuous motion.

Peirce's path was more logical than mathematical, but its turns and intentions are analogous. In a more general way than Descartes, he relieved the cognitive process of the need for steady, unchanging, clear-and-distinct perception of a fixed idea.

The presence of the sign is the reminder of the transience of consciousness: it is a mark of what *has been* experienced (the object) and also of the new interpreting sign *to come* (the interpretant). Moreover, precisely as a presence itself (the representamen), the sign's value is not mere appearance but its pointing backward and forward. The empiricist's basic epistemological object was a unit appearance present for the understanding, which observed the unit and assimilated it to and differentiated it from other unit appearances. The sign, by contrast, already channels mental motion from the object to the next sign. Each sign is a locale *between* its immediate neighbors, before and after. It is not quite correct to argue, then, that the sign can be made completely independent of its psychological references. Apart from the fact that the entire process of semiosis is driven by the "irritation of doubt," the description that Peirce gave of the generation of the sign function in the "New List" intrinsically counted (like Descartes with *intuitus*) on the variable and differential direction of mental attention. His argument concerning the *confusedness* of the substance that is seen not just as *stove* but as *black stove* is intrinsically dependent on the character of human psychological functions. "Function" does not necessarily imply "faculty," of course, nor does it necessarily imply introspection (a self-conscious looking inward at one's own powers and their presumed products), but it does imply a certain minimal psychology (just as Kant's transcendental psychology did).

Although Peirce was the consummate taxonomist of signs and created an elaborate technical vocabulary for classifying them, he did not fully explicate them as a unified field, the chief features and contrarieties of which would have supported (and been the source of) classificatory distinctions. The signs thus retain a degree of independence from one another that is reminiscent of the traditional empiricist conception of ideas as autonomous units of experience. Descartes's genius, by contrast, was to understand (a) geometric space as a field of figuration, (b) algebra as a field of symbols in manipulable proportionate formulas, and, (c) a jointure of the two: analytic geometry as a dual field of differentiable paths and motions expressible in formulas.

Thus, as with some other apparently strict-construction antipsychologisms (Wittgenstein's, for example), in Peirce there is less an elimination of psychology (and imagination in particular) than a rejection of some stock historical versions of it. There is less a rejection of the presence of images or imagelike experience than of certain assumptions about the ontology of images, their intrinsic qualities (e.g., their clarity and distinctness), and their location and function. And if, as there is every reason to think, signs have distinctive ways of appearance, they then have a certain image character quite apart from resemblance, not least insofar as they have distinctive ways of emerging into appearance.

8.7 Semiology: Signs as the Fusion of Imaginative Planes

At first glance, signs seem to be quite different from images. Signs share with images a degree of concreteness. Signs are intrinsically designed for communication, however, whereas the specific appearance and character of images can be so private as to be (nearly) incommunicable. Yet, as we have argued throughout this book, insofar as images and their aspects are differential and proportionable field phenomena—localized positions of appearance differentiated from remoter positions—images and their characteristics are far more articulable and communicable than is typically thought. Even many "private" images can be publicly produced, or at least analogized. A patient tries to convey to his analyst the events of a bizarre dream ("it was like being in a vat of molasses that at every moment produced dendritelike extensions that hardened and grasped at me"); a composer hums the different instrumental parts to an orchestral piece she has been mentally composing or scores it at the piano.

Wittgenstein and Peirce were inclined to deny less the existence of private experiences than their direct relevance to cognition. Both actually needed privacy at a certain originating level—in Peirce's case, for example, at the moment when the IT that is the beginning of consciousness comes to focus one's attention to the present. Moreover, antipsychologistic inclinations can get along famously with *social* psychology. A theory of signs like Peirce's provides us with the medium and element of a community's experience of the world. Indeed, individual psychology can all the more easily be scientifically probed if it is understood largely as the individual human being's acquisition of socially shared and communicated conceptions; a place even for uniquely individual psychology can be more easily defined by differentiating it from what is owed to the social. Perhaps, then, we could distinguish in human experience the form and content shared intersubjectively as sign, and the unique experience of private subjectivity as the pure image. Whether such a stark division is tenable is doubtful, however. This will become more apparent by considering the other great contemporary tradition of sign theory, Saussurean structuralism—or, to be more accurate, Saussure's semiology.

Saussure's historical influence has been due chiefly to the *Course in General Linguistics* published under his name in 1916. The three offerings of the University of Geneva course on which it was based were given over a relatively short period of time (1908–1912). The course surveyed the science of linguistics for advanced students and thus included a great deal of material not specifically related to semiology. The "book" published in 1916 was in fact a collation from student notebooks. Saussure's lecture notes were not extant, and he apparently destroyed all materials he had been preparing for an intended *opus magnum* on the phenomenon of language.

In 1996, however, materials written by Saussure, some in view of the *opus magnum*, were found in the attic of the Saussure family home in Geneva. They were published in 2002; an English translation followed in 2006.[79] Along with the earlier publication of different versions of Saussure's course lectures recorded by students, these materials have begun to transform the conception of Saussure's project. Saussure's motives, the immediate objects of his thinking, and his principal goals now appear rather different—and in some cases decisively different—from those of

[79] See Saussure 2002, 2006. They are known as the Orangery Manuscripts.

"author" of the 1916 *Course*. It can be argued that the 1916 work was a pastiche that concealed as much as it revealed of Saussure's philosophy of language.[80]

If there were already hints in the *Course* that it was undertaking a new approach to psychology—there is in particular his statement that the study of language is a part of the study of signs, called semiology, and semiology a branch of social psychology (Saussure 1916, 33)—the variant student notes and the new manuscripts give this conception real substance. If there were hints in the *Course* that language and signs are a form of imagining, it is the manuscripts that make this a virtual certainty. And if there was at least one passage in the *Course*, a moment of almost poetic and artistic grace in the collation from student notes, that hinted at Saussure's conception of language as the incipience of appearances not simply *within* a field but precisely *as* a complex *biplanar* field, it is the manuscripts and the notebook variants that provide an explanation of why this is so and how it provides a sounder foundation for Saussurean science than do the usual accounts of structuralism.

Our task is not to participate in the rehabilitation of structuralism or the revision of Saussure per se. Some background setting is nevertheless in order. In the next paragraphs I will present four basic Saussurean themes and problems they raise, then I will discuss how the manuscript finds allow us to reconceive the Saussurean ontology of language as imaginative.

(1) We mentioned at the beginning of the previous section the dual character of the Saussurean sign, represented by an oval divided into two halves, with the signification above and the signifier below: for example, a picture of a tree above the line, the Latin word *arbor* beneath (Fig. 8.1). In first approximation it looks as though idea–concepts and idea–images (above) are being joined to words (below). But this is not what Saussure is saying. To come closer to Saussure's conception (and a fundamental thesis of all structuralism), we must realize that what is below the line in the oval is a sound or some other mark or signal. In the first instance Saussure is concerned with words as spoken rather than written, so we can say that the oval contains an idea and a sound. The word–sign is a duality, the joining together of an idea and a sound. Contrary to what is often claimed, this is not a very new idea at all. In a sense it is only a slightly more rigorous form of Locke or Condillac, for whom the spoken word is a label for ideas in the service of communication. It can scarcely be the key to structuralism's influence.

(2) That this notion of sign is hardly revolutionary becomes clearer from a diagram in the *Course* that illustrates the *communication cycle* (see Fig. 8.2, after

[80]In their introduction to the recent discoveries, Bouquet and Engler claim that the influence of the 1916 work kept scholars from seeing the implications of the variants that began appearing in the 1950s and 1960s; see Saussure 2002, 8–11, and Saussure 2006, xii–xiv. Engler's earlier, magisterial collation (in six parallel columns) of variants of the 1916 *Course* provided the most abundant evidence of a different Saussure; see Saussure 1968. Agamben 1993 [1977] was brilliantly prescient in presenting him as not merely a great linguistic scientist but also a profound philosopher of language and the language-using mind. Two recent critical works that have helped transform the study of Saussure are Thibault 1997 and Maniglier 2006. The latter especially has allowed me to turn what had been occasional insights and intimations into a focused view of Saussure. The following account could not have been written without it.

Fig. 8.2 The communication cycle

Saussure 1916, 27). According to the diagram and the accompanying account, something happens in the brain of person *a* that signals the mouth to utter a word; the sound travels through the air to the ear of person *b*; the effects on the ear travel to *b*'s brain; from there a signal is sent to the mouth, from which sound travels through the air to the ear and brain of *a*; and the cycle begins once more. Combined with the notion that the sign is a fusion of idea and sound, it suggests in first approximation that from the mouth to the ear it is the sound that is effective, whereas from the ear to the brain and on to the mouth of the second person there occurs a physiological event with a psychological turn that leads to a new physiological event that, at the mouth of the second speaker, again turns into acoustic physics. The problem is that this seems to break up the sign into multiple, discrete components, thus more or less mocking the notion that there is a *fusion* of the two. Moreover, it resembles a notion common in the seventeenth and eighteenth centuries: that the sound elicits in the mind of the hearer the idea with which it is associated, and a new thought in turn is communicated to others by a new sound. From the 1916 *Course* it is hard to know how this inconsistency should be resolved, at least without accusing Saussure of being a very sloppy thinker.

If we remind ourselves that he conceived language as part of semiology and semiology as a branch of social psychology, we can make a second-approximation conjecture: that the sign is what unifies the apparently discrete phases of the analysis into a whole. At the social level, at least, the sign is a complex unity of idea and sound, though a unity that can be "parsed" into components in the course of acts of communication. But whether this is an improvement or simply divides and spreads around the problems is unclear.

(3) A third theme makes Saussure look more distinctive. The theme addresses what goes on in the mind of the speaker and listener, in terms not of psychophysiology but of the linguistic rules and structures that are acquired by the individual speaker through his interaction with existing speakers. These rules and structures govern all language use, and are reinforced in every act of communication. Thus they could be seen as an extension of Kant's conception that the human mind is intrinsically structured by functional rules. The *Course* illustrates this with a series of diagrams. For example, Fig. 8.3 (after Saussure 1916, 159) represents signs as fitting and coexisting with every other. Saussure grants that the diagram does not reflect very well the complication that each sign bears a duality of signifier/signified, which immediately leads to a discussion of a "paradoxical principle"

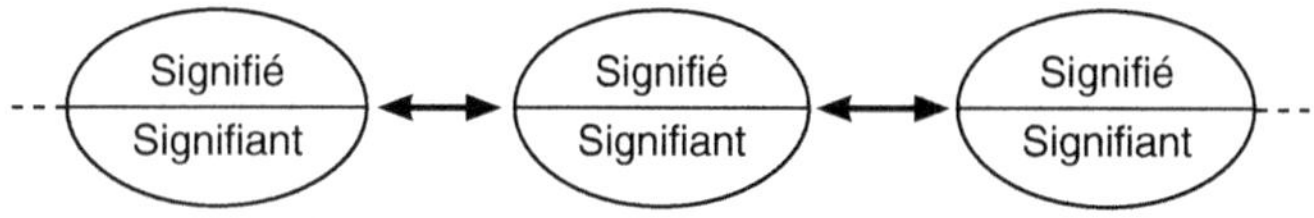

Fig. 8.3 Representing the interconnection of signs

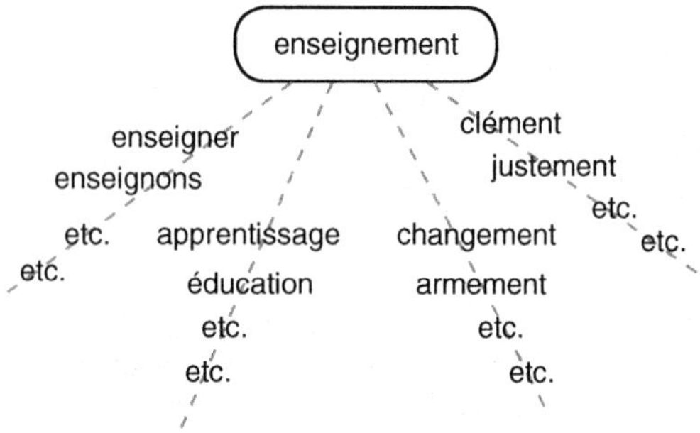

Fig. 8.4 The position of the sign *enseignement* in the network of signs

governing *values*. "They [values] are always constituted: (1) by a *dissimilar* thing susceptible of being *exchanged* for that whose value is to be determined; (2) by *similar* things that one can *compare* with that [thing] whose value is at issue." The value of a word does not depend simply on being exchangeable for other concepts (that is, for other significations or signifieds), it also "has to be compared with similar values, with other words that are opposed to it." The point is that the word's "content is not really determined except through the coming together [*concours*] of what exists outside of it" (Saussure 1916, 159–160).

A chapter later Saussure provides a more elaborate figuration of the interconnection of a sign with other signs; it develops further the complications produced by the various aspects of value and the corresponding signifier–signified duality (see Fig. 8.4, after Saussure 1916, 175). He places the word *enseignement*, "teaching," in an oval; the oval as usual represents the whole sign, but here he does not highlight the divided fusion of signifier and signified. From the oval extend broken lines leading down and away in different directions; along each of these broken lines are strung series of words. The leftmost series stands for different forms of the conjugation of *enseigner*, "to teach," from which the noun *enseignement* is derived. The second, third, and fourth series illustrate, respectively, synonyms of *enseignement*, other nouns ending like *enseignement* in *–ment* (thus they are also single-syllable unstressed noun rhymes), and adverbs ending with the letters *–ment* (which are also single-syllable unstressed adverb rhymes). These are only a few of myriad possible series of associations that depend principally on meaning (e.g., the synonym series), principally on sound (the adverb series), or on a combination of both (the conjugation series and the noun–rhyme series). They suggest the intricacy of networks that are constituted by relations between words, networks that are acquired and potentiated in a speaker's mind by being raised in and living in the community of speakers. The network relations are actualized when a person hears, thinks, and responds, and although hearing and speaking will settle on one possibility rather than others, in principle many possibilities of many series will enter into at least a low level of actuation or potentiation. One can hear a word in a certain relation to synonyms and at the same time anticipate its grammatical inflection and possible rhyming words. Someone who likes plays on words will quickly potentiate other, more unexpected series.

Within the mind (soul) of any individual there is a correspondence of the phonic (and written) signifiers to a mental content that is fluid but still more or less determinate. It is a gross oversimplification of this position to say that whenever we hear "tree" the sound is immediately associated with a given picture. Nevertheless, hearing the word triggers an already potentiated, deeply structured process that has various rules of form and formation in readiness. It is something like a multidimensional Kantian schema, which we recall is a rule joining a concept and an image. These Saussurean "schemata" join *meaning inflections* with *signifier inflections*. This is the level at which the sign's fusion of idea and signal is most intimate. Without the meaning component we do not have a sign at all, just noise, and without the rules of structured sound–formation we have no possibility of putting the meaning into communicable sign–form. Active formation, hearing, and utterance takes place at least in part in the individual's mind, psyche, or soul. That does not mean we will run into insuperable privacy problems, however. Traditional philosophical worries about the differences of mental content from person to person—even the possible incommensurability of ideas from one person to another—are overridden by the grammatical, lexical, phonic, and graphic formation rules that structure language. Every time "walk" is heard there is a potentiation of possible suffixes (*–ing*, *–ed*, *–er*, *–like*), verb-compounding auxiliaries (as in "walk up to," "walk down," "walk along," etc.), modality discriminators (I walk every day, I went for a walk), rhymes and near-rhymes (talk, Salk, balk, hawk, woke) and so forth. This is all part of a network of modulations that allows for a finely nuanced thinking and speaking about things.

Clearly each particular word appearing along the lines of Fig. 8.4 is not placed according to an absolute necessity. These are potentialities rather than actualities, so none is assumed to be actually conscious or even at the threshold of consciousness. The only exception to that principle is that once one is in process of using a word, some appropriately inflected version of it will be uttered in a way that is both immediately and mediately connected to other words according to the system of signifier-signified values. If you are about to tell someone what happened when you went walking yesterday, the *–ing* and the *–ed* and all the other possible endings appropriate to verbs will come to mind and speech by second nature, according to the grammatical and phonological principles one has inwardized and reinforced over a lifetime by speaking with and listening to others. One might (*à la* Chomsky) schematically conceive of many computational programs by means of which the potentials of the sign network are put into action and produce integrated and differentiated sounds, grammar structures, and meanings.

Native speakers of a language ordinarily have no trouble immediately distinguishing what words are susceptible of these transformations by association, fusion, and structure, and what others are not, even if they have not acquired the technical terminology to express it ("walk" and "run" feel alike because they are verbs, though they are different because you can't say "runned" unless you are a beginning learner; "walk" and "beautiful" are different enough that they cannot be assimilated in this way). That is, we already implicitly class together verbs, regular verbs, irregular verbs, nouns, adjectives, and the like, and keep track of the distinctions just fine, for the most part, and we share this acquired disposition with all other native speakers. There are also structures that are more directly related to sound, for example rhyming words, assonances, and alliterations. As individuals we have been assimilated to the social psychology of all these language structures. Notice in particular that, although the idea behind the word "walk" is not exhausted by all these structures, at least some aspects of its meaning are present in every use: it is an action, it is in primary usage intransitive (so as an act it is not acting upon an object), with *–ed* attached it is about the past, and so forth. Meaning and the forms of representing or signaling are intimately connected. The nature of the sign as a fusion of idea and signal thus acquires stronger support, and the distinctiveness of Saussure's approach appears more clearly. It is not just supergrammar with two-faced signs; it is an expression of the continuously dual inflectional character of linguistic experience. Both signal/signifier and idea/signified are inflected. That is to say as well that its character is doubly differential, in fact that the rules affecting both sound and idea are differentiators.[81] This, of course, is a trait of the biplanarity of incipiently differentiable, localized images that this book has been tracing in the occluded-occulted tradition of imagination. It also anticipates the fourth and perhaps most truly distinctive theme of semiology.

(4) Saussure says in the *Course* that the sign is nothing positive. In the first instance this strikes some people as the kind of nonsense on which postmodernism is built. A word is a sign, and a word is something positive; you can find it in dictionaries, for one thing, and in other books you will find it not just mentioned and defined but also used meaningfully. In speech it appears as a sound vibrating air and eardrums. There are easy defenses of Saussure's claim, and there are more difficult but richer ones.

It is possible to take a first step toward making the nonpositivity claim credible by calling to mind another famous thesis of the *Course*: that signs are arbitrary. As with most of Saussure's themes, that is not new. Almost from the moment that human beings began reflecting on the multiplicity of human languages, it has been

[81] Notice that most of the explanation here having to do with the signal or signifier have dealt more with the written forms of differentiation than with the spoken. Dealing with the spoken forms is more complicated, but in that respect more similar to the complications of the signified in use. An indefinite but still limited range of variations of idea/image elicits the same sound–signal; an indefinite but still limited range of sounds (yet all marked by the identical letters on a page, whether one says for "tuba" something that sounds like /too–buh/ or /tyu–buh/ or /too–bah/ or /too–bur/) elicits the same idea/image. One can try to define a social standard of pronunciation and meaning—after all, semiology is social psychology—but it will not for all that always be the authoritative standard for either, and over time any standard will inevitably shift, both in terms of meaning and of sound. See the following paragraphs.

recognized that the sounds uttered to name and describe things, actions, states, and events are not a matter of nature or necessity. The different names given to things, actions, states, and events make clear that convention and choice are at work. Although names *are* in some sense set down—that is, posited—they do not have the positivity of a real-world thing like a statue, a tree, or a squirrel.

Saussure is more radical than this, however. *Every* sign, that is, the union of idea and sound in every case, is arbitrary. But "arbitrary" has to be understood in a more determinate way than is typical. It does not mean uncaused or unmotivated, but determined by a choice or judgment (thus consonant with the root meaning of Latin *arbitror*)—at the level less of individual psychology, however, than of social psychology.[82] Why any particular sound or signal is associated with any particular meaning or signified is ultimately a matter of an anonymous social imposition that does not have to adhere to any particular kind of rationality or precedent, other than social consensus. An individual can try to coin a new word, or a new usage for an existing word, but in effect he is not so much inventing it as he is offering it to others for acceptance or rejection. Most of the time it will be rejected, not by committee decision but by, say, confused looks or uncomplimentary interjections. A few will be picked up in his circle; most of those will never reach larger circles of people, although once again a few will.

This kind of arbitrariness does allow the operation of motives and conditions, though in the last analysis they are never final. The introduction of words is, in any case, always subject to various considerations, for example, by what is considered pronounceable. If an English-speaker proposes the word *aphthschempdrü* (with the *ü* pronounced as in German), we can be fairly confident that it will be rejected by almost every other English speaker. That is a matter of acceptable/unacceptable sound.[83] But what is weird in one language may be conventional in another. If, on the other hand, someone proposed a word for the gleam of the sun at dusk off the hood of an old car whose red paint was fading, it is almost certain—apart from a scenario that would take some fictional construction[84]—that the larger society would see no need for anything more specific than the nineteen-word description I have just given. If someone proposes a new word whose meaning is already expressed or nearly expressed by many others, it too is not likely to be accepted because of the existing

---

[82]One has to wonder whether the usual interpretations of "arbitrary" in the *Course* is not in part a result of what French linguists call *faux amis*, linguistic "false friends" (words in different languages that appear to be cognate but are not, or were but have undergone semantic shifts). "Arbitrary" in English can scarcely be used in other than a pejorative sense, but that is not true of the French "arbitraire" even today, and even less so according to its usage a century ago.

[83]But let an unconventional-sounding word be introduced in a wildly popular movie and we will suddenly find ourselves (or our children) using a sound we never would have expected.

[84]Fictional, but not science fictional: that is, we would not have to postulate a race of Alpha Centaurans whose physiology was based on silicon rather than carbon. It is certainly imaginable that professional auto painters might have words of this general type in their argot. That would go some way toward establishing a Saussurean point: that language (*langue*) shapes mind to experience the world and to bespeak it (in *parole*) in the ways that language adumbrates. In that sense it is a form of anticipatory imagination.

words holding the field. A new word always corresponds to two differentiations, one in the field of the signal, the other in the field of the signified; but these two differentiations are not isomorphic.[85] Moreover, existing words and other signs fall into obsolescence and sometimes disappear entirely from the language when the circumstances of their use no longer hold or other signs displace them.

Any word that exists in a language might *not* exist—that is, its particular fusion of sound and idea might be absent from the language. The sound might be joined to some other idea, the idea to some other sound, or there might not be any such sound and any such idea as part of any word in the language. A new coinage that rhymes with "love" will change the history of English lyric poetry. In 1990 you could not google yourself; Google Inc. was still a gleam in the eye of Sergey Brin and Larry Page, and the spelling "google" did not even exist, except as a mistaken form of the fanciful "googol" that had been coined for the number 10^{100}. But as soon as the new name was formed—arbitrarily as far as the world of English-speakers was concerned, although perhaps strongly motivated in the direction it took by the personal psychology of the inventors of the Google search engine—it was absorbed into the network established by a myriad of rules of transformation. It began to develop its own associations and differentiations, and even the possibilities of "google" serving as a verb was foreshadowed by linguistic analogy.[86]

New fusions of sound and meaning can pop up "out of nowhere," existing ones can vanish. There is a positive side to the arbitrariness thesis, then: words exist by virtue of being used and reused (in *parole*, to use a familiar Saussurean concept), and every particular use is a new expression of a social choice, judgment, and affirmation (a matter of *arbitrage*, so to speak) that is mediated by the individual.[87] At the level of the individual everything seems to be dominated by the

[85] So, for example, if the new word corresponding to the gleam of the sun off the fading paint of a car hood were "sinteraze," it would not be the case that all the differentiations of the sound from related sounds in the language would closely parallel all the differentiations of the meaning from related meanings (it is not a sin to sinteraze). But as soon as the word is accepted, it becomes a node not just in two differential networks—of sound on the one hand and meaning on the other—but also in the doubly differential network of language as signs.

[86] One might speak of potentiation probabilities here, which will themselves always be differential. It was not inevitable that "google" would be accepted into the English language, although the success of the search engine made it more likely. So did the phonological and morphological characteristics of the word—for example, a word that is more euphoniously pronounced with the various suffixes that a verb will require is more likely to be adopted as a verb than another word that does not suffix as easily. If the Google search engine had been named Basis, we might well not be saying that we are "basising" ourselves.

[87] This concept may be annoying to atomistic individualists for whom the social is nothing but the sum of individual choices and acts. But even for atomistic individualists there is a difference in point of view between private and public acts. Hobbes's theory of language illustrates this: each person is free to reason according to the marks of similarity and dissimilarity of ideas that please him, but in order to communicate he must be willing to surrender this autonomy. (It is impossible, of course, that a real language could come into existence simply on this model.) This is in almost perfect parallel with Hobbes's conception of how legitimate government (that is, the Leviathan) is constituted. Language is, after all, the first social institution.

necessities of the lexicon and grammar (as *langue*, to use another familiar Saussurean concept). But over time the *social being* of language—which is to say simply the *being* of language—changes nevertheless, and sometimes quite rapidly, as new terms are introduced and old ones fall into desuetude. Widespread pronunciation changes (comparable to the great vowel shifts in early modern Germanic languages) are slower; they may not even require changing spellings in the written form of the language, but over decades and centuries they can transform the spoken language to the point that communication with speakers living at a century's remove would be impossible. Grammatical changes tend to be slower still, but they too shift as the social reality affirms changes to structures embedded even more deeply in the language.

The arbitrariness of individual signs is related to the nonpositivity of signs in general. If they were genuinely thinglike, they would continue to exist even after falling into disuse. Here we have to set aside our literate convictions that a word continues to exist because it is listed in a dictionary, perhaps marked as "obsolete." That would be an index of its still occasionally being heard and read. If it has fallen into complete disuse, however, it is no longer part of the synchronic language, the language as it is used in a given moment of time, no matter how many copies of the dictionary of obsolete words have been printed.

Understanding nonpositivity requires recognizing that a word is not just the fusion of a single sound with a single idea but an approximating node or expression in a system (language) that fuses two other systems (mental appearances and speakable sounds) that are not naturally isomorphic.[88] There is a sound–system, and an idea– or meaning–system. Because of the deficiency of isomorphism, "cat," "hat," "mat," and "sat" differ each from the others by a single sound, but that does not

[88]One has to acknowledge *three* systems here, precisely because the fused system is not a simple association of the sound system and the mental system. As the next few sentences make clear, the sound system and the meaning system are not and cannot be, overall, isomorphic to one another. But this is not to say that there are not localized isomorphisms in particular locales (/play/ used as the sound fused with the meaning of a certain kind of recreational activity sets up a locale for both sound forms and meaning forms like *played*, *player*, *playful*, *playing*, *displayed*, etc.), and schematized general isomorphisms across the entire field of sound–with–meaning (like the –ed suffix lending the sense of pastness to verbs—which is general but not universal, as *take–taken* exhibits). The recently discovered notes of Saussure emphasize that it is not simply sound per se that gets fused with meaning but a small subset of sounds that are significantly producible (phonemes)—thus this subset has already entered into the realm of significant mental experience and is not totally separate from the "mental system"—and, similarly but not isomorphically, it is only a subset of the blooming, buzzing confusion of mentality that is signified. Saussure's fundamental insight could be seen as an extension of Descartes' insight regarding the foundations of analytic geometry. The arbitrary, imaginative naming of geometrical lengths and areas can be done not just ad hoc but with the purpose of incorporating the labeling into formulaic representations of complex mathematical relationships; these in turn are a limited subset of possible formulaic uses, a subset systematically treatable in the more abstract imagining of algebra. Letters are not isomorphic to points and lines, yet they can be combined in a manner that nevertheless allows the mind to recognize and preserve features of the phenomena of interest, beginning with order and measure. But if one is blind to the imaginative character of mathematics, it is easier to be deaf to the dual- or even triple-aspect phenomenality of word–schemata.

mean that they are close in meaning; similarly "blue," "cerulean," "azure," and "ultramarine" are all close in meaning but not in sound. These examples also illustrate the principle of Saussurean linguistics that was at work in the diagrammatic structures discussed a short while ago: words (and other signs) are constituted not positively but rather as positions within networks of differentiations of sound/signal, meaning/idea, and grammar/inflection. If you erase the sign at the center of a figure representing networked relations, you have a gap surrounded by a structure. The structure has been partially hollowed out by erasing the word—which would of course have to be erased anywhere else it occurs in the network—but the hollow place would in a sense still be implicitly defined.

The nonpositivity of signs is, like the arbitrariness of words, a more complicated matter than it at first seems. One sense of "posit" that it does not exclude is the posit, or putting in place, that takes place with every use of a word. More particularly, tomorrow's language is the consequence of today's constant, massive, social-psychological "positing" that is the use of the system of language by all the native speakers communicating with one another in society at large. The sense of "posit" that is excluded is any positivity like that of the objects of the sciences of nature. If the region of the brain known as Broca's area were, in a science fiction scenario, destroyed by radiation, human organisms would continue (at least in the short term) to exist, positively; but because the fused system of sound and sense would have vanished, there would no longer be any words or language. Perhaps more to the point is that if overnight every English speaker in the world forgot "cerulean" and every occurrence in records was erased, the word would have been "disappeared" from the language. The sound of "cerulean" would still be possible, but it would mean nothing or come to be associated with a different meaning; but people who had used "cerulean" before might associate its meaning with a new sound. In neither case of reemergence of a sign would the same sign have been re-created. For that, one would need the unlikely situation of the same sound becoming spontaneously reassociated with the same former meaning. But since not even the sound and the meaning are simple positivities—consider that no two people pronounce words identically and that over time general pronunciation and applications of meaning both shift—the re-fusion would nevertheless almost certainly be at least slightly different from the original one.

The arbitrariness thesis and the thesis of the nonpositivity of signs are essentially implications of language as social psychology, although this point has often been insufficiently emphasized in English-language secondary literature. More important than either thesis for our purposes, however, is what the implications are of taking language seriously as social psychology. What is at issue is no less than the rationally imaginative character of human thought, whether social or individual.

8.8 Language as the Social Imagination of the World

We do not have to start from Saussure to make some essential points about the socially imaginative rationality of language. That it is intrinsically social we can take as settled; the notion of private language and the hoary, related conception that individuals who already possess a rational, fully articulated mental world can decide to invent words to begin communicating with one another are unsustainable. A child deprived of community will be language-deprived. Whether and how language is rational requires a bit more argument. That many irrationalities can be written and spoken hardly proves the contrary, nor that one can use logical and mathematical symbols to represent logical and mathematical falsities and impossibilities. The ancient Greeks' use of *logos* for both speech and proportion is suggestive but hardly conclusive. Yet it was precisely the Greek Aristotle who provided a conception of language use regularized by term–inclusion and –exclusion (All S are M, No M are P, therefore No S are P) and category type (according to his categories: e.g., "this dog is a mammal," "this dog is golden," "this dog is running," "this dog is large") that corresponded both to the material–and–formal being of things in the world and their being formally in mind. If a more epistemologically minded age has many more doubts about the veridicality and reliability of the various possible correspondences (mind–world, mind–language, language–world), today even more than in the past we are inclined to think that any correspondences we can establish have to be mediated by some kind of linguistic system. The very possibility of rationality requires language use, however regulated or restricted it might have to be; and even sloppy uses of natural language attempt to initiate rational engagement with things and/or people.

What about language as imaginative? When the model of imagining is holding a visual picture in mind, language appears to be different from imagining. But here, virtually at the end of this study, the time for thinking that that is an appropriate model is long past. It totally ignores what we can take as demonstrated: that an isolated image is at best an element or a place where imagination can be at work. A fixed, isolated image is prescinded from image–incipience and –formation and uniplanar at best (and not even uniplanar if it is conceived as an autonomous unit outside of any image field). Plato conceived *logoi* as images precisely because it is the everyday way in which we human beings constantly represent the world to ourselves and one another, and it is a medium plane (that of the dialogue, for example) in which we can conceive and incipiently reconceive what lies before our senses or that we find in various levels of intelligibility. Our representation and our re-representation of things is intrinsically an image making and image marking.

That words are intimately connected with image making is also implied by modifying Kantian schematism in light of the social character of language. A schema is transcendental imagination's device for moving back and forth between representations in the manifold of sensibility—sensory images, in less Kantian vocabulary—and concepts. After the linguistic turn in philosophy, however, conceptual thought of any significant depth and density requires language. So an updated version of schematism needs to incorporate words. One immediately plausible way would be to say that the word needs to replace the concept in the schema: that is, the schema is transcendental imagination's device for moving between *words* and images. Alternatively, one might make the word precisely the intermediating device—and that would bring us closer to Saussure. The word would then be a function of imagination in its transcendental use. Would it not, then, become a kind of image–function (with some sensory aspect) of the concept? Moreover, one thing we need to keep in mind is a consideration that troubled Saussure but seems hardly to have occurred to Kant. It is easy enough when thinking about the schema that it connects the real world and its appearances to the conceptual world: you think "dog" and start drawing one on paper, you see a Weimaraner talking down the sidewalk and think "dog." But it actually connects the manifold of sensibility to the understanding, and both of these are mental. So is the purely conceptual realm, if there is such a realm, as is also the manifold of sensibility in external sense/in the memory/in the produced and reproduced imagination. The sounds of the words we hear are in the experienced mental realm, as are the marks on paper of writing and the paper they are marked on. Schematism may lead to publicly observable behavior, but it is in the first place and in essence a structure of consciousness itself.

Moving back into the Saussurean topography proper, every time we utter a word we are imaging an established regularity of the community's psychology, and every time we see something and utter its name it we are reinforcing our thought pattern and our speaking according to the established template of language. Language, especially spoken language, exists or rather takes place on and between several planes, and it provides vectors and paths to guide the movement of our awareness from one to another. This is imagining and image making of the most complex kind human beings are capable of, though it is also the most commonplace way. That has no doubt made it easy to overlook.

Our interpretation of Saussure so far has remained within the ambit of the 1916 edition of the *Course*, which was a heavily redacted compilation from student notebooks. We have not yet appealed to the manuscripts that were discovered later or to variants in the notes. One thing that is clear from these other sources is that Saussure was troubled by an unbridgeable difference we have already mentioned between the objects of all other sciences (as they were known ca. 1900) and the object of linguistics. Even if some scientific objects (like atoms) cannot be directly exhibited, they can be inferred from other objects that are exhibitable. We cannot see much of the physics or chemistry or biology of the ordinary material world at the level of our senses or even our senses assisted by instruments like the microscope. But if they are insufficient for observing four atoms of hydrogen join with two of oxygen to form two water molecules, we do nevertheless attendantly experience phenomenal manifestations of the occurrence. There is little doubt that, with the help of theory, these invisible things explain the things that appear in the "manifold of sensibility" (as Kant would have put it), and so are real. They are *res*: they are real, positive things in the world, and they are the objects of these sciences.

What, comparably, are the objects of linguistics? This is what puzzled—one must almost say tormented—Saussure.[89] The history of his discipline was of little help in solving the puzzle. For most of the nineteenth century, linguistics was historical and comparative. Its first answer to the question about the proper objects of linguistics was that they were languages and words as historically developing

[89]This is a major point of Agamben 1993 [1977], and it is reinforced over and over again in Maniglier 2006.

entities: for example, *German*, which emerged from West Germanic languages at some time in the second half of the first millennium C.E. and then passed through various stages until becoming modern in the sixteenth century. The problem with this object, however, is that it seems to be many objects rather than a single one. If you could, in a science fiction scenario, bring together five native German speakers, one each from the years 1000, 1250, 1500, 1750, and 2000 C.E., each would probably find two or three others unintelligible. What is the essence that makes it the same language, then? To today's scientists of linguistics, the notion that a language has a historical essence, that it somehow appears with an essence that persists intact through centuries or even millennia, seems vaporous, even mystical.

In the last third of the nineteenth century many linguistic scientists began to search for verifiable facts of language here and now, under the influence of philosophical and scientific positivism. Spoken (and written) utterances seemed to fill the bill. Accordingly, the proper methodology of linguistics was to accumulate these types of facts, to use them to find inductive generalizations, and then to postulate laws in explanation of them. Saussure saw two major difficulties with this approach. The *first* was a remnant of the historical problem: given that languages change, over what length of time could one say that the facts one had accumulated still reflected the same language? It is to this question that his conception of language studied *synchronically* is an answer. Considered all at one moment, that is, synchronically, the language consists of a lexicon, or rather a set of signs, and all the structuring rules and principles that shape the minds and utterances of native speakers of a language; what the language (*langue*) is, is all possible sentences that could be actually uttered (in *parole*) according to those rules and the lexicon. The scientific study of language development over time, *diachronic* linguistics, has to presuppose the synchronic approach, in this sense: one takes a synchronic "snapshot" (as far as available sources allow) of the lexicon and structure of a language at a place and time, say High German in Westphalia around 1750, and snapshots again 50, 100, 150, etc., years later, and then one compares these linguistic structures regarding what changed and what stayed the same in order to write a history of what was spoken in Westphalia over the past 250 years. A scientific historical or diachronic linguistics, whether it is dealing with a "single" language or many, needs to compare synchronic moments; it compares one linguistic snapshot of an entire language at a given time and place with other such snapshots. The history of a language, insofar as it can be described, has to be written by comparing not just individual words or small sets of words over time but the sequence of all important synchronic structures in their succession. Where the history begins and ends is to a large degree arbitrary.

As Patrice Maniglier has argued in his magisterial reassessment of Saussure, *La vie énigmatique des signes*, historians of linguistics say that Saussure's crucial contribution to making linguistics a science was identifying the synchronic structure of language as the proper object of linguistic science. But Maniglier points out that, for all the merits of this claim, it misses the overriding problem that troubled Saussure and overlooks what Saussure himself thought about the adequacy (or rather inadequacy) of his own approach. What kept him from producing a *magnum opus* on general linguistics was not premature death or perfectionism but profound insight into a question that neither linguistics, nor psychology, nor philosophy could help him answer, because none of them had even begun to glimpse its possibility—a question that is nevertheless very much at home in the occluded-occulted tradition of imagination.

I mentioned earlier that Saussure saw *two* major difficulties with positivist approaches to language. The first was its historicist remnants. If that had been the only difficulty, distinguishing synchronic from diachronic linguistics and making synchronic linguistics the ruling type should have resolved it. But the second problem, more fundamental, was not at all addressed by the synchronic/diachronic division. This second problem pervades the Orangery Manuscripts discovered in 1996: that there is no positive phenomenon of language as an object of science *outside of the viewpoint taken by linguistic scientists*. This is a radicalization of the usual non-positivity thesis, in that it contends that there are *no* facts *at all* in linguistic science! If true, this would make linguistics totally different from the sciences of nature. Saussure spent the last decades of his career trying to conceive what this implied, about linguistics but even more about language per se. As Maniglier argues, despite all the inadequacies he acutely felt, Saussure nevertheless arrived at the elements of a new ontology of language that has hardly been addressed since, much less improved, perfected, criticized, or rejected. If he did not arrive at a fully satisfactory conception of what makes linguistics a science, the failure of his followers and his critics to address what he glimpsed and actually accomplished still plagues, or ought to plague, all the social sciences and philosophy.

Maniglier's ultimate aim is to reinvigorate structuralism by revising our understanding of Saussure. My goal is more limited: to explain Saussure's theory as a theory of the humanly social mind based on the biplanar structure of incipiently localized appearance—in brief, to explain it as the quintessentially human form of imagining. If Saussure did not see and resolve all the outstanding issues of the occluded-occulted tradition, he nevertheless brought it to a place from which any credible future theory must commence.

The 1916 *Course* has many simple, sometimes oversimple, figures and diagrams. One stands out from the others because it is so pictorial (see Fig. 8.5, after Saussure 1916, 156).[90] It occurs in part 2, chapter 4, at the beginning of the discussion of linguistic value. Without commentary this drawing is mystifying. As the text of the *Course* explains, in the first instance it represents wind or wind-driven clouds (*a*) above a stormy ocean (*b*). Waves are formed on the water by the motions of the wind. (In fact Saussure appeals not to the winds but to atmospheric pressure, differentials of which produce winds.) To begin to interpret: there are two systems present, the mass of air and the mass of water. When the two interact, when the wind moves over the waters, we get a structure that is a structure neither of the one nor the other but

[90] "Pictorial" because it is quite literally a picture rather than a diagram, though it is used analogically. There is one that is actually more pictorial near the end of part 1: a botanical drawing showing the anatomy of a plant stem (Saussure 1916, 125). But unlike the case of Fig. 8.5, interpreting it is quite straightforward: the difference between a transverse and longitudinal section of the stem is analogized to synchronic and diachronic approaches to language.

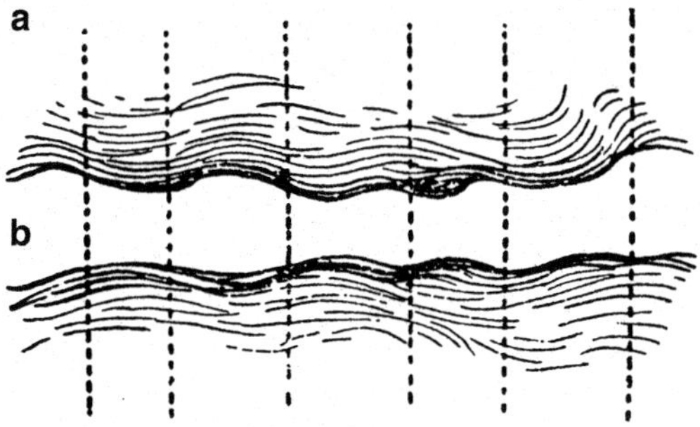

Fig. 8.5 The interaction of the “masses” of meaning and sound: wind over water

rather of both, of their interface. In an analogous way the mind/spirit is the waves at the interface of two masses, the mass of sound and the mass of meaning; it is at this interface that all the structural complexities of the language system are formed.

The passage is a rare moment of near-poetry in the *Course*; it implicitly evokes the account of creation in *Genesis*. It also has more significance than at first appears from the explanation provided in the 1916 *Course*. Albert Riedlinger, one of the *Course*’s two principal editors, recorded an alternate version during the second of the three general linguistics courses Saussure gave at Geneva.[91] The passage *in extenso* reads as follows:

> The characteristic role of language *vis-à-vis* thought is not to be a phonic, material means; but it is to create an intermediate milieu of such a nature that the compromise between thought and sound ends in an inevitable fashion at particular units. Thought, of its nature chaotic, is forced to make itself specific because it is broken up, it is distributed by language into units. But one must not fall into the banal idea that language is a mold: that is to consider it like something fixed, something rigid, when the phonic material is just as chaotic in itself as thought. It is not that at all: it is not the materialization of these thoughts through a sound that is a useful phenomenon; it is the somewhat mysterious fact that the thought–sound implies divisions that are the final units of linguistics. Sound and thought cannot combine except through these units. Comparison of two amorphous masses: water and air. If the atmospheric pressure changes, the surface of the water breaks up into a succession of units: the wave (=intermediate chain that does not form substance!). This undulation represents the union and so to speak the coupling of thought with the phonic chain that is in itself amorphous. Their combination produces a form. The terrain of linguistics is the terrain that one could call in a very broad sense the *common* terrain of articulations, that is to say the *articuli*, the small members in which thought takes consciousness through sound. Outside of these articulations, of these units, either one does pure psychology (thought), or phonology (sound).

There is a subtle but important difference between this passage and the parallel passage in the first-published version (Saussure 1916, 155–156). The two proceed for the most part in parallel. Yet the 1916 passage, more polished, also softens the

[91] I have translated the passage in the form given in Maniglier 2006, 278, which produces a smoothly readable version incorporating both the more episodic division of Riedlinger’s notes in Saussure 1968, 253–254 (column 2) and the corresponding, more continuous version of his notes in Saussure 1957, 37–38.

impression of what are quite striking claims for which the rest of the book, both what precedes and follows, scarcely prepares the reader. Then it ends with a paragraph that virtually takes back those claims: "One could call language [*la langue*] the domain of articulations, taking this word in the sense defined on p. 26: each linguistic term is a tiny member, an *articulus* where an idea fixes itself in a sound and where a sound becomes the sign of an idea" (Saussure 1916, 156). The ideas and the sounds do the "fixing" and the "becoming" here, rather than being themselves formed, even created, by the mysterious process. And if one looks back to p. 26 of the *Course*, one finds a very flattened account with no mystery at all: spoken language is divided into syllables, and significations into significant units, and "what is natural to the human being [is]...the faculty of constituting a language [*une langue*], that is, a system of distinct signs corresponding to distinct ideas." Both the mental and the sonic realms are articulated even before their encounter commences. Saussure's students could not simply change the words he spoke, but they framed them so that their bite was made harmless and far more traditional (not to mention that they talk of "distinct signs" when they should say "signifiers").

One of the essential insights that impel the notes found in 1996 is that neither mind nor sound is articulate before their encounter; their articulated formation and forms come precisely in and from that encounter. If it looks initially as though the objective physical realm of acoustic physics is being disturbed by a psychological impingement, the last sentence of the alternative passage strongly suggests differently. It implies what the Saussure of the Orangery Manuscripts insists on over and over: language as thought–sound is both psychological and phonological, it is an "impure" mixture. The thoughts and the relevant sounds are both *psychological–and–sonic manifestations*. He is not merely an opponent of private language, that one can have a fully articulated thought–mass that is linguistic without the sounds or words; he insists that, without the fusion of thought and sound in language, both human thought and human sound would be an unarticulated, chaotic, blooming and buzzing confusion.

The realm of sound with which the two-masses analogy is concerned is not the sounds of nature and technology studied by acoustical physicists but the sound that is articulated by the socialized psychology of the human being. *Both* meaning *and* sound are psychological or mental phenomena.[92] To put things in an idiom more congenial to the purposes of this study: both consciousness and sound are fields that need articulation, and out of their encounter, in consciousness, there arises a new field that is neither meaning nor sound but a psychological interface–fusion of the two. That fusion is the uniquely articulated social-psychological dynamism of language, which is the typically human form of incipient imaginative appearance.[93]

[92] In confirmation of this point, see Saussure 2002, 19; Saussure 2006, 4.

[93] In the *Écrits* Saussure says that if we were simply presented with a succession of simple colors (projected from slides onto a screen), "it would appear almost impossible to conceive of all these signs in their sequence, or 'as a synthesizable sequence, forming a whole'"; but when simultaneously displayed "we will have a pattern, which while not synthesizable for everybody, is at least beginning to become synthesizable and to be a pattern" (Saussure 2002, 109–110; Saussure 2006, 74–75). For Saussure, signitive articulation is a phenomenon that occurs with masses and wholes rather than particulars, with fields rather than objects or points. The language of synthesis suggests more a revised Kantianism than Humean associationism.

Langue, the social-psychological, synchronic structure of language, then amounts to what Aristotle called a first actuality; *parole* raises that actuality to utterance, which is full or second actuality.

This is one instance of a theme that recurs constantly in the Orangery Manuscripts: scientists of linguistics think that they can divide their subject into the objective and the subjective and sort real nonhuman, noncultural, nonlinguistic parts—e.g., physical sound waves—from the uniquely human and subjective. But instead those scientists are always already within the perspective of language when they take up the study of a language or any of its aspects. To be sure, a linguistic scientist can get a degree in physical acoustics and learn how to use physical detectors and recording devices, but to know what the phonemes of language are one must already be within the phenomenon of language. As example, Saussure points out that the variety of pronunciation of phonemes is so various, even for the same individual speaker, that a physicist with no previous experience of human language (an Alpha Centauran physicist?) would find it impossible to sort random noise (and silences) produced in speaking from phonemes and to identify the allophonic variants of each.[94] These do not exist as what they are before language and speaking. Once again, the objects of linguistic science are not real things like chemicals or plants.

In his book, Maniglier develops the psychological complexity of the "components" of signs to offer a more profound sense of what Saussure was after. He shows that there is a threefold, quasi-Hegelian "conceptual genesis of the concept of sign as a double entity, but also a *realist* representation of the acquisition of language" (Maniglier 2006, 299–300). (A) This conceptual genesis—or explanatory genesis, as I prefer to call it, insofar as it concerns the progressive sophistication of the model as it approaches explanatory adequacy—begins with two constitution–differentiations, that of the domain of sound and that of the domain of meaning. But that is only the beginning. There are differences in produced sounds, there are differences in thought appearances, and one cannot immediately overlay the one on the other to get language. (B) Differentiations can be infinitely nuanced, but language requires units, and to get those units there has to be opposition. Opposition arises in the second stage of genesis. The differentiated sound domain is not laid over the meaning domain; rather, the two interact, and in the process the differentiations within each domain become sharper, and some differentiations become oppositions. (C) These oppositions in the two domains finally give the appearance of the fully articulated system of signs, where despite the general arbitrariness and nonpositivity of language there arise relatively fixed, reproducible units of sound–thought.

A musical example, restricted in first approximation to voice, is helpful here. Imagine a tribe in which there is a tradition of wordless song. Traditional melodies get handed down from generation to generation and new ones are constantly improvised, without any notation to record it. Some songs are a nearly continuous stream of sound, others are segmented, percussive, or strongly rhythmic; some sound

[94] Today one might be able to design an algorithm and program a device to do some of this classificatory work, but one could hardly maintain that the algorithm and device had no experience of language insofar as their design would be predicated on the programmer–designer's linguistic competencies, both cultural and scientific.

earthy, others ethereal. Some people sing with a rasping sound to their voices (think of Bob Dylan), others with tremolo, others with elaborate melismas, and so forth. This musical tradition embraces not all sound but a quite restricted, though indefinitely various, subset. With the invention of notation, the written notes become the signifiers, the subset of sounds the signified. Imagine that at first a budding musicologist uses a continuous, (basically) horizontal line that more or less tracks the length of time the voice continues at a certain pitch and then shows the voice's rises and falls by an upward/downward shift of the line. Let us suppose further that this recording system very rapidly develops into a standard representation of pitch levels by lines and spaces and length of holding the pitch by differentiated marks (e.g., like the standard Western representation of the treble or bass clef with whole–, half–, quarter–, eighth–, etc. notes). We started with an established musical practice and now have developed a roughly parallel system of signifiers, by starting with a way of representing differences and then "hardening" it into sets of oppositions. Notice that the quarter– and eighth–notes are not merely different, they are now marked as exclusive (other, opposed): you hold a tone for a length of time or half that time. The variations in pitch recorded by a continuous line have become lines and spaces that exclude one another and thus are opposed.

So far the example has progressed in the genesis of signs through stages A and B, chiefly by considering the development of the notation–signifiers. But at the same time it is very likely that the performance tradition itself will be undergoing a movement as a result of the emergence of the signifiers. In fact that has been masked by my initial description of the available varieties in the performance tradition before notation: the description assumed the existence of some of the kinds of differentiations and oppositions I have noted in the signifier system. To oversimplify: at first the tribe simply sings; everyone recognizes certain characteristics of songs and voices in a *nonthematized* way that begins with liking or not liking the song and voice. People imitate one another, they are inspired to variation, they compete and want to differentiate themselves. Over time the cognoscenti among the tribe begin to thematize and name some of the practices—to begin with they might be named after the song or the singer rather than by abstraction. Gradually the differences "harden" into different and opposed practices, with or without anyone's trying to invent a notation system.

The progression of the genesis to stage C occurs when the signifiers and the signified, the notation and the singing, begin to fuse. Here "fusion" indicates something more and other than "association," indeed an opposition to it. Actual singing practices and a set of themes have already been associated by musicologists and cognoscenti. They become *fused* when in general the members of the tribe begin experiencing the music as a deployment of notes, pitches, modes, and genres. That is, the units that arise by the association of notation with singing come to be experienced as the element out of which the singing is composed (thus not just a description after the fact). The notion of styles and influences will be transposed accordingly. Although differences not directly recordable or expressible in notation will continue to be appreciated naïvely (e.g., the mellifluousness of a voice), everything that has been successfully translated into notation will be expressed in that medium. One expects, then, that the actual music will begin to be affected—for example, perhaps theoretical developments using terms of the notated music will encourage harmonic forms, and styles that use rhythms not commensurable with the notation might go out of fashion. Out of the oppositions of signs there will emerge a fully articulated domain of signs in which "music notes" (and their syntagmatic relationships) will prevail. They are not real like natural objects, but they exist for those who know the tradition, who speak the musical language. And perhaps there will also be an overwhelming temptation on the part of musicologists to write the history of this people's music as though it had always already been understood in the terms of the developed musicological categories and had developed as such.

With language, of course, we cannot trace the gradual historical genesis of its semiology through stages A, B, and C. The reason is what Saussure emphasized over and over in the recovered notes: one simply cannot remove oneself from language, and any attempt to do so will fail because the phenomena relevant to language can appear only to those who are already fully in language. With that caveat—that is, acknowledging that all our analytic efforts stand within the charmed circle of the synthesis that is language—we can deploy linguistic theories and terms to develop a conceptuality for the science of language. If we cannot give a historical account of how the difference and opposition arrived in (our) language per se, we can still note (for example) that the subset of all the natural sounds we can produce that are linguistically relevant are systematically differentiated. Certain different sounds are accepted as variants of a single phoneme, as allophones (the consecutive dental sounds in "went to" and "butter"), whereas in other cases sound differences no greater than this are taken as opposed, as different phonemes. This is all enforced by social consensus. It is the social construction of social reality with real bite to it.[95]

The conceptual genesis of signs throws light on Saussure's discussion of linguistic value. In explaining value and distinguishing it from meaning (which, in first approximation, is the signified of a sign) Saussure discusses a five-franc coin. It involves, he says, the two principles that *all* values do: (a) something dissimilar can be exchanged for the item of value, and (b) similar things can be compared with the item of value. Applying these specifically: (a) the five-franc coin can be used to purchase a certain quantity of noncoins, like bread (or chickens, or pencils, or stocks, etc.); (b) the five-franc coin can be exchanged for a determinate number of other coins (e.g., five one-franc coins, or at an exchange rate of five francs to a dollar, one U.S. silver dollar). By analogy, then, a word or words can be exchanged for something in experience (e.g., a dog runs through the classroom, and the teacher says: "A dog has run through the classroom!"), and can be exchanged for other words (e.g., for the teacher's sentence we can exchange "The Schnauzer entered at the door, sprinted along the first row of children, and leaped out the window!"). What the example/analogy conceals, however, is that it appeals not to two domains along with the interface between them (that is, the realm of the signifier, the realm

[95] One does not need to be a strong social constructivist to recognize that language is socially constructed. That does not mean, however, that everything about it and its use is irrational, unreal, and artificial. Quite the contrary!

of the signified, and the interface where the signs belong) but only to the signified domain (that of the bread purchase or the world events) and the interface plane (that of the coins or the words/sentences). Here, the properly psychological is left out of account.

Maniglier argues that the monetary example of value is a simple extension of the more apt *aesthetic value*. Aesthetic value concerns differentiations, inflections, and modifications in the use of an artist's medium insofar as they affect the overall quality of the work's aesthetic representation and expression. Thus it involves two differentiated domains or fields, that of the materials (their preparation, mixture, application, etc.) and that of the artist's thinking responsiveness to what he materially plans and sees. The interface between them is the picture that is actually drawn (and perhaps sketches and the like as well). Aesthetic representation always involves a unified two–fold, each part of which is psychological.[96] Each of these considered just by itself can be differentiated in manifold ways; the work of art is the result of the artist's negotiating a compromise between the two domains which in its final being, the interface, exceeds the analytic sum of these two parts. And this allows for the parallel with the threefold genesis of the sign.

So, for example, Monet's paintings of the Rouen cathedral are not objective snapshots but a series of evocations of the changing object of perception: changing according to the circumstances of light, ambiance, weather, and perspective, as well as according to the expressive possibilities of the artist's palette and (impressionist) method. The series of works constitutes a dual system of differentiations. No single representative (i.e., painting) is simply the equivalent of any number of the others, but each takes up a very clearly defined differentiating position with respect to the others. One can very easily locate this phenomenon in the work process of artists, even when he or she is not aiming to produce a series, by looking to the "work product"—all the sketches, sampling, cartoons, miniatures, retouchings, and even alternate paintings, drawings, or sculptures—that accompanies almost every major work.

This artistic example can be considered a kind of generalization of Descartes's discovery with respect to mathematical (and, in his earliest work, poetic) imagination, a discovery implicit already in the levels of ontological imaging in Plato. Imagining is not of things in themselves in a world in itself but commences with things and situations as they have been experienced, as we "take" them. In imaginative consciousness one simplifies and views aspects derived from the original experience by representing it differently: for instance, the plowed pattern in a farmer's field becomes a trapezoid, a sandy beach becomes an extended layer of tiny polyhedra, a dictionary becomes a representative of organized knowledge. The aspects one thinks relevant are thereby placed into a new, even fictional modeling domain different from the original. One then actively imagines (by variation and

[96] The materials are not simple physical entities but materials with established and discoverable expressivity in the art. That is the psychological, indeed social-psychological aspect of the materials. The resultant work of art is also both social-psychological and individually psychological. None of these aspects exists without the network of materials and practices.

iteration) the image–against–its–background in this new domain/plane. Eventually one may project the results of this imagining back into the domain/plane of the originating experience. If one finds an optimally efficient way of traversing a trapezoid in a euclidean or cartesian plane, one can apply this knowledge to the plowable field—as long as all the features that were not projected into the imaginative representation (rocks, trees, natural gas line easements) allow it. One might segment or alter the original's shape considered in the geometric plane and thereby hit upon other ways of patterning the path of the plow. These are ad hoc imaginings which one possibly never uses again once the practical problem is solved, although if they lead to a new mathematical theorem or innovative agricultural practice they will be generalized, systematized, and named. The imaginings of the Rouen cathedral in Monet's paintings are a relatively permanent *series* of imaginings that fuse what he noticed in appearance with the expressive possibilities of his paints and techniques. Few artists would be foolish enough to try to replicate the series, but its existence and success is an invitation to imaginative analogues.

Works in language (which can be more permanent than physical artifacts) involve the same kind of interaction between two realms of apparently independent experience that become jointly articulated and fused in the practices of mind. The difference is that, unlike with the development of both mathematics and visual art, almost all the practices and expedients are already fully shared socially. There is no doubt that an individual can innovate in language, but it is usually an isolated item, and it always has to be accepted and ratified by the speaking community. By contrast, the paintings remain what they are, and the mathematical theorems are valid, whether they are taken up by others or not.

This analysis of art helps to articulate the economic analogy between money and language in a way that perhaps more accurately illuminates Saussure's purposes. It is not so much that words are like coins and bills as that the invention and use of money presupposes a historical, social-psychological development of an imaginary field of "value." In the first instance the things of everyday life have each a distinctive character, both as brute presence and as part of everyday use. Speaking literally, there is and can be no equivalence between them: chickens are for laying eggs and for eating, knives are for cutting, bowls for containing, bracelets for decorating the wrist or proving a suitor's love. Human need inevitably leads to exchange, but every exchange made is at first ad hoc: it is of the moment, driven by the needs here and now of the participants. Setting up a market with standard exchange equivalents (two knives for one bowl, two and a half chickens for one bracelet) would have too many gaps and incommensurabilities (two half chickens are not the same as one whole) to offer a continuous field of universal valuation. Only with the invention of money does there come into existence a perfect fusion of the fields of goods and of exchange value, analogous to the fusion of the fields of meaning and of sound in language.[97]

[97]The comparison limps in the sense that the differentiation between prices (which are signs that fuse goods and exchange value) is purely linear, whereas the signs of language have all the types of differentiation (and more) that Saussure represents in his various diagrams.

8.9 Conclusion: The Ontology of Language

In the recovered Orangery Manuscripts, Saussure repeatedly brings up the "point of view" from which people approach language. For example, a researcher will discuss phonetics or phonology as an objective approach to language, so to speak an approach that in the first instance considers only sequences of sound as though they were objectively like any other sequences of sound. The problem he sees is that the researcher starts out knowing the phenomenon of language and what it means for sounds to be linguistic. The claim that the study is purely acoustic is belied by a prior selection of the sounds—the phonemes—that will be studied. This selection has always already been accomplished and has to be taken for granted. One thus never manages to step out of the point of view of the established language and its speakers; the distinction and the method that develop the acoustic approach to it are not prior to language, neither in general nor with respect to the specific language one is studying.

Even once one has entered into the particulars of the study of phonemes, the inability to step out of the linguistic point of view strikes again and again, for example with the so-called identity of linguistic sounds. Any attempt to define the sounds to be studied faces difficulties of pronunciation differences: not just of the variety of pronunciations of the "same" word or phoneme in different dialects, or the variety of sounds and allophones within the same dialect by different speakers, but also the variety of pronunciations by each speaker of the "same" word or sound in different "occurrences" (that is, in combination with different words or sounds in various orders)[98] and according to the speaker's momentary state (anger, fatigue, ennui, etc.). Saussure points out that this is a direct consequence of the profoundly differential character of signs, in particular of the sound/signifier aspect. Even if one can manage to identify for a given person an "average" sound for a phoneme around which most usages will occur, there will invariably be some that fall outside the expected range yet will be perfectly comprehensible to native speakers. Even to differentiate the sounds that are linguistic from "noise" you have to stand within the perspective of language. The subsequent study of these sounds according to the categories of, say, physiology or acoustics does not imply that from that point on the study is purely physiological or acoustical. Although the concepts, techniques, and devices are borrowed from the biological or the physical sciences, the objects being studied are not just nerve synapse firings or air vibrations, but synapse firings and air vibrations produced in the networked processes of speaking and hearing. The investigation is decisively conditioned by the initial standpoint within language. The sounds are studied as the utterances of speakers and listeners and the mental processes they share. It is a study of social psychology, as Saussure posited.

The immediate consequence that Saussure draws from these examples is that it is not possible to study language positivistically, as *facts* determined to be

[98] One small example: the subtle difference in the "p" sounds in pat, bump, bumper cars, and appear.

linguistic apart from the more basic phenomenon that both investigators and speakers/listeners are always already producing language and existing within its element. But it also points in the direction of the very strange ontology of the word as sign. If ordinarily we think of a word like "put" as a composite of three phonemes, that is fundamentally wrong, because that is to compose it out of three *identities*. This is the way someone assembling a voice synthesizer thinks of the problem, not a native speaker.[99] It would be closer to the truth to say that it is composed of three differences, even if that still suggests conceiving the word as an identity consisting of exactly specifiable, naturally occurring units. Native speakers learn not the phonemes per se but the words as a whole, as different in sound and meaning from others. They acquire language not by acquiring successive sound–unit identities and then learning to put them together, but rather by hearing complexes of language–meaning and learning to differentiate them, from the very first moment that they become part of the social universe of language use. The linguistic scientist, on the other hand (and that in essence includes any adult learners of language who start with the target language's alphabet, a dictionary, and grammatical rules), turns the natural language into an object of study by imaginatively imposing on it a network of linguistic concepts like phoneme and morpheme and applies techniques that analyze the language's unities of speaking and hearing into reproducible parts, which it then projects into various dimensions useful for study (acoustics, neurophysiology, logic, structural diagrams, etc.).

The original ontology of language is, by contrast, a matter of social psychology's differentially joining two imaginatively structured realms into articulate units, into the determinate signs that fuse sound and meaning. It is important to recognize that the sound in question is sound as heard and produced, not sound as the merely physical phenomenon of the transmission of vibrations through solids, liquids, and gases. This means that the realm of sound is just as psychological as is the realm of the signified meaning. Neither realm is stable in itself or stably structured; it is their fusion that accomplishes such structure and whatever stability it has.[100]

[99]Of course a cognitive scientist might argue that evolutionary physiology must already have solved the problem of combining the three sound–units into a complex sound. But this very statement reveals rather than conceals the point-of-view problem, and analyzing it would reveal the scientist's multiple acts of imaginative relocation of the original (e.g., reconceiving the situation in the field of physiology, then evolution; or conceiving "put" as /p–u–t/, projecting it into loci of physiological activity in nerves and brain, and projecting that into the framework of information processing). Of course the cognitive scientist might make a counterclaim that Saussureans do the same thing (Saussure agrees!). But that is less a counterclaim than a substantiation of the point this book has been making: that the human mind for the most part, and perhaps universally, thinks imaginatively; it produces fields by cross-sectioning the real world and possibilities of the real world, and it works in and projects to and from such fields, over and over again. The fields and projections of imagining are legion.

[100]This allows us to answer Wittgenstein's question about private language in a Saussurean fashion: no, there can be no private language, because the ideal realm is unstable until it is shaped by signs. Signs are stable because they are a socially enforced psychological fusion of sound and meaning and because as signs they have systematic value relations to one another that reinforce their stability.

If there is a realm of pure thought, it is inaccessible to language. Formally Saussure allows for it, but in his notes he says that, without signs, we will have to find some other way to access pure ideal forms as such; he ironically leaves that task to someone other than the linguist.[101] His deep conviction is that the account of language acquisition as a series of labelings of things is doubly or triply false: it substitutes two sets of identities for two fields of form possibilities, and it understands the creation of signs as punctiform, individual, and inductive, whereas that creation could not have come about except globally, as a comprehensive social phenomenon creating an interface between fields.

There is then a final irony in the Saussurean conception that helps explain the direction in which the structuralist movement took it. It looks, in the first instance, as though from the perspective of the structuralist the individual human mind is filled with words, rules, and other positivities, including differentiating functions. Yet language as semiological is part of social psychology. It is, as I remarked earlier, an Aristotelian first actuality. The very example Aristotle uses to illustrate the distinction of potentiality, first actuality, and second actuality—almost in anticipation of Saussure's conceptions—is the different respects in which we are "grammatical" as infants, as possessing the language without actually speaking or hearing it, and as hearing and speaking. This "semiological way" is structural and formal *as a whole*. It is not an assemblage of enumerable positive forms and structures but instead exists in the unified, exquisitely differentiated field of signs.

Each of the differential diagrams Saussure drew symbolizes a different dimensional field of value. He gave no more than a small sampling of the possibilities. This differentiation takes place in two fused elements (sound and meaning), with innumerable dimensions and no fewer differentiating factors than the system demands. The language is a social psychology that is acquired as first actuality by every native speaker, who then produces limitlessly *paroles* as second actualities. When Aristotle discussed the fields of sensation he differentiated them only by contrariety: dark–light, rough–smooth, sharp–blunt, and so forth. The Saussurean linguistic sign–fields, by contrast, can be differentiated by every linguistically relevant concept, by every inflection of meaning, not just linearly between one quality and its opposite but in a network of nodes that on the printed page of the *Course* is expressed in two dimensions but that, in principle, must be differentiated in multiple dimensions—in as many as are needed, in as many as correspond to the differentiations that the language makes. In that sense the differential space has to be represented in a corresponding number of dimensions; the space is n-dimensional, with n exactly as large or small as is needed to represent the being of the sign. Although order and measure in this space of language is not exactly orderable and measurable

[101] See Saussure 2002, 44, 64, 73, and 83; Saussure 2006, 25–26, 41, 48, and 57. Saussure 2002, 227 (in English, Saussure 2006, 159–160) suggests that there is nothing for psychology to study beyond what is semiological. This would open up a different kind of critique of psychologism than most existent types: ideas, images, concepts, and the like cannot be studied from the perspective of discrete soul powers because they are always already interactively semiological phenomena. Substitute "semiotic" for "semiological" and one gets the Peircean equivalent.

in the manner of cartesian space, its system is built on the principle that motivated Descartes's creation of analytic geometry: by fusing symbolic algebra with spatial relation, there results a new field of a higher order of complexity that is more distinctly conceivable precisely because it is more exactly imaginable as a result of the fusion.

To adapt Lacan's phrase: in this way the psyche of each human being is structured like a language. Indeed, it is structured by and with language, comprehensively. If, for Aristotle, the psyche is by definition the first actuality of an organized body, this means that the psyche of the human being who has been made part of a language community is an essentially linguistic first actuality. To inflect this idea more decidedly to Saussure's topography: each individual human being is, from infancy, quickly and progressively shaped as a speaker/listener, in the sense that the very appearances of the manifold he or she experiences are structured according to *langue*, the differentiated system of signs that has been socially instituted. We see, hear, touch, smell, taste, and locate that which *langue* articulates and linguistic and sensory imagination can discriminate.[102] The second actualities of this social psychology are *parole*, our acts of speaking and listening. They are how the first actuality is enforced, renewed, and, occasionally, freshly differentiated in individual, innovative ways. And it is through *parole* that the synchronic system *langue* is reaffirmed, reshaped, and transmitted to the future. *Parole* is how *langue* becomes diachronic.[103] It is the active imagining of the potential imagination of language.

References[104]

Agamben, Giorgio. 1993 [1977]. The barrier and the fold. In *Stanzas: Word and Phantasm in Western culture*, trans. Ronald A. Martinez, 152–158. Minneapolis: University of Minnesota Press.

Beiser, Frederick C. 2002. *German idealism: The struggle against subjectivism, 1781–1801*. Cambridge: Harvard University Press.

Boring, Edwin G. 1950. *A history of experimental psychology*, 2nd ed. New York: Appleton.

[102]To shift for a moment to Kantian terms: language is a schematism more comprehensive than the First Critique (even amplified by all the other critical writings) explains. The manifold of sensibility is organized not just aesthetically (in the presentations of space and time) and logically (in the forms of the pure concepts of the understanding) but as an accordance with all the forms of one's native language. This is perhaps what Kant was beginning to recognize with his examples of single schemata like triangle and dog, although the approach was too ad hoc and particular to reflect the systematicity of language that is key to Saussure. And to shift for an even briefer moment to Wittgenstein: if the *Tractatus* presupposes a kind of transduction of experience into logical form, Saussure's *langue* can be understood as the comprehensive transducing machine from which emerges even the forms of logic.

[103]This paragraph simply translates into Aristotelian idiom Maniglier's account of Saussure's semiology.

[104]If there is a second, square-bracketed date, it indicates the year the work first appeared in its original language.

Burch, Robert. 2010. Charles Sanders Peirce. *The Stanford encyclopedia of philosophy*, Fall 2010, ed. Edward N. Zalta. http://plato.stanford.edu/archives/fall2010/entries/peirce/. Accessed 30 Dec 2012.

Casey, Edward. 2000. *Imagining: A phenomenological inquiry*, 2nd ed. Bloomington/Indianapolis: Indiana University Press.

Castoriadis, Cornelius. 1987 [1975]. *The imaginary institution of society*. Trans. Kathleen Blamey. Cambridge: The MIT Press.

Castoriadis, Cornelius. 1997. From the monad to autonomy. In *World in fragments: Writings on politics, society, psychoanalysis, and imagination*, trans. David Ames Curtis, 172–195. Stanford: Stanford University Press.

Coleridge, Samuel Taylor. 1907 [1817]. *Biographia literaria*. Ed. J. Shawcross. 2 vols. Oxford: Clarendon Press.

Dumont, Richard E. 1965. The role of the phantasm in the psychology of Duns Scotus. *The Monist* 49: 617–633.

Eco, Umberto. 2000. *Kant and the platypus: Essays on language and cognition*. Trans. Alastair McEwen. New York: Harcourt Brace.

Engell, James. 1981. *The creative imagination: Enlightenment to romanticism*. Cambridge: Harvard University Press.

Ferrarin, Alfredo. 2001. *Hegel and Aristotle*. Cambridge: Cambridge University Press.

Fichte, Johann Gottlieb. 1970 [1794]. *The Science of knowledge [Grundlage der gesammten Wissenschaftslehre]*. Trans. Peter Heath and John Lachs. New York: Meredith Corp.

Fichte, Johann Gottlieb. 1984 [1795]. On the spirit and the letter in philosophy. Trans. Elizabeth Rubenstein. In *German aesthetic and literary criticism: Kant, Fichte, Schelling, Schopenhauer, Hegel*, ed. David Simpson. Cambridge: Cambridge University Press.

Fichte, Johann Gottlieb. 1994. Second introduction to the "Wissenschaftslehre". In *Introductions to the Wissenschaftslehre and other writings*, trans. Daniel Breazeale. Indianapolis: Hackett.

Frey, Hans-Jost. 1996. *Studies in poetic discourse: Mallarmé, Baudelaire, Rimbaud, Hölderlin*. Trans. William Whobrey. Stanford: Stanford University Press.

Hegel, Georg Wilhelm Friedrich. 1840–1845. *Enzyklopädie der philosophischen Wissenschaften im Grundrisse*. 3 vols. Georg Wilhelm Friedrich Hegels Werke, ed. Friedrich Förster et al., vols. 6–8. Berlin: Duncker & Humblot.

Heidegger, Martin. 1927. *Sein und Zeit*. Halle a.d. Saale: Max Niemeyer. English edition: Heidegger, Martin. 1962. *Being and Time*. Trans. John Macquarrie and Edward Robinson. New York: Harper & Row.

Heidegger, Martin. 1929. *Kant und das Problem der Metaphysik*. Bonn: Friedrich Cohen. English edition: Heidegger, Martin. 1997. *Kant and the problem of metaphysics*, 5th ed. Trans. Richard Taft. Bloomington: Indiana University Press.

Heidegger, Martin. 1979. *Prolegomena zur Geschichte des Zeitbegriffs*. Gesamtausgabe, vol. 20. Frankfurt a.M.: Klostermann. English edition: Heidegger, Martin. 1985. *History of the concept of time: Prolegomena*. Trans. Theodore Kisiel. Bloomington: Indiana University Press.

Johnson, Mark. 1987. *The body in the mind: The bodily basis of meaning, imagination, and reason*. Chicago: University of Chicago Press.

Kant, Immanuel. 1900 ff. *Kant's Gesammelte Schriften* (known as the Akademie Ausgabe, "Akad."). 29 vols. to date. Königlich Preußische Akademie der Wissenschaften. Berlin: de Gruyter (before 1922 Reimer).

Kant, Immanuel. 1987 [1790]. *Critique of judgment*. Trans. Werner S. Pluhar. Indianapolis/Cambridge: Hackett.

Kant, Immanuel. 2002 [1788]. *Critique of practical reason*. Trans. Werner S. Pluhar. Indianapolis/Cambridge: Hackett.

Kretzmann, Norman. 1974. Aristotle on spoken sound significant by convention. In *Ancient logic and its modern interpretations*, ed. John Corcoran, 3–21. Dordrecht: Reidel.

Lacan, Jacques. 2006. The mirror stage as formative of the I function as revealed in psychoanalytic experience. In *Écrits*, trans. Bruce Fink, 75–81. New York: W. W. Norton.

Laplanche, Jean, and Jean-Bertrand Pontalis. 1968. Fantasy and the origins of sexuality. *The International Journal of Psycho-Analysis* 49: 1–18.

Lonergan, Bernard J. F. 1997. *Verbum: Word and idea in Aquinas*. Ed. Frederick E. Crowe and Robert M. Doran. Collected Works of Bernard Lonergan, vol. 2. Toronto: University of Toronto Press.

Longuenesse, Béatrice. 1998. *Kant and the capacity to judge: Sensibility and discursivity in the transcendental analytic of the "Critique of pure reason"*. Trans. Charles T. Wolfe. Princeton: Princeton University Press.

Manetti, Giovanni. 1993. *Theories of the sign in classical antiquity*. Trans. Christine Richardson. Bloomington: Indiana University Press.

Maniglier, Patrice. 2006. *La vie enigmatique des signes*. Paris: Léo Scheer.

McManus, Denis. 2006. *The enchantment of words: Wittgenstein's "Tractatus logico-philosophicus"*. Oxford: Clarendon.

Mueller, Gustav E. 1958. The Hegel legend of "Thesis–Antithesis–Synthesis". *Journal of the History of Ideas* 19: 411–414.

Peirce, Charles Sanders. 1867. On a new list of categories. *Proceedings of the American Academy of Arts and Sciences* 7: 287–298.

Peirce, Charles Sanders. 1868a. Questions concerning certain faculties claimed for man. *The Journal of Speculative Philosophy* 2: 103–114.

Peirce, Charles Sanders. 1868b. Some consequences of four incapacities. *The Journal of Speculative Philosophy* 2: 140–157.

Perrin, Denis. 2007. Wittgenstein et le clair–obscur de l'image: Logique, psychologie, philosophie. In *L'image*, ed. Alexander Schnell, 91–113. Paris: Vrin.

Ricoeur, Paul. 1978. Imagination in discourse and in action. *Analecta Husserliana* 7: 3–22.

Saussure, Ferdinand de. 1916. *Cours de linguistique générale*. Ed. Charles Bally, Albert Séchehaye, and Albert Riedlinger. Lausanne/Paris: Payot & Rivage.

Saussure, Ferdinand de. 1957. Cours de linguistique générale, Cours II (1908–1909): Introduction (d'après des notes d'étudiants), ed. Robert Godel. *Cahiers Ferdinand de Saussure* 15: 3–103.

Saussure, Ferdinand de. 1968. *Cours de linguistique générale*, vol. 1. Ed. Rudolf Engler. Wiesbaden: Harrassowitz.

Saussure, Ferdinand de. 2002. *Écrits de linguistique générale*. Ed. Simon Bouquet and Rudolf Engler. Paris: Gallimard.

Saussure, Ferdinand de. 2006. *Writings in general linguistics*. Trans. Carol Sanders and Matthew Pires. Oxford: Oxford University Press.

Schelling, F. W. J. 1978 [1800]. *System of transcendental idealism (1800)*. Trans. Peter Heath. Charlottesville: University of Virginia Press.

Sepper, Dennis L. 2009. Goethe, Newton, and modern scientific imagination. *Revue philosophique internationale* 63: 261–277.

Thibault, Paul J. 1997. *Re-reading Saussure: The dynamics of signs in social life*. London: Routledge.

Walser, Martin. 1981. Romantische Ironie in Wirklichkeit. In *Selbstbewußtsein und Ironie: Frankfurter Vorlesungen*. Frankfurt a.M.: Suhrkamp.

Wittgenstein, Ludwig. 1922. *Tractatus logico-philosophicus*. Trans. C. K. Ogden. London: Routledge and Kegan Paul.

Wittgenstein, Ludwig. 1953. *Philosophical investigations*. Trans. G. E. M. Anscombe. Oxford: Blackwell.

Wittgenstein, Ludwig. 1998. *Notebooks: 1914–1916*, 2nd ed. Ed. G. H. von Wright and G. E. M. Anscombe. Oxford: Blackwell.

Chapter 9
The Ethos of Imagining

Philosophy guided by a sense of responsibility for everything should no longer lay claim to a mastery of the absolute, should in fact renounce all such notions, in order not to betray them in the event, without, however, sacrificing the concept of truth itself. (Adorno 1992, 22)

Cornelius Castoriadis, in an indispensable discussion of how speaking, making, and imagining are interrelated, argues that nature and the canons of modern logic and set theory (which he designates ensemblistic-identitary, or *ensidic*, logic) constitute a *basis* for human and social being.[1] Human and social being proper become determinate, however, only as the result of a more fundamental institution than ensidic logic, an institution that indeed "leans" on nature and accepts principles of logical inclusion and exclusion but that occurs through communally shared historical acts. Fundamental institution is expressed *in*, and *as*, the basic social imaginary of the community, and language is its chief repository. Language, the most fundamental of fundamental institutions, is in essence an imaginary creation.

In explication of this complex contention, Castoriadis offers a quasi-Kantian theory of *legein* (giving accounts by making determinations) and *teukhein* (making and doing—the word is etymologically related to *technē*) that elaborates several grounding schemata. As prerequisite for acts of *legein* he includes, among others, schemata for (1) separation/discreteness (or identity and difference), (2) union (or assemblage into

[1]"The Social-Historical Institution: *Legein* and *Teukhein*," in Castoriadis 1987 [1975], 221–272. This claim should make clear, all by itself, that Castoriadis is not proposing an epistemological relativism. But ensidic logic, though implicit in all human reasoning, is rarely as rigorous and thoroughgoing as it is in the forms of modern mathematical logic. The logic of inclusion and exclusion is universal, but precisely what classes or groupings a culture establishes and how and in what circumstances it enforces criteria of inclusion and exclusion can be determined only by the fundamental institution of language.

D.L. Sepper, *Understanding Imagination: The Reason of Images*, Studies in History and Philosophy of Science 33, DOI 10.1007/978-94-007-6507-8_9,

a whole), (3) decomposition (a combination of the first two), (4) the as (that is, the positing of something *as* something else, in some respect), and (5) designation (which presupposes the individuation and the collecting together implied by the previous four schemata; see Castoriadis 1987 [1975], 224–225). *Legein* is "the ability to distinguish–choose–posit–assemble–count–speak" (223); the most basic principles of ensidic logic are incorporated into it, though without ever, until recently, having been developed into set theory and mathematical logic (227). Castoriadis's explication of these as imaginative functions underlying all human thinking, making, and doing not only critically develops Kantian schematism in the direction of language but also provides a path for elaborating Saussure's semiology in a way that avoids pitfalls of later structuralism.[2]

Saussure did take seriously that there can be no semiological seeing without the seeing–as… and seeing–as–different of whatever presents to the senses and imagination; more accurately, semiological hearing and "seeing"—the hearing is literal, the seeing is figurative—involves hearing and seeing as and as different. In the first instance the sound is heard as the meaning, the sound makes the meaning emerge in the network of differential meanings; and the nascent appearance of meaning makes corresponding sounds and other signals emerge in the aspect of speaking. The sign is simultaneously a mental formation of the sound–signal and of the ideational content. Seeing–as, more generally the taking of something as something else, something more, or something beyond, in a particular respect, has been a basic trait of abstractional imagination from the beginning of conceptualization in Western thought, and so too has been concretizing imagination. In imagination the mind typically attends to two places at once, and the exact position of the attention comes into question precisely by way of the schema of seeing–as.

Especially in the Orangery Manuscripts found in the 1990s, Saussure returns over and over to the paradox of a scientific approach to language. In order to study language one must already be in its midst—thus one must already be seeing things A as things B (for instance, sounds are seen, or rather heard, as what they signify)—but the scientists of language are constantly intent on a radical separation of the domains or fields of A and B, as though there could be (for example) phonemes without meaning.[3] You can study physical acoustics without attending to meaning,

[2] Although Castoriadis is a harsh critic of structuralism, he is appreciative of Saussure. For example, in Castoriadis 1987 [1975], 216, he contrasts the later radicalization of the difference between synchronic and diachronic with Saussure's justified "reaction to a pseudo-historicism in the linguistic domain." On 253 Saussure is cited twice as an authority for important distinctions in the technical vocabulary of linguistics. On 244 Castoriadis offers a discussion of "sign" that he claims is different from Saussure's; in the end it closely resembles the account we have given of Saussure in Sects. 8.7 and 8.8, above, and differs instead from the later structuralism Castoriadis dislikes.

[3] This is, once again, the problem of prescission vs. abstraction (see Sect. 5.13, above, esp. n. 102). There is no doubt that one can study a naturally produced sound that is acoustically the same as a particular phoneme of a language, but it would not be a phoneme or even linguistic. Phonemes are singled out by the ear and produced by the human vocal equipment precisely insofar as they play their role in signifying language. Extraterrestrials that had no experience of signs (that is, of the fusion of sound or gesture with meaning—imagine that they "communicated" by the immediate perception of one another's thoughts) could without a moment's pause begin studying water as an earthling chemist does, or vibrations in the air as an earthling physicist would, but they would have no access whatsoever to the phenomena of language as such.

and you can apply the techniques learned in physical acoustics to study phonetics, and you can use general phonetics to study the phonologies of particular languages, but the phonology of a language has no existence or status prior to or apart from the existence of the language and the researcher's intralinguistic construal of how that language (and language in general) works.[4] The sound system of this or that language picks out, with an incredible flexibility in use that cannot be explained by physical acoustics, only a tiny subset of physically possible sounds that is also a small subset of the sounds that can be produced by the human vocalizing system. More fundamentally, however, there cannot be language–sound or language–meaning in any proper sense without human participation in the reciprocal self–and–world–organization (a social psychology) of sound and mental appearance that we call language.

Which came first for human beings, linguistic imagination or projective, biplanar imagination? Today they are so intricated and co-implicated that it seems impossible to answer. If we resort to evolutionary psychology, the more plausible answer would be that first came brute appearance, then appearance seen as part of a field, and then appearances in one field seen against the background of another field, and then the fusion of fields in language. We must, of course, also recognize that evolutionary psychology is itself a system of projective, biplanar imagining, and therefore it cannot escape the fundamental conditions of imagining. That means, for one thing, that in order to be credible evolutionary psychology must extend and deepen the imaginative structures it uses for the sake of such explanations. It must become more adequately comprehensive in what it tries to embrace and more comprehensively adequate to the phenomena it tries to explain. Its imagination of imagination can certainly begin with just-so stories made to fit, at least loosely, the appearances in their appropriate fields. Then it has to make the fit tighter; then it has to encompass with appropriate concreteness ever more aspects of the life of the mind and soul. And it must not claim the kind of certainty that has become customary in the age of supreme rationality; it has to get accustomed to being the kind of imaginative rationality and rational imagining that by its nature must put things in proportional connection with one another.

Saussurean semiology with a Castoriadian twist yields human imagination that is fundamentally linguistic. That would explain in large part why what we have called imagination is so important to us. Perhaps it would also allow both romantics and hyperrationalists to see that rationality without emergent, field-positioned images is blind, and imagination without taking proportional measures of what is imagined (that is, without rationality) is manic. And it would no longer be a question of just *adding* rationality to animal imagination to get *homo sapiens sapiens*—not if rationality, to exist in the human sense, requires the fusion of image fields that are indeterminate (chaotic) before the fusion. The divorce of rationality from imagination that has been a cultural commonplace for nearly three centuries would turn

[4]In first approximation, phonetics is the general science of the human production of linguistically significant sounds, whereas phonology studies the specific structures in a given language governing the significant use of its sounds.

out to be illusory. Rationality itself, insofar as definition is in question, would have to be conceived as a power of reckoning in images seen and marked against the background of other (kinds of) images. Even logic would be an abstract but still imaginative marking system that tracks and shows with sharp distinction an important subset of the informing structures of language. But not all.[5]

Those who want to hold on to a rationality devoid of all materiality—that would mean purged of everything that allows for differences of appearance in the element of thought where it took place—would have to prove that such an element exists. Such an element, however, would have to be without time, without space, without change, without movement of thought, even without different thoughts. It could have only one thought in eternal changelessness. More than a few philosophers, it appears, have entertained such a possibility. In our historical investigations, however, we have found at least some grounds for thinking that the greatest of the philosophers of imagination did not think so—and some of them were philosophers who have been thought to be the strongest supporters of eternal changelessness. The more one tries to think the possibility through, the more alien it is to the fundamental condition of being human.

9.1 Delimiting Imagination Rationally

I began this book with very general questions and considerations about how imagination is popularly and technically conceived: on the one hand, imagination as the power or source of creativity; on the other, imagination as forming and holding an image in mind. The two appeared to have little to do with one another. I put aside creativity in order to consider how forming and holding images became the default model of imagination, then began searching for alternative ways of conceiving the act of imagining. This led to the phenomenon of imaginative fields, which in turn led to the notion of the conceptual topology of imagination. All along I took my bearings, both affirmatively and negatively, from classic, or at least typical, modern thinkers. Then it began to dawn that there is a long and strange history of how the conception of imagination and images developed, a history that is marked by apparent missteps but also by hardly exploited resources. It turned out that, in several of the most classic thinkers of all, the role of imagination was central, yet the tradition of interpretation of those thinkers overlooked and deemphasized it. Moreover, although there are many connections joining later classic authors to earlier ones, it appears that few thinkers have been directly aware of the deep historical roots of the problems and the solutions they entertained. We ourselves are in a similar position: the "ordinary level" of learned discussion of imagination simply does not rise to the level of what past thinkers have accomplished. As I pointed out at the beginning,

[5] To see this point, one need only reflect on the intractable problems that logicians have incorporating time and modality into their systems.

contemporary philosophy and psychology have grown so natively and naïvely antipsychologistic that they have tended either to turn away from questions of imagination or to reduce imagining to a propositional or behavioral attitude. As I have tried to show through analysis and history, this is a monumental case of missing the point, of not even glimpsing the phenomenon, much less articulating it.

I have noted several times that it was Plato's Socrates who first explained that, when the issue is what to do or think next, we are never fully supplied with all the knowledge we need. When a knowledge claim is contested we rarely manage to make it fully consistent with all our other claims and actions and all the cases we can imagine. Our ability to accommodate what we (think we) know to other things and cases—past, present, future, or merely possible—is limited, no matter how much evidence we have gathered, how much mathematics we have applied, and how much formalism we have used. Indeed, it often happens that the more evidence we have and the more rigorously we reason, the more problems we discover. I see no reason to believe that this is about to change in the future. But Socrates' response to this situation, even a few moments before his death, was to keep facing up to questions by patiently tracing out the interconnections of things and all the considerations we ordinarily neglect. The only way to do this is to develop *logoi*, accounts, in ever greater detail—and for Plato and Plato's Socrates, *logoi*, strictly speaking, are *eikōnes*, images. To think is to imagine, now more concretely, now less.

Descartes, the putative father of modern rationalism, sought method as a response to the nearly random way in which people undertook the search for truth: they as it were wandered about, expecting to find answers where their aimless path took them. Implicitly recognizing the truth of Aristotle's dictum that there is no thinking without phantasms, he countered with a theory of invention based on our native psychological capacity (*ingenium*) for making simplified figures and images of what has been given in the field of any problem we encounter. The simplified forms allow us to see clearly basic facts and relations that we can link to one another according to elementary proportions. But furthermore, in what I have called the neglected rule of Descartes's method, the rule of enumeration, even once this method has led to a solution we must make sure that we have left nothing important out. Every day we must continue providing ourselves with the widest possible experience in search of all the forms of order and measure the world has to offer and all the ways in which order exists in things. Our knowledge must have amplitude as well as accuracy, it must be far-seeing and extensive as well as clear and distinct. Nothing is beneath our notice, no truth is too small to bother with. Rationality without vigorous, wide-ranging imagination does not know enough to take another step; rationality without amplitude will always misjudge how much it has accomplished and how universal its conclusions are.[6]

[6]Here, unlike in Chap. 6, above, we can set aside the question of whether Descartes's mature dualism ultimately sinned against the insight behind his fourth rule. For us, at any rate, tracing the history of thought does not mean that we have to accept everything we find there as true. We are better Cartesians (though not better cartesians: see Sect. 6.6, above, esp. n. 31) if we diverge from Descartes by thinking more rigorously and vigorously than he did.

In the historical chapters of this book I have shown that the classic philosophers of imagination offer well-articulated theories of the psychological economy of imagination and reason, even if no single theory is entirely satisfactory or addresses all questions. On the other hand, as Castoriadis has argued, even the best thinkers tolerated and sometimes encouraged the premature closure or occlusion of their understanding of imagination, especially when it posed a threat to the sovereignty of reason. Occlusion was often followed by concealment or occultation in the works of interpreters. As counterpoise, I have tried to reconstruct those well-articulated theories and their subsequent occlusions–and–occultations as a tradition. The tradition, though tacit, is greater than the sum of its parts. It not only sheds light on our own confusions but also points beyond confusion to a renewed understanding and appreciation of imagination and reason. But recognizing, much less revitalizing, is a task that is difficult for us, not simply because we tend to fall back thoughtlessly on a heritage from thinkers who turned their backs on crucial questions, but also because *we* in the meantime have largely ignored our everyday experiences of the relevant phenomena and thus lost the sensibility needed for articulating them well and accurately.

In the introduction to Chap. 2, I gave a long definition of imagination that, at the beginning of the investigation, scarcely made sense. This is how it went:

> Imagination is a (psychologically) evocative, anticipatory, abstractional-concretional activity that follows upon actual perception. It allows the imaginer to (1) dynamically (re)position herself and incipiently explore, place, vary, connect, and re–present appearances originating within a field of concern, (2) attend to and mark the field's potentials, and (3) exploit those potentials by projecting them to other fields (possibly new) in abstracted/concreted appearances.

I went on to question the value of definitions in philosophy: insofar as philosophy is a search for wisdom or an attempt at explanation, it can never be content with definitions. I do not intend to reverse myself now simply because I have reached a point where I can explain my definition in detail. Definitions are a logical and rhetorical means, not an end. They invite a reader or listener to conceive and reconceive what they mean, in logical consequence and in concrete application. That means, of course, that they have to be both abstractly and concretely imagined. Definitions are heuristically descriptive and summative rather than essential. That means that they can be useful for starting a discussion, for refining one's sense of things along the way, and for trying to mark at the end of discussion the point one has reached. If understanding and science are never fully at an end, marking the point one has reached as one leaves off discussion should be taken as a provisional act, as marking the point where future investigation can and should pick up, rather than as the sum of wisdom, philosophy, or science. So what follows here is less a definitive ending than an analytic reading of the definition that will try to point ahead, toward a few places we might go next.

The definition is divided into two sentences. (1) The first sentence places imagination in a genus—activity—with certain qualifications. (2) The three clauses of the second sentence then elaborate the specific character of imagination suggested by the qualifications.

(1) According to the first sentence, imagination is an activity. Or should we say instead that *imagining* is an activity, or perhaps an act; and that imagination is a potential, or a power, or a faculty? I mention the last as a term that, among philosophers and psychologists, is often treated as that–which–is–not–to–be–named. There is a taboo imposed on "faculty," a taboo that has become little more than an irrational tic of people uncertainly striving to be rational. It is, to begin, simply the Englishing of a Latin word that was used to render a Greek word. There exist alternative English renderings that are not regarded as problematic, even if, as mere lexical items, they in effect mean exactly the same thing.

Greek *dunamis* is what became *facultas* in Latin, though it also became "power" or "potential" (*potestas, potentia*). In the Chap. 5 discussion of Aristotle we identified a major historical reason for the faculty–taboo. Late ancient philosophy and various medieval reinterpretations proliferated soul powers and subpowers. In the course of the transmission of these innovations and interpretations, there was a tendency to reify the powers, to treat them (though Aristotle did not) as distinct things, as distinct modules (as we might say). So, for example, the proper sensibles of sensation that can be brought back to mind by imagination are stored in one part of the brain, the common sensibles in another, and in a third place there occurs their dis– and re–assembling. *Here* things remembered can be called upon, *there* the memory of events, and, at a certain juncture of this ever more complicated modular process, the phantasm–complexes we have built up receive their associated concepts and names. The term "interior/inward senses" became a cover for this multiplicity. If such an approach seemed justified in light of the medical knowledge of the day, in our more jaundiced view what the name covered was unjustified certainty—unjustified because the underlying conceptions were overprecise and too particularly specified, beyond any real, demonstrable knowledge. Insofar as "faculty" was intrinsic to the language used to express this certainty, one can easily understand that later thinkers felt a moratorium was in order.

One can accept this diagnosis yet feel the need to raise an objection: Were the medieval approaches really so different from those of today? If the spaces of the brain ventricles are nothing but spaces—thus by contemporary lights incapable of being the place of the exercise of a power—we today take a more organ- or suborgan-centered approach that could easily be regarded as more authentically Aristotelian. For Aristotle and 2000 years of philosophizing in his name, the soul was nothing more or less than the basic level of activity of a body that is divided into organs for the sake of living. This basic level of life bore the potential for all the more specialized and particularized organ–activities of the organized body of the organism. Today, we search for brain areas in which neurons metabolize more rapidly when a fly is buzzing across the perceptual field and then talk about having located the various elements of vision and hearing. With our fMRIs and PET scans and electron-flow scans and progress in miniaturization and less invasive detection we are beginning to isolate the activity of even single neurons and neural paths. Our science is calibrated to detect the workings of ever-smaller parts. So we have, first, a consciously describable experience, like vision; we have, second, gross brain scans that show relatively large but nevertheless particular areas more[7] activated than others; and we have, third, an increasingly detailed microscopic specification of the tiniest parts that are involved and what is going on in them.

I do not at all wish to be dismissive of the knowledge that this work represents. But every time experimentalists take previous work and try to analyze it more particularly, and every time theorists try to bring into a common field of explanation disparate results that lie before them, they are opening up new gaps at the same time they fill in old ones. A new theory, for example, typically unifies in a certain respect a range of phenomena that were considered disparate. The immediate impression it leaves is that "we" have "mastered" something. But then begins the hard work of filling in details, of taking other factors into consideration and asking whether their relation to the new theory is evident. Moreover, the new theory provides terms and concepts that need to be brought into explicit relation with neighboring fields, first near, then more remote. A God's-eye view of things, the realization of a total explanation of the phenomenon, therefore retreats as we advance in knowledge.[8] Here I will no more than mention that this phenomenon is explainable, at least in part, as a movement in and between imaginative planes and the establishment of new ones. Thus if we begin to look at "faculty" theories with a more historically alert and generous eye, we can see that what the ancients and the medievals were doing was not a world apart from what we do. They, too, tried to establish places and planes and to elucidate relations between them.

My definition does not call imagination a faculty or even a power or a potential, but an activity. Aristotle of course called imagination a power, and he even established a conceptual scheme for placing powers with respect to objects and activities of organs. Soul powers are expressed in acts, and those acts are directed toward appropriate objects: a visible object, say a blue sphere, is only potentially blue insofar as there is no light activating the transparent medium between it and an eye prepared (having the power) to perceive color. His most detailed explanation of this object–act–power schema concerned the nutritive power of animals and plants. The nutritive power of the animal or plant gives rise to and is expressed in acts of taking in nourishment, of eating; and those acts are directed to things that are, for that

[7]The "more" and the "less" here is crucial. Even relatively sophisticated accounts of neural functions tend to ignore the fact that individual neurons are not inertly awaiting outside activation but rather constantly "firing." Both the inhibition and the speeding up of the rate of firing are significant. Moreover, slower metabolism in a brain area can indicate a suppression of activity that is as necessary for the completion of a complex neural process as is the increase of activity in other areas.

[8]This is not always apparent at the moment of invention, when the inventor envisions a field as totally unified by a principle that opens up the prospect of a total reduction of all similar phenomena to the principle. But working out the details of the reduction always takes time and opens up unexpected complications. Quantum theory explains the hydrogen atom in exquisite detail, but a comparable understanding of the helium atom (next in the periodic table of elements) still eludes us—not to begin to mention the more than 100 other natural and artificial elements. Thus the claim that quantum theory is the best-confirmed theory ever has to be accepted with appropriate kinds (and fields) of qualification.

particular plant or animal, appropriate food. The sequence of investigation for a researcher, he argued, moves from the object to the act to the power.[9] Most conspicuous to the researcher are the things in the environment, like squirrels and nuts. When the squirrel picks up a nut and nibbles at it but doesn't do the same thing to a similar-looking piece of bark, we can draw a corresponding distinction between what is and is not food for the squirrel. The fact that the organism can repeatedly and appropriately move from a merely potential state (nuts lying on the ground all around it) to an active state (ingesting them) justifies our talking of the corresponding power. So Aristotle started his particular investigation of soul powers by asking about nutrition, thus about what food is.

I argued earlier that this schema, applied to imagination, was one of the fundamental sources of the tradition of (mis)understanding imagination. It put the image at the beginning. Insofar as the prototype of image was taken to be the fixed and easily reproducible visual image—that is, an image taken according to cognitive standards—the image was misconceived, and thus the acts of imagining and the power of imagination were misconceived as well. If Aristotle can sometimes be convicted by his own words, it is because at those places the words go plain contrary to his formal definition of imagination: it is a *motion* (*kinēsis*) originating in sensation—a motion that, in the proper places in the body of the sensitive animal, gives rise to appearances like those that originally showed in sensation. Aristotle does not clearly express the addition I have given after the dash; it is left implicit. But, as we saw in Chap. 5, it is a direct consequence of his physics and his understanding of the originating source of imagination, *sensation*, as involving the kind of motion he called *alloiōsis*, the qualitative change that is a repositioning of the specific appearance of the quality with respect to contraries (*enantia*) between which all the qualities of that kind take up a place (e.g., all the shades of color positioned between white and black, all the tactile qualities between sharp and blunt, etc.). When that motion is only potential the quality is of course unperceived; when it is actual, when the finger comes down onto the point of a needle or a seam on a baseball and touches it, the touch gives rise to a change that is simultaneously and correspondingly a change in quality–appearance. Imagining the sensation, then, means that in appropriately situated body organs there is the same kind of activity attended by very much the same kind of experience—though one not coming now from real-world objects like needles on the carpet or raised threads sewn into leather. And that makes a world of difference in how we need to conceive both the object and the activity—and, by extension, also the power, which the definition describes as "evocative" and "abstractional-concretional."

When I am trying to remember as exactly as possible the blue stain on my dining room's north wall the object is clear: it is the blue of that stain, not the stain itself. I cannot see either the stain or the blue at the moment when I am standing in the

[9] I would partially correct Aristotle here: what we often first see are acts of nutrition, like a squirrel nibbling on an acorn. We then isolate the object as a proper form of nourishment for the animal. This prepares us to analyze the activity of taking in nutrition, and then talking comprehensively of the squirrel's power of nutrition.

paint department of a hardware store. I may, of course, have taken a photo of the stain with my smartphone, and even used software to improve the color match of the photo to the original—but unless I have corrected perfectly, I do not at this moment see a perfect simulacrum of the color. Not being able to present myself with the original directly, I must evoke it instead—with or without the help of a photo. I must call it forth, in some meaningful sense I must produce it. I will be producing it as released or detached from the original (even when I use a photo and take the color there as exactly what is to be matched). What I produce, and what even the photo produces, is an abstracted color. Thus in the definition I refer to the "abstractional."[10] But as a produced appearance, as an emergent imaginative phenomenon—whether vivid or muted, distinct or blurry, private or public—it has a concreteness that is at least reminiscent of original sensation, so I call the evocation "concretional." Being exactly the kind of thing it is, the imagined thing is simultaneously both abstractional and concretional. The further advantage of these terms ending in *–tional* is that they imply an activity or process (of abstraction and concretion, respectively) that is involved in the production of this imagined color. In trying to find the right paint for my dining room wall, I look at the photos I've taken and the store's paint samples and try to picture more exactly the hue of the wall, its lightness or darkness, its gloss or its matteness. I vary my imagining in the direction of the more or the less (a brightening or an attentuation of the hue, a shift a little toward green or toward indigo) and say to myself, "No, it's not quite the way it is in the photo, especially as the photo appears in the garish light of this store; it's actually closer to row 7, column D, of the semigloss chart, but even that's not quite it, because there's a hint of aquamarine that I don't see in any of them; and the texture is not right."

If Aristotle was right to say that there is no thinking without images, then it is likely that pure acts of imagining (whatever they might be) would be rare, because imagining would typically be found along with or as part of other and more complex activities of mind. Untangling such complexities would then be one of the next orders of business in imagination studies. The paint-matching example in the previous paragraph shows something about the evocative, abstractional-concretional character of imagining "in the wild," in one typical (but not exclusive or prototypical) real-world situation. The example does not determine whether imagination is a function or a faculty or a module. It does not decide whether remembering is essential to imagining or reducible to a special kind of imagining. It does not settle the degree of involvement of reason with imagining, nor whether privacy and introspection are imprescindible characteristics. But it does allow us to gain some further (if only approximative) precision with respect to the abstraction, the concretion, and the evocation that imagination is and that imagination brings to mental events. The power of abstraction, it is often said, is distinctively human, a characteristic that allows some part of us—let us say "mind"—to be elsewhere than our immediate surroundings. Abstraction has long, and traditionally, been regarded as intellectual, a rational endowment. If imagination is not

[10] I use this word (unusual but attested already in the 1860s) to distinguish it from both colloquial and technical uses of "abstract." It also parallels "concretional" (first attested in the 1840s).

derivatively but fundamentally abstractive, however, it would appear that abstraction cannot be simply identified with intellection or rationality. So we need to be ready to start our psychologizing from the beginning.

If imagination often withdraws attention from our immediate surroundings,[11] this translocation or translocative power is already anticipated in animal vision and hearing, which in contrast to taste, aroma, and touch (to name three classic channels of sensation that require bodily contact or intimacy) extend the scope of the here and the now. For an insect with vision, a threat is not at the very next step (or wiggle, as it would be for a worm) but at several paces, say behind a rock up ahead. An owl scanning the woodland floor is already *with* the mouse at a hundred meters. This expansion of the field of spatial and temporal presence is not abstraction proper. Except in fables and in the dreams of pure associationism, there is no reason to think that the owl is consciously and directively comparing this situation to others like and unlike it. Even the first stage of such a limited ability of comparison presupposes considerably more-developed capacities, however. For an owl to expressly compare a mouse in this part of the woods on this night with a rabbit yesterday in a clearing—unless it is a random, flash association—would require that the owl already possess an at least minimal sense of the hunting situation: prey in environment. Any imagining (at least in an Aristotelian sense of an image's requiring intermediacy between extreme possibilities) that might take place would have to operate according to a few basic prey– and environment–possibilities. Thinking out the possibilities phylogenetically might lead us to conclude that this type of environmentally located hunting–comparison presupposes the evolution of an appropriate field awareness without deliberation or even an express consciousness of the field and its schematization by alternative possibilities. Deliberation requires being aware, in addition, of alternatives as such with respect to the same level(s) of the relevant field and probably also an at least minimal awareness of "self." To conceive the mental possession of alternatives located in ranges of perceptual-field possibilities would require taking further steps "upward" in a hierarchy of phylogenetic capacity.[12]

[11]I am not implying that imagination must always withdraw from immediacy. First of all, it is by withdrawing from one immediacy that it plunges us into another (for example, thinking intensively about what colors go well together in a room 5 miles away). And in the long run imagining vigorously can allow us to inhabit with greater awareness the immediate situations we find ourselves in. The active imagining of the basketball point guard immerses him more fully into the immediacies of the game he is playing.

[12]One point that needs to be made clear is that associationism is constantly faced with the problem that Saussure so strongly insisted upon with language: that one cannot have a first thing of the kind (word) without implicitly postulating that both speaker and listener already have the whole (language). Being able to recognize two mental events as two instances of the same thing already presupposes possessing some field or ground for the comparison. It makes no sense to talk about the possibilities for individual animals without first conceiving the possibility for the species. An individual organism with light sensitivity cannot suddenly progress from mere photosensitivity to focused images of discrete objects in its environment; but some individual many generations later may have progressed from light-dark discrimination to a basic kind of object discrimination if many more light-sensitive cells have in the meantime developed into an appropriately responsive organ in typical members of the species.

The simplest abstractional competency native to imagination is producing and having image–appearances without the corresponding originals. This kind of minimal abstraction is also immediately a concretion: it is an emergent appearance, however weak, that is evoked or at least evocable in the organism's awareness. This way of putting things unfortunately falls in the direction of confusing imagination with memory. One can imagine something never before experienced, an image without an experienced original other than the imagining itself (Hume's blue, for example). The point of examining Aristotle's definition, in Chap. 5, in light of his entire soul doctrine and his physics was to show that this "inclination" toward confusion was not inevitable. The definition Aristotle gives mentions only the continuation of the original motion of sensation beyond the sensation. Without putting it in the context of the more general theory of sensation as physical motion, it looks like each occurrence of sense perception is a discrete event that leads to a correspondingly discrete division of the appearance from the perceptive act. Understood in the larger context, however, that event is just one possibility of the sense organ's total activity; and any activation of the organ is already an activated potentiation of the organism's total capacity for color appearances (plural)—the potentiation of the organism's color field. This is a point of major division between Aristotle's empiricism and modern versions: in modern versions the image simply pops into consciousness, fully formed and detached from all other images and circumstances, whereas for Aristotle it is always the result of an activity in a sensory field corresponding to ongoing organic activity.

Being able to perceive a single color presupposes a capacity to perceive others. A physico-physiological explanation of why this is so would be quite different today than it was for Aristotle, but it would still appeal to the same basic scheme: discriminable possibilities between extremes (say a neuron firing faster or much faster, slower or much slower); or as a computation based on a network of neural outputs that issue in not just the binary possibilities of "see blue" or "not see blue" but in one of the seemingly limitless possibilities of hue and chromatic relationships. It has not been my aim, however, to discuss how field theories of imagination correlate to our best cognitive science and neuroscience. That there might be ways of correlating it to contemporary facts and theories works in its favor, and that the principle involved can be conceived as a continuation of basic Aristotelian insight reinforces the contemporary worth of the kind of historical-philosophical archeology I have undertaken in this book.

The analysis of the occluded-occulted tradition of imagination provides resources to forestall confusion between imagination and memory. The *detachment* from immediate sensation of an appearance–form that is originally concomitant with the activity of sensation is the very beginning of imagination. Without the possibility of such detachment, memory is not even conceivable. Indeed, this detachment–with–the–possibility–of–reevocation is the foundation of an organism's sense of temporality. It establishes a slight but pregnant division between the actual and what follows. There is a short but not insignificant step from that very slight division to the possibility of the organism's taking up the "viewpoints" of past, present, and future. The implicit translocation of attention is one of the most fundamental gestures of imagination we have discovered in our historical investigation: the ability to shift fields and then to see one field in light of the other. With the division of a real act from its *possible* reevocation, there is implied the organism's possession of past, present, and future. The motion from the actual to the reevocable is the anticipation of the future; the view of the actual from the perspective of the reevoked is the recognition of the past. The actual present ends up having a certain mobility and extendability as a result.[13]

My definition of imagination does not expressly mention time or temporality. This might be a defect, although in the first instance one might see it implied by the words "follows upon" in the first sentence and as implicitly treated as a variety of "field" or "positioning" mentioned in the three subclauses of the second sentence. One might also take this uncertainty about whether and where time should be placed in the definition as a task to be addressed by a new phase of inquiry into imagination. If, as I have repeatedly emphasized, a definition always occurs as a mark of transition, the imaginative status of time may be seen as something that is less a part of the historical archeology of philosophy (and philosophical approaches to psychology) than as a contemporary issue that has been occasionally but imperfectly addressed.[14]

(2) The last several pages have in effect begun the commentary on clause 1 of the second sentence of my definition. They have discussed fundamental ways in which imagination is about positioning individual appearances of sensation with respect to a field and positioning the subjective consciousness with respect to near and far (space, to put it in perhaps too abstract and unarticulated a form) and present–past–future (time, with a similar caveat). The further qualifications given in the second sentence's three clauses are more obviously and more directly related to the historical archeology of Chaps. 4, 5, 6, 7, and 8, with an emphasis on the fields within and between which imagination does its principal work.

[13] This brief account of emergent temporality is indebted to Gustave Guillaume's quasi-Saussurean explorations of linguistic tenses, in Guillaume 1965 [1929/1945] (for a very brief introduction, see Agamben 2005, 65–67). Kant was the first to argue that time was a function of imagination proper (albeit in its transcendental use). Unlike Guillaume, who begins with the gap, however slight, that opens up between things and events on the one hand and beginning to speak about them on the other, Kant simply postulates the temporal character of the inner sense; then, later and quite separately in the First Critique, he builds temporality into the spatial manifold (of external sense) by using schematism to enact there the pure concepts of the understanding by means of the pure principles of the understanding. Kant did not bridge the gap between these two temporalities, the first a time–differentiation, the second the express articulation of time flow. As for the mobility and extendability of the actual, one might recall Augustine's analysis of time (in Book 11 of *Confessions*) into three presents: the past present, the present present, and the future present. One could amplify this by extrapolating from his Book 10 discussion of reciting a line of verse. The meanings of words already-recited, currently-being-recited, and about-to-be-recited are differentially copresent at every moment of the utterance. There is no limit in principle to this extendability of the temporal sense. In particular, writing an autobiography like the *Confessions* suggests that in the present we can have a living sensibility for our distant past and even some part of our future.

[14] To mention only the first third of the twentieth century, one finds quite diverse approaches to the question in Bergson, in Heidegger around the period of publication of his first Kant book (Heidegger 1929), and Husserl throughout his career. But none of these was entirely satisfactory even to their respective authors. See also the immediately preceding note.

The first subclause of the second sentence says the imaginer (re)positions herself and then works upon the appearances within a field of concern. The second subclause emphasizes the imaginer's focus on a crucial aspect of the field, its potentials. The third subclause starts with these field potentials and envisions them as a source for projection—in particular for the abstractional-concretional projection of the potentials onto other fields, as appearance–possibilities appropriate to those fields. A virtue of the whole sentence is that it divides the activity of imagining into distinct phases, which is useful for analytic purposes. By the same token, this is a vice, for one might easily want to argue that all three of the aspects highlighted by the subclauses are, in human beings at least, interdependent and copresent.

By this stage of the inquiry the notion that imagination works primarily, even exclusively, in imaginative fields needs no specific defense, nor does the idea that imaginative consciousness by its nature is always ready to move between fields as it works them, although how long it can remain in a field (say the strictly geometric treatment of plane figures that land surveyors undertake in phases of their work) without moving to others is unspecified. The idea of field potentials, however, requires a little more explication. In a sense it combines a notion of differential geometry with Walter Benjamin's insight into the *Entstaltung*, de–formation, of the perceptual appearance as the moment in which imagination properly commences (see Sect. 3.3, above). Consider a mathematical example: With most curves there is a function that indicates the directional tendency of the curve at every point; it is called the derivative or differential of the function that describes the curve. This means that if one knows little about the entire curve but has some knowledge of what happens as one moves in the near vicinity of a point on it, one can begin to explore, (re)constitute, and understand the more distant parts of the curve and the space it traverses. To analogize: By imagining something, one has activated not just an isolated image but also affine images in a surrounding field. The imagined thing is less a fully determined, isolated entity than it is a labile or mobile appearance in a more or less definite locale. (Among other things, this is a field-appropriate way of expressing the typical instability of images as they first emerge.)[15] Spontaneous movements around this "point" of the field surface may be irregular at first, but they are not simply random insofar as they have this point of reference.

This mathematical analogy also describes the situation of someone who has an experience of an imaginable quality or character but has not yet become acquainted with its relevant fields. Such a starting point would be like the brute factuality of a Lockean idea before it has been compared or contrasted with others by conscious or subconscious understanding. Any random comparison would be unlikely to lead to recognition of a *coherent* field to which the idea might belong—but you never

[15]This would include the "flash appearances" that Alain described in trying to imagine the façade of the Panthéon. See Sects. 2.3 and 2.4, above.

know! Contrasts, per se, do not yield coherency, only exclusion of the new idea from being very similar to existing ideas or already existing sets of ideas that have been gathered coherently. Yet as soon as one comes across a comparison that has a natural-seeming similarity, one begins to constitute a sense of similar experiences "close" to the original one.

Empiricists prefer to talk of association and assemblage, but, as I have argued earlier, association is too weak a notion to articulate, for example, the similarity and coherency of colors in the color field. What the last paragraph describes in a Lockean way would be the beginning of the exploration of a *field*, though in first approximation it does little harm to think in terms of modern empiricist idea–units that are assembled into a set of assimilated/associated experiences under some common term. Benjamin's conception of a preceding deformation captures the dynamism of perceptual attention without reducing it to the mechanical processing and classification of idea–units. *As soon as one begins to attend to the appearance as such*, the release and detachment of the appearance–form has commenced—and that is the commencement of imagining.

Conceived according to Kant, it is true, the imagination in its transcendental functioning has already taken place before this de–formation; it has organized and schematized the manifold of sensibility in the appearance of the here-and-there, the now-and-then, the this-and-that of ordinary sense experience. The blind spot where the optic nerve enters the retina has been filled in with a look that fits its surrounding, the complex composition of the trillions of photons impinging on the rods and cones has been averaged, the boundaries of objects have been determined and adapted to the three-dimensionality of the field of visual perception, and the synthesis of the channels of the various senses into a single spatial-temporal world has been achieved. All of that is presupposed as already accomplished, before we take a look around and "simply" see what is there. Given the fact of memory, there is already in the background of our present awareness the prior detachment–deformation of what has previously appeared. But the emergent appearance of the imagining that is under our control begins with our noticing, with whatever minimal awareness, that things look a certain way. That moment already contains within itself the implicit "a certain way *and not others*" that is the properly Benjaminian (and Castoriadian) moment of imagination's start. It is the incipience of the *field character* of the appearance, of its variations and field potentials, of its abstractions and concretions and projections into less and more complex settings.

Whether an imaginative field is continuous-analog or discrete-unitized depends, on the one hand, on the imaginative character in question and is, on the other, one of the basic topics that future imagination studies will need to pursue. It is possible, for example, that all imaginative fields are ultimately quantized, so to speak. Although many of them appear to be continuous, that might just be an artifact of the limits of our perceptive and imaginative powers. That is, the discrete "points" of different appearance might be so densely packed that we do not notice the gaps between them. Alternatively, some fields may be quantized while others are continuous, or (at least as a mere possibility) all might be continuous though some might appear to be discrete. There is even the possibility of some kind of duality analogous to wave-particle duality. To give an account of the paths light takes we have to use probabilities describing the continuous progress of waves, but every interaction of light with matter (like striking a screen or being absorbed by an electron) is discrete. According to the analogy, every actually perceived character or imagined possibility of the imaginative field would be discrete, but the field of possibilities might be traversable as continuous.[16]

In any case, what the second and the third subclauses of the second sentence of my definition highlight is the kind of work that the imaginer does in fields and between them. Once one has experienced some characteristic of a phenomenon—it does not have to be sensory in any usual sense, it could be, for example, symbolic or signitive—it comes to be experienced in relation to other characters nearer and more remote. This dynamic experience or "viewing" implicitly establishes an underlying field that, once the imaginer has enough experience with it, begins to take on an independence from the originating experience precisely as field. This release from the conditions of one imaginative field into the circumstances of a new one is itself a case of imaginative *Entstaltung*. When Descartes realized that the signs with which he marked points, lines, line lengths, areas, etc., could be articulated in formulas expressing relational connections by means of arithmetic signs, and that such articulations have near and distant consequences because of the possibilities of calculation, it became possible to explore algebra as algebra. Of course it is also historically true that Descartes always considered this algebra to correlate closely with and express *geometrical* relationships, so that for him algebra did not have the kind of autonomy that we take for granted (and that has been taken for granted by most mathematicians since Descartes).[17] But that autonomy—which is always only relative—is precisely what imaginative fields have if they are genuinely coherent and, in that sense, humanly accessible. Even the purest of pure mathematicians sometimes take a moment or two to glance outside of the mathematical structure that is their immediate concern to what the structure can stand for, to fields the structure is affine to, to other problems and problem fields upon which it might be projected. It is by working the field of the structure, by understanding it as articulated according to field potentials that guide us from one discrete possibility of the field to another, that we perfect our experience of the field as such and come to be at home in it. This is what the mathematician does as he explores possible consequences, conceives new problems, and proves new theorems. The more comprehensively and densely we articulate the structured field, the more likely it is that we will find multiple ways in which it can be projected into other theoretical worlds or, most concretely, into the lifeworld. But remember: even the most abstract of fields has some distinctive appearance to consciousness. The field of all such related

[16]I am *not* arguing or even suggesting that a physico-physiological explanation of imagination should be quantized or that there is a basic image–field duality *because* of some quantum dependency. But notice that I am imagining one field (that of imagining) in terms of another (quantum duality of photon travel and interaction)—which, I think, is what always happens when one tries to explain scientifically. And it is more fundamental than hypothesis formation or model building.

[17]Timothy Lenoir has given a compact sketch of what this means; see Lenoir 1979.

appearances has an underlying basis that is matter with respect to the form of the appearance. This is the *abstract* or *formal materiality* of the appearances. Insofar as something has such abstract materiality, however abstract it might be, it is an image and imaginable.[18]

The definition of imagination I have given fails to do one important thing. To use an old-fashioned language: it fails to unify the genus and the specific difference(s) of imagination into a simple, one-sentence formula. The first sentence of my definition in effect presents the complex genus of imagination; the second sentence provides many differentiators. Imagination, says sentence 1, is a (psychologically) evocative, abstractional-concretional activity that follows upon actual perception. The three subclauses of the second sentence then elaborate imagination's work of placing appearances as field positioning, as a marking of fields and their potentials, and as a projection of the potentials from one field of appearances to another. As I noted at the beginning of this book, imagination has almost always been understood as being in a middle position, as a medial power between sensation and intellect (the rational power, *ratio*). But I also noted that the Latin *ratio* (which was used to translate the Greek *logos*), in its most basic sense, indicates that one thing is set into proportional relation with another. *Ratio* can do this proportion setting directly when the two things have community in a (common) field, or more indirectly (the Greek's *analogia*, the Latin-speaker's *proportio*) when a relation between two items in different "fields" or "planes" can be compared to the relation between a third item existing in the first field and a fourth item in the second field. But at the root of this comparative power, and thus more fundamental, is our ability to put one length in relation with another length (3 inches to 4 inches), or one length with an area (3 inches to 12 square inches), or one thing in relation to another (a horse is a mammal), or one thing in relation to an attribute (the horse is piebald). By expressing both predication and basic proportionalization with the same term, *logos*, the Greek language implicitly held to a theory of a commonality between the two, a commonality that was expressly articulated in the epistemologies and ontologies of their outstanding thinkers. If we do not agree with this theory, it is less because we have thought out its defects or limits than that we have neglected it, let it drop, overlooked or forgotten it. This is another, perhaps deeper case of occultation–occlusion–eclipse in the history of Western thought, not to say a type of nihilism that deeply shapes ordinary philosophical and even scientific life. As the historical part of this investigation has shown, however, this commonality was not neglected by the greatest thinkers of imagination even after Athens' philosophical glory was millennia past. If philosophy were nothing more than a game of symbol manipulation according to rule, then foregoing the past would be no great loss. But if past thinkers understood things that, to us, are terra incognita, then forgetting is our own misfortune and our own shame.

Perhaps, as a memory aid, there is need for a simpler definition that places imagination with respect to rationality. Given the likelihood that there is a very basic kind of imagination that animals other than humans possess, however, any alternative definition will still require two parts: viz., imagination is the emergence of the (re)apparent as such; and human imagination is reason insofar as it works with and produces images.

---

[18]This is simply a consequence of the nature of imagination as abstractional as well as concretional.

Unfortunately, despite its greater brevity, this new definition is just as complicated as the first. A careful comparison would uncover strong parallels, beyond the two-part structure, showing that the new one conceals in its abstracted brevity the more detailed specifications of the other. I will not undertake such a comparison here or explicate at length this new definition but will make three remarks. First, the new definition suppresses any mention of imaginative fields and leaves tacit the *where* of imaginative work and image production. That is less a flaw than an expression of the level on which this definition focuses (the human-animal distinction). Second, it identifies human imagination specifically as a function of reason. At first glance this looks like yet another attempt in the long history of philosophy to establish the hegemony of reason over imagination. Hegemony would not necessarily be implied, however, even if reason turned out to have "higher" functions than working with images. Only if the more encompassing psychological or anthropological theory held that reason works by domination would hegemony be a consequence—and "domination" is not a synonym of "mastery" when the latter term is used in the sense of, say, a mastery like that of an artist with respect to her materials. Such mastery does not so much impose a master will as demonstrate skill in helping all the elements that enter into the result achieve their utmost possibilities. The formulation also leaves open the quasi-Aristotelian possibility that reason always works with phantasms, or even that reason is nothing other than (that is, it is by definition) the working with images and placing them in relationship to one another.

The third remark about the new definition must be a bit longer than the first two. The phrase "emergence of the (re)apparent as such" marks several nuances. It distinguishes imagination from original appearance. So, for example, the original sensory appearance of a greening spring woodland is not an imaginative phenomenon: it is perceived by sense. It would belong to imagination, however, if we defined it as the "emergence of the apparent," because then every appearance of every kind (for example, the appearance of perceived things, of memories, and of concepts) would be included. The "as such" does the delimiting work here: if I create the slightest separation between the phenomenon of the greening woodland and a follow-up moment that incipiently takes in the situation, form, or placement of the appearance *as* appearance of some kind—of appearance *as such*—I will have entered the realm of imagining.[19] This separation is the reason for not insisting (by use of the parentheses around "re") on there actually being a *re*appearance: imagining is not always and only reproductive. If I look at the greening wood and think how I would photograph or paint it, I am not necessarily making the appearance reappear; the original appearance still persists in some fashion, but I have shifted the plane in or from which I am considering it. Another reason not to insist on reappearance is illustrated by Hume's example of the never-before-experienced shade of blue that nevertheless has a well-defined position in an orderly sequence of blues.

[19]This point combines and develops the insights of Aristotle, Guillaume, and Benjamin with respect to the very beginning of the phenomenon of imagining.

The appearance of the new blue is, strictly speaking, imaginative: it is neither sensory nor memorative, and it is not conceptual in the sense of abstracted from hue.[20]

Expressing the new, second definition in terms of the apparent or reapparent is also broader than may appear at first glance. It includes, for example, concepts and formal structures. Anything that can enter into awareness or consciousness becomes apparent, and thus is subject to the definition. (The reader who happens upon this section without reading the rest of the book may fear that this is an attempt to give imagination hegemony over reason simply by fiat.) One might well want to restrict the definition to sensory appearance. Horror at the possibility that reason and logic might be tainted by imagination is not an acceptable motive for the restriction, however. As the preceding chapters of this book have argued, there is good reason for thinking that the occluded-occulted tradition of imagination has always allowed for the possibility that the fact of appearance itself, appearance of any kind, has a concreteness that implies a formable and deformable matter, even if it is "logical" matter. Thus imagination is inadequately treated if it is restricted to the purely sensory.

Whether the implications of this definition hold or not cannot be decided by vote of members of a philosophical or psychological association or even by chairholders in the relevant departments of research universities. Nor is it to be determined simply by what currently appears or not in professional journals, or even in books like this one. Such things would be matters of the sociology of professional disciplines, and some even of academic fashion. Truth or even plausibility is never finally decided by opinion.

Adorno's epigraph to this chapter suggests why. What distinguishes philosophy, in particular, from other kinds of inquiry is a sense of responsibility for everything. Although philosophical argument and philosophical progress (if there is such a thing) require particularization and even specialization, in the first and the last analysis they need to be undertaken with what a Kantian might call a regulative sense of totality: of needing, ultimately, to account for everything in principle. If an investigation avoids this, it is a special science rather than philosophy. If it is thoroughgoingly specialized, it is no more philosophical than any other kind of specialization; nor do specialists become philosophers simply by cultivating interdisciplinarity—though it is probably better than nothing.

---

[20] As usual, drawing a clear line between memory and imagination obtrudes, or rather requires some additional analysis and distinctions. One might wonder whether recalling the discomfort one felt from last year's sunburn qualifies as the emergence of the (re)apparent as such—the problem being with the "as such," since the discomfort's coming back to one's attention is an emergence of the reapparent and thus also of the (re)apparent. This could be resolved by dividing cases: imagination is either the reemergence of the apparent, or the emergence of the apparent (which therefore contains the possibility of its later reappearance) as such. The first case is Aristotelian and includes memory, whether human or animal more generally; the second is solely about *human* imagining, since (plausibly) only human beings can take the appearance *as such*, which I note as implying that the appearance is being taken, at least incipiently, within the emergent framework of a plane or field. These are complications for the future to develop and resolve.

9.2 The Ethos of Imagining Found; or, Topological Topics of Placing Imagination

There is still one task to perform, one more valuable than scrutinizing a definition: to describe the topics and topologies of imagination. Presumably we have learned something about these from our investigations. That is, we have learned something about the way in which the phenomena of imagination hold together in fields. Given the claim that the hitherto occluded-occulted tradition understands imagination as a field phenomenon, as the rational work on images in and between fields, it seems plausible that we might be able to understand imagining itself as a kind of complex field, or perhaps a complex of fields.

At the outset of this book I argued that the relatively recent tradition of antipsychologism in professional philosophy and psychology has obstructed our ability to come to terms with imagination (in particular) and the human psyche (in general). My response was not simply to reject the conceptual and methodological concerns that underlie antipsychologism. One cannot simply reassert the supposed rights of privacy, intuition, and introspection; for the most accurate research in psychology one has to take very seriously the need for the verifiability and reproducibility of phenomena. One cannot reimpose folk psychology, a kind of *consensus gentium*, that is known to be deficient. Still, as my historical elaboration of the occluded-and-occulted tradition has attempted to show, we can have greater confidence in an ontologically and epistemologically based psychology that is a kind of *consensus sapientium*, especially where it evinces aspects of imagination and mind we have neglected.

The imaginary dialogue between a singer and a philosopher in Sect. 2.6 did argue for the rights of privacy in certain kinds of imagining, but it also implicitly suggested a way to avoid the Scylla of psychologism and the Charybdis of antipsychologism. It simply makes no sense—it is in fact nihilistic—to deny that a singer rehearsing at home for a recording session can imagine an instrumental accompaniment to her singing, though it might sound to a roommate like *a cappella* singing. There was of course a kind of "proof" to satisfy behaviorists: in the recording session the singer calls a stop and explains to the players how their performance did not match the expectation of what she had imagined. Where such "proofs" are available we need to take advantage of them. But the most recalcitrant antipsychologists will not want to take the further step of acknowledging *in general* that there is private experience in imagining that is not immediately accessible but nevertheless is the substance of the imagining.

I think that such recalcitrance is indefensible; it is based more on recent tradition and the contemporary sociology of academic disciplines than on sober analysis. Nevertheless, what Chap. 3 offered in an attempt at the reconciliation of positions was the notion of conceptual topology: places where the possibilities of concretely experiencing the world and its various aspects subsist in a fusion of the experiential, the qualitative, the quantitative, and the conceptual. For example: the articulated field of all possible colors, the articulated field of all possible tones, and the field of algebraic operations and the corresponding field of geometrical formations and transformations. Color and tones have an immediate sensory basis, whereas the mathematical fields are abstract and conceptual. Yet the former kind have not just qualitative but also quantitative character (at a minimum, the more and the less saturated, the brighter and the duller, the sharper and the flatter) and a degree of conceptualization (color contrast, major and minor keys), whereas the latter have typifying appearances that we count on, for instance in teaching them to novices (but even more importantly in originating and inventing them). Taking the field of sensation as a manifold of colors is in effect a cross-sectioning of the experienced world; so is taking the shapes and figures of the things of that world in a geometrical plane. Moreover, as we argued in Chap. 6, it was Descartes more than any previous historical figure who opened up the possibilities of "cross-sectioning" the field of geometry in order to produce the corresponding algebraic formulas in analytic geometry. Ever since, this kind of sectioning of the world, and imagined worlds, has proceeded apace.

Just as phenomenologists will point out that the distinction of objectivity from subjectivity must follow upon a prior experience of the lifeworld, it is possible to argue that the sciences of nature (e.g., physics, chemistry, and biology) are complex imaginative cross-sectionings of the original phenomenon called "world." This is not to deny the reality of chemistry, but rather to contest one's right to absolutize its status as existent in itself, apart from the original cross-sectioning that establishes its foundational conceptual topology. The sciences would thus be abstract concretions and concrete abstractions of founding imaginative acts; absolutizing their status would be prescission, with all the attendant problems that we explained in Sects. 5.13 and 8.6. The conceptual topology is the fusion of the qualitative, the quantitative, the experiential, and the conceptualized that underlies the science.

If this "fusion" goes against our analytic methodological instincts, that is perhaps the fault, and an index, of the very success of our sciences. Making this claim is not an "indictment" of the sciences from a "humanist perspective." Any economist who resents an attempt to reduce the reality he knows to genes or reproductive viability, any biologist who faces a chemist's claim that life is a mere consequence of valence bonds, any chemist who is told that it is just a matter of time before chemistry is totally deducible from quantum physics understands that too ruthless an analysis does away with most of the phenomena specific to the field. The scientist trying to reduce another science to his own can cavalierly sweep them away all the more easily for having just a passing acquaintance with them.

If the world as we encounter it is constantly experienced in the cross-sectionings of our attention and intentions, then conceptual topologies are not "add-ons" to our world experience but its very medium—but of course "medium" then has to be taken as essentially plural. The human psyche would thus be conceptually topological. Biplanarity and all the other characteristics we have traced out in this study would be intrinsic not just to imagination but to human consciousness per se. I have in principle allowed for the radical potential of such possibilities in this book, though without always emphasizing it. One radical consequence, for example, might be that we should feel there is something profoundly unsatisfactory in treating (say) sensation, imagination, and reason as sharply differentiated. "Reason," "intellect," "concept," "imagination," "image," and the like might all be ways of cross-sectioning our experience of ourselves. Which of these we notice would depend on which cross-sectioning we had taken. It will not do to imagine, as some cognitive philosophers do, that we are on the verge of discovering a truly scientific vocabulary that will displace old "folk-psychological" categories. First, because folk psychology is more complex than it seems; second, because its various forms are all cross-sectionings of human being and action, no more and no less than cognitive science. If the old categories need replacement, the only way we can know is by thinking them through, thinking through the alternatives that are proposed, establishing a kind of concordance of usage between the two, and then seeing where that leads—and then searching for other cross sections.

Cognitive philosophers have not gotten that far, at least not yet. But in this penultimate section I will take for granted that *we* have successfully established that imagination typically works in and between fields. I will elaborate various **topics**—that is, *topoi* or places—of the field of imagination, and the **topologies**, which are the conformations of the field with respect to the topics or places—conformations that implicitly express the field's potentials. The potentials are like the forces that accelerate motion one way or another as one moves across the field (think of rising and descending slopes); the topology is like a topographical map that represents the positions and potentials in a significant (if not exhaustive) and understandable (if not fully understood) way. This discussion will be, I hope, organized if not systematic. It will be part summary, part elucidation of structure and relationship, part speculation beyond anything the occluded-occulted tradition has openly presented.

Topic 1. Imagination begins with the emergence of appearance as appearance; appearance as appearance is the beginning of field placement. The former clause is something that dawned already in Chap. 2, the latter in Chap. 3. Imagination is not to be gauged by a fixed and reproducible image, visual or not, but rather by the flash of appearance that imagination initiates. In that sense it would appear that remembering and even recalling and forming concepts, insofar as they involve an emergence of appearance as such, have an imaginative moment. This aspect of imagination makes moot the "refutations" of imagination that prove psychological images do not exist because they are evanescent. Evanescence constitutes the temporal essence of images. The fixation of images in more permanent form is itself a work of imagination, but not so fundamental as evocation.

"Appearance as appearance" indicates that what appears in the imagining shows itself as something that has already appeared, or might have appeared. The latter qualification, the "might have," is more revealing than the "has already," which is chiefly directed to imagination's reproductive capacity. When we try to reevoke "shades of blue" as Hume did, we have begun to treat the various blues we have seen precisely as the *same kind* of appearance. In logical terms we might say that we are treating the various *instances* as belonging to a species, or various *species representatives* as part of a genus. That does not mean we have entered a logical realm, however, but rather that this type of reevocation exhibits the abstractive side of imagination that the occluded-occulted tradition has known all along. Abstraction begins to uncover a state within which the species can be seen, the species of the instances, and this species state is itself concrete (though not as concrete as the instances). Taking the appearance as such and looking ahead to more intensive imaginative work to come is to begin to uncover the field within which the instances are related to one another as positions/possibilities of a more, or less, abstract field.

The "might have appeared" qualification is nevertheless more revealing of what is most characteristic of human imagining. It characterizes, for example, Hume's blue, never before experienced but sensorially evident from where it stands in a well-organized array of actually experienced hues. These possibilities are the most basic kind of imaginative innovation, the imprescindible starting point for creativity. These are new (even if only slightly new) positions in the field that are determined by potentials defined at positions we already know. There is no reason that a casually imagining person should be unable to come across a previously unexperienced position in the field—a color never before seen, a shape never before experienced—whether as a flash of awareness or in a fixed and easily reproducible image ready for clear cognition. But the field, to begin with, is only nascent. The first person to see color was the first to exercise the field of color perception, but he or she had not by that fact alone grasped the field as field. The field of analytic geometry has been expressly known for less than 400 years, but decades of mathematical work before that began to grasp at least a few of its features. In any case, the reliably fixed, reproducible image, whether it is a word, a curve, or a patch of blue, is far easier to achieve for someone who has consciously taken possession of the field of the phenomenon and begun to recognize field structures and their overall topology.

Topic 2. If the image is the proper object of imagining, it is as a labile, flexible, inchoate object, with its lability, flexibility, and possibility of emergence determined by field potentials in place, in context. This is for the most part a restatement of Topic 1, with focus on image rather than imagination. Yet the last four words add a complication: the image is *in place, in context.* What does this mean? First, it relativizes or relationalizes what the image–object is. The simplest examples we ordinarily use are items like a mental picture of a blue patch, or of a bluebird. The patch or the bird, imagined out of the blue, perhaps with fuzzy limits and obscure contours, is primed to be more specifically situated in a context: the blue patch placed next to or succeeded by a green one, the bluebird about to take flight in the backyard from the branch of tree. But then does not the object change, into the two patches taken together, or the whole backyard scene? This is less an objection than the beginning of an elucidation of imagination, both epistemologically and ontologically.

The epistemological aspect was clearly present in Plato and further developed by Descartes as the ground of mobile imaginative problem solving. Imagination is one of our freest powers: whatever it is currently entertaining it may add to, diminish, alter, efface. As we explained in introducing the field concept in Chap. 3, what is the focal object at one moment (a squirrel, a curve) can become simply part of the scene or scenario (animal activity in the backyard, the path followed by one of many moving objects). The scenario then becomes, at least for the time being, the new object, and one cannot (except in special circumstances) reductively claim that it is intrinsically composed of all the "elements" that appear in it. The object or the scenario can be analytically/synthetically schematized as a field potential (or perhaps a potentiator) and thus taken as alterable (with as many dimensions/degrees of freedom as there are recognized variables); or it can be schematized abstractively and/or projectively into a field that displays structural similarities or isomorphisms. All these moments are part of the imagining, and part of imagination's progress or sequence. And throughout, images, no matter how simple or complex, *are* precisely as they *appear*—even if any particular description or other portrayal of them seems inadequate. This "precise" being but limited portrayability is less a problem than it is the basic situation in human imagining. Those who see it as chiefly a problem with respect to knowledge tend to overlook that the localized finitude of imagination is a fundamental condition for the existence of human knowledge. I speak of human knowledge, not divine or angelic knowledge or the knowledge of some fantastic being resident in another world, whether imagined as real or as logical, much less in all possible worlds. Imagination is able to change its mind: it is the formative element in which mind–changing takes place.

The inclination of those who hold rationality to be distinct from animal powers is to exclude imagination from cognition. But imagination usually has its revenge, or at least its return, as in the case of Plotinus, who doubled imagination so that intellect could have the forms of lower imagination translated into higher imagination without the taint of materiality (see Sect. 6.1, above). In the seventeenth century, mathematics (in the form of *mathesis universalis*) was taken as the standard for determining everything (including images) with respect to quantity and magnitude, and Descartes unleashed its universal, regularizing power to solve every physical problem (whether practical or theoretical); by the beginning of the eighteenth century, mathematics had become an analytical instrument of formalization by rational principles that allowed scientists to reduce and even eliminate any reference to perception and the ordinary things of the world. But of course that was "accomplished" without anyone having realized that one needed more than tacit historical development to show that mathematized natural science had really left imagining behind. As I argued in Chap. 6, modern mathematics is the most rigorous use of field-shifting imagination ever conceived, with a radical deployment of its abstractive character and a simplified deployment of its concretional one. Imagination's secret return, if not its revenge, has been evidenced by philosophy and psychology's failed struggle ever since to locate forms without concreteness that might nonimaginatively account for human logical and rational abilities.

Topic 3. Imagination has a basic and a developed aspect, historically distinguished as "animal" and "human" imagination. Human beings are animals, of course, so they have both "animal" and "human" imagination, but in light of the flexibility of purposive activity that higher animals (especially mammals) can display it is better to call these "basic" and "developed." The former are strongly dependent upon the sensory and even motor[21] capacities of animals. Only hearing animals could

[21] I include motor capacities not just because of the existence of mirror neurons in higher animals but even more basically because of the imaginative character of certain routinized motor activities (culminating in dance and gesture—but also in speaking). It might also capture the kind of inventive activity exhibited by lower animals (like wasps) when their "purely instinctive" routines are blocked.

have aural imagination, a reemergence of the heard as hearing. But it is improbable that *all* hearing animals have the ability to imagine sounds. If dogs have dreams in which they hear the barking of other dogs, the howling of cats, and the voices of their masters, those dogs are imagining sounds as such. That would not, however, immediately translate into an abstractive power constituting an imaginative field in which dog consciousness could work as such—for example, by composing dog songs. It might be possible, by some theory of image combinatorics, to simulate an animal imagination capable of arriving at new images without appealing to *ratio*—by which I mean the express, conscious establishment of ratios and extended image proportions in one field or between different fields. This *rationalizing imagination* is reason's work with images and imaging—and one of the questions for the future is whether this is reason's highest, or even only, work.

Topic 4. Human imagination is the mobile working of and with images taken comprehensively. The dual directedness implied by "working *of* and *with* images" is deliberate and significant. "Working of images" is, by itself, already twofold, in fact an amphiboly with a point. In the first place it means (as it has throughout this book) that the imaginer works and elaborates images, just as an artist works and elaborates paintings. But it can also serve to attribute the working activity to the images themselves. It is possible that neurobiological and cognitive studies will in future provide insight into how this might function. That images are per se dynamic and not just autonomously generated by self, ego, or reason seems certain. They are not just flexible and labile, they often possess an internal principle of change or development. By internal principle I do not mean, of course, that images exist apart from the natural perceptive fields intrinsic to the psychophysiology of the human being. Any color patch is a realization of a possibility of the human visual field; any imagined future event is temporalized and spatialized by means of the actuation of our nervous system and brain. These activations of basic psychophysiological fields constitute the ground level of imaginative function, according to the basic articulations these fields establish. Ontologically speaking, then, an image is an actualization of the field. Thus it is virtually certain that any imagination-possessing animal beyond the human has some similar psychophysiological grounding fields that lend to the animal's images their own proper dynamism and development, which of course will often be dissimilar to the human (for example, insofar as the compound eye of the fly or the honeybee produces a light pattern with many different focal points rather than one).

"Working with images" can even suggest that images work with (other) images, but primarily it indicates the human ability (and any comparable ability of other imagination-possessing animal) to take a position outside an original image, or even outside the apparent plane of that image. Comparing image to image, or viewing an image according to some character that appears or is suggested in its original plane but is made conspicuous and effective in another—for example, the geometric shapes of things drawn from the plane of the lifeworld—is thus the "working with images" that I have in mind.

Topic 4 therefore does not immediately imply that there must be an independent (or quasi-independent) power—say, reason—that does this work. If there is such a power, it can indeed be something "outside" the image. But it can be imaginatively effective only insofar as it works within and between the planes of imagination. Fichte, whose work inspired the hyperbolic Romantic inflation of imagination to the level of world creation, argued that it was the outward drive to experience–and–cognition that produced a first multiplication and division of fields, that of the I and the not–I. This division was followed in the subsequent outward movements of the drive by further and deeper articulations of the not–I, which in the first instance amount to new (and therefore multiple) planes against which to project the I's experience. Dialectically, as the world of the I becomes more complex, so does the I. Moreover, and most important for the interpretation of Topic 4, the appearance that resulted from the success of the outward-moving drive was creatively produced by imagination out of the tensions in consciousness.

We do not have to be Idealists or Romantics to appreciate this theory. Ratcheting down its (to us) excessive epistemological and metaphysical commitments, we might adapt it to conceive the human organism and at least some other animal organisms as having the power of complicating and dividing appearances and of seeing or recognizing different levels of appearance in relationship to one another. In the human case, one presumes, there would be a more explicit taking possession of complicated-and-divided perception in a conscious way. To a greater or lesser degree, accordingly, certain images would provide a (relative) standard for others and thus set an intrinsic principle of some possible ordering, even measurement.

Topic 5. Images, when multiplied, have a tendency to stabilize one another by being situated; alternatively, imagination is self-stabilizing insofar as images bear intrinsic structure by virtue of fields and scenario placement. This is a development especially of the concluding part of the previous paragraph. The realization that there was a concretely abstract field of fundamental mathematics based on simple perceptive comparison—in its fully developed form called *mathesis universalis*—underlay the universality of Descartes's imaginative method of problem solving. One world–thing appeared longer, or brighter, or sharper than another. One could accordingly create a matrix, with each thing ranked with regard to comparable characteristics, from longest to shortest, brightest to dimmest, etc. One might, applying this ordering method, eventually find a unit that allowed one to refine this ordering into measurement: two units of length, six of brightness. One would thus fix the experienced things into a regularized field sequence (which is schematized and abstract in comparison to the level of ordinary experience, where things with their lengths and brightnesses and sharpnesses are lying about or being used) and would translate these sequenced things into a language that borrows geometric and arithmetic determinations from the fields of geometry and arithmetic. Of course geometry and arithmetic are themselves fields standing in a complex near-isomorphism of field structure, which led to Descartes's discovery of the new fused field called analytic geometry.

Both Descartes and Kant understood rulelike principles to be behind these field structures. Descartes called them "mathematical truths" and said they were created by God to govern the things of the world; the human mind in turn can recognize them, and when it imagines the physical world truly it observes and obeys them.

Kant regarded space to be a pure intuition that unified (synthesized) the manifold of sensibility according to imagination in its transcendental function, and this intuitable space could be further articulated in accordance with the pure concepts and principles of the understanding synthesized with the manifold by further acts of transcendental imagination. For Plato the structures were due to the abstractional and concretional projections of the forms, ontological projections on the one hand and epistemological on the other. For Aristotle the structures were due chiefly to the metaphysics of motion and the fact that geometry was naturally abstractable from formed matter in nature. As *explanations* all these principles have their advocates and their critics—and for a variety of reasons. Some of them seem plausible, others possible but problematic, yet others highly suggestive but uncertain. As far as the occluded-occulted tradition of imagination goes, however, what I am counting on as more impressive than one or another particular explanation is the commonality in how these thinkers grasp the features and processes of imagining, in a kind of *consensus sapientium*, the consensus of those most learned about imagination. If in the last analysis nothing in philosophy can be settled simply by vote, perhaps a vote of those who have most profoundly considered matters is worth more than others—even if the vote is filtered through the mind of a twenty-first century writer.

The deeper point is this. With notable exceptions, philosophers and scientists have long suspected imagination of being too arbitrary and unstable for the purposes of knowing. Philosophers and moralists have long suspected it to be too willful and unstable for purposes of morally, ethically, or religiously good action. But now it looks as though the source of these problems may be quite the opposite: it is pursuing imagining too little, thinking too little about the possible variations of what we have encountered, that routinizes our imagination in ways that limit our knowledge, encourage complacency, and narrow our experience so that we overlook the self-stabilizing capacities of imagining. Putting an image against backgrounds brings out specific imaginative characteristics that are invisible and ineffective when we try to isolate it.

Vigorous imagining is a cure for one of the weaknesses of traditional empiricism. If facts are disconnected units of experience, then the accumulation of facts makes it harder to do all the comparisons needed for finding the truth. Francis Bacon conceived of dealing with "instances" logically and linguistically, by creating elaborate charts and schemas of correlations and contrasts of all the various instances. These correlations and contrasts are conceived according to the natures that we see exhibited in the instances and that are conceived by the words ("heat," "motion," and the like) we use to express phenomenal categories. The facts of the instances are thus schematized and organized, then reschematized and reorganized for the sake of higher-level generalizations and relations. Instead of simply accumulating instances under terms, the field topology of imagining suggests also searching for field relations and field structures that are due, ultimately, to manifest similarities in appearance.

Topic 6. Imagination is abstractive with respect to perception, and this constitutes the first step toward rationalization. Rationalization requires seeing what presents to consciousness *as something* or *in a certain respect*, although such "seeing as" arguably commences before rationality proper. By virtue of "seeing as" there is opened a field of comparability (the minimal field would begin with two things: what currently presents and what it is seen as). According to the oldest philosophical traditions, seeing–as is seeing a thing as an instance of a form or kind, or as belonging (or not belonging) to another kind, or as an aspect of the larger presentation. All of these acts of the mind are abstractive in that they, along one or several paths, retreat from the full set of affordances of the thing in order either to focus on a few or to set the thing in a foreground or against a background. If the thing in its totality is conceived as a set of possible predicates (this is a philosopher's seeing–as),[22] one inevitably leaves out most of them in abstracted seeing. Even if the first moments of perception and recognition are not properly imaginative, they immediately open onto the imaginative realm of field recognition and field operations.

Topic 7. Imagination is abstractively projective with respect to images when it is searching for structural features and trying to fit them to new planes; when it achieves a certain control and predictability of images and image planes it can be concretely projective. More than 2000 years ago, Aristotle's way of characterizing the basic work of intellect with images—that intellect sees the forms in the phantasms—already noted the complex character of human imagination. Although his statement might at first glance be taken to imply that imagination quickly finishes its work and turns the form over to intellect (which is what the medieval theory of abstraction ultimately claimed), examples he gives (like snub noses, or lights moving at night in the valley below the city's ramparts that are recognized as the movements of enemy troops) make clear that the recognized appearance–form is immediately projected by the mind into new concrete formations. The appearance–form is the presentational form of a thing; it can be analyzed as physical, as mental, as having aspects, as existing in a scenario. The mind, when it abstractively recognizes the form, sees it precisely in its formally topological character. In the cross-sectional planes that the mind thus opens it can place, re–place, and vary the thing as presented in accordance with the possible ways of imaginatively placing the appearance–form. The appearance–forms occupy and form the "matter" of that plane. The matter of the plane is, with respect to ordinary awareness, less real, less a natural or physical *res*, than the matter that shows in sense perception, but it still has degrees of concreteness, lent to it by human imaginative psychophysiology.

Topic 8. A physical or mental "thing," aspects of such a thing, scenarios with many such things, and anything else that presents (vorstellt) **itself to consciousness are imageable, and insofar as the experience can be cross-sectioned by a relevant plane of concern they are imaginable (that is, variable in situation and appearance) in that plane.** This is a direct corollary of topics 6 and 7. We do not yet possess an understanding of human psychophysiology sufficient to explain how and why this happens, but anything that enters

[22] An example would be Kant's *prototypon transcendentale* in the First Critique; see Kant 1996 [1781, 1787], A571/B599. "Affordances" in the previous sentence should call to mind J. J. Gibson's ecological theory of perception.

consciousness has formed the "matter" of consciousness, and if this matter has once been originally formed it can be formed again in a more immanent way, that is, in a way that the human being can initiate in imagination. This does not mean that every human being can produce at a moment's notice a perfect and perfectly vivid reproduction of what has once presented itself, no matter its kind—although every person can train herself to do better than she already does and can expand her repertory. Yet the very fact of reproducibility of what appears to consciousness is evidence of the original producibility (in the relevant "matter" of consciousness) of these appearances as appearances—that is, evidence of imagination and its underlying powers. If temporality itself is produced in or by imagination, as was suggested already by Aristotle and argued fundamentally by Kant, then reproducibility is doubly an imaginative power—and triply so if one goes beyond Kant and says that even the forms of general logic and language are subject to the synthetic appearance–power of imagination.

Topic 9. The mobility of imagination calls into question standard interpretations of consciousness, for example the homunculus model and the functional model. None of the four great representatives of the occluded-occulted tradition of imagination, Plato, Aristotle, Descartes, and Kant, held to a strong notion of ego, self, or foundational consciousness.[23] This is not always apparent to us, insofar as we bring our modern and postmodern concerns into our reading of them. It is in any case easy to fall into homunculus language in following out the contours of the topology of imagination that they share. For instance, one might be tempted to conclude that this book presupposes something like an indefatigable little consciousness–being who constantly shifts attention as he looks or leaps from imaginative plane to imaginative plane. Yet the "mobility" of human consciousness is one of its most conspicuous features, and any acceptable theory of consciousness needs to accommodate it fundamentally. Mobility without homunculi is an open possibility.

The problem with contemporary functional interpretations, on the other hand, is that one never knows in advance which neurocellular functions are tacit, which enter into consciousness, and why. Does the neural subnetwork that detects vertical boundaries in the field of vision show itself directly? Apparently not, but only at a late stage of visual processing, in the finished visual presentation. Similar questions

[23] Plato's theory of ideas requires the mind to be in more than one place at a time if it is to see things clearly; in terms of the cave allegory, to see truth comprehensively one needs to see the shadows on the wall as social projections, but in turn one must see the situation of the cave analogically to the basic situation of the forms projecting themselves into things. Aristotle's theory of perception argues that there is some awareness at the level of the activity of each organ that has perceptive power, so experience is multiply layered. For the ancients, apart from the Stoics, it is hard to find a strong or magisterial ego. Kant's transcendental unity of apperception, the "I think X" that supervenes as the final synthesis of every subordinate synthesis, is not a substance but a weak unifier that assembles awareness from other levels of synthesis. Descartes appears to be the odd man out. Yet the real distinction of the thinking ego from the physical world is evident only after a laborious process of untangling consciousness from its objects, and this ego is affirmed (in the *Meditations*) not as master and possessor of nature but as an insistent presence that we tend to overlook, even when we are aware of it, because of our engagement with the things of the world.

and concerns arise every time one discovers or invokes a brain region or process in order to explain what is experienced.[24]

In the first instance, then, it might be best to say that consciousness is as you find it, and where you find it. Suppose I am a movie director sitting in a café and watching a woman walk along the street, and wondering whether or not she would be appropriate for the role of a passerby in the film's final shot: she will do a double-take of the heroine and hero kissing passionately. Look back at the previous sentence. Is it not a complexly nuanced articulation of where and how the imagined movie director's mind is placed, and where and how my mind is as well (not to mention yours)? Are not our minds in several places—in several places of the imaginatively projected worlds of coffee–drinking, movie–making, and philosophical example–making? These projections and places almost immediately suggest many other planes of concern that an interested thinker, reader, or listener could subsequently provide and explore.

Hegel in the preface to the *Phenomenology of Spirit* already noted that in uttering a simple sentence of the form "S is P," the mind starts with S, moves to P, and then moves back to S in order to see S and P in their predicated unity. This is less a paradox than an everyday fact that expresses *in nuce* the nature of dialectic. If so simple a sentence can display complicated temporal and local movements of the mind, why should anyone expect more complicated sentences and scenes to have any fewer? It seems implausible—unbelievable—that any reasonable human being might think such phenomena can occur without appearance (including private and intimate appearance) to the psyche or that these can be captured by static logical form. Is this psychological movement and successive placement intellectual or imaginative? By now the reader knows my preferences. Nevertheless, the question must be a matter and place for future inquiry, evidence–gathering and –presentation, and argument. It is toward such matters and places that we finally turn.

9.3 Conclusion: Toward a New Beginning

This study aims to be a new beginning rather than an end, a beginning that commences with a de–occlusion and de–occultation of the occluded-and-occulted tradition of imagination. Even if someone disagreed with virtually everything I have said in this chapter or in this book, its claims can be analyzed into elements and questions that any adequate theory of imagination *must* address. An adequate theory, in this sense, is one that aims to elucidate all the phenomena of imagining and approaches the task of explanation and interpretation in as ample and comprehensive a way as possible.

[24] It is high time that researchers of every pertinent field recognize that in addressing this question they are in essence facing the pineal-gland paradox: why is there consciousness *here* but not elsewhere? Having scores of semiconscious and several fully conscious subnetworks just distributes and multiplies the question.

One way to amplitude would apply lessons we have learned from our historical investigations. If imagination works in and between fields of phenomena, and each field is articulated by oppositions (Aristotle's *enantia* in an underlying matter), then one must try to grasp imagining and images with respect to these oppositional articulations. At the beginning of this study we saw that images are typically conceived as static, according to a model of visual fixation, and we found that this tendency appeared already at the beginnings of the tradition of imagination. Yet we also found in many of the very same early authors signs of labile and mobile images, and ultimately discovered modern thinkers developing an intrinsic dynamics of images (for example, in the analytic geometry of Descartes and the schematism of Kant). It is not a question of who is right and who wrong, however, but that the topology of imagining indicates stability and lability (an aptness to change) as opposing poles. I say "indicates" because it is not immediately apparent how this particular opposition should be implemented. Certainly it is possible to argue that many or most images (not just visual ones) are fairly determinate as images, so that changes to them have to be imposed from the "outside"—say a deliberate act of imagining a change to the image. Yet if I ask you to imagine a ball on a steep hillside, there will be a tendency to continue the imagining in its rolling down the hill. An imaged ball bears an intrinsic tendency to roll. This is not completely absent from imagining the ball in any situation whatsoever; disengaging this tendency would require a focused effort of mind. An imagined cube or even a point, all things being equal, has more stability than a ball. But a cube or line segment that comes into imaginative being "in a flash" can be easily seen as rotating in space in a way that a point cannot—unless, that is, you conceive the point as a miniature ball! Even imagining simple qualities can bring out labilities. A yellow that is pure, appearing exactly midway between orange and green, probably has for most people a greater stability as an imagined visual quality than does a yellow that is closer to orange, because the orange can exercise an imaginative "pull" on the yellow.

It will not do to object that such "facts" about images are due to vagaries of human capabilities, whether individual or culturally induced. One is of course free to hypothesize (imagine) a different kind of imagining, given any particular act; one can even imagine a different kind of imagining *being* than the human, one (say) for whom all images are completely fixed and separate from one another, for whom the balls and colors it imagines stay put. Yet the counterfactual hypothesis is itself predicated on the field opposition between stability and lability. It shifts the imagining to a plane in which one has projected variant psychological capabilities by setting the stability/lability parameter to a maximum of stability. Rather than refute the existence of a set of paired opposites it confirms it. It also confirms that in imagining it is difficult, if not impossible, to get a grip on the nature of images without a correlative grip on the capacities of the one doing the conceiving of the images (that is, imagining the imaginer). Imagining will tend to follow the "vectors" of the tendencies in images in a field, all things being equal; it requires some additional effort on the part of the imaginer to resist. Uncovering such relations is not detecting a fault in the analysis, it is an essential aspect of the method of investigating imagination. Conceiving and articulating imagination as a field phenomenon requires exploring such oppositions and relations. To assume that analysis requires conceiving all the elements of imagining in ontological isolation from one another is to commit a common but fundamental offense against philosophy: to fail to make the distinction between prescission and abstraction.[25]

Looking back to our historical investigation, we can with very little effort find many other oppositional pairs that structure images and imagining. For example, the contrast between productive imagination and reproductive imagination, which has always been at least latent in philosophical and psychological theory and which exploded into the foreground of the conceptual topology with Kant and his successors, is implicit in every act of imagining. It implicates both the imagining power and the images (the essential provenance of the latter is fundamental to the very possibility of productive imagination); it also opens onto the question of memory's role in imagining. The degree to which an image is complete in itself or intrinsically part of one or another scenario is another opposition that can be brought to the analysis of images; it is no more than partially reducible to the stable/labile contrast (if an image is part of a static tableau the question of its lability is peripheral). Some imagining focuses on a specific object, some conceives many objects together. When one reemphasizes the temporality of this contrast, the opposition between conceiving something in its aspects (a kind of imagination focused on an image–object and its internal composition) and conceiving it in a scenario provides a useful distinction in the possibilities of imagining. Imagining can foreground and background both fixed and dynamic objects and situations; it can be carefully directed and focused or unregulated and diffuse (one might venture here the possibility of invoking the contrast between conscious and unconscious); it can be guided by resemblances or proceed in variation and differentiation. It can work on objects in a field or project between fields; it can be uniquely individual or socially formed; it can be purely imagistic or primarily linguistic. These are some of the more obvious contrastive pairs.

The demand that future theories strive for amplitude commensurate with the topological richness of the hitherto occluded-occulted tradition applies not just to philosophy but to psychology and all other disciplinary approaches to imagination. The demand cannot be satisfied just by hypothesis or modeling, however. The problem is not so much that making hypotheses and models requires the use of imagination as that hypotheses and models work by *simplification*. Achieving *amplitude* and *adequacy* requires different strategies. Amplitude is a theme hardly addressed by contemporary theories of science, and adequacy is taken for granted where the prevailing paradigms of science understand both evidence and explanation as precise (i.e., expressing prescission), formalized, and quantitative. Contemporary scientific

[25] See Sects. 5.13 and 8.6, above. Perhaps the objection with which this paragraph begins is further evidence that philosophy and psychology have always tended merely to postulate by prescission a faculty of mind that rises fully and entirely above the "human, all too human," or "animal, all too animal" limits of perception, memory, and imagination. I have suggested throughout this study that philosophers who have put their mind to the task of conceiving this arrive at it quite precisely by a rigorous kind of imaginative projection of the abstractive tendencies of imagining. To conceive pure reason clearly and distinctly it is necessary to imagine one's way to it.

understanding has learned to exploit with unprecedented skill the abstractive capabilities of imagination and (in applied science and technology) the power of the concretional schemas of projective imagination. But it has not risen to the challenge expressed by Plato at the beginning of our philosophical—and therefore also our scientific—tradition: that being itself, as well as all the sciences of what is, have to be measured by the ontological projection of what exceeds being, a projection that is in effect the image production of being itself. Not every philosopher has to face up to this question every day, much less every scientist; but not being willing to face it or, worse, simply denying its relevance or meaning is, in essence, nihilism. One cannot in the last analysis (or synthesis) escape metaphysics, or something like it.

Here at the end, I want to identify four major areas of fundamental concern that must be addressed in a future philosophy and science of imagination. In some cases my own preferences are clear, though I freely admit that, however well motivated and plausible they may be, they are by no means certain. The discussions are brief but, I hope, sufficiently indicative.

Area A. The depth psychology of imagination. Several times I have remarked that this study emphasizes the cognitive and proto-cognitive powers most crucial to the occluded-occulted tradition of imagination, and that therefore it leaves out of primary consideration will, desire, passion, emotion, and feeling. From certain points of view, therefore, one could argue that I have been "cognitivizing" imagination just as much as it has been in the past, that I sometimes abandon even more irrevocably the "subjective" elements that enter into imagining and that, according to some theories at least, might constitute its essential ground. In so doing I would be giving license to a new kind of naïve but totalizing claim about imagination. Every computer programmer could proclaim himself an artist and scientist because he is situating things in fields, re–presenting them in other fields, and working out the morphological structures that shape the fields per se.

I can scarcely deny that much of what I have done here lends itself to such use; indeed, in some sense I wish to encourage it. Imagination does, I believe, play an essential role in cognition *of all kinds at all times*. Of the early philosophers in the occluded-occulted tradition, only Descartes genuinely began, but only began, to address the deeper psychological dimensions, in his correspondence and in the *Passions of the Soul*. In Chap. 6, I explained that in the *Passions* he stated categorically that the *thinking* of the *res cogitans* is better understood according to its active rather than its passive side, the passive side being perception (including intellectual perception), the active being *voluntas* or will. He stated further that imagination is an action and thus a form of will more than a form of perception (although there is a corresponding perception to this imaginative act of will). Sense perception, by contrast, is not an action of will but simply a passion. Perhaps even more surprising is that several times in the *Passions* he gives examples showing how imagination has more direct access to the passional and emotional side of the human being than reason and will do, and other examples indicating that *hidden passions* can express themselves in our actions and imaginings.

Earlier, of course, the ancients and medievals were aware of the incursions of passions and affections in both waking and sleeping life, but these were understood as irrationalities that needed control rather than as constitutive of a significant part of human psychological life. Later rationalism and empiricism were neither of them very favorable to a deeper exploration of the positivity of affect and affect's role in imagination. Spinoza and Hume were exceptions. But as far as conventional historiography is concerned, Spinoza wanted to subordinate everything to intellect, while Hume was interpreted less according to the sophisticated psychology of the *Treatise* than the more simplified and popularized *Enquiries*. Kant attempted to establish impassable territorial boundaries between cognition, desire, and feeling, but his theory of the transcendental functions of imagination—and in particular its "play" with the understanding in aesthetics—inspired Fichte and Schelling to develop a more powerfully unified and interactive metaphysics of subjectivity. Thus it was not until the emergence of depth psychology, in particular that of Freud, that the affective realm came to be seen as *the* fundamental source of images and their power. Despite many mentions of *Phantasie* in his writings, however, Freud never actually made imagination a thematic object of focused research or theorizing.

Most psychologists today have little confidence in Freud, and even less in the image archetypes of the "Freudian renegade" Jung, though both played not insignificant roles in popular twentieth-century theories of imagination. Yet the idea that human imagining is actuated at least in part from deep psychophysiological levels of personality to which we do not have direct access does not appear to be prima facie implausible. Quite the opposite! We have scarcely developed ways of probing these things, however, even at the level of first-approximation qualitative description. It is likely that any account of imagination that incorporates these factors will have a long and very complicated road to follow. Perhaps in the first instance we are likely to find better cues about how to proceed from literature, music, and the visual arts than from experimental psychology or neurophysiological probes.

Area B. Anthropology and imagination. The depth psychology of imagination would go a long way toward extending our understanding of human being and thus of (philosophical) anthropology. Yet there is also need of studies that are anthropological in a more conventional disciplinary sense. Lévi-Strauss's structuralism offers, for example, a conception of the deep structuring of fables and other imaginative narratives. Following the anthropological traces indicated by Kantian schematism, researchers like Lakoff and Johnson have theorized the existence of basic human metaphorical functions that provide a foundation for human language and image making. Beyond theory, of course, there is also need for comparative, cross-cultural studies of imagination. Although Berlin and Kay's research into color naming in scores of languages was designed as a test of the Sapir–Whorf hypothesis—that languages and the concepts they express are to a significant degree incommensurable—it provided insight into surprising cross-cultural invariants of color experience and naming. Insofar as the research was intended to be a first step toward comparing languages in their expression of simple sensory experiences, one can only hope that future investigators will find ways (and funding) to extend the approach to the other modalities of sensation.[26]

[26]See Lévi-Strauss 1964, Lakoff and Johnson 1999, and Berlin and Kay 1969.

There is, of course, also need for cultural-anthropological inquiry into how people of different cultures experience the full variety of phenomena we have been calling imaginative. This is more difficult than the Berlin and Kay type of study. For one thing, it presupposes that the investigator is already in possession of the kind of data that Berlin and Kay accumulated. For another, effective inquiries of this kind would probably require investigators to develop beforehand more acute and more creative ways of researching our (and their) own experience and basic conceptions of imagination. The psychological investigations pursued by Shepard and Kosslyn, many of which are based on timing the performance of mental tasks like rotating a statue in imagination, were a first step. Undoubtedly there will be surprises in store for anyone trying to arrive at a genuinely culture-neutral articulation of imagination.

There is another approach, until now little pursued, that might allow an amplified understanding of imagination as placement and might also incorporate the affective aspects of imagination mentioned under Area A. In *Being and Time* Martin Heidegger broached a network of issues by interpreting the "being–in" of being–in–the–world in terms of the existentials *Befindlichkeit* and *Stimmung*, state–of–mind/attunement[27] and mood. Although Heidegger rejected Sartre's view of *Being and Time* as a kind of anthropology, it nevertheless seems hardly credible that nothing at all can or should be borrowed from it for anthropological purposes. As Heidegger makes clear from the beginning of his discussion of space and placement, the human being (*Dasein*, to be more accurate) does not exist in the world the way a gift exists inside a box. *Dasein* is not a relationship of one object to other objects, or of subject to objects. The world is arranged not just in terms of countable time and measurable space (what Heidegger criticizes as the world concept founded by Descartes) but also articulated as a network of meanings. Meaning articulation includes place, in the sense that, say, the Bastille is not just at a determinate longitude and latitude or the intersection of a certain few designated streets but a location with complexly articulated social, historical, and even personal meaning. One might call this a fundamental orientation to place that we typically but wrongly think of as subjective. In addition, Heidegger desubjectivizes *Stimmungen*, moods/attunements, by understanding them as typical ways of experiencing this orientation; they are not species of orientation, but modulations of it.

If we disregard Heidegger's qualms about interpreting his work anthropologically and introduce more conventional language, we could express this by saying that

[27]"State of mind," used in the Macquarrie and Robinson translation (Heidegger 1962), is too generic and bland. Stambaugh uses "attunement," which she also uses for *Gestimmtheit* and *Gestimmtsein* (Heidegger 1996); *Stimmung* she renders, like Macquarrie and Robinson, with "mood." Alternatives like "disposition" or "disposedness" have been proposed for *Befindlichkeit*; they are better in etymological terms, but the average English-speaker would not hear in them the peculiar balance of abstractness and concreteness conveyed by the German. John Haugeland has suggested the neologism "sofindingness" (Haugeland 2000), but it needs more than a little explaining. A virtue is that it expresses the *befinden* in *Befindlichkeit*. *Wie befinden Sie sich?* is a somewhat formal question–greeting to another person: roughly "how are you doing," more literally "how do you find yourself," "how do you find yourself situated."

human beings exist in the world with certain fundamental intersubjective possibilities of orientation to surroundings, and that these possibilities are lived out as mood, which, even before we talk about it in private subjective terms, has structural characteristics that "attune" us to the world, to its notably meaningful places and to the other people and things there. Turning even more decisively toward conventional philosophical vocabulary, we might say that there is no conscious being in the world without fundamental orientations and moods. There is no consciousness without disposition, mood, and mood tonality. The clinching thought here for our purposes is this: that insofar as imagination works with and by placement, studying it is not just psychology but even more an articulation of fundamental human dispositions and attunements to the world through imaginative fields and of human aptness and adroitness in moving between fields in ordinary experience. The study of imagination would therefore turn out to be a pathway to the fundamental analysis of *Dasein* as being in the world, a kind of ontological anthropology.

Kant provides a different kind of guidance to how we might approach feelings and moods as implicit in imagination. The Third Critique understands inquiry and reflection as involving the interactive cooperation of different powers. When the aim is knowledge and is successful, the activity comes to rest in a concept or a proposition. But when knowledge—that is, subsumption under a concept or principle—fails or is not possible, the interaction or play of powers does not necessarily produce a null result. The first, aesthetic half of the Third Critique discusses two specific ways in which the play between imagination and understanding takes settled dynamic form. The first produces a special kind of disinterested sentiment or feeling. "Disinterested" here means: having no *direct* interest in the *existence* of the object, whether past, present, or future. The feeling, because it is produced in a harmonious interplay, is stable enough to be nameable: it is the feeling of the beautiful. The second kind of play that takes settled form occurs when the power of imagination is overwhelmed by what it experiences. In this case, the play of powers redounds to the advantage of the understanding. The sensory and imaginative powers may be taxed beyond their limit, but through the power of understanding the subject experiences himself as beyond even the most awesome and dangerous natural forces. Understanding owes this to reason's infinity: that is, to reason's infinite aspiration to total and unconditioned knowledge (which, Kant says, it never achieves). This stable form of experience and feeling is called *sublime*. Beyond the beautiful and the sublime, Kant presents the "intellectual feeling" of respect for the law in the Second Critique, *The Critique of Practical Reason*.[28]

There is no *a priori* reason to think that Kant exhausted the possible outcomes of the play of powers in his Critiques. In comparing Byzantine Christian painting and the painting of the Baroque, one can hardly avoid noticing a difference between the static and the dynamic in art. In comparing periods and styles one can often distinguish other aesthetic categories and subcategories (the magisterial and

[28] Kant ascribes respect to reason alone—which, as I explained earlier, is a category mistake due to Kant's characteristic attempt to draw sharp, impermeable boundaries between the basic psychological powers.

the quietist might be subcategories of the static style, the dramatic and the turbulent of the dynamic).[29] Given the harmonious play of the psychological powers, the aesthetic phenomenon leads not only to feeling but also to concepts and language, and the principal medium of their production is the imagination. Because this feeling and naming is stable, based on the interaction of fundamental human powers, and because it is directed toward an object, or at least the appearance of an object, a Kantian can rightly deny that the judgments associated with the feelings are merely subjective.

We of course are very far from understanding how to spell out further such concepts and judgments, much less knowing what to search for neurophysiologically. But an approach that draws on the play of powers might allow us to express more fully and adequately some of the basic phenomena experienced by people who seriously engage an imaginative *field*, rather than just imaginatively play with respect to an object or a scene. The Kantian play of powers might well in some cases *constitute* the field as field. The object or object–appearance is not the aim but the starting point of an implicit and largely unconscious exploration of how the object/appearance fits the various fields in which it evokes an aesthetic response. This approach might, in addition, provide insights into matters like *style* in the arts.

In cognitive fields, where Kant allows for the "intellectual feeling" of the ethical respect for law, the approach might lead us to other cognitive attunements. For example, different mathematicians or theoretical physicists often approach the same subject field with different attitudes and scientific styles. One geometer might always go to work analytically, whereas another might resort first to synthesizing transformations. An atomist/particularian will attack problems by looking for theories and conceptions that emphasize discrete unit–entities; a field theorist will start with a higher-order vector space and interpret entities as determinate functions therein.[30] It is conceivable that one might be able to identify an associated sensibility or "feel" of such approaches.

Area C. Ontology and imagination. By asking what imagination and its basic phenomena are, this book has engaged the ontology of imagination, images, and signs. "Ontology of X" is a conventional way for a philosopher to signal that she is talking about what makes X the kind of thing it is. In a reductionist system, that would imply that X, which appears to be of kind A, is really some explicable organization of a set of things y_i in kind B. A table looks like an assemblage of solid wood, glue, and fasteners, but it is really mostly empty space because it consists of atoms organized by electromagnetic forces.

I note only in passing that this kind of scientific reduction requires the imaginative articulation of one field in terms of another. The more immediately relevant but

---

[29] These concepts do not need to be conceived as oppositional, though doubtless one can identify oppositional pairings.

[30] The earliest theoretical attempt I know to account for styles and types of approaches within any given science is Johann Wolfgang von Goethe's historical studies of the sciences of color and optics, beginning ca. 1795. He appears to have in fact understood this approach as a development of Kant's theory of the transcendental functions of imagination.

related point is that, at crucial moments, fundamental ontological explanations look elsewhere to give an accounting of what they are trying to explain. The *explanans*, what does the explaining, is located in another field than the *explanandum*, the thing to be explained. To explain how things of ordinary experience are we resort to things beyond that experience: the wetness of water we explain by the bonding of atoms, the motions of matter by mathematical principles, the nature of mathematical truth by the forms of intelligibility. This is as old as metaphysics/ontology/philosophy of being itself; Plato did it as much as a modern physicist does. It is necessary but not always harmless, especially when the abstractions that allow us to constitute the new imaginative field (e.g., that of fundamental physical forces) are absolutized.

In the *Phaedo*, Plato's Socrates tells the story of how he came to philosophy. He was especially attracted by the doctrine of Anaxagoras of Clazomenae (ca. 500–428 B.C.E.), who declared that everything was ruled by mind (*noûs*). Socrates soon turned away, however, because he found that Anaxagoras resorted to explanations employing mind only when his more usual materialist explanations failed. Why explain the justice of an action in terms of an assemblage of matter, or quantities, or qualities, or sets, Socrates asked. Such explanations lose sight of what is to be explained. So he turned instead to trying to explain just actions in terms of justice, beautiful things in terms of beauty, and so forth. The Platonic interpretation of this kind of explanation led to the ideas or forms, and in later Platonism this was reified into the realm of the forms (that the philosopher strives to visit or revisit), of absolutized abstractions, of the mind of God. Whether Plato's irony was also Socrates' irony is arguable, but there are reasons to think that both would have found these later Platonistic developments as the self-deceptions of philosophers, and perhaps even comical insofar as they claimed to be authoritative interpretations of what Socrates and Plato had meant.

The contemporary relevance of this anecdote is not far to seek. The cognitive neurosciences and neurobiological approaches to mind explain psychological phenomena according to the activities of neurons and brain regions. This kind of work and this kind of explanation, which employ the projectively abstractional-concretional imagination that I have taken pains to make evident, has to be undertaken if we intend to have good science. The problem comes when the most reductionistically minded researchers explain psychological phenomena precisely *as* such activities, and nothing more. One might imagistically portray the situation this way. The reductionist begins with the phenomena to be explained, projects them in abstractional-concretional form into a new plane, explains the workings and interactions of the elements on that plane, and then says that this plane and its working constitute a perfect template or overlay that exactly matches the original plane. But of course such claims almost never turn out to be thoroughgoingly correct—something in the original plane blurs under the overlay, some phenomena simply do not match what is expected in the explanatory plane, some features of the original plane are ignored or distorted in the plane of explanation. In traditional philosophical theories of explanation that means that there is something wrong with the explanation plane, not with the phenomena to be explained. Sometimes—and this is one of the most remarkable things about scientific progress—the explanation does help us to rearticulate the plane of the phenomena in a way that shows the latter more clearly. But often it leaves out what does not suit it, perhaps with a promissory note that at some indefinite time in the future what is left out will finally be included, occasionally with the assertion that what is left out is not important.

I am perhaps less disturbed than Socrates was by the only occasional express resort to mind in explaining things. But perhaps that is because this book has taken pains to show that, by way of projectively abstractional-concretional imagination, things of mind are actually present in explanations that reductionists think are resolutely material-physical. Rather than play the game that begins by setting mind over against matter—itself an act of imagination—I ask, if someone wants to be a pure materialist (or a pure mentalist, for that matter), that he demonstrate his right to the claim. It is not enough to vaunt the success of the sciences, since success is never as perfect as would be required for the claim to be true. The question is what principle is at work, and how the application of the principle accords with the world in its totality—all the way from naïve lifeworld to scientifically explained world.

Perhaps we will never fully understand how human psychology enters into scientific explanation and the claims of reduction. But perhaps there is also a different kind of ontological thinking to bring to bear that both acknowledges the merits and necessity of rigorous scientific approaches and allows for the articulation of our experience "close to the ground." Perhaps we can see the task of philosophy in particular as a different kind of *reductio*, see it in a sense that resonates with an older, more traditional meaning: a bringing back of one thing or state to another to show fundamental relationships, rather than to dismiss the importance or even the reality of the reduced thing.[31]

Here I will not so much *explain* as *make an image of* what I mean, in the mode of a reminder. With the cascade of figures and images in Books VI and VII of the *Republic* (discussed in Chap. 4), Plato proceeded as follows. Socrates, asked to explain the good itself, laughs and pleads incapacity. He does, however, offer an image, or rather two metaphors: on the one hand he calls what he has to offer the offspring of the good—insofar as the good/image relationship is like the father/son relationship—and on the other he calls it interest (like the interest earned on the investment of principal). He uses these figures because they express a resemblance between an original and what derives from it while preserving a superiority of the original. Then he offers the image proper: the good itself is like the sun. The sun makes grow, gives *being* to, things like plants, and it also makes them all *appear* to other beings. He elucidates this image further by presenting the image of the divided line, which extends from the ideas/forms, through the geometrical things and the physical things, to the images and shadows produced by the physical. The ideas and geometrical things are, both of them, accessible to thought, while the physical

[31] The medieval scholastics typically used *reduction* to describe the induced or caused movement from a state of potentiality to a state of actuality, a conception that is fundamentally ontological. In this conception, being in the full sense is *actual*; every actualization is the actualization of a potentiality; and the actuality of every actual thing (short of divinity) necessarily contains further potentialities. Thus beings are actively potentiated fields of articulated possibilities.

things and their images are accessible to visual perception. Then, in the allegory of the cave, a scenario–image that is structured to image analogically the divided line, he presents the basic situation of human life, starting with the slavish devotion to mere appearances—that is, to shadows on the wall that are not understood by the enchained slaves to be projections from elsewhere but are taken as ultimate realities. The slave whose chains are broken can turn around to see that the flickerings on the wall are social projections. If he lets himself be led from the cave he may perceive the realities of nature beyond the social names and appearances, and finally even glimpse the source of nature and of light: the sun, from which everything ultimately derives its being and appearance.

Rather than take this as proof that the realm of the forms is ontologically *superior* to everything beneath it, what Chap. 4 argued is that the entire sequence of images in Books VI and VII is a complex image of the total being and possibility of the good itself, a complex image that derives from the good itself and therefore has to be understood as also good and as showing the ways of the good. Every level of the line and cave and both aspects (being and appearance) of the sun's reality *are*, and as such they cannot be denied without falsity. The forms exist insofar as they deploy the dynamism of the good; the forms themselves resemble the good, and they do their work by progressively, level by level, imaging themselves. Each level "below," whether mathematical truth or physical reality or image/shadow, images itself in each of the others. This is an ontological and epistemological vision. Plato thereby provides us with an ontology that not only explains appearances but also understands being and appearance as dual and reciprocally reflexive rather than oppositional. In this sense, the tales historians of thought tell about the Greek origins of the *opposition* between being and appearance blind us to a juster reading of what can be found in the texts; a juster reading might have led to a different philosophical tradition.

Our ontology, whatever it ultimately turns out to be, has to be as rich as being.[32] Even appearance presents itself in being, so it, too, is being. Accommodating the ontology of imagination therefore requires that our metaphysics have as many principles as are needed—though not necessarily more—to explain everything that shows itself in being and possibility. This means, too, that we need to resist the ordinary philosophical and scientific temptations to nihilism. I mentioned near the outset of the historical part of this book the possibility that, when Parmenides

[32] In his late work, Paul Feyerabend was concerned with the gap between typical scientific and philosophical explanation on the one hand and the ontological and phenomenological abundance of the world on the other; see Feyerabend 1999. Philosophy and science long ago mastered the art of abstracting and prescinding evidence so that it might be well adapted to the available conceptual mechanisms of explanation. The conception of imagination I am fostering, an imagination that is essential to rationality and works by both abstraction and concretion, presents an implicit rebuke to prevailing methods as one-sided. It is a fundamental matter of human finitude that all investigators are limited in abilities, interests, and preparation, and that all approaches must come from one side or another. But in the long run we have the obligation (philosophers above all) of trying to put it all together. If inconsistencies, problems, and gaps remain, we must acknowledge them and learn how to accommodate them rather than ignore, or worse deny, their existence.

made his radical claims about being, he was less establishing the pattern for later philosophers (that is, asserting the superiority of the ideal unchanging realm to the realm of mere appearance) than postulating the indefeasibility of being for all that appears to thinking. In particular: where there is being one may not assert nonbeing; whatever is is; and when one tries to account for something by explaining it away (whether absolutely, or for the sake of explanatory reduction, or for the sake of interpretation) one falls short of both being and thought. All being must be, indeed is, preserved in thought. Accounting for that is the basic task of the philosopher, the task which s/he never really completes but to which s/he must constantly return.

Area D. The ethics of imagination. The ethos of imagination is the inhabitable place of imagination. This book has been an exploration of its ethos. The ways of inhabiting a place become a way of life, and those ways take on a shape that, in articulated form, becomes an ethics. Rather than just word play, these matters express fundamental truths. As such they have consequences. Too many, in fact. Since here I am indicating directions for future work, I will be aphoristically brief.

The most abstract of rationalisms hold that ethics can be expressed in propositions or commands. The problem with formulas is that they do not tell us how to apply them, nor what we should do when they conflict with other formulas. In any case we need to develop the art of making our ethical propositions conform to the fields of practice in which they apply and to learn how to educe more adequate ethical accounts of good practice. We need to think imaginatively in the places where we live.

There is more, however. If the ontology of imagination suggests that the being of the world has to be rich enough to account for all the cross-sectionings of it we make, then we must also think about whether and how we need to exercise our imaginative powers responsibly. Some imagined worlds are merely fantastic; some are possibilities we should want to realize; yet others might be evil. About most imaginings, in any case, we might want to be careful, at least once we go beyond the first few steps of imagining in a given direction. Yet we have very few guidelines and even less experience in negotiating such matters. And we have wagered heavily on creating economic and political arrangements that amplify and accelerate the introduction of new goods, new practices, new entertainments, new ways of life. We are committed unlike any previous generation to scientific and technological imagination. It is urgent that we think and imagine the consequences. That is an ethical necessity.

We can *share* thoughtful imagining, fortunately, but we must cultivate it first. That requires of us a new kind of pedagogical ethics. We have become (especially in the United States) obsessed with inculcating measurable knowledge and skills in our children. We believe that imagination is proper to the arts, and that rationality is chiefly formal and procedural. The obsession is disastrous, and the beliefs about imagination and rationality are false. Rationality is the art of putting things and appearances into proportionate relation, and thereby developing our sensibilities for the best placement of those things and appearances in coherent fields. Most fields cannot be coherently reduced to others, however, so pedagogically we must arrive at some agreement on the fields of experience that are indispensable for our children to know with well-developed imaginative articulation: not just as means, but also as ends in themselves. That means that it is a question of the fundamental and comprehensive good of being human. Thus it cannot be left to the decision of those who have narrow economic, political, or professional interests. To put it in a formula: the only adequate way of developing rationality is to develop our ability to imagine comprehensively; we must start with ourselves, or we will inevitably fail our children and the future world.

****

The danger that sometimes occurs when professional academics address a topic is that they appropriate and subordinate it for their professional purposes. They enframe phenomena and, even against their best intentions, reduce it to the terms of the framework. The problem is universal, whatever the academic specialization. Theory, a way of seeing that attempts to leave behind the inessential for the underlying or overarching, in effect decides what is inessential simply by leaving it behind and subjecting it to neglect.

The effort to gain a naturally authoritative perspective, to penetrate to the heart of things, always runs the risk of mistaking the character of things and misinterpreting its own standpoint. There is a permanent danger in theory, whether it is academic or everyday. We have no choice, however, but to run the risks, which come with our human nature, especially our curiosity. The risks become even greater when our inquiry touches most nearly who and what we are. The danger is at its maximum in philosophy—at least in philosophy that sees no subject matter as alien and has as ultimate goal leaving nothing out of account.

Easier said than done. If I ask myself now what I have done, might I not have to answer, "I've presented a new theory of imagination, to be added to the neverending story of such theories"? What value does it have?

What I have presented is a way of looking at and conceiving things, to be sure, but all that means is that it stands ready for future application, specification, and amplification. It is not new, at least insofar as it explicates the topology of imagination underlying our traditions. It is therefore not a theory in the sense that it can be precisely predictive and precisely confirmed or disconfirmed. It is historical in that, as amply as its author could make it, it gives an account of past theoretical and practical experience with imagination. It is practical in that, at least a little, it asked you, the reader, to encounter imagination in its proper fields and on its own ground. It is theoretical in that it describes the conceptual topology of imagination not just as it has existed in the past but also as it exists today, as the source of most of what we do and say about imagining. The more clearly we see this, the better we can decide where to head next.

This is work in progress on a matter that may have no proper end. By its very nature and activity imagination seems always to be in the middle of things; images are far more often landmarks and stopping places than they are ends. Yet that is to reaffirm imagination's elemental and environmental character: it is as basic to human being as is air for breathing, light and sound for sensing, and the cycles of waking and sleeping for life's basic rhythm. It is more a function than a faculty; the function is both sensitive and cognitive; it is the central place and the center of gravity of thought's activity. It is rooted in the stepping back from any sensitive or cognitive presence for the sake of focusing, attending, configuring, conceptualizing, enjoying. As a sensitive power it is not far removed from the inner powers of sensibility that we call emotion and mood; as moved by forms of desire, not far from will.

These matters taken all together would require a study far more ambitious than this one. When all is said and done, treating them would demand that we grasp, far better than we do, the unity–and–diversity, and the ground for the unity–and–diversity, of being human. Perhaps these matters are, for the time being at least, too unwieldy, not just for our scientific theories but also for our philosophy and culture, which have not proved very good at recognizing the right questions and the grounds for responding to them as amply and adequately as possible.

It is high time, then, to get serious, even if part of that seriousness will require us to learn many forms of imaginative play. It is time to find adequate questions, to ask them in a well-directed sense, and to cultivate ample awareness and focused attention, so that our science, our philosophy, and our culture might finally rise to the level that our thoughtfully imaginative engagement with the world demands.

References[33]

Adorno, Theodor W. 1992. Why philosophy? In *Critical theory: The essential readings*, ed. David Ingram and Julia Simon-Ingram, 20–30. New York: Paragon House.

Agamben, Giorgio. 2005. *The time that remains: A commentary on the letter to the Romans*. Trans. Patricia Dailey. Stanford: Stanford University Press.

Berlin, Brent, and Paul Kay. 1969. *Basic color terms: Their universality and evolution*. Berkeley/Los Angeles: University of California Press.

Castoriadis, Cornelius. 1987 [1975]. *The imaginary institution of society*. Trans. Kathleen Blamey. Cambridge: The MIT Press.

Feyerabend, Paul. 1999. *Conquest of abundance: A tale of abstraction versus the richness of being*. Ed. Bert Terpstra. Chicago: University of Chicago Press.

Guillaume, Gustave. 1965 [1929/1945]. *Temps et verbe: Théorie des aspects, des modes et des temps; suivi de L'Architectonique du temps dans les langues classiques*. Paris: Librairie Honoré Champion.

Haugeland, John. 2000. Truth and finitude: Heidegger's transcendental existentialism. In *Heidegger, authenticity, and modernity: Essays in honor of Hubert L. Dreyfus*, vol. 1, ed. Mark Wrathall and Jeff Malpas, 43–77. Cambridge: The MIT Press.

Heidegger, Martin. 1927. *Sein und Zeit*. Halle a.d. Saale: Max Niemeyer. English editions: Heidegger, Martin. 1962. *Being and time*. Trans. John Macquarrie and Edward Robinson. New York: Harper & Row; and Heidegger, Martin. 1996. *Being and time*. Trans. Joan Stambaugh. Albany: State University of New York Press.

Heidegger, Martin. 1929. *Kant und das Problem der Metaphysik*. Bonn: Friedrich Cohen. English edition: Heidegger, Martin. 1997. *Kant and the problem of metaphysics*, 5th ed. Trans. Richard Taft. Bloomington: Indiana University Press.

Kant, Immanuel. 1996 [1781, 1787]. *Critique of pure reason: Unified edition*. Trans. Werner S. Pluhar. Indianapolis/Cambridge: Hackett.

[33]If there is a second, square-bracketed date, it indicates the year the work first appeared in its original language.

Lakoff, George, and Mark Johnson. 1999. *Philosophy in the flesh: The embodied mind and its challenge to western thought*. New York: Basic Books.

Lenoir, Timothy. 1979. Descartes and the geometrization of thought: The methodological background of Descartes' "Géométrie". *Historia Mathematica* 6: 355–379.

Lévi-Strauss, Claude. 1964. *Le cru et le cuit. Mythologiques*, vol. 1. Paris: Plon. English edition: Lévi-Strauss, Claude. 1969. *The raw and the cooked*. Trans. John and Doreen Weightman. Mythologiques, vol. 1. New York: Harper & Row.

Index

A
Abstractio, abstraction, 12, 25, 33, 39, 61, 62, 64, 78, 79, 89, 90, 104, 107, 116, 147, 152, 179, 197, 202, 208, 241, 245–248, 251, 254–256, 258, 261, 263, 272, 315, 316, 333, 358, 370, 417, 418, 424, 426–428, 442, 445, 446, 448, 450, 472, 484, 492–494, 499, 503, 504, 510, 514, 520, 522
Abstractional, 19, 73, 484, 488, 491, 492, 494, 496, 499, 509, 520, 521
Actuality, 191, 195, 196, 204, 209, 219, 233, 258, 310, 376, 411, 412, 447, 471, 478, 479, 496, 521
Adeimantus, 147, 149, 150, 157, 192
Adorno, Theodor, 501
Aesthetics, 5, 342, 352–361, 373, 380, 383–396, 516
Agamben, Giorgio, 21, 33, 34, 37–39, 456, 466, 495
Aisthēsis, aesthetic, 2, 5, 91, 134, 188, 190, 213, 274, 281, 342, 352–361, 367, 373, 376, 380, 383–396, 404, 407, 409, 474, 479, 516, 518, 519
Alain (Emile Chartier), 27, 28, 30–32, 34, 36–38, 40, 41, 43, 47, 49, 53, 63, 65, 71, 122, 408, 496
Albert the Great, 271, 272
Alexander the Great, 185
Alloiōsis, 222, 231, 233, 291
Amplitude, 8, 12, 13, 99, 106, 136, 174, 175, 267, 279, 345, 487, 513, 514
Analogia, analogy, 39, 75, 82, 90, 105–107, 116, 121, 124, 132, 133, 147, 151, 154, 155, 158, 166, 176, 207, 242, 244, 281, 283, 302, 339, 345, 355, 365, 367, 369, 370, 378, 383, 389, 400, 405, 408, 414, 462, 470, 473, 475, 496, 498, 499
Analytic geometry, 5, 94, 268, 280, 296, 310, 313, 328, 427, 454, 463, 479, 503, 505, 508, 513
Anaxagoras, 520
Anaximander, 106–108
Anaximenes, 105–108, 112
Animal
  imagination, 9, 61, 215, 425, 485, 507
  psychology, 1, 5, 346
  spirits, 199, 323, 326
Antipsychologism, 27–31, 284, 285, 376, 407, 408, 429, 430, 432, 433, 439, 453, 454, 502
Aphairesis, 179, 245, 247, 251, 254, 258, 450
Appearance-bearer, 104
Aristotelianism, 204, 226, 417
Aristotle
  Categories, 76, 192, 193, 252
  conventional Aristotelian psychology, 197, 198, 202, 204, 205, 210
  On Generation and Corruption, 231
  On Interpretation, 76
  On Memory and Recollection, 200, 206, 207, 236, 240, 245, 258
  Metaphysics, 187, 192–194, 230–232, 243, 245, 249, 250, 261, 271
  Nicomachean Ethics, 213, 248, 259, 261
  Organon, 192, 270, 277
  On the Parts of Animals, 199, 225
  Physics, 187, 189, 194, 205, 207, 216, 219, 220, 222–226, 229–233, 255, 261, 262

D.L. Sepper, *Understanding Imagination: The Reason of Images*, Studies in History and Philosophy of Science 33, DOI 10.1007/978-94-007-6507-8,

Aristotle (*cont.*)
Posterior Analytics, 243, 246, 248, 273
Prior Analystics, 243
Rhetoric, 212, 270
On Sense and Sensible Things, 230, 241
On Sleep and Sleeplessness, 225
On Sophistical Refutations, 76
On the Soul, 103, 134, 188, 190, 195, 196, 200, 205–216, 220–222, 225, 227, 229, 230, 235–237, 240, 241, 243–248, 253, 256–258, 261, 263, 274, 417, 420, 422, 426
Topics, 50, 76, 77, 186, 208
Arnauld, Antoine, 340, 341, 347
Art, 2, 18, 61, 111, 187, 275, 332, 400, 518
Associationism, 38
Atomism, 105, 111, 113
Augustine, 302, 339, 416, 495
Averroës, 189, 204, 245, 271, 272
Avicenna, 189, 245, 271, 272

B
Bachelard, Gaston, 66, 67, 82
Bacon, Francis, 509
Bacon, Roger, 271
Baudelaire, Charles, 404–407
Baumgarten, Alexander, 359
Beeckman, Isaac, 274, 280
Beethoven, Ludwig van, 46
Behaviorism, 29
Benjamin, Walter, 66–68, 70, 73, 82, 251, 496, 497, 500
Bergson, Henri, 32, 33, 35
Berkeley, George, 343
Bespeaking, 187, 200
Biplanarity, 87–96, 276, 342–344, 414, 417, 428, 460, 503

C
Camus, Albert, 53
Cartesianism, 301, 338
Castoriadis, Cornelius, 1, 89, 174, 180, 208–210, 256, 268, 409, 483
Cephalus, 140, 147
Cézanne, Paul, 95
Chomsky, Noam, 415
Christina, Queen of Sweden, 268, 328
Cinnamon, 21, 22, 35, 36, 43, 49, 57, 65, 71, 93, 221
Coconstitution, 183
Cogito, 303, 317, 318
Coimbran commentaries, 277
Coleridge, Samuel Taylor, 92, 403
Common sensation, 198–200, 202, 203, 206, 209, 211, 215, 217, 219, 220, 225, 227, 229, 232, 233, 235–237, 248, 253, 256, 270, 272, 273, 286, 339, 344, 373, 377
Comprehensiveness, 19, 20, 48, 62, 198, 221
Conceptual scheme, 63, 64, 490
Conceptual topography, 108, 189, 251, 252, 352, 360, 368, 383, 400, 401, 451
Conceptual topology, 8, 10, 14, 15, 50, 51, 78, 79, 83, 96, 99, 106, 139, 154, 175, 176, 179, 181, 197, 204, 226, 228, 232, 233, 245, 247, 250, 251, 255, 263, 264, 267, 268, 270, 273, 276, 290, 316, 323, 325, 331–396, 399–479, 486, 502, 503, 514, 524
Concomitant sensible, 200, 209, 217, 218, 233, 239, 247, 253, 257
Concretional, 19, 73, 428, 488, 491, 492, 496, 499, 506, 509, 515, 520, 521
Condillac, Etienne, 456
Contemplation, 155, 167, 172, 242, 277, 428
Contexture, 39, 121, 249
Contrarieties, 57, 229–232, 235, 251, 253, 259, 261, 263, 423, 454, 478
Core phenomenon, 11
Creativity, 1, 2, 4–6, 8, 17, 18, 58, 95, 96, 372, 391, 401, 404, 486, 505
Cross section, 33, 34, 87, 90, 96, 162, 251, 260, 414, 477, 503, 504, 510, 523

D
Deduction, 76, 290, 291, 297, 298, 317, 355, 371, 374, 377, 378, 399, 452, 453
Definition, 9, 19, 128, 196–197, 488, 500
Delacroix, Eugène, 405
Deleuze, Gilles, 33, 34
Democritus, 110
Descartes
Compendium, 274, 275, 279, 283, 287, 290
Discourse on Method, 268, 284–286, 303, 310, 312, 313, 321
Geometry, 268, 269, 275, 280, 290, 294–296, 310, 311, 313–315, 328
Meditations, 268, 277, 285, 301, 302, 304, 307, 312–321, 324, 327
Notebook C, 279–283
Optics, 268, 292, 297
Passions of the Soul, 268, 319, 321, 322, 326, 328
Principles of Philosophy, 268–272, 276–278

Regulae, 284–287, 289–291, 293, 295, 297–299, 303, 305–308, 310–313, 315, 316, 318, 324
World, 268, 270, 276, 278, 279, 298, 303, 306–313, 316, 317, 320, 325
Descartes, René, 5, 267, 348
Diairesis, 128
Dianoia, 155, 158, 207, 212, 213, 344
Diotima, 177, 178, 180
Divided line, 155, 158, 160, 161, 163, 164, 167, 168, 170, 171, 175, 176, 179, 182, 216, 240, 242, 383, 416, 521, 522
Division (method of), 128–130, 132, 133, 136, 161
Doxa, 134, 152, 157, 169, 212, 213
Dunamis, 194, 489

E
Effluence, 113–116, 136
Eidetic memory, 59
Eidōlon, *eidōla*, 113, 131, 133, 137, 167, 172, 270
Eidos, idea, 138
Eikastic art, 131, 132, 216
Eikōn, icon, 130, 150, 152, 158, 160, 487
Eikāsia, 4, 158, 162, 170, 177, 182, 216, 240
Einbildungskraft, 409, 420, 424, 440
Elisabeth, Princess of Bohemia, 321, 328
Emblem, 279, 283, 328
Empedocles, 112–116, 136, 270
Empiricism, 32, 33, 56, 58, 62, 68–70, 186, 214, 229, 252, 257, 339, 341–344, 367, 368, 378, 399, 443, 451, 494, 509, 516
Energeia, 191, 194
Enlightenment, 3, 56, 406
Entstaltung, 67, 70, 496, 498
Epagōgē, 202, 243, 253, 450
Epicurus, 110, 113, 270
Epistemology, 25, 35, 37, 42, 58, 61, 68, 69, 99, 252, 289, 320, 339, 414
Ethics, 77, 99, 172, 213, 248, 259, 261, 320, 337, 339, 354, 383–396, 402, 403, 414, 435, 523
Ethos, 40, 427, 483–525
Evolution, 22, 94, 308, 311, 365, 477, 493

F
Feyerabend, Paul, 522
Fichte, Johann Gottlieb, 400–403
Field, 4, 19, 55, 125, 197, 274, 337, 409, 484
Flash presentations, 37
Folk psychology, 198, 502, 504
Frege, Gottlob, 29
Freud, Sigmund, 409, 411, 452, 516

G
Galilei, Galileo, 31, 224, 280, 307, 324
Game, 53–100, 129, 156, 177, 260, 291, 431–434, 440, 493, 499, 521
Gassendi, Pierre, 341
Genius, 10, 81, 95, 96, 189, 289, 319, 359–361, 388, 391, 404–406, 419, 430, 431, 454
Glaucon, 143, 147–149, 155, 157, 160, 165–169, 192, 193
Goethe, Johann Wolfgang von, 125, 391, 392, 406, 519
Google, 462
Gorgias, 115, 116

H
Hamlet, 23
Harmonia, harmony, 165
Hegel, Georg Wilhelm Friedrich, 419
Heidegger, Martin, 125, 188, 378, 399, 400, 495, 517
Heraclitus, 107–109, 112, 166, 175, 181, 284
Herder, Johann Gottfried, 142, 143, 419
Hēgemonikon, 270
Historiography, 12, 264, 268, 339, 342, 406, 516
Hobbes, Thomas, 4, 302, 341
Hugh of St. Victor, 277
Humboldt, Wilhelm von, 419, 431
Hume
Enquiry Concerning Human Understanding, 55
Hume's blue, Humean blue, 36, 55–58, 62, 494, 505
Treatise of Human Nature, 33, 56, 344, 413
Hume, David, 33, 36, 55, 56, 320
Humming, 42, 44, 45, 48, 91
Hupokeimenon, 224, 232, 249, 250
Hupolēpsis, 212–214, 216, 251
Husserl, Edmund, 24

I
Idealism, 33, 156, 169, 361, 362, 396, 400, 403, 407, 419, 431
Ignatius Loyola, 277
Image

field, 90, 91, 303, 465, 485, 498
location of, 262
model, 166
object-image, 26, 326
perception, 158, 162, 168, 177, 262
perseverance of, 53, 54
scenario image, 150–152, 156, 159, 163, 170, 173, 260, 522
Imagination
eikastic, 183, 276
intelligent, 61
model of, 20, 23, 27, 486
ontological, 71, 90, 100
para-imagination, 326
productive, 38, 57, 63, 182, 424, 514
reproductive, 38, 182, 282, 394, 424, 514
thinking, 155, 252–261, 396
Inception, 50, 70, 214, 229, 376
Inceptive, 53–100, 381
Incipience, 19, 64, 65, 71, 82, 213, 214, 229, 381, 402, 456, 465, 497
Ingenium, 285–287, 331, 487
Institution, 89, 180, 323, 371, 403, 406, 414, 420, 462, 483
Intelligible species, 202, 203, 205, 245, 247, 273, 321
Intentions (first and second), 11, 21, 24, 29, 64, 65, 89, 112, 123, 220, 240, 255, 387, 418, 424, 445, 452, 453, 503
Internal sense, 198–204, 229, 272, 276, 278, 286, 317, 318, 336, 341, 377, 390, 420, 423
Intuition, 5, 64, 278, 284, 285, 290, 291, 297, 298, 347, 369–371, 374, 375, 377, 379, 381–383, 385–389, 392, 400–402, 408, 409, 418, 420–424, 426, 438, 445, 452, 453, 502, 509

J
Jesuits, 277–279, 283, 327
John of the Cross, 278
Jones, Sir William, 431
Jung, Carl, 516

K
Kant
Anthropology from a Pragmatic Standpoint, 396
Critique of Judgment, 383, 389
Critique of Practical Reason, 394
Critique of Pure Reason, 362, 364, 368, 379
Metaphysical Foundations of Natural Science, 383
Kant, Immanuel, 226, 267, 276, 359, 361
Kierkegaard, Søren, 406
Kinēsis, 222, 233
Kosslyn, Stephen, 31

L
Lacan, Jacques, 411
Language-game, Sprachspiel, 440
Lavoisier, Antoine, 97
Leibniz, Gottfried Wilhelm, 336
Lifeworld, 97, 498, 503, 507, 521
Local memory, 212, 213, 282
Local motion, 207, 222, 223, 225, 226, 231, 233, 234, 237, 244
Location of images, 262
Locke, John, 252
Essay Concerning Human Understanding, 252
Logic, 28, 29, 50, 54, 70, 76, 78, 79, 97, 99, 106, 111, 128, 132, 167, 168, 179, 181, 186, 193, 243, 244, 270, 277, 307, 337–341, 347, 353, 354, 358, 376, 380, 387, 414, 419, 428, 435, 437–440, 443, 446, 448, 452, 477, 479, 483, 484, 486, 501, 511
Logos, 77, 99, 105, 108, 109, 112, 127, 128, 133, 151, 160, 163–171, 173, 175, 181, 185, 186, 197, 215, 284, 416, 430–440, 449, 465, 499

M
Maimonides, Moses, 189, 271
Malebranche, Nicolas, 331
Maniglier, Patrice, 467–469, 471, 474
Mathematics
mathematical truths, 307, 309–312, 508, 520, 522
mathematization, 268
Mathesis universalis, 304, 305, 307, 506, 508
Matrix, 4, 8–10, 13–15, 24, 77, 78, 84, 292, 294, 345, 372, 508
Meditation, 5, 268, 277, 285, 301, 302, 304, 307, 312–321, 324, 327, 331, 340, 341, 347–351, 362, 370, 511
Melissus, 111
Memory
art of, 282, 283
eidetic, 59
local, 212, 213, 282

Meno, 114–124, 126, 127, 162, 171, 174, 197, 313
Mersenne, Marin, 306, 307, 310
Metabolē, 222
Middle ages, 4, 186, 191, 199, 243, 245, 271, 272, 342
Mimesis, 130, 133, 142–144, 146, 147
Mirror stage, 411–415
Monet, Claude, 474, 475
Mozart, Wolfgang Amadeus, 42
Muybridge, Eadweard, 34, 38
Myth of Er, 146–148, 154, 178

N
Neo-Platonism, 163, 270
Newton, Sir Isaac, 56, 57, 60, 62, 204, 226, 251, 297, 360, 367, 370, 391–393
Nicole, Pierre, 340, 341
Nietzsche, Friedrich, 28
Nihilism, 47, 110, 408, 427, 499, 515, 522
Noûs, 207, 213, 244, 344, 520
Noēsis, 25, 155, 158, 190, 215
Nutmeg, 21, 22, 35, 36, 43, 48, 49, 57, 93

O
Occluded and occulted, 2, 50, 72, 502, 512
One-sidedness, 25, 97, 98, 390
Ontological imaging, 164, 166, 174, 175, 416, 474
Ontology, 40, 61, 68, 100, 104, 127–137, 148–154, 178, 183, 185, 219, 227, 289, 303, 304, 338, 359, 382, 400, 424, 454, 456, 468, 476–479, 519, 520, 522, 523
Ontology of imagination, 519, 522, 523
Orangery Manuscripts, 455, 468, 470, 471, 476, 484
Ousia, 191

P
Painting, 92, 93, 131, 135, 158, 171, 213, 293, 304, 305, 361, 404, 405, 434, 436, 474, 475, 507, 518
Panthéon, 17, 27, 28, 31, 32, 34–42, 47–49, 53, 63, 65, 70, 122, 408, 496
Parmenides, 109–112, 127, 128, 134, 181, 522
Pascal, Blaise, 335
Peirce, Charles Sanders, 354, 431, 442
Phantasia, 4, 5, 103, 185–264, 271, 277, 287, 298, 299, 302, 325, 422
Phantastic art, 130–133, 141
Phenomenology, 14, 19, 24–27, 29, 33, 65, 66, 233, 273, 364, 420, 445, 448, 512
Phusis, 105, 187, 222
Physico-mathematics, 274, 280
Pineal gland, 273, 299, 309, 314, 317, 318, 323–326, 512
Pistis, 157, 158, 215
Plato
  Apology, 150, 171
  Cratylus, 416
  Crito, 171
  Meno, 114–121
  Phaedo, 171, 172, 174
  Phaedrus, 416
  Republic, 115, 128, 130, 136–143, 147, 154–164, 170, 174, 176, 178, 183
  Sophist, 113, 115, 127–130, 132–134, 136–141, 147, 149, 150, 158, 161, 163, 165–168, 171, 174, 178
  Symposium, 174, 177, 180
  Theaetetus, 127, 129, 130, 134
Plotinus, 182, 271, 506
Polemarchus, 140, 147
Port-Royal Logic, 340, 347
Potentiality, 67, 69, 195, 196, 209, 217, 244, 478, 521
Praescissio, 254, 345
Precision, 35, 124, 132, 197, 254, 255, 261, 314, 322, 347, 411, 492
Prescission, 254, 316, 317, 442, 446, 448, 450, 484, 503, 514
Primitive notion, 322, 327
Proper sensible, 198–200, 211, 212, 214, 217, 218, 227, 228, 232, 233, 239, 247, 253, 256, 257, 360, 373, 489
Proposita, 345
Propositional attitude, 25
Propositionalism, 30, 345
Psuchē, 103, 185–188, 190, 376, 426
Psychologism, 27–31, 376, 432, 478, 502
Psychophysiology, 189, 204, 233, 253, 259, 273, 321, 323–325, 457, 507, 510

Q
Quaestio, 284, 291

R
Ratio, 9, 108, 151, 169, 203, 230, 281, 283, 289, 290, 312, 326, 360, 499, 507

Rationalism, 5, 32, 68, 109, 111, 112, 156, 169, 170, 186, 259, 312, 327, 328, 336, 339, 342, 343, 347, 354, 356, 359, 361, 378, 399, 442, 487, 516, 523
Rationality, 2, 3, 7–9, 49, 92, 138, 139, 156, 160, 170, 178, 180, 181, 183, 186, 188, 191, 210, 245, 248, 263, 270, 290, 291, 299, 313, 327, 335, 342, 369, 403, 461, 464, 465, 485–487, 493, 499, 506, 510, 522–524
Recombination, 23, 59, 203
Recontextualization, 8, 246, 257, 263, 277, 297, 303, 352, 403, 434
Recursion, 26, 49, 75, 215, 405
Reevocation, 42, 256, 257, 494, 495, 504
Reification, 34, 98, 254, 345–352
Rejectionism, rejection, 12, 415
Remotion, 300, 316, 318
Res, 5, 194, 312, 316, 350, 414, 466, 510, 515
Resemblant instances, 57, 58
Revolutions, 11, 12, 20, 96, 97, 171, 220, 226, 273, 323, 331, 338, 361, 373, 456
Richard of St. Victor, 277
Ricoeur, Paul, 415
Romanticism, 1, 3, 92, 361, 396, 400, 401, 404, 407, 430
Rousseau, Jean-Jacques, 419
Runge, Philipp Otto, 60

S

Sapir-Whorf hypothesis, 516
Sartre, Jean-Paul, 24, 25, 37, 68, 70, 216
Saussure, Ferdinand de, 354, 431
Scenario-image, 150–152, 156, 159, 163, 170, 173, 260, 522
Schelling, Friedrich Wilhelm Joseph von, 47, 403
Schemata, 379, 381–383, 389, 392, 403, 410, 419, 424, 425, 449, 459, 463, 479, 483, 484
Schematism, 370, 378, 379, 381, 382, 385, 387, 392–395, 400, 403, 409, 410, 418, 427, 439, 447, 449, 465, 466, 484, 495, 513, 516
Schools, 13, 24, 67, 68, 73, 125, 189, 268, 270, 272, 277, 279, 284, 405, 446
Schweben, 402
Semiology, 454–464, 473, 484, 485
Semiosis/semeiosis, 442–444, 446, 449, 451, 453, 454
Sensus communis, 388, 390, 392
Series, 10, 17, 28, 36, 41, 57, 60, 61, 63, 92, 93, 96, 116, 128, 158, 165, 201, 203, 208, 214, 254, 264, 267, 278, 283, 290–296, 300, 306, 307, 316, 332, 333, 348, 359, 378, 405, 413, 452, 457–459, 474, 475, 478
Severino, Emanuele, 109, 110
Shakespeare, William, 2, 19, 51
Shepard, Roger, 30–32
Simplicius, 106
Simulacrum, 7, 36, 130–133, 163, 169, 172, 174, 176, 309, 415, 492
Snubness, 247, 248
Socrates, 12, 114, 192, 282, 416, 487
Spinoza, Baruch (Benedict), 2, 95, 342, 343, 347, 354, 371, 407, 516
Spiritual exercises, 277, 278, 318, 320, 327, 453
Squirrel, 40, 64, 71, 138, 152, 153, 155, 200, 202, 203, 244, 246, 249, 250, 253, 255, 258, 259, 379, 380, 385, 424, 425, 427, 450, 461, 491, 505
Stimmungen, 517
Stoicism, 326
Substrate, 123, 163, 224, 230, 232, 242, 247, 249–253, 255, 256, 260, 263, 372, 421

T

Teresa of Avila, 278
Texture, 9, 13, 17, 22, 39, 93, 95, 121, 131, 135, 232, 247, 250, 252, 345, 346, 492
Thales, 104–108, 223, 447, 448
Theōria, 154, 172
Theseus, 19, 51
Thomas Aquinas, 197, 202–204, 254, 272, 377
Thrasymachus, 140, 148
Topography, 10, 14, 15, 82, 83, 108, 189, 198, 250–252, 337, 338, 344, 345, 347, 352, 360, 368, 383, 394, 400, 401, 429, 451, 466, 479
Topology, 8, 10, 14, 15, 50, 51, 53–100, 106, 133, 139, 154, 168, 175, 176, 179, 181, 182, 197, 201, 204, 219, 226–229, 231–233, 245, 247, 249–251, 255, 263, 264, 267, 268, 270, 272, 273, 276, 281, 290, 316, 323, 325, 331–396, 399–479, 486, 502–505, 509, 511, 513, 514, 524
Topos, 12, 50, 75, 76, 418
Tradition, 1, 19, 54, 103, 186, 267, 336, 400, 486

Transcendental psychology, 361–372, 375, 376, 387, 394, 401, 407, 419, 454
Typology, 395

U
Universal characteristic, 353, 354, 358

V
Ventricles, 199, 272, 273, 323, 351, 489
Visual model, 23, 27–37, 166
Vivacity, 56, 65

W
Wittgenstein
  On Certainty, 431
  Philosophical Investigations, 432
  Tractatus logico-philosophicus, 435
Wittgenstein, Ludwig, 431
Wolff, Christian, 354

Z
Zebras, 58, 59
Zeno, 111

Printed by Printforce, United Kingdom